Gravitational Measurements, Fundamental Metrology and Constants

NATO ASI Series

Advanced Science Institutes Series

A Series presenting the results of activities sponsored by the NATO Science Committee, which aims at the dissemination of advanced scientific and technological knowledge, with a view to strengthening links between scientific communities.

The Series is published by an international board of publishers in conjunction with the NATO Scientific Affairs Division

A **Life Sciences** B **Physics**	Plenum Publishing Corporation London and New York
C **Mathematical and Physical Sciences** D **Behavioural and Social Sciences** E **Applied Sciences**	Kluwer Academic Publishers Dordrecht, Boston and London
F **Computer and Systems Sciences** G **Ecological Sciences** H **Cell Biology**	Springer-Verlag Berlin, Heidelberg, New York, London, Paris and Tokyo

Series C: Mathematical and Physical Sciences - Vol. 230

Gravitational Measurements, Fundamental Metrology and Constants

edited by

Venzo De Sabbata

Department of Physics,
University of Bologna and Ferrara, Italy

and

V. N. Melnikov

Gosstandart, Moscow, U.S.S.R.

Kluwer Academic Publishers

Dordrecht / Boston / London

Published in cooperation with NATO Scientific Affairs Division

Proceedings of the NATO Advanced Study Institute on
Gravitational Measurements, Fundamental Metrology and Constants
(10th Course of the International School of Cosmology and Gravitation
of the Ettore Majorana Centre for Scientific Culture)
Erice, Italy
2–12 May 1987

Library of Congress Cataloging in Publication Data

NATO Advanced Research Workshop on Gravitational Measurements,
Fundamental Metrology, and Constants (1987 : Erice, Sicily)
Gravitational measurements, fundamental metrology, and constants /
edited by Venzo de Sabbata and V.N. Melnikov.
p. cm. -- (NATO ASI series. Series C, Mathematical and
physical sciences ; vol. 230)
"Proceedings of the NATO Advanced Research Workshop on
Gravitational Measurements, Fundamental Metrology, and Constants,
Erice, Italy, 2-12-May 1987"--T.p. verso.
"Published in cooperation with NATO Scientific Affairs Division."
ISBN 9027727090
1. Gravity--Measurement--Congresses. 2. Physical constants-
-Measurement--Congresses. I. De Sabbata, Venzo. II. Melnikov,
V. N. III. North Atlantic Treaty Organization. Scientific Affairs
Division. IV. Title. V. Series: NATO ASI series. Series C,
Mathematical and physical sciences ; no. 230.
QB330.N38 1987
526'.7--dc19 88-4393

ISBN 90–277–2709–0

Published by Kluwer Academic Publishers,
P.O. Box 17, 3300 AA Dordrecht, The Netherlands.

Kluwer Academic Publishers incorporates the publishing programmes of
D. Reidel, Martinus Nijhoff, Dr W. Junk, and MTP Press.

Sold and distributed in the U.S.A. and Canada
by Kluwer Academic Publishers,
101 Philip Drive, Norwell, MA 02061, U.S.A.

In all other countries, sold and distributed
by Kluwer Academic Publishers Group,
P.O. Box 322, 3300 AH Dordrecht, The Netherlands.

Printed in The Netherlands

TABLE OF CONTENTS

P R E F A C E

The Tenth Course of the International School of Cosmology and Gravitation of the "Ettore Majorana" Centre for Scientific Culture is concerned with "Gravitational Measurements, Fundamental Metrology and Constants". The choice of this subject was made because the increasing role that relativistic and quantum metrology play in practical applications and in the perspective of the unification of all fundamental interactions.

Practical applications are caused by the increase in precision of measuring devices, the spread of precise measurements over large distances and time intervals with the progress of space technology, the laser interferometer and the improvement in gravitational wave experiments.

In this respect and in order to have a global view of these problems and a correct perspective of the current state of the science, arguments as the variability of the fundamental physical constants, the relations between them, in particular the status of the Newtonian gravitational constant have been extensively developed and discussed.

A great excitation has been generated by the discussion of a new force in Nature and by the exposition of the various experiments that are now in progress over the world. Other arguments as quantum electrodynamics, quantum Hall effect (always in their relation with fundamental constants) tha status of gravitational antennas including laser interferometer, the physics at high densities and also gravity research in the solar system complete the collection of these interesting papers.

We have added an Appendix where we have liked to reproduce an old and beautiful paper by Quirino Majorana (the oncle of Ettore) on the absorption of gravitation by matter.

We have preferred to arrange all contributors alphabetically, by (first-named) author.

venzo de sabbata
v.n.melnikov

CAN WE CALCULATE THE FUNDAMENTAL DIMENSIONLESS CONSTANTS OF PHYSICS?

A.O. Barut *
International Centre
for Theoretical Physics
Trieste
Italy

ABSTRACT. We review some dynamical models to calculate the dimensionless constants $\alpha = e^2/4\pi\varepsilon_0\hbar c$, $\beta = M_p/m_e$ (or $\beta' = m_\mu/m_e$), $\gamma = G_{Fermi}\, m_e^2/c\hbar^3$ and $\delta = G_{Newton}\, m_e^2/(e^2/4\pi\varepsilon_0)$ which are associated with the four different manifestations (electromagnetic, strong, weak and gravitational) of a possible single interaction.

1. WHAT WE CAN CALCULATE IN PHYSICS AND WHAT NOT

There are only four units in physics which determine the scales of all measurable phenomena: length scale ℓ, time scale t, mass scale m and charge scale q. Since all charge in the world is a multiple of the electronic charge, q = ne, the last unit may be reduced to a pure number. (This point will be discussed further later). There are no generally accepted theories with a fundamental length ℓ_0, fundamental time t_0, fundamental mass m_0; and the last one is surely not additive. Thus we must keep these three units and establish some standards for them. Other basic physical constants may be used to set the scales instead of the man-made standards like meter, sec and kg. We must choose three quantities as uncalculable and try to calculate and measure everything relative to these quantities, which can then be set equal to unity.

One such system is based on the electron, perhaps the most basic elementary particle, presumable absolutely stable. We can take as uncalculable the mass of the electron m_e and from special relativity the velocity of light, and set

$$m_e = 1, \quad c = 1 \quad \text{and} \quad e^2/4\pi\varepsilon_0 = 1 \tag{1}$$

In these units, $\hbar = \lambda\!\!\!^{-}_{compton} = \alpha^{-1} = 137.03$, classical electron radius $r_e = (e^2/4\pi\varepsilon_0 m) = 1$, $t_e = r_e/c = 1$, Bohr radius $a_{Bohr} = \alpha^{-2} = (137.03)^2$. The system (1) is based on "classical" quantities, and is different than the familiar units in particle physics: $c = \hbar = 1$.

V. De Sabbata and V. N. Melnikov (eds.),
Gravitational Measurements, Fundamental Metrology and Constants, 1–14.

Another system of units is based on the assumption that the gravitational constant G, and c and $\hbar$ are both uncalculable and fundamental, and unchanging:

$$G = 1, \quad c = 1, \quad \hbar = 1 \tag{2}$$

In this system the scales for length, time and mass

$$\ell_P = \left(\frac{G\hbar}{c^3}\right)^{1/2} = 1 \quad (\cong 10^{-35} \text{m}); \quad t_p = \left(\frac{G\hbar}{c^5}\right)^{1/2} = 1 \quad (\cong 10^{-43} \text{sec});$$

$$m_p = \left(\frac{\hbar c}{G}\right)^{1/2} = 1 \quad (\cong 10^{-5} \text{gr})$$

are called the Planck scales and are independent of any properties of matter like m_e or e.

The choice of a system, such as (1) or (2), is not really arbitrary or final. It depends what the theory considers to be more a priori or basic, an electron or light, or gravitation, or the quantum of action. Only a successful unified theory based on one such system can tell us that the chosen system is a good one. And once we have assumed three such uncalculable quantities, all other quantities become dimensionless and should be in principle calculable (dimensionless quantities are distinct from dimensionless constants - see next section).

One should keep in mind that setting $c = 1$, for example, excludes the possibility of explaining it further, in terms of the Young modulus of a relativistic deformable ether, for example; setting $\hbar = 1$ excludes explaining the scale of atomic energy levels; $m_e = 1$, trying to understand the nature of mass, $G = 1$, the nature of gravity, etc. One might think that with the advance of physics we shall be able to calculate eventually everything. One might think, for example, that the mass of the electron is of electromagnetic origin and could be calculated from the energy of the electromagnetic self-field. But we see that this is not the case. Even if we give up to calculate any two of the three constants c, $e^2/4\pi\varepsilon_0$, or $\hbar$ (all three are related) we still need a scale for mass. This could be a basic mass m_e, or any other quantity, a length or a magnetic moment, etc. For example, the classical calculation of the static Coulomb energy of the electron requires a cutoff r_0:

$$\frac{1}{8\pi} \int_{r_e}^{\infty} \frac{e^2}{4\pi\varepsilon_0 r^4} r^2 \, dr \cdot 4\pi = m_e c^2 \quad \text{gives} \quad r_e = \frac{e^2}{4\pi\varepsilon_0 m_e c^2}, \tag{3}$$

We see that the choice of the system (1) or (2) is equivalent to the choice of an elementary length ℓ_0, elementary time t_0 and elementary mass m_0, all set to unity. But unfortunately they do not imply that they are smallest possible units of length, time and mass. The question of a possible ultimate smallest discrete values of length, time and mass is still open.

2. FUNDAMENTAL DIMENSIONLESS CONSTANTS

The fundamental dimensionless constants are numbers in <u>any</u> system of units. Hence they must be calculable using any system. They are four fundamental dimensionless constants that can be directly associated with the strength of four basic category of phenomena:

1) $$\alpha = \frac{e^2}{4\pi\varepsilon_0 \hbar c} = (137.0359)^{-1} = 7.297 \times 10^{-3} \quad \text{(electromagnetic)} \quad (4)$$

2) $$\beta = \frac{M_p}{m_e} = 1.836 \times 10^3 \quad \text{(strong)} \quad (5)$$

$$\left\{\beta' = \frac{m_\mu}{m_e} = 206.7694\right\}$$

3) $$\gamma = \frac{G_F m^2}{\alpha \hbar^3 c} = 3.04 \times 10^{-12} \quad \text{(weak)} \quad (6)$$

$$\delta = \frac{G m_e^2}{e^2/4\pi\varepsilon_0} = 2.3997 \times 10^{-43} \quad \text{(gravitational)} \quad (7)$$

One could also discuss another dimensionless number which involves however a model of the universe, based on the relation

$$G\mathcal{M} = c^2 \mathcal{R} \quad (8)$$

where $\mathcal{M}$ is the total mass of the universe and $\mathcal{R}$ the radius of the universe. Eq. (8) is of the same nature as the model of the electron, Eq. (3). The dimensionless numbers r_e/R and $m_e/\mathcal{M}$ are related by δ.

$$\frac{r_e}{\mathcal{R}} = \frac{e^2/4\pi\varepsilon_0}{m_e c^2} \left(\frac{G\mathcal{M}}{c^2}\right)^{-1} = \frac{1}{\delta} \left(\frac{m_e}{\mathcal{M}}\right) \quad (9)$$

or, $$\frac{\mathcal{R}}{r_e} \cdot \frac{m_e}{\mathcal{M}} = \delta .$$

The so-called Eddington number $\frac{m_e}{\mathcal{M}}$ is sometimes <u>assumed</u> to be of the order of δ^2, so that

$$\frac{r_e}{\mathcal{R}} = \delta . \quad (10)$$

There are two other dimensionless constants associated with two other strengths of phenomena which we have not listed because they have been already calculated. One is associated with the chemical force between neutral atoms and is essentially calculated by the Heitler-London theory of the H-bond. The other is associated with the strength of α-decay and is calculated by the tunnel-effect theory of Gamow and of Condon and Gurney. They are both very weak manifestations of electromagnetic interactions. By reducing these phenomena to electromagnetism we have understood them, and understanding comes from the calculation of the dimensionless numbers involved. These examples are very instructive of what we mean by the calculation of fundamental constants which we discuss in the next section.

Although it is true that there is no magic about the numbers in Eqs.(4)-(7), α^{-1} is for example not an integer as Eddington thought it was, and they can be multiplied by numbers like 2π or so, they really set the scale of the phenomena, the sizes, times and inertia of things. They can be defined entirely phenomenologically or observationally:

1) $\alpha = \frac{\lambda\!\!\!^{-}_{compton}}{a_{Bohr}} = \frac{r_{magnetic}}{\lambda\!\!\!^{-}_{compton}}$

2) β is the ratio of two observable masses

3) $\gamma \cong \frac{\Gamma_{\beta\text{-decay}}}{\Gamma_{2p\text{-}1s}} = \frac{\tau_{2p\text{-}1s}}{\tau_{\beta\text{-decay}}} \cong \frac{10^{-9}\,\text{sec}}{10^{3}\,\text{sec}} \sim 10^{-12}$

This is the ratio of observable lifetimes in atoms to radioactive lifetimes.

4) $\delta = \frac{\text{gravitational force between two electrons (extrapolated)}}{\text{electrostatic force between two electrons}}$

3. THE MEANING OF CALCULATION OF DIMENSIONLESS CONSTANTS

The calculation of dimensionless constants is a delicate matter, and we must be careful not to engage in pure numerology.

No theory can explain its postulates. The four strength parameters α, β, γ, δ enter as basic input constants in four separate theories or models of phenomena. Thus, in attempting to understand or calculate the dimensionless numbers we must realize that we have to go beyond a given theory. Here lies the difficulty. One cannot just go beyond a theory haphazardly. The generalization of a theory must proceed from the open problems of the theory and from the inevitable connections between the four separate different theories mentioned above. Any meaningful calculation of fundamental dimensionless constants must involve unification of theories. We then speak of understanding a theory when we succeed in calculating the dimensionless constant , and understanding means reducing one theory into another, as the examples of chemical interactions and α-decay discussed in the previous section show.

Thus the calculation of α, which is equivalent of calculating $\hbar$ in terms of the classical quantities $e^2/4\pi\varepsilon_0$ and c, can mean the understanding of quantum theory in terms of classical electrodynamics and some new input into the internal structure of the electron. Note that in the case of the chemical forces and α-decay we have to put some new input for the structure of the atom (and of the nucleus).

The calculation of β (we mean a dynamical calculation) implies that hadron masses have something to do with the electron mass. If hadron (or heavy lepton) masses have nothing to do with m_e, then β would be just an accidental number with no meaning at all. One could calculate the ratios of hadron masses among themselves from some composite model of hadrons, just like calculating the ratio of the masses of excited states of atoms to ground state. We would then need a new mass scale for hadrons. But of course, hadrons are not entirely independent of the electron, as the decay of the neutron into proton, electron and neutrino shows. Thus, the calculability of β goes beyond the standard complete separation of strong and electromagnetic interactions and points to a possible electromagnetic origin of strong interactions.

Same remarks apply to the constants γ and δ. In the case of γ the question is whether the Fermi coupling constant G_F is a fundamental new constant of nature, or is determined again by electromagnetism. In the electroweak standard model G_F is still a fundamental constant, but G_F and $e^2/4\pi\varepsilon_0$ are combinations of some other coupling constants g and g'. The model still has two coupling constants; it is not a unified theory. Alternatively one can introduce e and the masses of Z^0 and W instead of G_F and $G'_{Fi} = G_F \cos\theta_i$, θ_i = mixing angles; the weak interactions strength vary with θ_i and are not universal.

Finally, the gravitational constant G may be entirely independent of electromagnetism, in which case the Dirac's large number $\delta^{-1} \sim 10^{43}$ is just an accident, or perhaps having to do with the age of universe, hence cosmology, see Eq. (10). Or, there is a relation between gravitation and electromagnetism in which case the large number δ^{-1} should have a dynamical explanation.

4. ELECTRON THEORY AND ELECTROMAGNETIC ORIGIN OF INTERACTIONS

We have seen that a meaningful calculation of dimensionless constants requires some new extensions of the present theories, some new input into the structure of matter, or new hypotheses. I discuss in this paper two such slight extensions of present ideas. They are really some conservative follow-up of the elements which exist in the present theories

1) Internal structure of the electrons

The method is to take seriously the helical internal motion of the electron, called "Zitterbewegung", which is inherent in the Dirac theory of the electron. Any internal structure to an assumed point-like elementary object is bound to have profound effects, as the example of chemical force discussed before shows. The helical structure of the pointlike charge of the electron gives the electron its spin and a certain size and inertia, although, we emphasize, the charge is still point-like. This will give us some possible mechanisms to understand α and δ.

2) The self-energy anomalous magnetic moment forces at short distances

Again we know that anomalous magnetic moment of the electron exists and contributes to the interactions in atomic physics. We shall argue further that these interactions become very strong at short distances which then have far-reaching consequences concerning the elementary particle interactions at short distances or high energies. If this is the case then strong and weak interactions can be different manifestations of the electromagnetic interactions, hence the dimensionless ratios β and γ become meaningful.

5. THE FINE STRUCTURE CONSTANT α

Consider an atom making a transition between two levels of energies E_2 and E_1 an emitting an amount of energy $\hbar\omega$ in the form of radiation. By energy conservation $E_2-E_1 = E = \hbar\omega$. This relation between E and ω is the fundamental hypothesis of quantum theory, hence cannot be explained by quantum theory alone. We have to make a certain extension.

The first question is: why is the radiated energy from an atom proportional to the frequency ω of the emitted radiation and not to ω^3?

The energy radiated by the electron per unit time during an acceleration a is given by

$$\frac{dE}{dt} = \frac{2}{3}\frac{e^2}{4\pi\varepsilon_0 c^3} a^2 \tag{11}$$

If the charge performs oscillations such as $x = A\sin\omega t$, then $\dot{x} = A\,\omega\cos\omega t$, $\ddot{x} = -A\omega^2\sin\omega t$. Hence $a^2 = A^2\omega^4\sin^2\omega t$ and the energy emitted during one period is

$$E = \frac{2}{3}\frac{e^2}{4\pi\varepsilon_0 c^3} A^2\omega^4 \int_0^{2\pi/\omega} dt\,\sin^2\omega t = \frac{2}{3}\frac{e^2}{c^3} A^2\pi\omega^3 . \tag{12}$$

We could understand linear dependence of E on ω if each component of the velocity of the charge had the form $\dot{x} = c\cos\omega t$ with maximal velocity being independent of ω, say equal to light velocity. Then the energy radiated during N periods would be proportional to ω:

$$E = \frac{2}{3}\frac{e^2}{4\pi\varepsilon_0 c}\pi N\omega \tag{13}$$

Indeed, in the Dirac theory, the velocity of the electron has the eigenvalues $\pm c$. In the Heisenberg representation of the Dirac electron, $\vec{x}$, $\dot{\vec{x}}$ and $\vec{p}$ are independent dynamical variables, and velocity operator $\dot{\vec{x}} = c\vec{\alpha}$ commutes with both $\vec{x}$ and $\vec{p}$, where $\vec{\alpha}$ are the Dirac matrices. This leads to a larger phase space than a point particle. One can visualize the internal structure of the electron by a helical motion of the point charge around a center of mass, called Zitterbewegung [1]. The spin of the electron can then be evaluated as the orbital angular momentum of the helical motion around the center of mass. In Dirac theory the dynamical variables are operators and it is difficult to draw a picture. However, there is a classical theory of the electron with a Zitterbewegung for which we can actually draw the helical orbit of the charge. This theory is based on the Lagrangian [2]

$$L = -\frac{\lambda}{2i}(\bar{z}\dot{z} - \dot{\bar{z}}z) + p_\mu \dot{x}^\mu + (p_\mu - eA_\mu)\bar{z}\gamma^\mu z\ , \tag{14}$$

where λ is a constant of the dimension of an action; dots are with respect to a proper time τ. The mass m does not occur in Eq.(14); it comes in as a constant value of the integral of motion $\mathcal{H} = \bar{z}\gamma^\mu(p_\mu - eA_\mu)z = m$. There is the notion of antiparticles in the classical theory (14), and its quantization by a variety of ways leads exactly to the Dirac electron [2)-4)].

Let us now apply the radiation formula (11) to the Zitterbewegung with velocity $\dot{\vec{x}} = c\vec{\alpha}$. We obtain for the radiated energy per unit time

$$\frac{dE}{dt} = \frac{2}{3}\frac{e^2}{c^3}(c\dot{\vec{\alpha}})^2\ . \tag{15}$$

How can we interpret the operator $\dot{\alpha}^2$ in this formula? For the transition between two energy levels E_2 and E_1 of an atom, we could set

$$\dot{\alpha}^2 \to |\langle E_2|\dot{\vec{\alpha}}|E_1\rangle|^2\ . \tag{16}$$

In quantum theory, the absolute value squared of a matrix element is a probability. Hence with (16), Eq.(15) is the probability of emitting the energy dE/dt per unit time. The Einstein A-coefficient is the probability of emitting a photon per unit time, hence dividing (15) by $\hbar\omega = E_2 - E_1$, we have

$$A = \frac{(2/3)\,e^2/c\,|\langle E_2|\dot{\vec{\alpha}}|E_1\rangle|^2}{\hbar\omega} \tag{17}$$

Indeed, using the Heisenberg equation $\alpha(t) = e^{-iHt}\alpha(0)e^{iHt}$ it is possible to calculate [5] Einstein's A-coefficient relativistically from Eq.(17). So far this is exact relativistic quantum theory. Can we calculate the radiated energy itself, i.e. $\int_0^T dE/dt\, dt$? The velocity $\dot{\alpha}$ in (15) is

$$\dot{\vec{\alpha}} = \frac{i}{\hbar}[H, \vec{\alpha}]$$

$$\langle E_2|\vec{\alpha}|E_1\rangle = \frac{(E_2-E_1)}{\hbar}\langle E_2|\vec{\alpha}|E_1\rangle = \omega\langle E_2|\vec{\alpha}|E_1\rangle \tag{18}$$

Since we are not interested in probabilities but rather in the actual value of acceleration, we make the heuristic assumption of replacing the matrix element $|\langle E_2|\vec{\alpha}|E_1)|^2$ by its eigenvalue squared $\vec{\alpha}^2 = 3$ in calculating the energy radiated. The is where we go beyond quantum theory. We have then from (15)

$$E = \int_0^{T=2\pi/\omega N} \frac{dE}{dt}\,dt = \frac{2}{3}\,\frac{e^2}{4\pi\varepsilon_0 c}\,\omega^2 \cdot 3 \cdot \frac{2\pi}{\omega} N = 4\pi N \frac{e^2}{c}\,\omega \tag{19}$$

Eq.(19) agrees with (13), and is equivalent to an estimate of $\hbar$ or α. For writing (19) as $E = \hbar\omega$, we identify $\hbar = 2\pi\, e^2/4\pi\varepsilon_0 c\, N$ or

$$\alpha^{-1} = \left(\frac{e^2}{4\pi\varepsilon_0 \hbar c}\right) = 4\pi N \tag{20}$$

This mechanism reduces the calculation of $\hbar$ to the number of periods N in a photon in the laboratory frame. How long is a photon of frequency ω? We see also why this calculation goes beyond quantum mechanics, because such a question about the number of waves in a single photon cannot be answered by quantum mechanics.

In an earlier attempt [6] we have assumed that the atom consists of a continuous set of oscillators in every direction giving an additional factor 4π in Eq.(11), we have also calculated the energy emitted in one period in the proper frame of the oscillating charge $T' = 2\pi/\omega\ \sqrt{1-\beta^2} = 2\pi/\omega\ \sqrt{3}/2$ (assuming an average $\bar{\beta}^2 = (1/2)^2$) so that the value for α^{-1} obtained was

$$\alpha^{-1} = 8\pi^2\sqrt{3} = 136.7572$$

and with a Doppler correction a value $\alpha^{-1} = 137.03$ was obtained which is very close to its experimental value 137.0359. Alternatively, using (20), we need an estimate on the number of periods N in a single photon. Perhaps this question can be answered when experiments can be made on a single photon.

One should remark that the frequency of the Zitterbewegung for a free electron is fixed $\omega_0 = me^2/\hbar$ and the free electron does not radiate in spite of the oscillations. The Zitterbewegung is the natural non-radiating state of the electron as determined by the Lagrangian (14). The electron radiates in the presence of external fields which superimposes to the Zitterbewegung another oscillation of frequency ω as we have assumed.

6. SELF-ENERGY OF THE ELECTRON AND MUON-ELECTRON MASS RATIO

The muon-electron mass ratio can be calculated if we can think of the muon (and eventually τ) as some kind of an "excited state" of the electron. It cannot be an ordinary electric excited state, like an atom, decaying as $\mu \to e\gamma$. It can however be an excited state of a different kind decaying like

$$\mu^- \to e^- + (\bar{\nu}_e \nu_\mu)$$

The muon can also be thought to be a resonance in the channel $e\nu_e\bar{\nu}_\mu$ of three particles. Heavy lepton spectroscopy differs from atomic spectroscopy in that the quantum of transitions between e, μ and τ is not the photon but the new quantum $(\nu\bar{\nu})$.

The self-energy of the electron may give an answer to the question of excited states of the electron.

The classical electron equation including the self energy is the nonlinear Lorentz-Dirac equation. Witten in proper time τ, this is

$$m\ddot{x}_\mu = e\, F^{ext}_{\mu\nu}\, \dot{x}^\nu + \frac{2}{3}\frac{e^2}{4\pi\varepsilon_0 c^3}\,(\dddot{x}_\mu + (\ddot{x})^2\, \dot{x}_\mu) \tag{21}$$

The last term can be incorporated into a variable mass term [6b)] so that

$$\frac{d}{d\tau}(m\dot{x}_\mu) = e\, F^{ext}_{\mu\nu}\, \dot{x}^\nu + \frac{e^2}{4\pi\varepsilon_0 c^3}\, \dddot{x}_\mu . \tag{22}$$

First we show the analogy between the quantum Dirac equation and the classical Eq.(22). The former can be written using covariant Heisenberg equations of motion as [6b)]

$$\frac{d}{d\tau}(\mathcal{H}\dot{x}_\mu) = e\, F^{ext}_{\mu\nu}\dot{x}^\nu + i\,\frac{\hbar}{2}\, \dddot{x}_\mu \tag{23}$$

where $\mathcal{H}$ is the mass operator $\mathcal{H} = \gamma^\mu p_\mu$. For a free particle, Eqs.(22) and (23) become $d/d\tau(m\dot{x}_\mu) = 2/3\, e^2/4\pi\varepsilon_0 c^3\, \dddot{x}_\mu$ and $d/d\tau(\mathcal{H}\dot{x}_\mu) = i\,\hbar/2\, \dddot{x}_\mu$. Apart from the factor (i), the two equations coincide for $\mathcal{H} = (3/2)mc^2/\alpha$ which is equal to the muon mass. It is not yet clear what that means: it only compares self energy of the electron with the quantum effect of the muon. The self-energy effects have a characteristic size $2/3\, e^2/4\pi\varepsilon_0 mc^4$, whereas quantum effects have a characteristic size $\hbar/2mc$, 137 times larger. For a muon the Compton wave length becomes comparable to the classical radius of the electron.

We must still add to the quantum Dirac electron its self-energy. In QED the self-effects are treated perturbatively. There is however a way to write the self energy contribution for the Dirac equation in the form of a non-linear equation analogous to Eq.(21). This is obtained from Maxwell-Dirac equations by eliminating the vector potentials A_μ and reads

$$[\gamma_\mu(i\partial_\mu - eA^{ext}_\mu) - m]\psi(x) - e^2\gamma^\mu\psi(x)\int d^4y\; D(x-y)\bar{\psi}(y)\gamma_\mu\psi(y) \tag{24}$$

This brings a new term to Eq.(23) in Heisenberg representation

$$\frac{d}{d\tau}(\mathcal{H}\dot{x}_\mu) = eF^{ext}_{\mu\nu}\, \dot{x}^\nu + i\,\frac{h}{2}\dddot{x}_\mu + R \tag{25}$$

It is difficult to write the self energy term in Heisenberg representation corresponding to the nonlinear and nonlocal term in the Schrödinger representation (24). Its effects however would be to modify the Zitterbewegung, to add oscillations of an amplitude α-times smaller. The resulting change in the spin or magnetic moment is the anomalous magnetic moment of the electron which now finds an intuitive explanation. Its value for a weak field is $(\alpha/2\pi)\mu_0$ and exists also for the classical relativistic equation.

One can get a number of old and new information from the nonlinear equation (24). Old information are radiative effects like Lamb-shift, spontaneous emission, (g-2), etc. New information is the possible new localized solutions of (24) at short distances. Guided by equations (24) and (22) we can write a nonlinear wave equation which has two exact solutions, electron and muon, or the muon as an excited solution of the electron. The equation is ($c = \hbar = 1$)

$$(\gamma^{\mu}p_{\mu}-m)\psi - \frac{2}{3}\frac{\alpha}{m}(\gamma p-m)^2\psi = 0 \tag{26}$$

The first solution satisfies $(\gamma p-m)\psi_1 = 0$, for which the second term vanishes; so we can choose m as the electron mass. The second solution satisfies $(\gamma p-m)\psi_2 = K\psi_2$, where K is a constant to be determined. Or

$$(\gamma p - (m+K))\psi_2 = 0 \tag{27}$$

Inserting (27) into (26) gives $K\psi_2 - 2/3\ \alpha/m\ K^2\psi_2 = 0$, using $\psi_2^+\psi_2 = 1$, $K = (3/2)m/\alpha$. Hence from (27)

$$m_{\mu} = m + \frac{3}{2}\frac{m}{\alpha} \tag{28}$$

which is a rather accurate formula for the mass of the muon.

Another version of Eq. (26) has been given [8]. We postulate a second order equation

$$(\gamma^{\mu}p_{\mu} - ap_{\mu}p^{\mu} - \kappa)\psi = 0 \tag{29}$$

where a is so chosen that the anomalous magnetic moment of the particle has the classical value $2(2/3)\alpha$. The two masses from (29) are m_1 and $m_2 = 1/a - m_1$, and with $1/a = (3/2)m/\alpha\ (1 + 4\alpha/3)$ we obtain again (28). Is muon an "excited state" of the electron, or is it a bound state of $(e\nu\bar{\nu})$? These statements may be equivalent, for we can think also the excited state of an atom to be a bound state of the ground state with a photon. In fact a simple model of a magnetic bound states gives the formula (28) and one for the τ-lepton [9] which relates $\nu\bar{\nu}$-quanta to magnetic interactions.

These calculations of the muon-electron mass ratio make use of some heuristic assumption, but in view of the enormous importance of the μ-puzzle and generation problem they are worth considering.

7. FERMI COUPLING CONSTANT AND THE DIMENSIONLESS CONSTANT γ

The constant G_F has been first introduced in the β-decay: $n \to pe^-\bar{\nu}$, later with slight modification in the μ-decay: $\mu^- \to e^-\bar{\nu}_e\nu_\mu$ and then $e\nu$-scattering. For other new unstable particles the weak coupling constant is written as $G_F \cos\alpha$ or $G_F \sin\alpha$, where α's are the so-called mixing angles. Fermi's matrix element for $n \to pe^-\bar{\nu}$ contains simply an annihilation operator for n and creation operators for p, e^- and $\bar{\nu}$, all multiplied by the phenomenological constant G_F. It does not tell us what is going on dynamically in the decay process.

It is however possible to think of neutron as a dynamical long lived resonance in the 3-particle channel $(pe^-\bar{\nu})$ bound together by anomalous magnetic moment force [10]. Magnetic moment spin-orbit forces produce deep short-ranged potential wells which can support narrow resonances in the continuum which can then slowly tunnel. In this way the β-decay is reduced to a mechanism similar to α-decay. The model qualitatively accounts for the properties of the neutron. But we must assume a small anomalous magnetic moment for the neutrino of the order $a_\nu \cong (10^{-9} - 10^{-10})\mu_0$. This is the experimental upper limit for free neutrinos. We can then calculate in such a model the Fermi coupling constant G_F in terms of the magnetic moment of the neutrino. One can say that this model calculates one constant in terms of another. The point is however that the anomalous magnetic moment of the neutrino is an electromagnetic property, hence one embeds the weak interactions into the electromagnetic interactions. Moreover, a_ν enters into many other processes, and could be measured or calculated independently.

8. THE PROTON-ELECTRON MASS RATIO β

Continuing the idea that all particles can be built from the two basic stable spin-1/2 particles, the electron and the neutrino, a model for the proton is $p = (e^+e^+e^-)$ bound by short-ranged magnetic forces. It is difficult to perform a fully relativistic three-body Dirac problem. Moreover, the anomalous magnetic moment interactions have to be included in a self-consistent way non-linearly. But in order to see that such a mechanism can give the correct order of magnitudes we may perform a relativistic WKB-method by finding the extremum of the energy.

First for a 2-body (e^+e^-) magnetic problem, we assume an effective magnetic potential and a relativistic Hamiltonian of the form [11]

$$H = 2p - e\mu/r^2$$

where the magnetic moment is $\mu = e/2m$. We quantize the circular orbits by $rp = n$, and set $\partial H/\partial p = 0$. This gives $p_n = n^2/e\mu$ and $E_n = 2\ m/\alpha\ n^2$. For $n = 1$, we get the pion mass, $n = 2$, the η-mass. The same method applied to the $(p\bar{p})$-system with the proton magnetic momentum $\mu = (e/2M)2.79$ gives the mass formula $E_n = 2\ M/\alpha\ (1/2.79)n^2$ which for $n = 1$ with the Z^0-mass of 92.8 GeV found in $\bar{p}p$-collisions.

Now for the 3-body magnetic problem $p = (e^+e^+e^-)$, we assume the stable configuration of a Lagrange equilateral triangle [12] with sides $a = \sqrt{3}r$ and a Hamiltonian $H = 3p - e\mu/(\sqrt{3}r)^2$. The same quantization and extremum procedure, $pr = n$ and $\partial H/\partial p = 0$ gives the mass formula $E_n = (27/2)m/\alpha\, n^2$, which for $n = 1$, is close to the proton mass. (The Coulomb corrections brings the number to $M_p/m_e \simeq 1840$).

9. THE LARGE NUMBER δ^{-1}

The dimensionless constant most difficult to understand is perhaps δ, Eq.(7), the ratio of the gravitational to electrostatic forces between two electrons. The number δ^{-1} is so large compared to all physical processes that Dirac [12] tried to understand it in terms of the age of the universe and the variability of the constants of physics. If we set

$$\beta^2\delta = \frac{GM_p^2}{e^2/4\pi\varepsilon_0} \simeq \frac{t_e}{T} \tag{30}$$

with electronic time scale $t_e = r_e/c = (e^2/4\pi\varepsilon_0\hbar c)\,\hbar/m_ec^2 = \alpha\,\hbar/m_ec^2$, then T becomes of the order of the age of the universe : $T \simeq (\alpha/\beta^2\delta)(\hbar/me^2) \sim 10^{19}$ sec. And if Eq.(30) had always been true, then as T increases some of the constants must have changed with time. This is a hypothesis. The limits on the variation of G, α, $\alpha^2\cdot\mu_p\cdot m_e/M_p$, are $\dot{Q}/Q \leq (10^{-12} - 10^{-17})\ \mathrm{yr}^{-1}$.

We shall examine here the connection between gravitation and electromagnetism in order to understand the number δ. A possible macroscopic indication for such a connection comes from the Schuster-Blacket law between the magnetic moment, spin and gravitational constant G of the celestial bodies [13),14)]. We consider now this problem from the point of view of the internal structure of the electron. Any internal structure of a quantum system is found to have an effect on the force between atoms. We shall now apply this to the electron with the internal motion of the charge around a fictious center of mass. The only difference with the H-atom is the absence of the central charge. Consequently the force between two electrons consists mainly of the Coulomb force, but there are corrections.

Let the coordinates of the two charges be $\vec{x}_1$ and $\vec{x}_2$. Charge 1 is centered around $\vec{a}$ and charge 2 around $\vec{b}$, with relative coordinates $\vec{x}_1-\vec{a} = \vec{\xi}_1$ and $\vec{x}_2 - \vec{b} = \vec{\xi}_2$ and $\vec{a} - \vec{b} = \vec{r}$, the distance between the two centers of mass. The instantaneous Coulomb potential then gives

$$\frac{e_1e_2}{|\vec{x}_2-\vec{x}_1|} = \frac{e_1e_2}{|\vec{\xi}_2+\vec{b}-\vec{\xi}_1-\vec{a}|} = \frac{e_1e_2}{|\vec{r}+(\vec{\xi}_2-\vec{\xi}_1)|}$$

$$= \frac{e_1e_2}{r} - \frac{e_1e_2}{r^3}\,\vec{r}.(\vec{\xi}_2-\vec{\xi}_1) - \frac{1}{2}\frac{1}{r^3}\,[(\xi_2-\xi_1)^2$$

$$-3(\vec{r}\cdot(\vec{\xi}_2-\vec{\xi}_1))^2] + \ldots \tag{31}$$

The first term is the Coulomb potential as if the electron were point particles. The additional terms can be identified with spin forces, because the spin is precisely the orbital angular momentum of the relative coordinate $\vec{\xi}$ around the center of mass. Thus static spin forces are the analogues of van der Waals' forces. In addition, we have the extremely weak radiation forces. A single electron does not radiate in spite of the Zitterbewegung of its charge. But if we bring another electron at a distance r, then the Zitterbewegung is a bit distorted, the charge will accelerate and radiate. There is now another force coming from these radiative processes even if the centers of masses of the electrons are at rest. If we assume a force proportional to acceleration and going like $1/r$ on each particle

$$\vec{F}_{rad} \sim \frac{\ddot{\vec{x}}_1}{|\vec{x}_1-\vec{x}_2|} = \ddot{\vec{\xi}} \left[\frac{1}{r} - \frac{\vec{r}}{r^3} \cdot (\vec{\xi}_1-\vec{\xi}_2) \ldots\right], \tag{32}$$

where we have neglected the acceleration of the centers of mass, we obtain for the time-average of this force over the rapidly oscillating Zitterbewegung of both charges

$$<F^i> \sim \frac{r_k^j}{r^3} \lim \frac{1}{T} \int_0^T \ddot{\xi}_1^i \xi_1^j \, dt$$

$$= - \frac{r_k^j}{r^3} \lim \frac{1}{T} \int_0^T \dot{\xi}_1^i \dot{\xi}_1^j \, dt = - \frac{r_k^i}{r^3} \frac{1}{3} <\dot{\xi}_1^2> . \tag{33}$$

This is always an attractive force going like $1/r^2$ which is proportional to the average of the square of the velocity due to the presence of the other particle. It is important that the radiative force (32) be proportional to e^2 and not e. The argument of Eq.(33) has been given by Cook [15] in the case of gravity to connect $1/r^2$-law with the Einstein-Sciama force. In our case, we attempt to have a mechanism to calculate G itself. This requires the knowledge of the coefficients in (32) and the equations of motion of Zitterbewegung in external field. But the order of magnitude of G seems to come out correctly, since the radiative force in question is of the order of $(\Delta E/M)^2 \sim 10^{-30}$, where ΔE is Lambshift in an atom (a characteristic energy for radiative processes) and M the mass of the atom.

This is only an indication for a mechanism to calculate G on the basis of electodynamics; a lot more needs to be done.

REFERENCES

1) E. Schrödinger, Sitzungsberichte Akad. d. Wiss. 63 (1931); A.O. Barut and A.J. Brachen, Phys.Rev. D 23, 2454 (1981); D24, 3333 (1982).
2) A.O. Barut and N. Zanghi, Phys.Rev.Lett. 52, 2009 (1984).

3) A.O. Barut and I.H. Duru, Phys.Rev.Lett. 53, 2355 (1984).
4) A.O. Barut and M. Pavšić, Class. and Quant. Gravity 4, L41 and L131 (1987).
5) A.O. Barut, in Old and New Questions in Physics, Cosmology, Philosophy and Theor. Biology, Ed. A. van der Merwe (Plenum, 1983), p. 53.
6) A.O. Barut, Z.f.Naturf. 33a, 993 (1978).
7) A.O. Barut, in Quantum Optics Relativity and Theory of Measurement Ed. P. Meystre (Plenum, 1983) p. 155.
8) A.O. Barut, Physics Letters 73B, 310 (1978).
9) A.O. Barut, Phys.Rev.Lett. 42, 1251 (1979).
10) A.O. Barut, Surveys in High Energy Physics 1, 113 (1980).
11) A.O. Barut, Lett. N. Cimento 34, 1 (1985).
12) A.O. Barut, in Bull.Techn.Univ. Istanbul (Jan. 1987).
13) V. de Sabbato, These Proceedings.
14) A.O. Barut, in Proc. Second Marcel Grossman Meeting on General Relativity, Ed. R. Ruffini (North-Holland Publ. Co., 1982), p. 163.
15) R.J. Cook, N.C. 35B, 25 (1976).

* Permanent address: Physics Department
University of Colorado
Boulder, CO 80309
USA

OBSERVABLES IN GENERAL RELATIVITY*

Peter G. Bergmann
Departments of Physics, Syracuse University, New York University, and Queens College of the City University of New York.

The notion of observable in general relativity differs profoundly from the corresponding concept in special relativity and in Newtonian physics. This is because in general relativity fixation of a frame of reference and gauge transformations are intertwined in a manner not encountered in any other area of physics. In other field theories an observable is any (local) component of a field that is gauge-invariant, such as a component of the electric field strength at a stated instant in time and at a given space point. Of course, functions, or functionals of such observables are themselves observables.

Given a fully specified physical situation, and given the appropriate frame of reference, in classical field theory each observable has a definite value. In quantum field theory an observable has a definite expectation value, and a definite variance (which of course may be infinite). These predicted values can in principle be confirmed by experiment ("measured"); that is what we mean by the term "observable". It is to be noted that the fixation of a frame of reference (= choice of a four-dimensional pseudo-Cartesian coordinate system) though a necessary preliminary to any measurement, carries no dynamical implications and involves no Heisenberg relations.

* The author's participation in the 10th Course of the International School of Cosmology and Gravitation was made possible by the generous support of the Ettore Majorana Centre for Scientific Culture. Research support by the National Science Foundation (U.S.A.) under Grant PHYS-8541793 to Syracuse University is also gratefully acknowledged.

V. De Sabbata and V. N. Melnikov (eds.),
Gravitational Measurements, Fundamental Metrology and Constants, 15–18.

In a general-relativistic theory, that is to say a theory whose formulation does not give rise to a preferred set of coordinate systems that are related to each other by a finite-dimensional symmetry group of transformations, an observable is not that easily defined. Consider, as an example, the metric field in general relativity. If we know the metric tensor in some coordinate system, we have essentially complete information about the gravitational field. The information at hand enables us to determine, by calculation, whether the space-time is Ricci-flat or there is a non-vanishing energy-stress tensor; whether or not there is a Killing field; what Petrov-type field we have, etc. But because there are no preferred coordinate systems, which could be identified by means of a finite set of data, it is meaningless to predict the outcome of a measurement of the component $g_{\mu\nu}(x^0, \ldots, x^3)$, with given nemerical values for the subscripts μ, ν, and for the four coordinate arguments. In this respect, our field behaves much like a gauge-sensitive quantity behaves in a Poincaré-invariant theory, such as a component of an electrodynamic potential. Curvilinear coordinate transformation now play the role of gauge transformations. Just as an instrument cannot take into account our choice of a gauge frame, it cannot be programmed to act within a definite (curvilinear) coordinate system, as these choices are made within our minds, or on a piece of paper, not by interaction with the physical universe.

One can characterize an observable formally by drawing an analogy with the gauge groups that occur in Poincaré-invariant field theories. As infinitesimal gauge transformations are generated by certain variables, usually appearing in the form of constraints, a gauge-invariant variable, hence an observable, is a variable whose Poisson brackets with the gauge generators vanish. Now the generators of infinitesimal coordinate transformations are Dirac's Hamiltonian constraints, hence a general-relativistic observable is a quantity whose Poisson brackets with all of the Hamiltonian constraints are zero. But such a quantity is a constant of the motion! We must conclude that in a general-relativistic theory the observables are the constants of the motion.

This result brings but cold comfort: Comceptually, one hesitates to accept a situation in which the only observables do not change with time; practically, constants of the motion are hard to come by, particularly if we deal with generic solutions of, say, Einstein's field equations. One would like to deal with quantities that closely resemble the data that experimentalists collect as a matter of course.

A rather unwieldy scheme was proposed many years ago by A. Komar and myself. Suppose one can construct a number of scalar fields that are algebraically independent of each other, and which depend entirely on the metric field. Then a world point can be characterized by the numerical values of four such scalars. Such a construction is known as intrinsic coordinates. It happens that generically four scalars can be constructed algebraically from the components of Weyl's tensor; they are the Géhéniau-Debever scalars. With their help the components of any tensor field (such as the metric field) may be converted into observables. Intuitively, this approach can be likened to the establishment of coincidences: The occurence of (at least) five causally independent events at the same world point represents a concatenation not common to all physically distinct solutions of Einstein's field equations, but only to a certain subset of solutions. The statement of such a coincidence of events at a world point is entirely free of the accident of choice of a particular coordinate system. Formally, such a coincidence may be cast into Hamiltonian constraints. In principle, an instrument may be conceived that "observes" the coincidence. Such an instrument could, for instance, be programmed to search for the world point in which the four Géhéniau-Debever scalars take certain preassigned values, and to show on a digital read-out the value of a fifth variable then and there.

This scheme has a number of flaws. First, it is applicable only to the "generic" situation. It does not work in space-times that possess Killing fields, or that are algebraically special. Second, the Géhéniau-Debever scalars have only the remotest relationship to the quasi-Lorentzian coordinate systems that are used in practice.

It would seem that a more attractive construction of observables might restrict itself to space-times that are asymptotically flat at infinity, or that satisfy some other boundary conditions compatible with some cosmological model. It has been suggested that a preferred set of space-like hypersurfaces is available in the form of so-called minimal surfaces, that is to say surfaces for which the trace of the second fundamental form vanishes. A foliation of such minimal surfaces might serve as the basis of a globally well-defined coordinate system, provided existence and uniqueness theorems can be formulated and proved that relate the coordinate systems thus obtained to current or at least feasible observational procedures.

Obviously, the whole question of observables is intertwined with the possibility of constructing four-dimensional coordinate systems on the basis of some initial (Cauchy) fixation. One way consists of the arbitrary choice of some space-like hypersurface, and the adoption of the congruence of time-like geodesics normal to that hypersurface through each of its points as the local 'time axes'. This congruence of geodesics is surface-forming; these surfaces are equidistant from each other. The whole system of mutually orthogonal time-like geodesics and space-like hypersurfaces represents the solution of an eikonal equation (or Hamilton-Jacobi equation) having the form

$$g^{\mu\nu} \frac{\partial S}{\partial x^\mu} \frac{\partial S}{\partial x^\nu} = 1 .$$

The scalar quantity S will serve as the time coordinate, whereas the three remaining space-like coordinates may be chosen so that they remain constant along each of the time-like geodesics. Every hyper-surface S = constant serves as a defining surface for simultaneity in the infinitesimal Lorentz frame associated with the local time axis, and this definition of simultaneity is integrable over the space-like hypersurface. In the coordinate system just sketched four components of the metric tensor have trivial values,

$$g_{\mu 0} = \delta_{\mu 0} , \quad g^{\mu 0} = \delta^{\mu 0} .$$

This construction is assured to work only in a finite patch of space-time, not globally, as the time-like geodesics may converge to caustics or foci at some finite distance from the original hypersurface that underlies the whole construction. Intuitively one may think of this four-dimensional coordinate system as a collection of "good" clocks, i.e. clocks that indicate proper time, which are lined up on the initial hypersurface, at rest relative to it, there set to show time zero, and then permitted to fall freely (to travel along geodesics). The three spatial coordinates serve to identify each clock once and for all.

It is by no means trivial to proceed from such a fixation of a four-dimensional coordinate system to the theoretical construction of observables; but it is a step in the right direction. The construction just outlined narrows the "gauge" freedom of coordinate choice from four arbitrary functions of four arguments (including a time-like coordinate) to four functions of three arguments, the fixation of the initial hypersurface and its coordinatization. It is not difficult to invent similar coordinatization schemes, and one can hope that some will approximate what experimental physicists and observing astronomers actually do.

THE QUANTUM HALL EFFECT PART I: BASIC EXPERIMENTS

E. Braun
Physikalisch-Technische Bundesanstalt
Bundesallee 100
3300 Braunschweig
Federal Republic of Germany

ABSTRACT. In a short historical review, some of the prerequisites which were essential for the discovery of the quantum Hall effect are discussed. Subsequently, von Klitzings experiment on silicon MOSFET's is described and a quantitative description of the quantum Hall effect is given. The impact of the new quantum effect on electrical units and fundamental constants is described.

1. HISTORICAL RETROSPECTIVE

The history of the quantum Hall effect begins in 1879 in Baltimore, when Prof. Rowland's student Edwin Herbert Hall put a gold foil in a magnetic field and passed a current through the foil (Fig.1). He observed a voltage perpendicular to the direction of the current and to the magnetic field /1/. This voltage was afterwards called the Hall voltage. When the 100th anniversary of this discovery was celebrated in Baltimore in 1979, none of the guests suspected that only a few months later Klaus von Klitzing would make a 2nd discovery with the Hall effect, a discovery which would be of such importance that it would be honored with the Nobel Prize.

The explanation of the Hall effect was one of the highlights of the theory of a free electron gas for electrical conduction. The electrons were supposed to move through a conductor with a mean free path before they are scattered. The basic idea behind this explanation of the Hall effect was that the Lorentz force deflects the electrons to one edge of the sample, where they set up the so-called Hall field (Fig.2). Equilibrium is achieved when the force in this Hall field cancels the Lorentz force. Under this condition a Hall voltage U_H can be measured perpendicular to the current I and to the magnetic flux with the density B:

$$U_H = \frac{1}{ne} \cdot \frac{1}{d} \cdot I \cdot B \tag{1}$$

(n is the density of the electrons, e the elementary charge and d is the

V. De Sabbata and V. N. Melnikov (eds.),
Gravitational Measurements, Fundamental Metrology and Constants, 19–38.

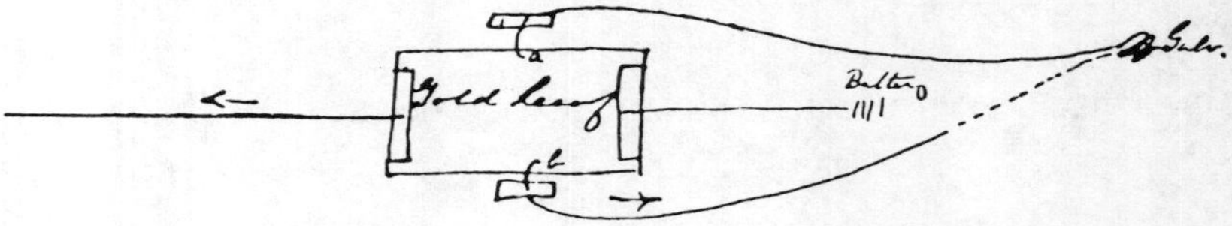

Fig. 1 Edwin H. Hall's sketch of his experiment in 1879 /1/. When he connected a gold foil to a battery and put it into a transverse magnetic field, he observed a voltage between the probes a and b. This effect is now called the Hall effect.

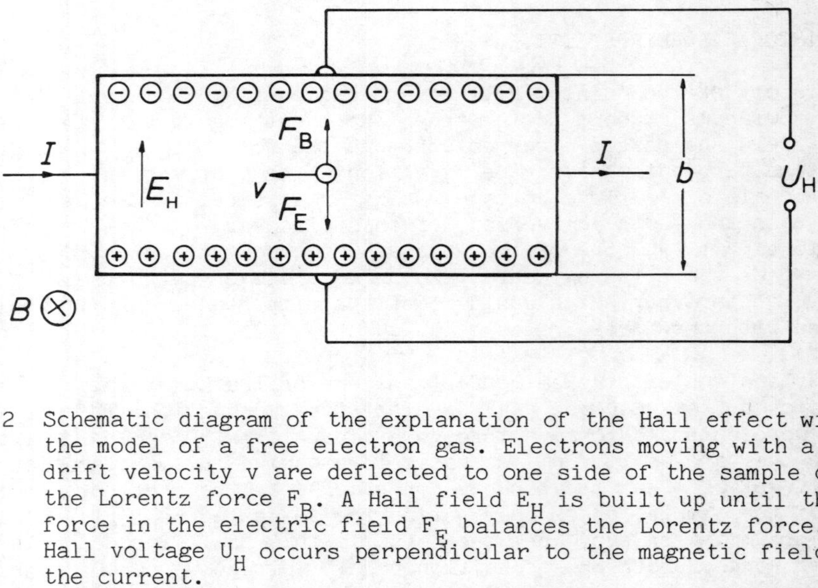

Fig. 2 Schematic diagram of the explanation of the Hall effect with the model of a free electron gas. Electrons moving with a drift velocity v are deflected to one side of the sample due to the Lorentz force F_B. A Hall field E_H is built up until the force in the electric field F_E balances the Lorentz force. The Hall voltage U_H occurs perpendicular to the magnetic field and the current.
(I current, B magnetic flux density, b width of the sample).

thickness of the sample in the direction of the magnetic field).

The measurement of the Hall coefficient A_H

$$A_H = \frac{1}{ne} \qquad (2)$$

became a very useful tool for the growing production of semiconductors: measurements of A_H where used to determine the sign of the carriers (electrons or holes), their density, and, in combination with the measured conductivity σ, the Hall mobility μ_H could be calculated as $\mu_H = \sigma A_H$. Moreover, Hall devices had been built making use of the fact that the Hall voltage is proportional to the product of two currents, one passsing through the sample and the other producing the magnetic field.

The technique of liquifying helium opened up the possibility of working at very low temperatures in the range of about 1 K. It also became possible to work in very high magnetic fields either by producing them with much effort in Bitter coils, available mainly in special high magnetic field laboratories or by producing them for a brief period of a few microseconds by discharging energy storages, e.g. capacitor banks. In very recent years it has become possible to produce steady fields up to 15 T and even more in superconducting coils. The combination of superconducting coils with Bitter coils extended the availability of steady fields to 30 T.

From a physical point of view, it was important, to reach field ranges in which the carriers cycle several times in closed orbits before they are scattered. This condition is expressed by

$$\mu B \gg 1. \qquad (3)$$

Not only the techniques to produce low temperatures and high fields made progress, but also the methods of growing semiconductors improved to such an extent that samples with high mobility values became available.

According to Bohr's law

$$\oint p\,dq = nh \qquad (4)$$

(p momentum, q coordinate of the carrier, n integer, h Planck constant) quantum effects occur under this condition.

One of the quantum effects occurring in semiconductors at low temperatures and in high magnetic fields are oscillations in the magnetoresistance, called Shubnikov-de Haas oscillations (Fig. 3) /2/. Their investigation leads to the band structure of the material and to the shape of the surfaces of constant energy in k space.

In 1957 Schrieffer mentioned for the first time the possibility of inversion layers on the surface of semiconductors where the motion of the

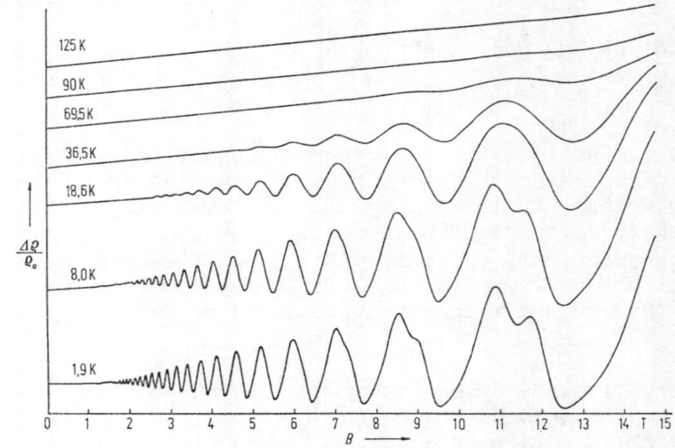

Fig. 3 Magnetoresistance curves of a HgSe sample. With decreasing temperature pronounced Shubnikov-de Haas oscillations occur /2/.

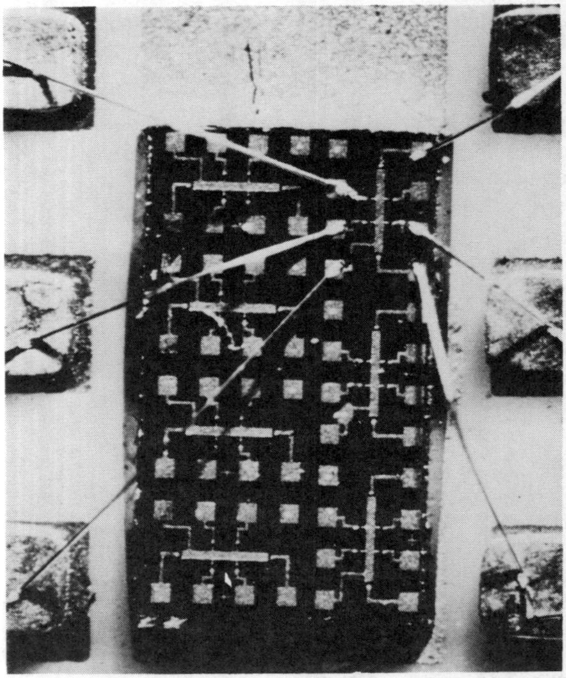

Fig. 4 Silicon chip with seven MOSFET's for resistance and Hall resistance measurements. Only one is bonded. This sample is one of the samples used by von Klitzing during the discovery of the quantum Hall effect.

carriers should be restricted in such a way that they could only move parallel to the surface /3/, and they should form a two-dimensional electron gas (2 DEG). This 2 DEG was first realized in MOSFET's by Fowler et al. /4/. In Japan, theories on transport in a 2 DEG in the presence of a high magnetic field were developed /5/.

All these prerequisites, listed from a personal point of view and not comprehensive, provided the essential background when Klaus von Klitzing went to the High Field Magnet Laboratory in Grenoble with a couple of silicon transistors in order to measure the Hall effect and made his famous discovery.

2. THE DISCOVERY OF THE QUANTUM HALL EFFECT

From 1979 to 1980, Klaus von Klitzing investigated the Hall effect in silicon MOSFET's (Metal Oxide Semiconductor Field Effect Transistor, Fig. 4) using the high steady magnetic fields of the High Magnetic Field Laboratory in Grenoble.

A typical silicon MOSFET is shown in Fig. 5. It works like a capacitor with an aluminium electrode on one side and p-type silicon on the other. Between these two sides, a gate voltage U_G is applied and carriers are influenced on the interfaces of the insulator and the semiconductor. The number of carriers is proportional to the applied gate voltage. When a voltage source is applied to source and drain electrodes of the sample which are also made of aluminium, a current passes through a conducting channel at the interface of the semiconductor and the oxide. This current can be controlled by the gate voltage, especially as higher the gate voltage, the higher the conductivity of the channel will be, since the number of carriers increases with increasing gate voltage. The dependence of the Hall voltage - which is inversely proportional to the carrier concentration - as a function of the gate voltage should therefore be of the shape shown in Fig. 6. Von Klitzing measured this dependence and the experimentally observed curve was in principle as expected, but under the conditions which had been present in Grenoble, (B = 19.5 T,T = 1.5 K), superimposed structures which had been previously observed /6/ turned out to be steps at which the Hall voltage as a function of the gate voltage did not change (Fig. 7). Moreover, his careful measurement with a digital voltmeter showed that the ratio of the Hall voltage to the source-drain current, the Hall resistance, at these steps had values like 12 906 Ω or 6 453 Ω. The same values could be measured on different samples with different shapes. Von Klitzing was very surprised at this, since it was well known that Hall voltage measurements depend on the individual geometry and carrier concentration of a sample. As can be seen from his notebook (Fig.8) /7/ it took him only 3 lines to uncover the mystery of these numbers. He showed that they are integer fractions of h/e^2 where h is the Planck constant and e the elementary charge. As can be seen from Fig. 9, the longitudinal voltage drop becomes zero within the resolution of the measuring system whenever the Hall voltage shows a plateau /8/.

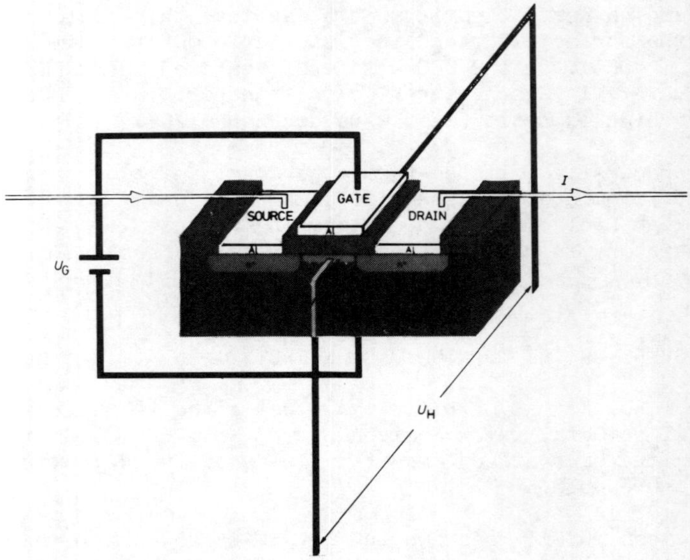

Fig. 5 Scheme of a MOSFET.
(Metal Oxide Field Effect Transistor)
Source and drain are the probes for the current I. The gate voltage U_G is applied between the gate probe and the p-Si substrate. The Hall voltage U_H occurs perpendicular to the current. The two-dimensional electron gas (2 DEG) is built at the interface of SiO_2 and p-Si.

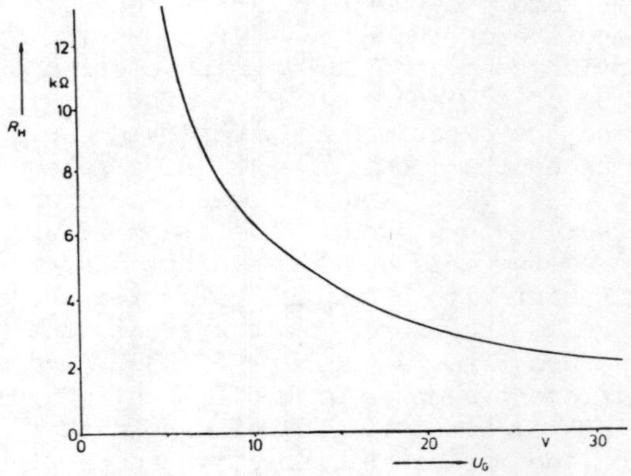

Fig. 6 Expected dependence of the Hall resistance R_H of a MOSFET on the gate voltage U_G according to the model of the free electron gas. This dependence can in reality only be observed at higher temperatures, at which the quantum Hall effect does not develop.

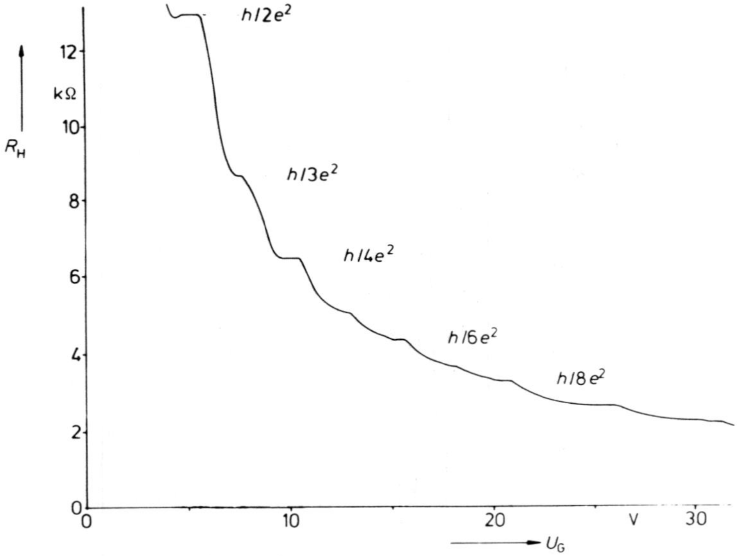

Fig. 7 Dependence of the Hall resistance R_H on the gate voltage U_G as measured by von Klitzing at B = 18.9 T and T = 1.5 K. The plateaus occur at integer fractions of h/e^2 /8/.

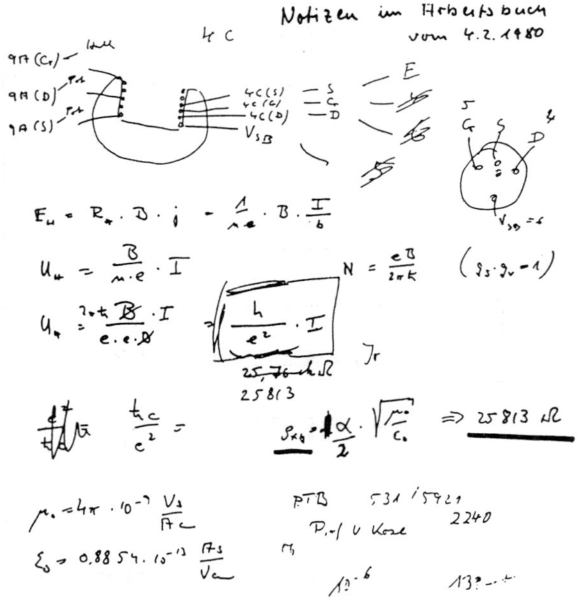

Fig. 8 Part of a page from von Klitzing's notebook with his remarks during the discovery of the quantum Hall effect /7/.

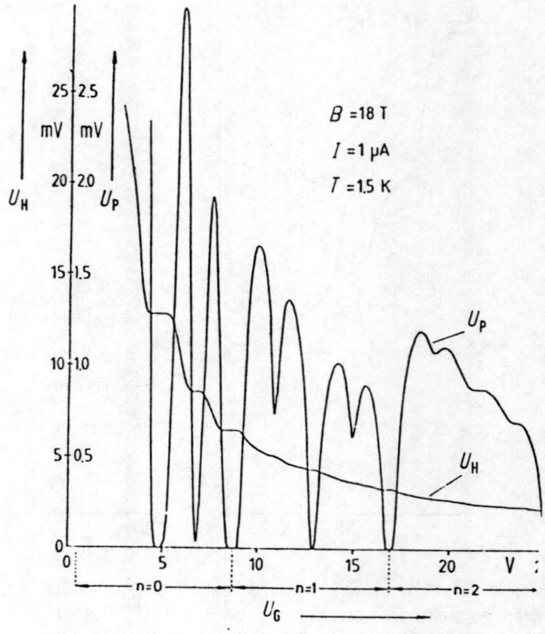

Fig. 9 Von Klitzing's originally published curves for the dependence of the Hall voltage U_H and the longitudinal voltage drop U_P as a function of the gate voltage U_G measured with a silicon MOSFET as shown in Fig. 4 /8/.

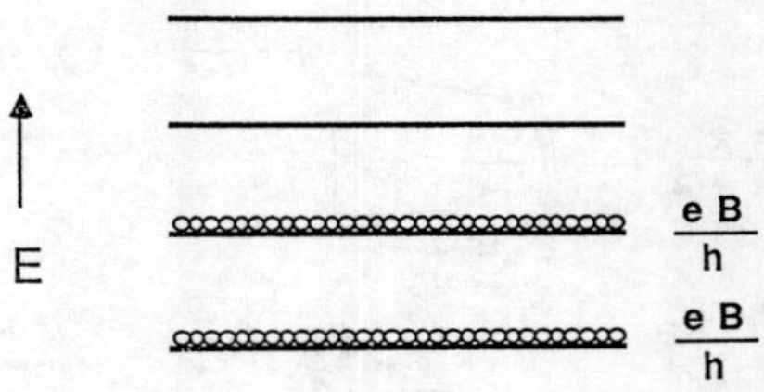

Fig. 10 Scheme of Landau levels.
Each of them can be occupied by eB/h electrons.

3. THE OCCURRENCE OF QUANTIZED RESISTANCE VALUES

The gate voltage applied to a MOSFET leads to an n-conducting inversion layer in the p-type semiconductor close to the insulating silicon oxide layer. Since the width of the occurring triangular potential well is in the order of the wavelength of the electrons, the energy is quantized with respect to this direction into electrical sub-bands E_i according to /9/

$$E_i = \left(\frac{\hbar^2}{2m_3^*}\right)^{1/3} \cdot \left(\frac{3}{2}\pi eF\right)^{2/3} \cdot \left(i + \frac{3}{4}\right)^{2/3} \qquad (5)$$

$$(i = 0,1,2,\dots)$$

where $m_3^* = 0.1\ m_o$ and F is the electrical field which is in the order of 10^5 V/cm. If only the lowest sub-band is occupied, the motion of the electrons is restricted to a plane parallel to the interface of the semiconductor and the insulator. The electrons form a 2 DEG.

The application of an additional strong magnetic field ($\mu B \gg 1$) forces the electrons to move in closed orbits. The energy of each electrical sub-band is once again quantized into the Landau levels, separated by $\hbar\omega$, where

$$\omega = \frac{eB}{m_c^*} \qquad (6)$$

is the cyclotron frequency (m_c^* is the effective cyclotron mass). Due to the valley and spin degeneracy in silicon, each of these Landau levels is again split into 4 sublevels. By means of quantum mechanics /10/ it can be shown that each of these levels is degenerated by the number n_Φ of flux quanta h/e. With the restriction to a unit of area, the number of electrons in one energy level is therefore given by

$$n_\Phi = B/(h/e). \qquad (7)$$

If i energy levels are fully occupied the number of electrons n_i is given by

$$n_i = i \cdot n_\Phi = i\,\frac{eB}{h}. \qquad (8)$$

The classical derivation of the Hall voltage in a two-dimensional electron gas where the density of the current I is given by $j = I/b$, b being the width of the sample, leads to the following expression

$$U_H = \frac{1}{ne} \cdot B \cdot I. \qquad (9)$$

(n is the density of the 2 DEG).

In contrast to the situation in a three dimensional electron gas, this voltage is independent of the geometry of the sample.

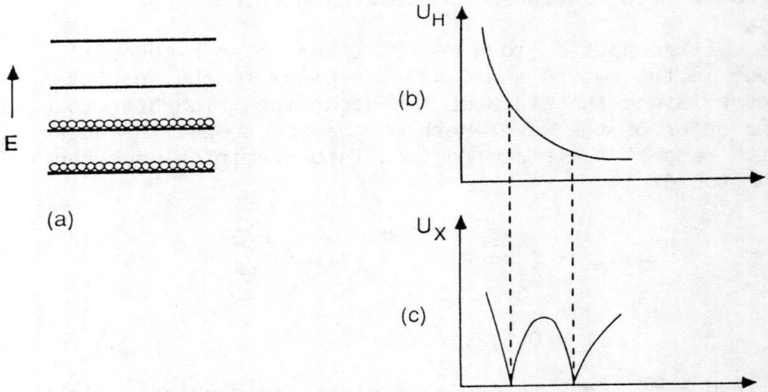

Fig. 11 (a) Energy levels for the electrons in Fig. 10 on the basis of the model of a free electron gas. (b) According to this model no plateaus occur in the dependence of the Hall voltage U_H on the gate voltage U_G. (c) At certain values of the gate voltage U_G - where i (i= 1,2,..) Landau levels are totally occupied by electrons - the longitudinal voltage drop U_x would become zero, but no plateaus would develop. At exactly this gate voltage the Hall resistance should have the quantized value $1/i \cdot h/e^2$.

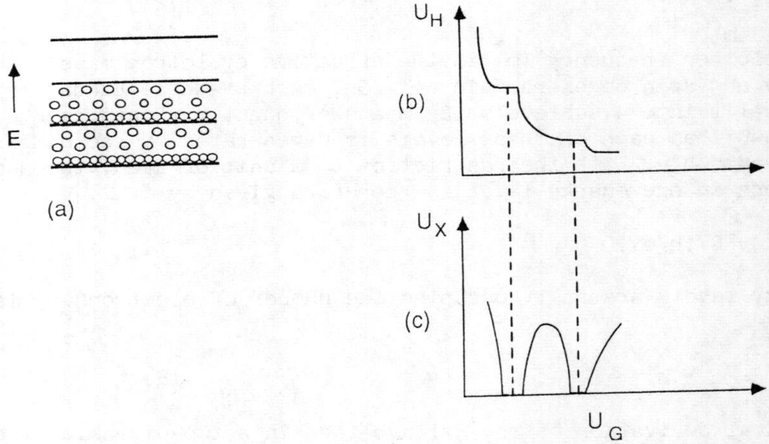

Fig. 12 (a) The model which is proposed for the electrons is now expanded. Besides the extended states of the free electron gas, additional states for electrons exist which are not extended but localized. (b),(c) In this case the Hall voltage U_H and the longitudinal voltage drop U_x show plateaus with $U_x \approx 0$ and $U_H = 1/i \cdot h/e^2 \cdot I$. (I current through the sample).

The explanation for the observation of certain distinguished Hall resistance values is as follows. A MOSFET is put into a steady magnetic field B. Each energy level of the energy spectrum of the 2DEG can be occupied by eB/h electrons. If the number of carriers is now increased by increasing the gate voltage, at certain gate voltage values exactly one, two or three and so forth, energy levels will be totally occupied (Fig. 10). At these gate voltage values the carrier concentration n_i is given by

$$n_i = i \cdot \frac{eB}{n} \qquad (10)$$

At this carrier concentration the Hall voltage $U_{H,i}$ is given by

$$U_{Hi} = \frac{1}{n_i e} B \cdot I = \frac{1}{i\frac{eB}{h}e} \cdot B \cdot I = \frac{1}{i} \cdot \frac{h}{e^2} I \qquad (11)$$

Since $\frac{h}{e^2} \approx 25\ 813\ \Omega$, 12 906 Ω is close to $\frac{1}{2}\frac{h}{e^2}$ and 6 453 Ω is close to $\frac{1}{4}\frac{h}{e^2}$. The distinguished resistance values observed by von Klitzing occur in the case of totally occupied energy levels. In this case - on the assumption that $kT \ll \hbar\omega$ - a scattering of the electrons is not possible and therefore the longitudinal resistance drops to zero. However, these considerations do not explain the occurrence of plateaus, which means that the Hall voltage remains constant while the gate voltage increases. The model described above would lead to the kind of experimental situation which is shown in Fig. 11. According to this model the gate voltage would be adjusted to that value at which the longitudinal voltage drop has its minimum. Exactly at this gate voltage, the Hall voltage should have the quantized value. There should be no indication of a step.

Although up to now there is no quantitative theory on the quantum Hall effect, most of the theories explain the occurrence of plateaus by localized states between the energy levels for the extended states (Fig. 12).

4. THE IMPACT OF THE DISCOVERY OF THE QUANTUM HALL EFFECT ON METROLOGY IN SUMMER 1980:

When von Klitzing reported his results for the first time at an international conference in the summer of 1980 at the CPEM 80 in Braunschweig /11/, the situation with respect to the main electrical units was as follows:

1.) The base electrical unit is the ampere. The definition of the ampere includes the definition of the permeability of vacuum $\mu_o = 4\pi 10^{-7}$ Ωs/m. The realization of the ampere in 1980 was only possible with an uncertainty of $6 \cdot 10^{-6}$.

2.) The volt is a derived unit. Its definition in base units is:

$$1\ V = 1\ W/A = 1\ kgm^2s^{-3}A^{-1}. \qquad (12)$$

its realization - depending on the ampere - was also only possible with an uncertainty of $6 \cdot 10^{-6}$.

However, it was possible to reproduce a certain voltage created with a Josephson element, with a relative uncertainty of 10^{-8}. This voltage is given by

$$U = \frac{h}{2e} \cdot \nu \qquad (13)$$

Where ν is the frequency applied to the Josephson element. Since 1972 the as-maintained unit of voltage, e.g. at PTB, has been related to a Josephson voltage obtained with a frequency of 483 594 GHz. This value had been recommended by the Comité Consultatif d'Électricité (CCE) in order to achieve international uniformity in the dissemination of the volt. It was well known that this Josephson effect-based volt did not agree with the SI volt, but the possibility of reproducing this voltage with an uncertainty of two orders of magnitude lower than the one for the SI value justified the international agreement.

3.) The ohm is also a derived unit:

$$1\ \Omega = 1\ V/A = 1\ kgm^2s^{-3}A^{-2}. \qquad (14)$$

However, its realization was possible with a lower uncertainty than that of the ampere. The reason for this was that the realization of the ohm was not based on a balance, but made use of the defined value for the permittivity of vacuum. Thompson and Lampard had shown that a capacitor can be built with a capacitance C which depends only on a length l /12/

$$C = \frac{\varepsilon_o \ln 2}{\pi} \cdot l \qquad (15)$$

$$(\varepsilon_o = \frac{1}{c_o^2 \mu_o}, \text{ permittivity of vacuum})$$

At a certain freqency ω an impedance $1/\omega C$ is obtained. A decimal fraction of this impedance can be compared with a dc resistance (at PTB this comparison is done at 1 kΩ). From this dc resistance a chain of resistors can be built up, finally leading to a realization of the ohm. This complicated method led in 1980 to a realization of the SI ohm with an uncertainty in the order of 10^{-7}.

Once derived, the ohm was maintained by a group of selected 1 Ω standard resistors - the PTB group consists of 7 standard resistors. Standard resistors from various laboratories were regularly compared either bilaterally or in the course of an international intercomparison at BIPM. The differences in the as-maintained units were in the order of a few micro-ohms (Fig. 13). The intercomparison itself was made with relative uncertainties of 10^{-8} to 10^{-7}. Although the differences in the as-maintained units between the various laboratories had been known at the date of an intercomparison, extrapolations gave rise to additional uncertainties.

In this situation, von Klitzing showed that similiar to the Josephson effect, a quantum effect exists which allows the reproduction of certain distinguished resistance values independent of space and time.

With respect to the precision measurement of fundamental constants, the situation was as follows:

h/e^2 could easily be derived from the fine-structure constant α:

$$\frac{h}{e^2} = \frac{\mu_o c_o}{2} \cdot \alpha^{-1} \qquad (16)$$

α could be determined by two significant experiments:

The experimentally determined anomalous magnetic moment of the electron and a quantum electrodynamic analysis led to /14/

$$\alpha^{-1} = 137.035\,993 \pm 0.000\,011. \qquad (17)$$

On the other hand, the measurement of the gyromagnetic ratio of the proton, the Josephson frequency-voltage ratio and the conversion factor of the ohm at NBS led to /15/

$$\alpha^{-1} = 137.035\,963 \pm 0.000\,015. \qquad (18)$$

As can be seen, the two values do not agree with each other. At that time this initiated a discussion about the possible validity ranges of quantum electrodynamics.

Many questions arose during CPEM '80, most of them concerned with the following points:

1.) To what extent do the measured values R_{Hi} agree with those derived from the fine-structure constant α,

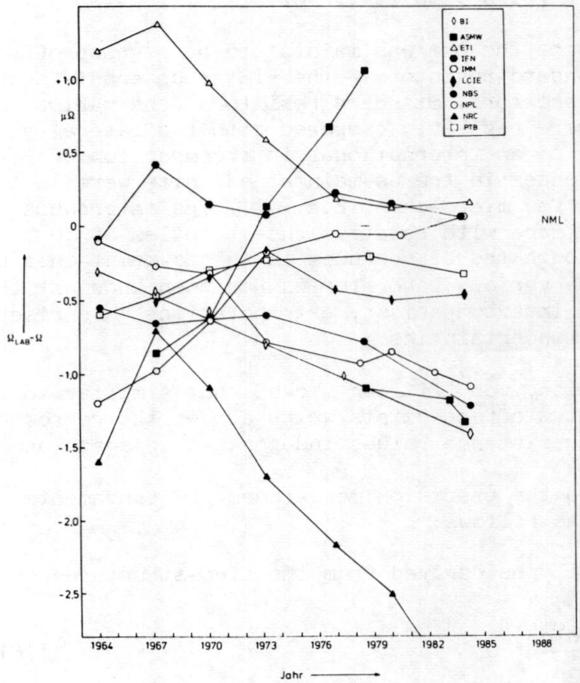

Fig. 13 Changes with time of the as-maintained units of resistance in various national institutes relative to the one at NML, (National Measurement Laboratory, Australia) which is supposed to be identical with the SI ohm to within $2 \cdot 10^{-7}$ /13/.

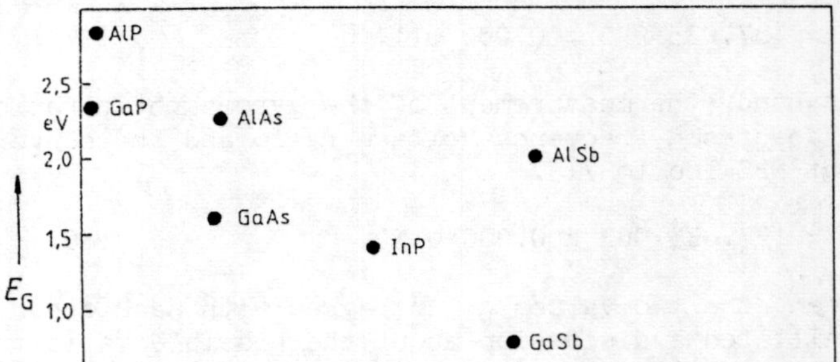

Fig. 14 Energy gap E_G as a function of the lattice constant a of various III/V semiconductors /17/.

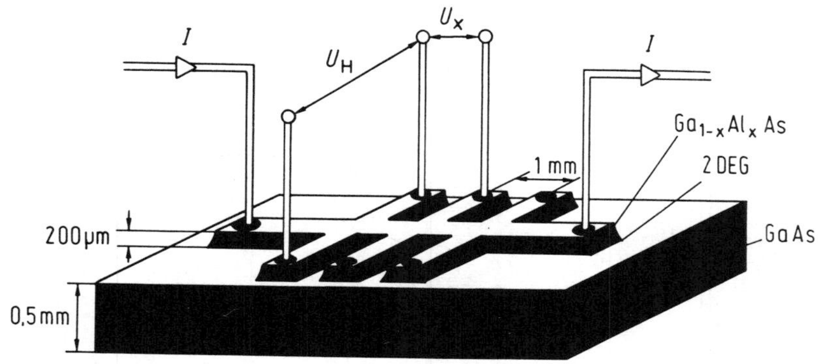

Fig. 15 Sample prepared from a GaAs-$Ga_{1-x}Al_xAs$ heterostructure. (I current through the sample, U_x longitudinal voltage drop, U_H Hall voltage, 2 DEG two-dimensional electron gas at the interface of the GaAs and $Ga_{1-x}Al_xAs$).

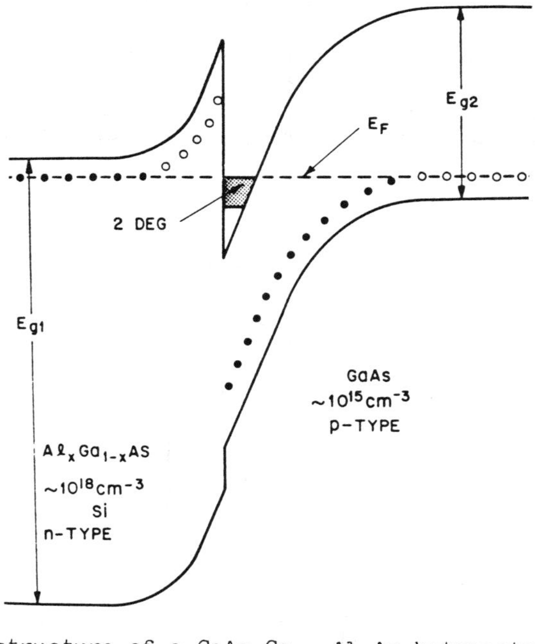

Fig. 16 Energy structure of a GaAs-$Ga_{1-x}Al_xAs$ heterostructure at the interface of the two layers. The electrons of the donators which are above the Fermi energy E_F form the two-dimensional electron gas (2 DEG) occurring in the triangular potential well. E_{g1} and E_{g2} are the energy gaps in $Ga_{1-x}Al_xAs$ and GaAs, respectively /18/.

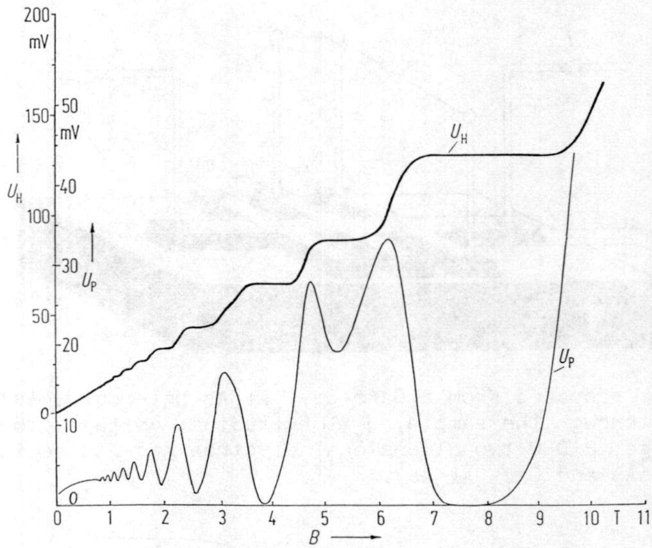

Fig. 17 Hall voltage U_H and longitudinal voltage drop U_P of a GaAs-$Ga_{1-x}Al_xAs$ heterostructure as a function of the magnetic field B at a temperature of 1.5 K, as measured at PTB and as typically used in high-precision measurements.

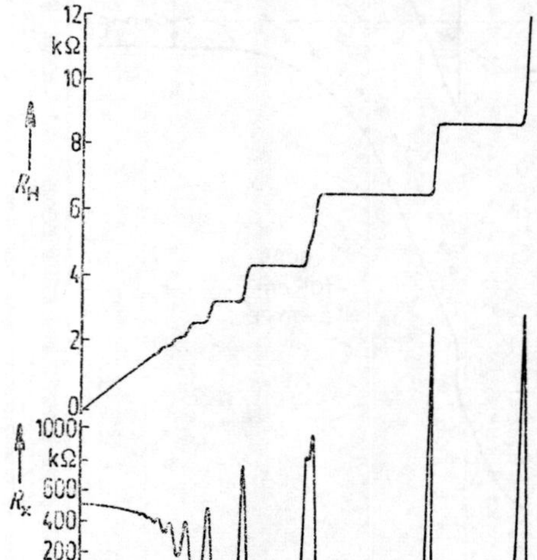

Fig. 18 Hall resistance R_H (upper part) and longitudinal resistance R_X (lower part) of a GaAs-$Ga_{1-x}Al_xAs$ heterostructure as a function of the magnetic field B at a temperature of 0.008 K, as measured by von Klitzing et al /19/.

$$R_{Hi} = \frac{\mu_o c_o}{2 \cdot i} \cdot \alpha^{-1} \quad ? \qquad (19)$$

2.) How sensitive are the measured values to changes of parameters like temperature, magnetic field, material and sample properties, step number etc?

3.) Will it be possible to lower the relative uncertainty of the reproducibility of a quantized resistance value to such a value ($\sim 10^{-8}$) that the effect could be used in a way similar to the Josephson effect?

Experimental results which provide an answer to these questions are given in part II.

5. THE QUANTUM HALL EFFECT IN HETEROSTRUCTURES

Shortly after the publication of von Klitzing's discovery, Tsui et al. /16/ reported the observation of the quantum Hall effect in a GaAs-$Ga_{1-x}Al_xAs$ heterostructure. This material is quite different from a MOSFET. As shown in Fig. 14, AlAs and GaAs have almost the same lattice constant but different energy gaps. Single crystals can therefore be obtained by substituting atomic layers of GaAs with AlAs or $Ga_{1-x}Al_xAs$ layers. A heterostructure (Fig. 15) which can be produced, for instance, by molecular beam epitaxy (MBE) consists of a GaAs substrate with epitaxial layers of GaAs which are superimposed by $Ga_{1-x}Al_xAs$ layers with $x \approx 0.3$. Due to the difference in the energy gap of the two materials (Fig. 16) a triangular potential well is built up at the interface, and under conditions similar to those ocurring in the MOSFET a 2 DEG is created /18/. The electrons in this case stem from Si donators in the AlGaAs layer. The mobility of the electrons of this 2 DEG can be tremendously increased by a spacer layer. This is formed by a thin layer of GaAlAs without Si donators. Mobilities of up to 200 T^{-1} have been reported in the meantime (usual mobility values of Si MOSFET's are 2 T^{-1}). Due to these high mobilities the quantum Hall effect can be observed at much lower magnetic fields than in silicon. Typical dependences of the Hall voltage and the longitudinal voltage as a function of the magnetic field obtained at PTB and at very low temperatures by von Klitzing's group /19/ are shown in Figs. 17 and 18. In contrast to the situation in MOSFET's, in GaAs heterostructures the carrier concentration is usually given by the concentration of the silicon donators and is not varied by a gate voltage.

Although the quantum Hall effect is very pronounced in these materials at extremely low temperatures, the optimum conditions for high precision measurement are at temperatures around 1 K, because a certain current through the sample is needed in order to get a high signal-to-noise ratio.

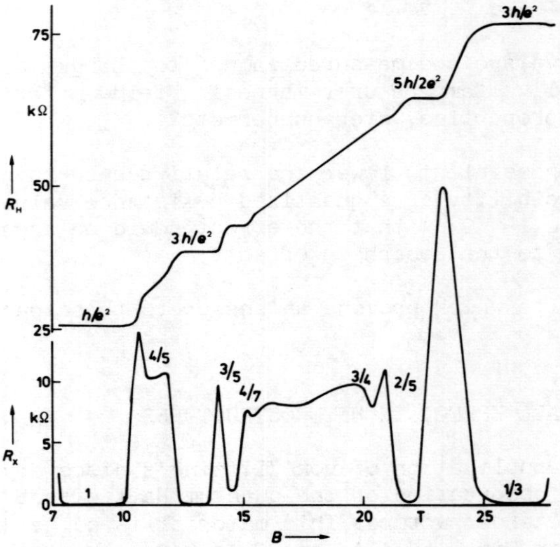

Fig. 19 Hall resistance R_H (upper part) and longitudinal resistance R_X (lower part) of a $GaAs$-$Ga_{1-x}Al_xAs$ heterostructure as a function of the magnetic field B in the quantum limit. Plateaus occur not at integer numbers of occupied energy levels, but at fractional ones. This effect is called the fractional quantum Hall effect /21/.

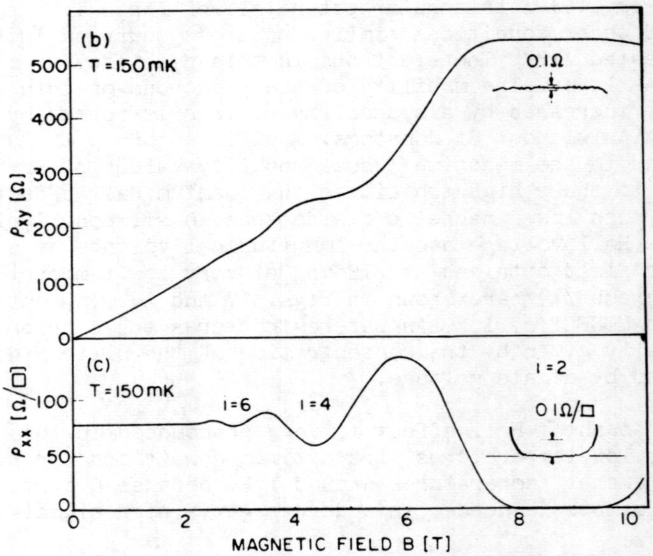

Fig. 20 Hall resistivity ρ_{xy} (upper part) and longitudinal resistivity ρ_{xx} (lower part) as a function of the magnetic field B of a special device consisting of 30 heterostructures /22/.

6. THE FRACTIONAL QUANTUM HALL EFFECT

Tsui et al. /20/ first reported the observation of Hall plateaus and minima in the longitudinal resistance which occurred at values of $i = 1/3$, $2/3$ and other fractions (Fig. 19). The observation of these steps is restricted to high-mobility samples, high magnetic fields and very low temperatures. Up to now a great many fractional values have been observed, all of them with odd denominators. The explanation of the fractional quantum Hall effect is based on many body effects, and seems to be different from the explanation of the integer effect.

7. LOWER QUANTIZED RESISTANCE VALUES

Störmer et al. /22/ showed that in a sample with 30 heterostructures in parallel, a superlattice, lower quantized values such as $\frac{h}{48e^2} \approx 598\ \Omega$ could be observed (Fig. 20). Obviously only 24 heterostructures were active.

REFERENCES

/1/ K.R. Sopka: *The Discovery of the Hall Effect: Edwin Hall's hitherto unpublished Account. The Hall Effect and its Application*. Edited by C.C. Chien and C.R. Westgate, Plenum Press, New York, 1980, p. 523.

/2/ E. Braun and G. Hein: Wiss. Ber. HMFA Braunschweig 3, 27 (1975).

/3/ J.R. Schrieffer: *Semiconductor Physics*. Edited by R.H. Kingston, Univ. of Pennsylvania Press, Philadelphia, 1957, p. 55.

/4/ A.B. Fowler, F.F. Fang, W.E. Howard and P.J. Stiles: Phys. Rev. Lett. 16, 901 (1966).

/5/ e.g. T. Ando, Y Matsumoto and Y. Vemuva: J. Phys. Soc. Japan 39, 279 (1975).

/6/ Th. Englert and K.v. Klitzing: Surface Science 73, 70 (1978).

/7/ I. Lemmerich: *Maß und Messen*. Katalog zur Ausstellung aus Anlaß des 100-jährigen Jubiläums der PTB. Physikalisch-Technische Bundesanstalt, Braunschweig und Berlin, 1987.

/8/ K.v. Klitzing, G. Dorda and M. Pepper: Phys. Rev. Lett. 45 (1980), 494 - 497.

/9/ G. Landwehr: Festkörperprobleme XV p. 49 (1975).

/10/ e.g. Ch. Kittel: Einführung in die Festkörperphysik, R. Oldenbourg Verlag München, Wien 1968, S. 349.

/11/ K.v. Klitzing: Conference on Precision Electromagnetic Measurements (CPEM '80), Braunschweig, Fed. Rep. of Germany, June 24, 1980.

/12/ A.M. Thompson and D.G. Lampard: Nature 177, 888 (1956).

/13/ H. Bachmair, private communication.

/14/ T. Kinoshita and W.B. Lindquist: Phys. Rev. 47, 1573 (1981).

/15/ E.R. Williams and P.T. Olsen, Phys. Rev. Lett. 42, 1575 (1979).

/16/ D.C. Tsui and A.C. Gossard: Appl. Phys. Lett. 38, 550 (1981).

/17/ N.T. Linh: Festkörperprobleme XXIII, 227 (1983).

/18/ H.L. Störmer: Festkörperprobleme XXIV, 25 (1984).

/19/ G. Ebert, K.v. Klitzing, C. Probst and K. Ploog: Sol. State Comm. 44 (1982), 95 - 98.

/20/ D.C. Tsui, H.L. Störmer, and A.C. Gossard: Phys. Rev. Lett. 48, 1559 (1982).

/21/ A.M. Chang, P. Berglund, D.C. Tsui, H.L. Störmer, and J.C.M. Hwang: Phys. Rev. Lett. 53, 997 (1984).

/22/ H.L. Störmer, J.P. Eisenstein, A.C. Gossard, W. Wiegmann, and K. Baldwin: Phys. Rev. Lett. 56, 85 (1986).

THE QUANTUM HALL EFFECT PART II: METROLOGICAL APPLICATIONS

E. Braun
Physikalisch-Technische Bundesanstalt
Bundesallee 100
3300 Braunschweig
Federal Republic of Germany

ABSTRACT. At first a general view of the problems concerned with the precision measurement of quantized Hall resistances is given. Experimental results obtained in various laboratories are presented. They include the influence of parameters like temperature, different materials, sample preparation etc. Among the applications of the quantum Hall effect are new possibilities for the determination of the fine-structure constant, the realization of the ohm, and the monitoring of the as-maintained units of resistance. Finally, the proposed recommendation of the Comité Consultatif d'Électricité is explained. This recommendation attempts to achieve international uniformity in the as-maintained units via commonly accepted numerical values for the quantized Hall resistances.

1. INTRODUCTION

In Part I it was shown that the quantum Hall effect is observable in a two-dimensional electron gas (2 DEG) realized in a MOSFET or in a GaAs-$Ga_{1-x}Al_xAs$ heterostructure in a strong magnetic field: Ranges of the applied gate voltage or the applied magnetic field exist where the Hall resistance becomes an integer fraction of h/e^2 and the longitudinal resistance diminishes within the resolution of the experimental set-up.

In this paper we shall first present some general aspects of the precision measurement of quantized Hall resistances. Afterwards we will describe experimental methods including necessary tests of the samples. After presenting experimental results including the influence of parameters like temperature, different materials, sample preparation etc, the possible metrological applications will be described. They include a high precision determination of the fine structure constant, the determination of other fundamental constants and the conversion factor of the ohm, or make use of the high reproducibility of quantized Hall resistances. Finally, the proposed recommendation of the Comité Consultatif d'Électricité aimed at international uniformity in resistance calibrations will be discussed.

V. De Sabbata and V. N. Melnikov (eds.),
Gravitational Measurements, Fundamental Metrology and Constants, 39–58.

2. GENERAL ASPECTS OF PRECISION MEASUREMENT OF QUANTIZED HALL RESISTANCES

The measurement of a resistance R_X means that the current I_X through the resistor and the voltage drop U_X across the resistor are measured at the same time t_o:

$$R_X = \frac{U_X(t_o)}{I_X(t_o)} \qquad (1)$$

The current through the resistor is usually measured through the voltage drop U_N across a standard resistor with known resistance R_N, which is connected in series with the unknown resistor as shown in Fig. 1. Generally, the same instrument is used and there is a time difference Δt between the two measurements. The unknown resistance is then measured as

$$R_X = \frac{U_X(t_o)}{U_N(t_o+\Delta t)} \cdot R_N \qquad (2)$$

This simple model already permits discussion of most of the problems associated with the resistance measurement if a high resolution like one part in 10^8 is aspired to:

1.) The insulation resistance of the circuit must be such high ($\geq 10^{12}\ \Omega$) that the total current flowing through R_X also flows through R_N.

2.) The current must be stable over a period of time Δt between the two measurements.

3.) If the values of the two resistances differ, the linearity of the instrument must be checked.

4.) No additional voltages such as thermal emf's should occur in the measuring circuit. However, if they do occur they should not change with time over one measuring cycle, which includes measurements with reversed current.

5.) Care should be taken that no rf noise is picked up in parts of the circuit and rectified.

6.) Noise arising from resistors or other sources should be avoided reduced by operating at cryogenic temperatures.

7.) Since the value of the standard resistor depends on temperature, pressure, dissipated power and probably on time, the appropriate coefficients for a correction must be known.

It is clear that the importance of these disturbing influences increases with decreasing uncertainty. While a relative uncertainty of 10^{-5} for the resistance ratio could easily be obtained using a commercial digi-

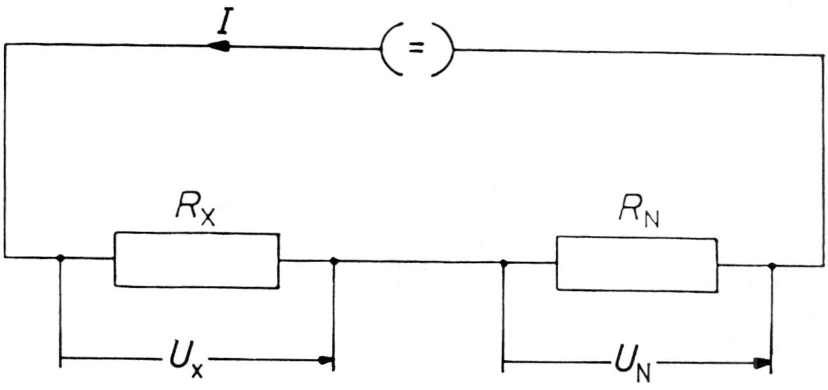

Fig. 1 Scheme of a dc-resistance measurement. The voltage drops U_x and U_N across the unknown resistor R_X and across the standard resistor R_N are measured (I current).

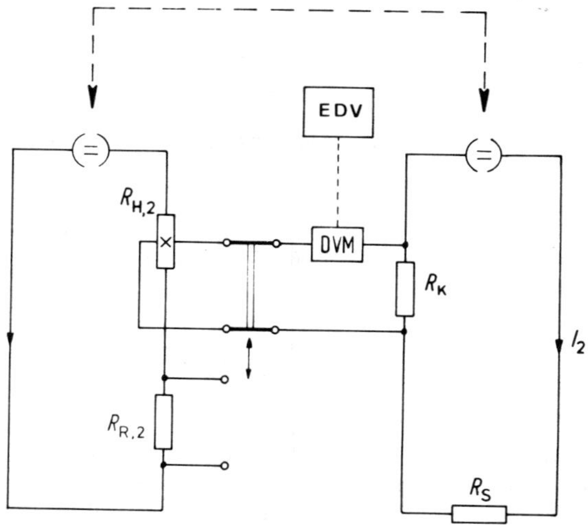

Fig. 2 Schematic diagram of the early PTB measurement circuit: The current I flows through the sample with a Hall resistance R_{H2} and through a reference resistance R_{R2}. An almost equal current I_2 flows through a potentiometer resistance R_K and a serial resistance R_S. The two current sources are in close thermal contact. (DVM digital nano voltmeter, EDV electronic data processing). $R_{H2} \approx R_{R2} \approx R_K \approx R_S \approx 12\ 906\ \Omega$.

tal multimeter with sufficient high resolution, a relative uncertainty of 10^{-8} involves a lot of care. The realization of the SI ohm via a calculable capacitor in any case limits the relative uncertainty of a quantized Hall resistance in SI units to about 10^{-7}. Any further accuracy relates only to the ratio of a Hall resistance to the resistance of a certain standard resistor.

Measuring methods for resistance ratios with very low uncertainties had been developed before the discovery of the quantum Hall effect, preferable for ratios close to 1 and powers of 10. High quality resistance standards had been developed only for decimal resistance values. However, even their resistances change with time and they also depend on temperature, pressure and dissipated power. These resistors were developed to be used at room temperatures.

Since the quantized Hall resistances occur at odd numbers like 12 906.4 Ω or 6 453.2 Ω, their measurement has often been split into several steps:

1.) A 1 : 1 comparison of a quantized Hall resistance with the resistance of a reference resistor with the same nominal value. It is of advantage to use a cryogenic reference resistor in order to avoid thermal noise.

2.) The comparison of this reference resistance with the closest decimal standard resistance value.

3.) If necessary, a recalibration of this standard resistor.

As an example we describe the 1 : 1 method used at PTB at the beginning of the quantum Hall effect experiments, with a reference resistor kept at 20 °C.

3. EXPERIMENTAL METHOD FOR 1 : 1 RESISTANCE RATIO COMPARISONS

Precision measurements of the Hall resistance were carried out using a potentiometric substitution method, shown schematically in Fig. 2. The two circuits were made as symmetrical as possible. In particular, the current I_2 was chosen equal to I_1, and for R_R, R_K and R_S resistors of the same type were used with nominal resistance values almost equal to $R_{H,2}$, i.e. 12 906.4 Ω. Great care was taken to avoid leakage currents and to keep both current sources and all reference resistors at one uniform temperature. Readings of the small differences between the voltages were taken from a digital nanovoltmeter by a desk computer.

4. EXPERIMENTAL RESULTS

4.1. Reproducibility

With the method described above we showed that the ratio of a quantized

Hall resistance to a reference resistance of the same nominal value could be reproduced with a standard deviation of less than $3\cdot10^{-8}$ independently of the Hall sample used (Fig. 3) /1/. Both reference resistances were corrected for a linear increase with time which is known to occur for this type of resistor. They had been kept at a constant temperature within 1 mK. However, a possible pressure dependence had not been corrected at the time.

4.2. Temperature dependence

On a sample with its i = 2 step at a magnetic field of about 9 T, we did not observe any systematic deviations within an (1 σ) uncertainty larger than $3\cdot10^{-8}$ when changing the temperature from 4.2 K to 1.2 K /2/.

Other groups did observe temperature dependences /3/, but at lower magnetic fields. Cage et al. /4/ showed that this temperature dependence not only occurs in the Hall resistivity ρ_{xy} but also in the longitudinal resistivity ρ_{xx}. They found that

$$|\Delta\rho_{xy}| = |\rho_{xy}(\rho_{xx}^{MIN} = 0) - \rho_{xy}| = s \cdot \rho_{xx}^{MIN} \qquad (3)$$

where $\rho_{xy}(\rho_{xx}^{MIN} = 0)$ is the value of ρ_{xy} linearily extrapolated to $\rho_{xx}^{MIN} = 0$ and ρ_{xx}^{MIN} is the minimum value observed for ρ_{xx}. The constant s depends on the sample and its preparation. Von Klitzing and coworkers showed that /5/

$$\rho_{xx}^{MIN} \sim \exp\left(-\frac{\hbar\omega}{2kT}\right) \qquad (4)$$

This behaviour shows that in higher magnetic fields (higher values of ω), the temperature dependence in a certain temperature range is lower than it is in lower magnetic fields.

4.3. Dependence on different samples and step numbers

We compared the i = 2 step of a GaAs-$Ga_{1-x}Al_xAs$-heterostructure with the i = 4 step of a Si MOSFET by the following procedure /2/:

1) comparison of $R_{H,2}$ of a GaAs sample with a reference resistance $R_{R,2}$;

2) comparison of $R_{H,4}$ of a Si MOSFET with a reference resistance $R_{R,4}$;

3) comparison of the resistance of a series connection of two auxiliary resistors, each with nominally 6453.2 Ω, with the reference resistance $R_{R,2}$;

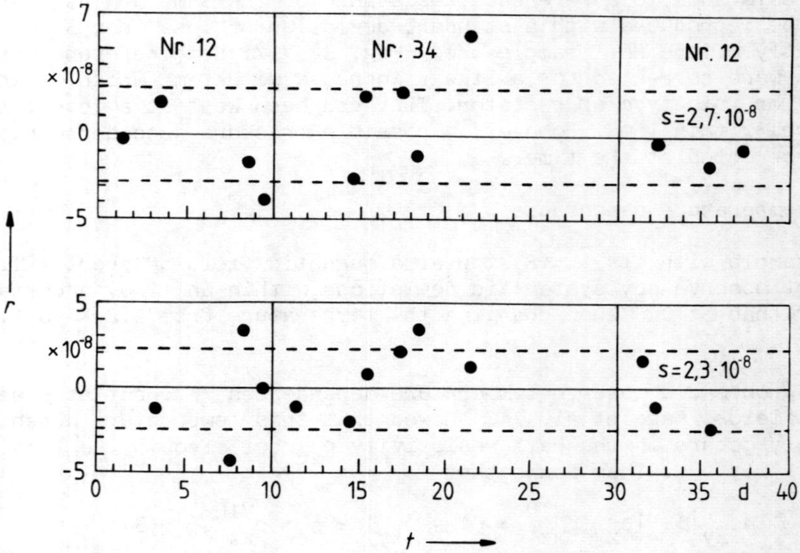

Fig. 3 Ratio of the Hall resistance R_{H2} (realized in two different GaAs heterostructures Nr. 12 and Nr. 34) to two reference resistances (upper and lower data) as a function of time. A least-squares-fitted straight line according to a linear change of the reference resistances has been subtracted in both sets of data. The remaining data sets are normalized to their mean value. The (1 σ) standard deviation is in both cases less than $3 \cdot 10^{-8}$.

4) comparison of each of these auxiliary resistances with the reference resistance $R_{R,4}$.

We obtained

$$\frac{R_{H,2}}{R_{H,4}} = 2 \cdot (1 + 3 \cdot 10^{-8} \pm 3 \cdot 10^{-8}). \qquad (5)$$

Our results on the reproducibility, the temperature dependence, the material and the step number have been verified by various other groups, also at other national institutes.

4.4. Comparison of results in different laboratories

The values for R_H obtained in different laboratories and hitherto published are summarized in Fig. 4.

As can be seen from the figure, the (1 σ) standard deviation of these results and the mean error bars of the individual results are all in the order of 10^{-7}, which is due to the limited knowledge of the ratios of the as-maintained units of resistance to the ohm.

The experimental results presented show that Hall resistances can be reproduced with an uncertainty 10 times lower than the uncertainty with which their value can be given in the SI unit.

These experimental results make possible a number of metrological applications of the quantum Hall effect.

5. METROLOGICAL APPLICATIONS OF THE QUANTUM HALL EFFECT

5.1. Determination of the fine-structure constant

The quantum Hall effect provides two possibilities to determine the fine-structure constant α. Since the inverse fine structure constant is given as

$$\alpha^{-1} = \frac{2}{\mu_o c_o} \cdot \frac{h}{e^2} \qquad (6)$$

where $\mu_o = (4\pi \cdot 10^{-7} \pm 0)\ \frac{Vs}{Am}$ according to the definition of the ampere and $c_o = (299\ 792\ 458 \pm 0)\ \frac{m}{s}$ according to the definition of the meter, a determination of $\frac{h}{e^2}$ is therefore a determination of the fine-structure constant as long as the ratio of the as-maintained unit of resistance Ω_{LAB} to the ohm is known. As already mentioned, the lowest obtainable uncertainty is limited by the uncertainty with which this ratio is known. However, there is a further possibility of determining the fine-structure constant by the simultaneous measurement of a quantized Hall resistance, the Josephson frequency - voltage ratio and the gyromagne-

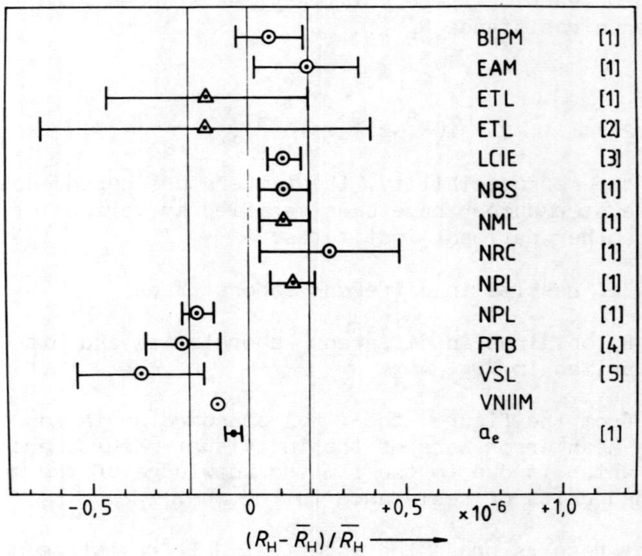

Fig. 4 Quantum Hall resistance measurements of various national institutes normalized to the mean value of all results compared. The vertical lines show the (1 σ) standard deviation of the mean value. These values are compared with the value calculated from the magnetic moment of the electron (a_e).

Δ result based on an Ω-realization in the laboratory; O result based on an 1 Ω comparison with BIPM or NML.

BIPM: Bureau International des Poids et Mesures, Paris.
EAM: Eidgenössisches Amt für Meßwesen, Switzerland.
ETL: Electrotechnical Laboratory, Japan.
LCIE: Laboratoire Central des Industries Electriques, France.
NBS: National Bureau of Standards, USA.
NML: National Measurement Laboratory, Australia.
NRC: National Research Council, Canada.
NPL: National Physical Laboratory, Great Britain
PTB: Physikalisch-Technische Bundesanstalt, Fed. Rep. of Germany.
VSL: Van Swinden Laboratory, Netherlands.
VNIM: Vsesojusnyi Nautschno Issledowatel'skij Institut Metrologii im D.I. Mendeleewa, Leningrad.

tic ratio of the proton $\gamma'_{P,low}$ /18/. All three do not have to be measured in SI units, but only in laboratory units:*)

$$\frac{h}{e^2} = \{\frac{h}{e^2}\}_{LAB} \cdot \Omega_{LAB} \qquad (7)$$

$$\frac{2e}{h} = \{\frac{2e}{h}\}_{LAB} \cdot V^{-1}_{LAB} \cdot s^{-1} \qquad (8)$$

$$\gamma'_{P,low} = \{\gamma'_{P,low}\}_{LAB} \cdot T^{-1}_{LAB} \cdot s^{-1} \qquad (9)$$

By making use of a further equation for the inverse fine-structure constant:

$$\alpha^{-2} = \frac{c_o}{4R_\infty} \cdot \frac{\mu'_P}{\mu_B} \cdot \frac{2e/h}{\gamma'_{P,low}} \qquad (10)$$

where the Rydberg constant R_∞ and the proton magnetic moment in Bohr magnetons μ'_P/μ_B are known with very low uncertainties, one obtains for α^{-1}:

$$\alpha^{-1} = (\frac{\{c_o\}}{4\{R_\infty\}} \cdot \{\frac{\mu'_P}{\mu B}\} \cdot \{\frac{2e}{h}\}_{LAB} \{\gamma'_{P,low}\}^{1}_{LAB} \cdot \frac{2}{\{\mu_o c_o\}} \cdot \{\frac{h}{e^2}\}_{LAB})^{1/3} \qquad (11)$$

The uncertainty with which α^{-1} can be obtained in this way is in practice one third of the uncertainty with which γ'_P is obtained.

If the uncertainty of γ'_P is less than 3 times the relative uncertainty of the Ω realization this second way gives a value for α^{-1} with a lower uncertainty than the one solely based on the quantum Hall effect.

The value for α^{-1} can be compared with another value which is derived from the experimentally determined anomalous magnetic moment of the electron by Dehmelt et al /19/:

$$\frac{g_e}{2} = 1.001\ 165\ 159\ 652\ 193\ (1 \pm 10 \cdot 10^{-12}) \qquad (12)$$

By use of quantum electrodynamics it is shown that α^{-1} can be calculated from this experimental result via the following series expansion:

$$\frac{g-2}{2} = C_1\ (\frac{\alpha}{\pi}) + C_2\ (\frac{\alpha}{\pi})^2 + C_3\ (\frac{\alpha}{\pi})^3 + C_4\ (\frac{\alpha}{\pi})^4 + \ldots\ldots \qquad (13)$$

The coefficients have been calculated by Kinoshita and led to /17/:

$$\alpha^{-1} = 137.035\ 989\ 8\ (1 \pm 20 \cdot 10^{-9}) \qquad (14)$$

The Kinoshita calculation uses a point-charge model for the electron. A model with a finite radius for the electron would lead to a value which

---

*) The symbol $\{x\}_{LAB}$ gives the number of a phyisical quantity in a laboratory unit.

is different by $2 \cdot 10^{-7}$ /20/. This implies that, a small transistor of less than a few mm^2 in size could contribute to the structure of an elementary particle like the electron if the experimental uncertainty is further decreased.

5.2. Development of new resistance measuring methods

Since the quantized resistance values occur at "odd" values, the comparison of these values with decimal ones is often the weak link in the whole measurement. To overcome this difficulty new methods based on cryogenic devices have been developed with the aim of a direct comparison of e.g., 12 906.4 Ω with 10 kΩ.

Several laboratories are using with great success cryogenic current comparators where the resistance ratio which can be measured is given by the ratio of the windings in the comparator /10/, /14/.

Another method makes use of a potentiometer based on a series array of Josephson junctions. At PTB resistance ratio measurements with 1440 Josephson junctions integrated on one microchip (Fig. 5) have been performed /22/. According to the equation

$$U = n \cdot f \cdot h/2e \qquad (15)$$

voltage steps differing by 140 μV up to more than 1 V can be obtained at a frequency of 70 GHz. After rough adjustment of the potentiomter by selecting a step n (e.g. n = 1 000 for 140 mV) any voltage between two steps can be obtained by a fine adjustment which changes that frequency by not more than ± 70 MHz (Fig. 6).

Since the drift of the current through the two resistors now becomes the limiting factor for the uncertainty, we are planning to use two of these Josephson potentiometers simultaneously. If a SQUID with a current resolution of $5 \cdot 10^{-14}$ A/√Hz is used as a null detector, a total uncertainty for the ratio of R_{H2}/10 kΩ lower than 10^{-8} should be obtainable.

5.3. Observations of drift rates of resistances

As mentioned above, our previous experimental set-up allowed the measurement of the ratio of a Hall resistance against a reference resistance of nominally the same value with a relative uncertainty of $3 \cdot 10^{-8}$. Drift rates of resistors can therefore be obtained in much shorter time than was possible before the discovery of the quantum Hall effect. Fig. 7 shows the drift rates of two resistors measured over a period of more than 30 days. They change by about $5 \cdot 10^{-8}$ per day /1/.

5.4. Monitoring the legal resistance units

If the experimental set up allows the determination of the Hall resistance in terms of the as-maintained unit with a sufficiently low uncertainty, drift rates of the national standard can be obtained. This has been done successfully at NBS for instance (see Fig. 8).

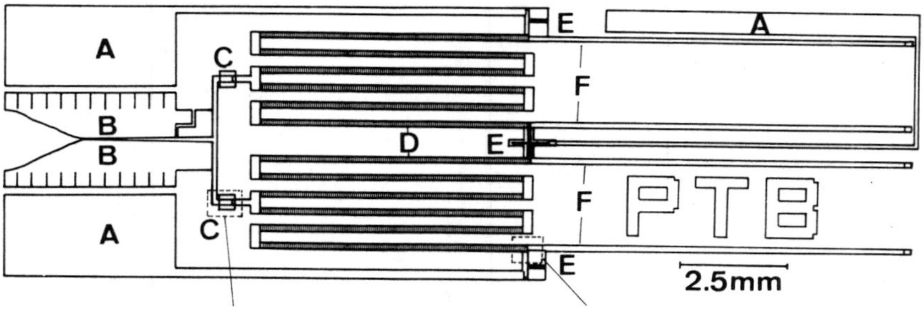

Fig. 5 Microchip with a series connection of 1440 miniaturized Josephson tunnel junctions which provides direct voltages of up to 1 V /21/.

A: constant voltage electrodes; B: fin line; C: dc blocks; D: tunnel junctions; E: filters F: matched loads

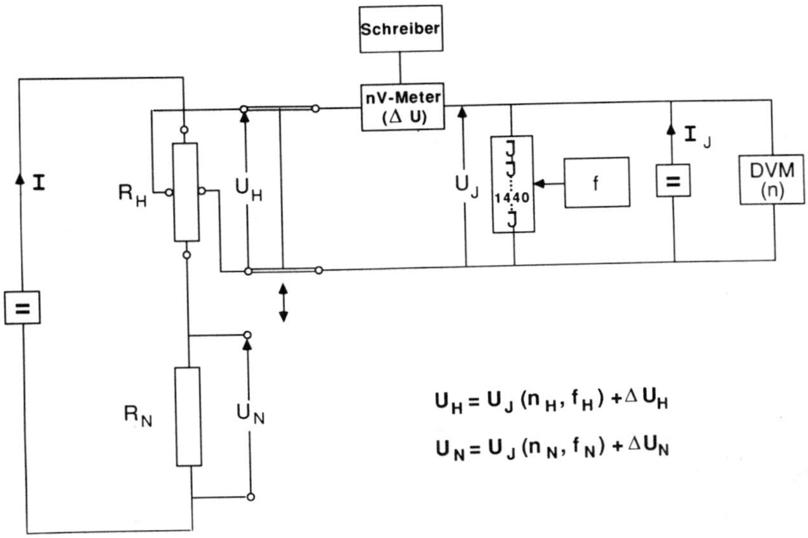

Fig. 6 Schematic diagram of the new PTB measurement circuit: The voltage drops at R_{H2} and R_N = 10 kΩ are compensated by a voltage U_J which is produced by the microchip shown in Fig. 5. (nV-Meter Nulldetector, f Frequency generator, DVM Digital volt meter to control the step number n)

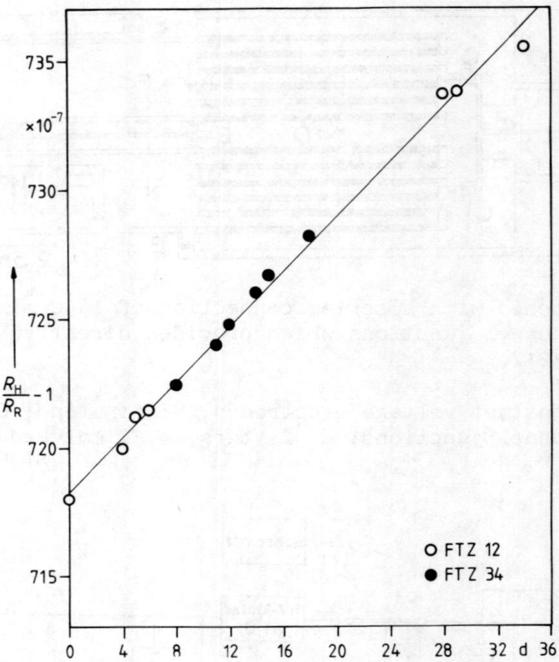

Fig. 7 Ratio of the Hall resistance to a reference resistance of nominally the same numerical value as a function of time. The samples (FTZ 12 and FTZ 34) are changed twice. The linear increase is due to a change of the reference resistance. (based on the same experimental data as one set of Fig. 3).

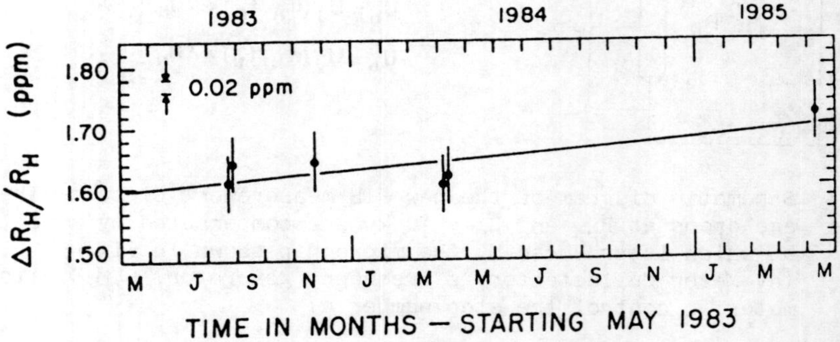

Fig. 8 Monitoring as a function of time the value $R_{H,4}$ which is calibrated in terms of Ω_{NBS}. These data imply that Ω_{NBS} may be decreasing by 50 μΩ/a /23/.

5.5. Determination of the SI ohm

As shown in 5.1 the simultaneous measurements of $\frac{h}{e^2}, \frac{h}{2e}$ and $\gamma'_{P,low}$ in terms of as-maintained units Ω_{LAB} and V_{LAB} allow the determination of the fine-structure constant. Equations 6 and 10 can also be solved for the unknown value Ω_{LAB}/Ω /24/:

$$\frac{\Omega_{LAB}}{\Omega} = \left[\frac{c_o^3 \mu_o^2}{16\, R_\infty} \cdot \frac{\mu_e}{\mu_B} \cdot \frac{\{\frac{2e}{h}\}_{LAB}}{\{\frac{h}{e^2}\}_{LAB}\ \{\gamma_P\}_{LAB}}\right]^{1/3} \qquad (16)$$

This method is in competition with the calculable capacitor. The ratio Ω_{LAB}/Ω can be obtained via the fundamental constants with a lower uncertainty compared with the calculable capacitor if the uncertainty of the $\gamma'_{P,low}$ measurements is lower than three times the uncertainty of the ohm determination with the calculable capacitor. This is because the relative uncertainties of the determination of h/e^2 and $h/2e$ are less than 10^{-7}.

5.6. Determination of the Avogadro constant

If the Faraday constant F is known in laboratory units according to:

$$F = \{F\}_{LAB} \cdot A_{LAB} \cdot s \cdot mol^{-1}, \qquad (17)$$

it is possible to determine the Avogadro constant with approximately the same relative uncertainty as the one of the determination of the Faraday constant if a quantized Hall resistance and the Josephson frequency - voltage ratio are measured in terms of as-maintained units:

$$\frac{h}{e^2} = \{\frac{h}{e^2}\}_{LAB} \cdot \Omega_{LAB} \qquad (18)$$

$$\frac{2e}{h} = \{\frac{2e}{h}\}_{LAB} \cdot V_{LAB}^{-1} \cdot s^{-1} \qquad (19)$$

Since the Avogadro constant can be written as

$$N_A = \frac{F}{e} = F \cdot \frac{h}{e^2} \cdot \frac{2e}{h} \cdot \frac{1}{2}, \qquad (20)$$

one obtains:

$$N_A = \frac{1}{2} \cdot \{F\}_{LAB} \cdot A_{LAB} \cdot s \cdot mol^{-1} \cdot \{\frac{h}{e^2}\}_{LAB} \cdot \frac{V_{LAB}}{A_{LAB}} \cdot \{\frac{2e}{h}\}_{LAB}\ V_{LAB}^{-1} \cdot S \qquad (21)$$

$$N_A = \frac{1}{2}\ \{F\}_{LAB} \cdot \{\frac{h}{e^2}\}_{LAB} \cdot \{\frac{2e}{h}\}_{LAB} \cdot mol^{-1} \qquad (22)$$

5.7. International agreement to obtain uniformity in the as-maintained units:

It has already been pointed out that the limitation in the uncertainty of the measurement of a quantized Hall resistance is at present given by the uncertainty of the ohm determination based on the calculable capacitor. It has also been shown that a quantized Hall resistance can be reproduced with a relative uncertainty of about one order of magnitude lower than the relative uncertainty of the realization of the ohm. This situation has led to a discussion of whether there should be a change in the definition of the ohm.

However, the SI system is a system of units where base units are defined and the other units are uniquely derived according to physical laws (Fig. 9). The electrical base unit is the ampere, while the volt and the ohm like all other electrical units are derived. Although it is possible to change the definition of a base unit (e.g. the change of the meter definition in 1983), it is not possible to change the definition of a derived unit. Possible but still speculative changes of the base units of the SI system are discussed in the Appendix.

About 15 years ago, a similar situation arose producing a similar discussion with respect to the Josephson effect and the volt. Voltages could be reproduced with a relative uncertainty of $6 \cdot 10^{-8}$, whereas their values could be given in the SI unit only with a relative uncertainty 100 times larger. The result of the discussion was not a change of a definition in the SI system, but a recommendation passed to the national laboratories by the Comité Consultatif d'Électricité for the common use of the same value for the Josephson frequency - voltage ratio for the maintenance of their volt.

It turns out that the value recommended in 1972 for the Josephson frequency - voltage ratio ($483\,594 \cdot 10^{14}$ Hz/V) was about $7 \cdot 10^{-6}$ too high, as can be seen, for example, from the 1986 adjustment of fundamental constants /25/. There is thus a need for a new recommendation on the maintenance of the volt.

The Comité Consultatif d'Électricité therefore decided at its last session in June 1985 to draft a new recommendation on the maintenance of the volt and, for the first time, a recommendation on the maintenance of the ohm in 1988. These recommendations are planned to become effective on January 1,1990.

The advantages of this recommendation for the as-maintained ohm at the various national laboratories which accept the recommendation will be that their as-maintained units of resistance agree with one another with an uncertainty which depends only on the measurement of a quantized Hall resistance in as-maintained units.

The lowest obtainable value for the uncertainty of this measurement is at present about 10^{-8}. National institutes are making every effort to

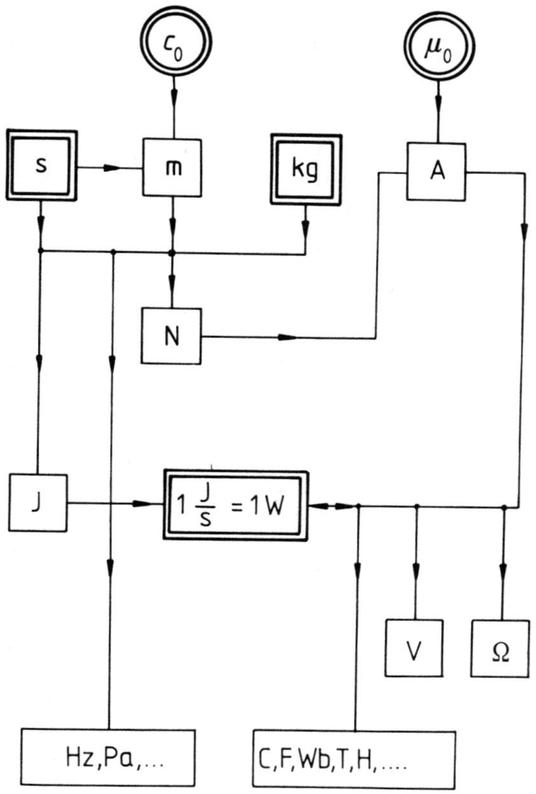

Fig. 9 Main mechanical and electrical units according to their definition in the SI system. c_o and μ_o are the only fundamental constants defined at present in the SI system.

decrease this value still further. The uncertainty of the ratio of the as-maintained units of resistance of two national laboratories based on a comparison of travelling resistance standards is usually not lower than $5 \cdot 10^{-8}$. As can be seen from Fig. 10 /26/ the extrapolation of the drift rate of an as-maintained unit after such a comparison produces an additional uncertainty.

It is clear that these as-maintained units, although they are in good agreement with one another, do not agree with the SI ohm. But the industrial need regarding resistance calibrations is international agreement on the one hand and constancy with time on the other. For this purpose agreement with the SI ohm is of minor importance.

Measurements based on the correct SI value are carried out chiefly in basic research, e.g. precision measurements of fundamental constants.

6. APPENDIX

The base electrical unit in the SI system is the ampere. Its definition is equivalent to the definition $\mu_o = 4\pi \cdot 10^{-7}$ Vs/Am.

This fundamental constant was the only one to be defined when the SI system was adopted. Meanwhile, the definition of the meter has been changed and the velocity of light in vacuum c_o is the second defined fundamental constant. With Maxwells equation $\varepsilon_o = 1/c_o^2 \mu_o$ is the third fundamental constant having zero uncertainty.

As shown by Kose and Wöger, for example, /27/, the derivation of the definition of SI units from fundamental constants is desirable and is the tendency in the development of the present system of units. However, it is not possible to realize the ampere via a macroscopic quantum effect on the basis of the fixed value of μ_o. On the other hand, as has been shown, it is possible to realize the units of voltage and resistance with extremely low uncertainty if h/2e and h/e^2, respectively e and h, were defined fundamental constants. From a practical point of view, a system of units would be of advantage based on the same units as present only instead of defining μ_o, e and h would be defined fundamental constants. The units of voltage and resistance could then be realized with a much lower uncertainty.

The price to be paid for this is that μ_o and ε_o would no longer be defined constants and mechanical and electrical energy could not be measured with the same unit.

If h/e is supposed to be a quantum of magnetic flux, the electrical units in this system are based on the quantum of electrical charge e, the quantum of magnetic flux h/e and on the time or frequency ν. Fig. 11 is a configuration of these three quantities and their relation to macroscopic quantum effects. While the existence of the Josephson effect and the quantum Hall effect are well established first experi-

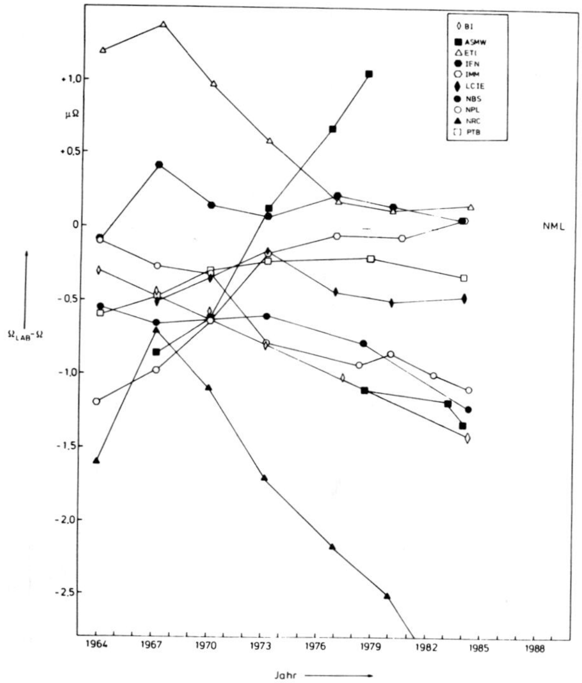

Fig. 10 Changes with time of the as-maintained units of resistance in various national institutes relative to the one at NML, (National Measurement Laboratory, Australia), which is supposed to be identical with the SI ohm to within $2 \cdot 10^{-7}$ /25/.

BI: Bureau International des Poids et Mesures, Paris
ASMW: Amt f. Standardisierung, Meßwesen und Warenprüfung, DDR

(see Fig. 4 for the other abbreviations)

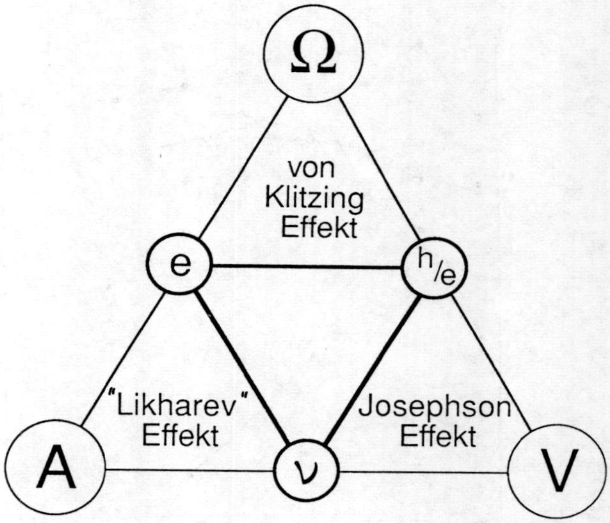

Fig. 11 The quanta of charge e and magnetic flux h/e and their relation to the electrical units via hitherto observed or proposed macroscopic quantum effects.

ments /28/ indicate that the quantization of eν proposed by Likharev /29/ is measurable.

Metrologists and not only metrologists would be happy if additional macroscopic quantum effects are discovered in the future.

REFERENCES

/ 1/ E. Braun, P. Warnecke, and H. Leontiew: Metrologia 22, 226 (1986).

/ 2/ L. Bliek, E. Braun, F. Melchert, P. Warnecke, W. Schlapp, G. Weimann, K. Ploog, G. Ebert, and G.E. Dorda: IEEE IM-34, 304 (1985).

/ 3/ W. van der Wel, K.J.P.M. Harmans, R. Kaarls, and J.E. Mooij: IEEE IM-34, 314 (1985).

/ 4/ M.E. Cage, B.F. Field, R.F. Dziuba, S.M. Girvin, D.C. Tsui, and A.C. Gossard: Phys. REv. B 30, 2286 (1984).

/ 5/ K.v. Klitzing: Rev. Mod. Phys. 58, 519 (1986).

/ 6/ T. Witt, T. Endo, and D. Reymann: to be published in the Proc. of CPEM 86 in IEEE Instr. and Meas.

/ 7/ W. Schwitz, L. Bander, H.-J. Bühlmann, M.A.Py, and M. Illegems: to be published in the Proc. of CPEM 86 in IEEE Instr. and Meas.

/ 8/ K. Shida, T. Wada, H. Nishinaka, M. Kobayashi, G. Yonezaki, T. Igarashi, and T. Nemoto: to be published in the Proc. of CPEM 86 in IEEE Instr. and Meas.

/ 9/ K. Yoshihiro, I. Kinoshita, K. Inagaki, and Ch. Yamanouchi: Phys. Rev. B 33, 6874 (1986).

/10/ F. Delahaye, D. Dominguez, F. Alexandre, I.P. Andre, I.P. Hirtz, and M. Razeghi: Metrologia 22, 103 (1986).

/11/ M.E. Cage, R.F. Dziuba, B.F. Field, Th.E. Kiess, C.T. van Degrift: to be published in the Proc. of CPEM 86 in IEEE Instr. and Meas.

/12/ B.W. Ricketts and M.E. Cage: to be published in the Proc. of CPEM 86 in IEEE Instr. and Meas.

/13/ B.M. Wood and M. D'Iorio: to be published in the Proc. of CPEM 86 in IEEE Instr. and Meas.

/14/ A. Hartland, R.G. Jones, B.P. Kibble, and D.J. Legg: to be published in the Proc. of CPEM 86 in IEEE Instr. and Meas.

/15/ E. Braun, P. Warnecke, G. Hein, G. Weimann, and W. Schlapp:

Document CCE/86-52 presented at the 17. session of the Comité Consultatif d'Électricité, Paris (1986).

/16/ Yu.V. Tarbeyer, N.V. Studentsov, and V.S. Tuminsky: Document CCE/86-4 presented at the 17. session of the Comité Consultatif d'Électricité, Paris (1986).

/17/ T. Kinoshita: to be published in the Proc. of CPEM 86 in IEEE Instr. and Meas.

/18/ D.C. Tsui, A.C. Gossard, B.F. Field, M.E. Cage, and R.F. Dziuba: Phys. Rev. Lett. 48, 3 (1982).

/19/ R.S. van Dyck, jr, P.B. Schwinberg, and H.G. Dehmelt in *Atomic Physics 9* edited by R.S. van Dyck, jr, and E.N. Fortson, World Scientific Publishing, Singapore, 1984, p. 53.

/20/ K.v. Klitzing: private communication

/21/ J. Niemeyer, L. Grimm, W. Meier, J.H. Hinken, E. Vollmer: Appl. Phys. Lett. 47, 1222 (1985).

/22/ P. Warnecke, J. Niemeyer, F.W. Dünschede, L. Grimm, W. Schlapp, and G. Weimann: to be published in the Proc. of CPEM 86 in IEEE Instr. and Meas.

/23/ M. Cage in *The Quantum Hall Effect* edited by R.E. Prange and S.M. Girvin, Springer Verlag New York, 1987, p. 49.

/24/ V. Kose: PTB Mitteilungen 92, 249, (1982).

/25/ CODATA Bulletin 63, Nov. 1986.

/26/ H. Bachmair: private communication

/27/ V. Kose and W. Wöger: Phys. Bl. 43, 227 (1987).

/28/ K. Yoshihiro, I. Kinoshita, K. Inagaki and Ch. Yamanouchi: to be published in: Proc. of 2nd International symposium on Foundations of Quantum Mechanics - In the Light of New Technology, Tokyo, 1986.

/29/ K.K. Likharev and A.B. Zorin in Proc. of the 17th International Conference on Low Temperature Physic, LT 17, Karlsruhe, 1984, edited by V. Eckern, A. Schmid, W. Weber and H. Wühl, North Holland, Amsterdam, 1984, p. 1153.

FUNDAMENTAL PHYSICAL CONSTANTS

E. Richard Cohen
Rockwell International Science Center
Thousand Oaks, California 91360
USA

ABSTRACT. The periodic review of the physical constants is important for two very different reasons. One is the need to have a set of numerical values that can be used reliably in all branches of science and technology. The other is the verification of physical models inherent in the consistency of those values. This paper discusses the definition of 'fundamental constants' and the reasons for choosing the specific set of constants for the 1986 adjustment. It summarizes the input data included in this adjustment, the results of the analysis, and the conclusions to be drawn from them. Of particular importance is the question of the consistency of QED and the status of the experimental verification of the validity of the perturbation expansion for the electron anomalous moment and the numerical evaluation of the multidimensional integrals associated with the sixth and eighth order terms.

An important aspect of modern metrology is the attempt to establish atomic, rather than artifact, standards for the base units of the *Système International.* The 1986 adjustment of the fundamental constants has implications with respect to the establishment of standards for electrical units. The 1986 CODATA recommendations based on the new adjustment will be compared with the previous (1973) recommendations and the major differences will be discussed and, to some extent, explained.

1. INTRODUCTION AND BACKGROUND

Prior to the twentieth century, the properties of matter could be unified only to the extent that the general principles of thermodynamics provided relationships between, and limitations upon, certain physical quantities. The development of atomic physics and the quantization of matter and energy introduced a new concept – fundamental constants. It has not been emphasized enough that quantum mechanics is fundamentally more a theory of precision and exact values than it is a theory of uncertainty. While there is indeed an indeterminacy about the outcome of the measurement of a microscopic system that has not been prepared in a pure quantum state, it is more significant that in some instances unlimited precision is conceptually possible. Whereas it is probably meaningless to determine the density of a sample of steel or aluminum to much better than 1 part in 10^4 and expect the result to have a meaning

V. De Sabbata and V. N. Melnikov (eds.),
Gravitational Measurements, Fundamental Metrology and Constants, 59–89.

for anything other than that specific sample at a specific temperature and pressure, the proton-electron mass ratio has been determined with a precision of better than one part in 10^8. Furthermore, this ratio is presumably the same throughout the universe (at least at locations where the density of matter and the curvature of space are sufficiently small).

The fundamental physical constants are those quantities that provide the basis for our models of the physical universe. The group of constants known as the 'Fundamental Constants' is, however, somewhat restricted. It should obviously include all of the independent properties of all of the particles considered today to be 'fundamental'. However, only the electron and the muon are included in the 1986 Adjustment, not the tauon (Cohen and Taylor, 1986). The zero mass of the photon can be considered to play a part but the masses of the heavy bosons do not. The baryons are represented by the proton and the neutron; quarks and pions are omitted. Only the electromagnetic force is involved in any direct sense, not the other three (or four) forces.

These omissions are a reflection of the state of our knowledge of the values of the properties concerned and are not basic constraints. As defined here, the set of fundamental constants is determined by those constants whose numerical values are experimentally interrelated so that the value of one cannot be found independently of the values of the others. Not only does this operational categorization of the fundamental constants separate them into hierarchical levels based on the levels of precision with which they are known, but it also requires the inclusion of calibration constants and conversion factors into the analysis with a standing equal to that of the 'fundamental constants'.

The analysis of the fundamental physical constants and the determination of their values to the greatest possible precision are important for two very distinct reasons:

(a) The atomic constants are important in all branches of physics because they are related to basic concepts and quantities. Their influence extends to such widely different fields as (for example) chemistry and electronics. Furthermore, to an increasing extent, atomic physics is becoming the basis for our entire system of measurement. Rather than continuing the pattern of the original Metric System, which defined units in terms of artifact examples of the quantity to be measured, the present *Système International* is, except for the kilogram, based on atomic standards.

(b) Knowledge of the values of the physical constants is the basis for determining the consistency and validity of the structure of physics. It is impossible to prove that an accepted description of a physical phenomenon or a physical process is correct; it is only possible to show that it is wrong. This is done by making predictions of the outcome of a measurement based on a given theory. If the result of a measurement does not agree with the prediction, then the theory must be rejected --- but only if the prediction has been correctly deduced from the theory and the measurements have been correctly carried out and interpreted. Any disagreement between measurements must be viewed in terms of the uncertainties inherent in the measurements as well as the uncertainties in the theory of their interpretation. There is an indeterminacy in choosing the precise dividing line between theory and measurement because every measurement must be interpreted and every interpretation requires some theoretical basis. One therefore looks for consistency, and in particular for instances of unexplained inconsistency, in order to evaluate the validity of physics. For this purpose the discrepancies rather than the mean values of the data are important.

A least-squares analysis provides precisely this separation. The least-squares criterion divides the abstract space of the observational data into one subspace containing the 'best' values of the unknown parameters and a second subspace containing the discrepancies among the data. The variance matrix of the observational data may be interpreted as the metric tensor of the space; with this metric the two spaces are orthogonal, reflecting the two orthogonal objectives of the analysis.

We can divide the constants into the 'universal' constants that describe the world as a whole and the 'atomic' constants that describe matter in particular. The categories need not be exclusive unless we want them to be, and a constant could well appear in more than one category. On a practical level, and in order to be able to determine their values, it will also be necessary to include some secondary quantities which are not 'fundamental' in the strict sense of the word, but which are metrological artifacts resulting from an earlier lack of knowledge, or which represent arbitrary metrological standards, 'local' units of measurement, or conversion factors.

1.1. Universal Constants.

The universal constants may be defined as those that are required for the description of the universe as a whole. In its broadest aspects the universe is dominated by gravitation, light – electromagnetic energy – and matter. The details of the structure of matter are unimportant. The associated fundamental constants are the Newtonian constant of gravitation G, the speed of light or of electromagnetic radiation c, and (perhaps surprisingly) the Planck constant h.

The Planck constant (or another constant equivalent to it) is required by thermodynamic and dimensional analysis arguments based entirely on classical considerations. The thermodynamics of a black-body shows that the energy distribution as a function of the wavelength of the radiation is a function of λT, and that the total electromagnetic energy density per unit volume is proportional to the fourth power of the temperature ($u = aT^4$). It then follows directly from dimensional arguments that the spectral distribution of black-body radiation is a function of the dimensionless quantity $(a/k)^{\frac{1}{3}}\lambda T$, where k is the Boltzmann constant. The appearance of k here is of trivial significance; it is only a conversion factor that relates an arbitrary temperature unit to the SI energy unit. The important result is that without specifically introducing Planck's constant, classical physics was able to yield an equivalent set of fundamental constants using G, c and $\xi = (k^4/a)^{\frac{1}{3}}$.* In

* From Planck's expression for the Stefan-Boltzmann constant, it follows that $\xi = (15/\pi^2)^{\frac{1}{3}}\hbar c$, but the derivation of this relationship requires more than purely thermodynamic considerations.

fact, twenty-five years before Planck, G. Johnstone Stoney (Barrow, 1983) introduced a system of units that essentially duplicated the Planck units at a scale different by the factor $\sqrt{137}$, using e^2 (in electrostatic units) in place of $\hbar c$. This gives a 'Stoney' mass, $m_S = e/\sqrt{4\pi\epsilon_0 G} = 1.86\,\mu g$.

1.2. Atomic Constants.

The atomic constants are generally recognized as those that define atomic (but not specifically nuclear) properties. Thus they comprise the elementary charge e, electron, proton and neutron masses m_e, m_p and m_n, magnetic moments μ_e, μ_p, and such related quantities as μ_B, μ_N, the Rydberg constant $m_e c\alpha^2/2h$, the Bohr radius $\alpha/4\pi R_\infty$, etc. If, in addition, one defines an atomic constant to be any constant that can be formed from a combination of atomic constants, then h and c must also be included as atomic constants as well as universal constants.

1.3. Secondary Constants.

As described above, the secondary constants are not fundamental, nor are they permanent members of the set. Their importance and their inclusion in an analysis is an accident of the state of metrology at the time. In various adjustments over the years, these constants have included the maximum density of water and the density of silicon, the calibration factors for various national and international representations of the basic electrical standards of the ohm and the volt, and the conversion factor from angstroms to 'x-units'.

2. INPUT DATA FOR THE 1986 ADJUSTMENT

An extraordinary amount of experimental and theoretical work relating to the fundamental constants has been published in the past dozen years. The measurement of frequency at infrared and visible wavelengths reached a level of development that resulted in the redefinition of the metre in terms of the distance traveled by light in a given time. Another advance was the direct linking of atomic lattice spacings to optical wavelengths, making possible a significant improvement in the determination of the Avogadro constant. Impressive progress has also been made in the precision of the numerical evaluation of the quantum electrodynamics of the electron anomalous moment as well as in its experimental determination. The most striking metrological advance occurred when von Klitzing (1980) observed the quantization of electrical conductance and achieved not only a direct macroscopic measurement of the fine-structure constant but a Nobel prize as well (von Klitzing, 1986).

In a tradition that goes back almost sixty years to the first major evaluation of the best numerical values of the then-recognized atomic constants by R. T. Birge (1929), the available data are divided into two categories: the more precise data (auxiliary constants) that are not subject to adjustment because of their relatively low uncertainties, and the less precise or stochastic data that are subject to adjustment.

There is no formal basis for separation into these two categories except that a variable with an uncertainty much smaller than that of any other variable to which it is connected will not be altered by the adjustment and hence can be treated as a constant. For example, at the start of the analysis for the 1986 adjustment, it was intended that the proton-electron mass ratio would be one of the unknowns of the analysis. However, as data collection proceeded, the precision with which m_p/m_e was known increased to the point that the weight of the direct observational equation was much larger than any weight contributed by any other relationship (such as the proton magnetic moment measurements that were of such importance in earlier adjustments).

2.1. Auxiliary Constants

The auxiliary constants of the 1986 Adjustment are given in Table 1. These quantities in general have uncertainties less than 0.02 ppm, and in no case is the uncertainty of an auxiliary constant greater than one-tenth the uncertainty of the stochastic variables with which it appears.

2.1.1. *The Speed of Light and the Definition of the Metre.* The new definition of the metre in terms of the speed of light (BIPM, 1983) establishes c as an exact auxiliary constant in the analysis,

$$c = 299\,792\,458 \text{ m s}^{-1}$$

but it does not make wavelength metrology obsolete, nor does it assure that a given laser will provide a valid length standard. The definition of the metre is conceptual rather than operational; the practical working standards for its representation are contained in a *'mise en pratique'* that defines the operating conditions under which specified hyperfine-structure features of selected absorption-stabilized lasers will constitute a consistent set of frequencies and wavelengths that reproduce the metre to within the specified uncertainties (Hudson, 1984).

Absorbing molecule and transition[a]	Frequency ν/MHz	standard deviation (parts in 10^9)[c]	Wavelength[b] $\lambda = c/\nu$ λ/fm
1. $CH_4, \nu_3, P(7), F_2^{(2)}$	88 376 181.608	0.044	3 392 231 397.03(15)
2. $^{127}I_2, 17-1, P(62), o$	520 206 808.51	0.20	576 294 760.27(12)
3. $^{127}I_2, 11-5, R(127), i$	476 312 214.8	0.34	632 991 398.09(21)
4. $^{127}I_2, 9-2, R(47), o$	489 880 355.1	0.37	611 970 769.76(22)
5. $^{127}I_2, 43-0, P(13), a_3$	582 490 603.6	0.43	514 673 466.22(22)

[a] Lasers 1 – 4 are He-Ne; laser 5 is Ar^+.
[b] These wavelengths are given to additional places beyond those appearing in the CCDM document in order to avoid inconsistencies in rounding.
[c] The precision indicated is achievable only when the lasers are operated under specified standard conditions.

2.1.2. *Proton-Electron Mass Ratio.* The value $m_p/m_e = 1836.152701(37)$ is the most recent result reported by R. S. van Dyck and his collaborators at the University of Washington (1986) from measurements of the cyclotron frequency, $\omega_c = eB/m$, of electrons and protons in the same magnetic field in a Penning trap. An almost identical, but less precise, value has been reported by Gräff *et al.* (1980), $m_p/m_e = 1836.1527(11)$.

In previous adjustments this ratio, or the equivalent ratio

$$\mu_p'/\mu_N = (\mu_p'/\mu_B)(m_p/m_e),$$

has been an adjustable stochastic variable. The data of Mamyrin *et al.* (1983) give a value for μ_p'/μ_N equivalent to $m_p/m_e = 1836.150\,90(79)$, while the measurements of Petley and Morris (1974) give 1836.1521(13). These earlier measurements are not in gross disagreement with the University of Washington data but they are clearly of such lower precision (and hence, lower weight) that they need not be considered in our analysis.

2.1.3. *Relative Atomic Masses and Mass Ratios.* The relative atomic masses of the nuclides are taken from the 1983 Atomic Mass Table of Wapstra and Audi (1985). The mass ratio m_μ/m_e is required in the evaluation of the reduced mass factor and QED terms for muonium (μ^+e^- atom) and for the QED calculation of the electron g-factor. This ratio (or equivalently, the magnetic moment ratio μ_μ/μ_p) is a variable in the adjustment, but for the auxiliary constant $1 + m_e/m_\mu$ the uncertainty is reduced by a factor of 200 and it may therefore be treated as a constant.

2.1.4. *Rydberg Constant.* The techniques of Doppler-free spectroscopy, yielding an increase in precision of one to two orders of magnitude compared to that available in 1973, have rendered all earlier measurements of the Rydberg constant obsolete. The present value is based on measurements at Stanford (Hänsch *et al.*, 1974; Goldsmith *et al.*, 1979) and Yale (Amin *et al.*, 1981, 1984; Lichten, 1985) reevaluated on the basis of the new definition of the metre and a revision of Erickson's spectroscopic energy-level calculations (Erickson, 1977; 1983) that incorporates the newer values of m_p/m_e and α. A determination of R_∞ carried out at NPL (Petley and Morris, 1979; Petley, Morris and Shawyer, 1980) giving a result that is in agreement with the Yale and Stanford results was not included in determining the recommended value because it is quoted to have a precision poorer by a factor of six than the Yale datum and hence, were it to be included, it would make no significant contribution to the final result.

More recently, Lichten and his collaborators (Zhao *et al.*, 1986, 1987) have repeated the measurement of the Balmer-α transition in H and D, and have also extended the technique to the Balmer-β transition. The results of these newer measurements give $R_\infty = 10\,973\,731.569(7)\,\mathrm{m}^{-1}$ from Balmer-α and $R_\infty = 10\,973\,731.573(3)\,\mathrm{m}^{-1}$ from Balmer-β.

2.1.5. *g-Factor for the Free Electron and Muon.* The g-factor for the free electron, $g_e = 2\mu_e/\mu_B = 2(1 + a_e)$, where μ_e is the magnetic moment of the electron and μ_B is the Bohr magneton, contributes to the adjustment in two ways: as an input variable a_e, the electron magnetic moment anomaly, from which a value of the fine-structure constant may be derived; and as a fixed auxiliary constant, g_e. The value in Table 2 for g_e is the one most recently reported by van Dyck, Schwinberg and Dehmelt (1984) at the University of Washington from measurements on a single electron stored in a Penning trap cooled to 4.2 K. The uncertainty assigned to this value has been increased from the value 4×10^{-12} of the experiment itself to take account of systematic experimental errors. The largest of these is the theoretically-estimated shift in the electron cyclotron orbit arising from the finite geometry of the apparatus, leading to wave functions that are confined to a region of the order of 1 cm (van Dyck, 1985; Brown *et al.*, 1985).

The g-factor for the free muon, $g_\mu = 2\mu_\mu/(e\hbar/2m_\mu) = 2(1+a_\mu)$, enters the adjustment only as an auxiliary constant. In contrast to the electron anomaly, the muon anomaly a_μ is not known with sufficient precision either experimentally or theoretically to allow the calculation of a competitive value for the fine-structure constant. The value used is the latest from a series of muon measurements that have been carried out over the past twenty years at CERN (Bailey *et al.*, 1979).

2.1.6. *Electron and Nuclear Magnetic Moment Ratios.* The ratio μ_e/μ_p is obtained from the g-factor measurements in hydrogen of Winkler *et al.* (1972) which are by far the most accurate available. These workers report the value

$$g_j(\mathrm{H})/g_p(\mathrm{H}) = 658.210\,7063(66) \quad (0.010\ \mathrm{ppm}). \tag{1}$$

The g-factors must be corrected to the free particle values using the theory of the bound state corrections due to Faustov (1970), Grotch and Hegstrom (1971), and Close and Osborne (1971) :

$$\mu_p(\mathrm{H}) = \mu_p\left[1-\alpha^2\left(\frac{1}{3}-\frac{m_e m_p}{2(m_p+m_e)^2}-\frac{(m_p+3m_e)m_e a_p}{6(m_p+m_e)^2(1+a_p)}\right)+\cdots\right] \tag{2}$$

where a_p is the proton magnetic moment anomaly and $\mu_e(\mathrm{H})$ is found by simply interchanging the rôles of the electron and the proton. One then obtains :

$$\frac{g_j(\mathrm{H})}{g_p(\mathrm{H})} = \frac{g_e}{g_p}(1+27.7\times10^{-9}).$$

When this is combined with eq. (3.1), we find

$$g_e/g_p = \mu_e/\mu_p = 658.210\,6881(66) \qquad (0.010\ \mathrm{ppm}). \tag{3}$$

The ratio μ_e/μ_p' is obtained from the measurements of Phillips, Cooke and Kleppner (1977) who give the value $g_j(\mathrm{H})/g_p = 658.216\,0091(69)$ for a spherical, pure H_2O NMR sample at 34.7 °C. The 0.10 ppm correction required to convert this result to the 25 °C reference temperature for proton NMR measurements is derived from the work of Petley and Donaldson (1984); hence

$$g_e/g_p' = 658.227\,5970(72) \quad (0.011\ \mathrm{ppm}). \tag{4}$$

2.1.7. *'As-maintained' Volt and Ohm Standards.* In the SI, electrical units are defined by assigning a fixed value to μ_0, the permeability of vacuum. The formal definition of the base unit 'ampere', involving 'straight parallel conductors of infinite length, of negligible circular cross-section' is obviously impossible to realize in the laboratory. Instead, the electrical units are represented in the national standards laboratories by sets of electrolytic cells and precision wire-wound resistors held in controlled environments and calibrated by international inter-calibrations.

Since 1973, most national standards laboratories have used the Josephson effect to provide a time-independent, reproducible standard of electrical potential to monitor and calibrate the banks of standard cells. Therefore, all measurements of electrical potential difference can now be related unambiguously to the BIPM standard characterized by the Josephson frequency-voltage quotient, 483 594.0 GHz/$V_{76\text{-BI}}$. The use of this standard is still limited by the precision of the calibration of the reference cells that are used as local standards.

The noncoherent unit $V_{76\text{-BI}}$ is related to the coherent SI unit through the relation

$$V_{76\text{-BI}} = K_V \, \mathrm{V}, \tag{5}$$

and the quantity K_V is a variable in the present adjustment. It is related to the physical constants by the definition

$$(2e/h) K_V \equiv 483\,594.0 \ \mathrm{GHz/V}. \tag{6}$$

Resistance standards have been compared in the BIPM triennial intercomparisons as well as in several bilateral international comparisons. The Australian CSIRO National Measurement Laboratory has used the Thompson-Lampard calculable capacitor (Thompson and Lampard, 1956) to maintain its standard of resistance in terms of the coherent SI ohm since 1964, so that there is more than a twenty-year data base tracing the time dependence of the BIPM resistance standard, $\Omega_{BIPM} = \Omega_{69\text{-BI}}$. These data indicate a surprisingly constant linear drift of −0.0566(15) μΩ/a. Because of this drift, all data are expressed in terms of Ω_{BI85}, the value of $\Omega_{69\text{-BI}}$ on 1 January, 1985:

$$\Omega_{BI85} \equiv \Omega_{69\text{-BI}}(1 \text{ Jan } 1985),$$

and in analogy with eq. (5),

$$\Omega_{BI85} = K_\Omega \, \Omega. \tag{7}$$

The drift rate is used as an auxiliary constant in the adjustment. On the other hand, K_Ω, the value of the BIPM maintained ohm in SI units, is taken as an unknown because of uncertainties (Type-B errors) associated with the transfer of capacitor measurements to resistance standards; a typical transfer chain might involve the series :

$$0.5\,\mathrm{pF} \rightarrow 10\,\mathrm{pF} \rightarrow 100\,\mathrm{pF} \rightarrow 10\,\mathrm{k}\Omega \rightarrow 1\,\Omega.$$

More than 100 individual bilateral comparisons together with the BIPM 1-Ω triennial intercomparisons, assuming that each national standard has its own linear drift rate, provide the means for determining the time-dependent offset between each national standard and the as-maintained unit Ω_{BIPM}.

Although the ampere is a base unit in SI, it is realized as a derived unit from the maintained standards:

$$A_{\mathrm{BIPM}} = V_{\mathrm{BIPM}}/\Omega_{\mathrm{BIPM}} = V_{76\text{-}\mathrm{BI}}/\Omega_{69\text{-}\mathrm{BI}}$$
$$A_{\mathrm{BI85}} = V_{\mathrm{BI85}}/\Omega_{\mathrm{BI85}} = (K_{\mathrm{V}}/K_{\Omega})\,\mathrm{A} = K_{\mathrm{A}}\,\mathrm{A} \qquad (8)$$

2.1.8. *Acceleration Due to Gravity.* The value of the acceleration due to gravity at the location of the measurement is required for the determination of the ampere and the volt, and in the high-field determination of the proton gyromagnetic ratio. The development of the transportable absolute laser gravimeter (Hammond and Faller, 1967; Faller, 1967) has released experimenters from the problem of relying on a series of relative gravity differences to relate the laboratory site to a national or international Fundamental Gravity Station. The transportable gravimeter yields a measurement precision of 0.01 – 0.03 ppm. The International Gravity Standardization Net, IGSN71, based on 11 absolute gravity determinations and approximately 25 000 gravity differences, provides a reference network of 1854 gravity stations with overall uncertainty of the order of 0.1 ppm for the evaluation of gravity at a laboratory site (Morelli, 1974). For those few measurements for which a modern gravity value is unavailable, the older value based on the Potsdam System has been used with a correction of $-14.0\times10^{-5}\ \mathrm{m\,s^{-2}}$ recognizing that the resultant gravity value may well be in error by 1 ppm.

2.2. Primary Stochastic Data

The 38 items of stochastic data considered in the adjustment and an estimated effective value of the degrees of freedom of each determination (Cohen, 1984) are listed in Table 2. The uncertainties of the stochastic data lie in the range from 0.05 ppm to 10 ppm. Of the 12 data types represented in Table 2, only the Faraday, the molar mass of Si, and the muonium hyperfine interval are represented by a single measurement. Unfortunately, this redundancy is not as useful as one might hope because not all of the data are of comparable precision.

Type 1. *Direct Ohm Determinations.* The Thompson-Lampard calculable capacitor (Thompson and Lampard, 1956; Lampard, 1957; van der Pauw, 1958) has been used not only at CSIRO where it was developed, but in several other National laboratories as well. The measurement constitutes a determination of the permittivity of vacuum ϵ_0 in units that are related to the laboratory-maintained ohm. Since $\epsilon_0 = 1/\mu_0 c^2$ is exactly defined in terms of the SI ohm, this allows a direct calibration of the laboratory standard.

The five measurements of the ohm (*1.1*, Thompson, 1968; *1.2*, Cutkosky, 1974; *1.3*, Igarashi *et al.*, 1968; Igarashi *et al.*, 1978; Igarashi, 1983, 1984; *1.4*, Dahake *et al.*, 1983; *1.5*, Jones and Kibble, 1985) are reasonably consistent; the mean value is

$$\Omega_{\mathrm{BI85}} = \Omega - 1.533(69)\ \mu\Omega. \qquad (9)$$

Type 2. *Direct Ampere Determinations.* The calibration of the maintained ampere in terms of SI (referred to in paragraph 1.3.7) is accomplished by comparing the measured and computed forces or torques exerted by a current-carrying coil on a second coil. The force depends upon the intensity and the geometry of the currents; the difficulties in the measurement lie in defining the precise geometry of the current paths, and in measuring the relatively small forces generated by the single-layer coils that are required in order to be able to define that geometry.

The direct ampere determinations (*2.1*, Driscoll and Cutkosky, 1958; *2.2*, Vigoreux, 1965; *2.3*, Gorbatsevitch, 1973; *2.4*, Driscoll and Olsen, 1968; *2.5*, Bender and Schlesok, 1974; *2.6*, Vigoreux and Dupuy, 1980) yield a mean value,

$$A_{BI85} = 1\,\mathrm{A} + (2.07 \pm 2.53)\,\mu\mathrm{A}. \tag{10}$$

If the relatively high NPL value (item *2.6*) is deleted, the mean is decreased to $A_{BI85} = 1\,\mathrm{A} - (0.48 \pm 3.22)\,\mu\mathrm{A}$.

Type 3. *Direct Volt Determinations.* The volt can be measured by determining the force on the plates of a parallel plate capacitor when a potential (known in terms of a laboratory standard volt) is applied. The energy stored in a capacitor per unit area of the plates (neglecting end effects) is given by $W = \frac{1}{2}CV^2$, where $C = \epsilon/z$ is the capacity per unit area and z is the separation. The force per unit area on the plates is

$$F = \mathrm{d}W/\mathrm{d}z = \tfrac{1}{2}V^2\epsilon/z^2.$$

At LCIE, Elnékavé and Fau (1965) have determined K_V using a Kelvin electrometer. The major uncertainties are associated with the accuracy of the calculation of the end-corrections and the precision with which the geometry can be defined and measured.

At CSIRO, Clothier *et al.* (1965) and Sloggett *et al.* (1984, 1985) have developed a novel variant of the Kelvin electrometer in which the lower electrode is a pool of mercury and the upper electrode is a semitransparent optical flat that is also one face of a laser interferometer. The level of the mercury under the electrode rises when a potential is applied until the gravitational forces on the mercury balance the electrical forces. The position of the mercury surface is measured interferometrically. Measurements are made at a succession of voltages and spacings chosen so that the height of the mercury is approximately constant.

The mean of the French and Australian measurements is

$$V_{76\text{-}BI} = \mathrm{V} - 7.86(58)\,\mu\mathrm{V}, \tag{11}$$

but there is a difference of a factor of 4 in the uncertainty of the two measurements, which translates to a factor of 16 in their relative statistical weights. The difference between the two values is (4.80 ± 2.46) ppm, or approximately two standard deviations.

More recently, Sloggett *et al.* (1986) have reported the revised value, $V_{76\text{-}BI} = \mathrm{V} - 8.07(31)\,\mu\mathrm{V}$, corresponding to the value $2e/h = 483.597\,90(15)\,\mathrm{THz\,V^{-1}}$. This is to be considered as the final result of their measurements; the decreased uncertainty is the result of a better understanding of the sources of systematic experimental shifts.

Type 4. *The Faraday Constant.* The Faraday constant was an important input to the 1973 adjustment, but it was deleted in the final analysis because of discrepancies with other data and the possibility of unsuspected systematic errors in the determination of the molar mass of the silver sample (Cohen and Taylor, 1973). The 1960 NBS experiments were repeated in a series of measurements extending over a nine-year period (1975-1984). The electrochemical equivalent of Ag was measured by Bower and Davis (1980) while the isotopic analysis and hence the determination of the molar mass of the sample of Ag used in the electrolysis was due to Powell, Murphy and Gramlich (1982).

Because one can write

$$\gamma_p' = \frac{\mu_p'}{\mu_N}\frac{e}{m_p} = \frac{\mu_p'}{\mu_N}\frac{F}{M_p},$$

and since $\mu_p'/\mu_N = (\mu_p'/\mu_B)(m_p/m_e)$ is an auxiliary constant in the 1986 analysis, a measurement of the Faraday is equivalent to a measurement of γ_p'. An electrolytic cell measures F in terms of the unit C/mol and the measurement is equivalent to a high-field γ_p' determination. On the other hand, the physical chemist who uses F to relate electrochemical and thermodynamic measurements usually requires the Faraday in units $\mathrm{J\,mol^{-1}}/V_{\mathrm{LAB}}$. In a unit system for which electrical energy is not equal to mechanical energy, the distinction between 'coulomb' and 'joule/volt' is significant. The Faraday measured in terms of J/V is a determination of $\gamma_p K_V$, which is neither a low-field nor a high-field determination.

Type 5. *Gyromagnetic Ratio (Low Field).* The low-field γ_p' data are in disagreement; the NPL measurement (item *5.2*) (Vigoureux and Dupuy, 1973, 1980) and the VNIIM measurement (item *5.5*) (Studentsov, Khorev and Shifrin, 1981) are significantly lower than the NIM measurement (item *5.3*) (Chiao, Liu and Shen, 1980), the NBS measurement (item *5.4*) (Williams and Olsen, 1979) and the ASMW measurement (item *5.6*) (Schlesok and Forkert, 1985). The ETL measurement (item *5.1*) (Hara *et al.*, 1968) is not in disagreement with the three high values only because of its relatively large uncertainty.

Of all the gyromagnetic ratio data, the most glaring discord comes from the NPL low-field value. The measurements of the proton resonance frequency were completed in December, 1975 after which the coil dimensions were measured, but no verification was made (by repeating the frequency measurements), to ensure that the measurement process did not affect the coils. Because the measurements were forced to terminate prematurely and because this result is so discrepant, it was not included in the final analysis.

Type 6. *Gyromagnetic Ratio (High Field).* The high-field γ_p' data are less discrepant, but they are also less precise. The Kharkov measurement (item *6.1*, Yagola, Zingerman and Sepetyi, 1962, 1966) appears to be discrepant; it differs from the mean of the other three values (item *6.2*,

Kibble and Hunt, 1979; item *6.3*, Chiao, Liu and Shen, 1980; Wang, 1984; item *6.4*, Schlesok and Forkert, 1985) by 2.4 standard deviations of that difference.

When the NBS Faraday measurement is expressed as a determination of the gyromagnetic ratio, it becomes 26751.716(36)$\times10^4$ C_{BI85}/kg; this is (2.8 ± 1.6) ppm higher than the mean of the direct measurements and (15.5 ± 5.6) ppm higher than the Kharkov value.

Type 7. *Silicon Lattice Spacing.* The first measurement of an atomic lattice spacing directly in terms of optical wavelengths was carried out at NBS (Deslattes, 1974, 1976) (item *7.1*). That value is, however, inconsistent with the later measurement at PTB (item *7.2*) using a somewhat different realization of the same concept (Becker, 1981; Seyfried, 1984; Becker and Seyfried, 1984; Siegert and Becker, 1984). The two results differ by more than 7 times the uncertainty of their difference, so that it is unjustified to keep both of them in the same adjustment. Additional measurements at PTB (Becker, Seyfried and Siegert, 1982; Siegert, Becker and Seyfried, 1984) have confirmed the quoted precision, while work at NBS has verified that the Si crystals used by PTB and by NBS have the same lattice spacing to within 0.4 ppm (Deslattes and Henins, 1984). The 0.23 ppm uncertainty implied by this limit of error is the major contributor to the uncertainty assigned to the PTB determination, although it actually applies not to the measurement itself but to the precision with which the result can be related to the NBS molar volume determination. Deslattes has indicated (private communication, 1985) that significant corrections to his early pioneering measurements with the X-ray-Optical Interferometer have been identified, but that it is premature to give a new value.

Type 8. *Molar Volume of Silicon.* The molar volume of Si has been measured in only one laboratory, NBS (Deslattes, 1980a,b). This value is based on the determinations of the isotopic composition (Barnes *et al.*, 1975) and the density (Bowman, Schoonover and Carroll, 1975) of pure silicon single crystals, corrected for known impurities and expressed in terms of the density at 22.5 °C in vacuum.

With the elimination of the NBS determination of the d_{220} lattice spacing from consideration in the analysis, there is no need to introduce the molar volume and the lattice spacing as separate entities in the analysis. The Avogadro constant N_A may be calculated directly from $N_A = nV_m(Si)/v$, where V_m is the molar volume and v is the volume of the unit cell containing n atoms of Si. The volume v is given by $(\sqrt{8}d_{220})^3$ so that $N_A = V_m/\sqrt{8}d_{220}^3$ and hence d_{220} may be dropped from the fitting process.*

Type 7 and Type 8 combined, with the NBS lattice spacing data

---

* The number 8 appearing in the expression for the volume is only indirectly related to $n = 8$ atoms in the unit cell. The lattice spacing of the (h,k,l) planes in a cubic lattice is given by $d_{hkl}^2 = a^2/(h^2+k^2+l^2)$. However, the (220) planes are strong x-ray reflecting planes because of the geometrical arrangement of the 8 atoms in the cell.

omitted, is then equivalent to the single input datum

$$N_A = 6.022\,1298(81)\times10^{23}\,\mathrm{mol}^{-1} \tag{12}$$

with an uncertainty of 1.34 ppm and $\nu = 4.8$ effective degrees of freedom.

Type 9. *Quantum Hall Effect.* The measurements of the quantum Hall effect have demonstrated over the past few years that the relation $R_H = h/e^2$ is valid at a precision better than 1 in 10^7. (The measurements only confirm that R_H is the same for different materials, and that, to this precision, it is independent of specific operating conditions. However, it is difficult to imagine a nonzero correction term that would be independent of the solid-state environment or of the operating conditions to this level.) The six measurements (*9.1*, Bliek *et al.*, 1985; *9.2*, Cage, Dziuba and Field, 1985; *9.3*, Wada *et al.*, 1985; *9.4*, Hartland and Davies, 1985; *9.5* van der Wel *et al.*, 1985; *9.6* Delahaye, 1986) give a mean value,

$$R_H = 25\,812.8461(16)\,\Omega_{BI85}.$$

From this, one finds

$$\alpha\Omega_{BI85} = [1/137.036\,2044(85)]\,\Omega.$$

The precision of 0.062 ppm is surprising for a physical measurement in a many-body system!*

Type 10. *Fine-Structure Constant.* The fine-structure constant can be calculated from the measurement of the electron magnetic moment anomaly carried out at the University of Washington by van Dyck *et al.*(1984) combined with the extensive quantum electrodynamic calculations of the theory, which culminate in Kinoshita's numerical evaluation of the eighth-order terms (Kinoshita and Lindquist, 1981a, 1981b, 1983) :

$$\tfrac{1}{2}g_e - 1 = a_e = \tfrac{1}{2}\alpha/\pi + C_2(\alpha/\pi)^2 + C_3(\alpha/\pi)^3 + C_4(\alpha/\pi)^4 + \cdots + \delta a$$

where $\delta a = 1.69(4)\times10^{-12}$ is the sum of the non-QED contributions for hadronic vacuum polarization and weak interactions, and

$$C_2 = -0.328\,478\,444, \qquad C_3 = 1.1763(13), \qquad C_4 = -0.8 \pm 2.5.$$

A measurement of the fine structure in atomic He (Lewis, Pichanick and Hughes, 1970; Frieze *et al.*, 1981; Kponou *et al.*, 1981) can yield a value for α only if the calculation of the energy levels (Lewis and Serafino, 1978) can be carried out with sufficient precision. Unfortunately, the computational uncertainty is relatively large, and uncalculated terms in the perturbation expansion of the wavefunctions of the two-electron system may contribute at the part in 10^7 level. The uncertainty assigned

* See also the contribution of E. Braun in this volume for additional information on recent developments in this fast-moving area of solid-state physics.

to the result is in fact dominated by the uncertainty of the theory with the result that the two measurements of the fine-structure constant differ in assigned weights by a factor of 85. Although the He fine-structure result is not in disagreement (within its stated uncertainty) with the anomalous electron moment value, the latter carries so much more weight that little is gained by including the former, and it is therefore justified to give this datum no further consideration in the analysis.

During the past year, too late for inclusion in the adjustment, measurements of the quantized Hall resistance at CSIRO (Australia) directly in terms of SI units (Sloggett *et al.*, 1986) yield the value $\alpha^{-1} = 137.036012(11)$. This result is (0.13 ± 0.11) ppm higher than the value derived from the electron g-factor.

Type 11. *Muon Magnetic Moment.* The magnetic moment of the muon has been determined from the hyperfine-structure splitting in a magnetic field by Mariam (1981; Mariam *et al.*, 1982) and from the precession frequency measurements of Klempt *et al.*(1982). These measurements are similar to the corresponding hyperfine measurements for hydrogen and to the anomalous muon precession, respectively. The mean of the two values is 3.183 345 47(95).

Type 12. *Muonium Hyperfine Splitting.* The measurement of the hyperfine structure in muonium will yield a value of $\alpha^2\mu_\mu/\mu_p$ if it is combined with the QED corrections to the original Fermi theory. This has been provided in large part by Sapirstein, Terray and Yennie (1984) and Bodwin, Yennie and Gregorio (1985), but still uncalculated higher order terms may be significant at the level of experimental precision that is now achievable. Because of this, a 0.13 ppm uncertainty, corresponding to ± 1 kHz limit of error, was included in the uncertainty of the theory of the experiment.

2.3. Secondary Stochastic Data

In addition to the data of Table 3 that form the basis for the least-squares analysis, there are three other stochastic quantities that must be considered in a general survey of the fundamental physical constants but which have such large uncertainties that they appear in a separate category. In essence, the results of the multivariate adjustment become auxiliary constants with respect to these data.

2.3.1. *Molar Gas Constant.* The 1973 recommended value for the gas constant R was based on measurements of the molar volume of oxygen and nitrogen. Measurements of volume are beset with problems of sorption of gas on the walls of the vessel, an effect that was not fully appreciated when the original experiments were carried out (1924 – 1952). In contrast to such extensive measurements, the speed of sound is an intensive measurement that avoids the necessity of an absolute volume determination. Quinn, Colclough and Chandler (1976) have used an acoustic interferometer to find the speed of sound in argon at the temperature of the triple point of water (Colclough, 1979, 1984;

Colclough, Quinn and Chandler, 1979) :

$$c_0^2 = 94\,756.75(58)\ \mathrm{m^2 s^{-2}} \qquad (6.1\ \mathrm{ppm}) \qquad (13)$$

where the uncertainty is Type A only.

The gas constant is given by

$$R = M(\mathrm{Ar})c_0^2/\gamma T, \qquad (14)$$

where $M(\mathrm{Ar})$ is the molar mass of argon and $\gamma = 5/3$ is the specific heat ratio for an ideal monatomic gas.

The isotopic abundances of atmospheric argon measured by Nier (1950) (which have been adopted by the International Commission on Atomic Weights and Isotopic Abundances of IUPAC (Holden and Martin, 1984; deBièvre *et al.*, 1984) as representative of the composition of atmospheric argon), the relative isotopic abundance measurements of the NPL argon, and the nuclidic masses of Wapstra and Audi (1985) lead to a molar mass for the argon used in the NPL measurements,

$$M(\mathrm{Ar_{NPL}}) = 0.039\,947\,753(75)\ \mathrm{kg/mol} \qquad (1.9\ \mathrm{ppm}) \qquad (15)$$

and hence to

$$R = 8.314\,510(70)\ \mathrm{J\,mol^{-1}K^{-1}} \qquad (8.4\ \mathrm{ppm}) \qquad (16)$$

where the uncertainty is expanded to include a contribution for evaluated systematic errors. The value for R given here is 3.6 ppm higher than that given by Colclough *et al.* who used

$$M(\mathrm{Ar_{NPL}}) = 0.039\,9476\ \mathrm{kg/mol}$$

for the isotopic composition of argon based on the results of Melton *et al.* (1971). Although the change we have introduced into the Quinn, Colclough result is within the range of its stated experimental uncertainty, eq. (15) has the advantage of being based on values directly related to the present IUPAC adopted reference abundances and its recommendation $A_r(Ar) = 39.9478$ for the relative atomic weight of atmospheric argon.

2.3.2. *Stefan-Boltzmann Constant.* The most accurate determination of the Stefan-Boltzmann constant available today is the one obtained by Quinn and Martin (1985) from the electrical calorimetric measurements of the radiation emitted by a black body at the temperature of the triple point of water. They give

$$\begin{aligned}\sigma &= 5.669\,67(76)\times10^{-8}\ \mathrm{W_{NPL}\,m^{-2}K^{-4}} \\ &= 5.669\,59(76)\times10^{-8}\ \mathrm{W\,m^{-2}K^{-4}} \qquad (134\ \mathrm{ppm}) \qquad (17)\end{aligned}$$

where $\mathrm{W_{NPL}}$ is the watt as maintained in electrical units at NPL at the time of the measurement:

$$1\ \mathrm{W_{NPL}} = 1\ \mathrm{V_{NPL}^2/\Omega_{NPL}} = 0.999\,985\,92(62)\ \mathrm{W}.$$

Although this represents a significant increase in precision over the Blevin and Brown (1971) determination (with an uncertainty of approximately 500 ppm) discussed in the 1973 adjustment, a still more accurate value may be calculated from the theoretical expression for σ,

$$\sigma = \pi^2 (R/N_A)^4/60\hbar^3c^2, \tag{18}$$

with the 1986 recommended values of N_A and $\hbar$ and the measured molar gas constant, eq. (5.4):

$$\sigma = 5.67051(19)\times10^{-8}\ \mathrm{W\,m^{-2}\,K^{-4}}. \quad (34\ \mathrm{ppm})$$

Thus the direct measurement is 4 times less precise than the indirect value. Since the difference between the two is 1.1 times the standard deviation of that difference, the two values are not in disagreement. However, there is presently insufficient precision in the direct measurement for it to influence a determination of the gas constant significantly. [A *pro forma* weighted mean of R from acoustic interferometry, eq. (16), and $R = 8.314175(280)\ \mathrm{J\,mol^{-1}K^{-1}}$ from the Quinn-Martin measurement of σ yields $R = 8.314490(68)\ \mathrm{J\,mol^{-1}K^{-1}}$ (8.2 ppm)].

2.3.3. *Newtonian Constant of Gravitation.* There is no well established relationship between the gravitational constant G and other physical quantities; it stands completely uncoupled from the remainder of the adjustment. Dimensional analysis allows one to write $G = \hbar c/m_{\mathrm{Pl}}^2$ where m_{Pl} is the Planck mass, 21.77×10^{-9} kg, but this is of no help in determining the value of G since there is no independent determination of m_{Pl}. The attempts to relate G to other forces in Grand Unification are, in the sense of the 1986 Adjustment, at best, only qualitative. Even if a theoretical expression existed for the gravitational constant, the present level of precision of the experimental determinations of G is such that it can have no effect on our knowledge of the values of any other constants.* It may be possible to obtain expressions relating the Planck mass to the masses associated with other forces, but it will probably be a long time before the precision of those expressions will rank with the current, not to mention the projected, precision of QED.

The 1973 recommended value was based on the measurements of Heyl (1930) and of Heyl and Chrzanowski (1942) (*see also* Cohen and Taylor, 1973):

$$G = 6.6720(41)\times10^{-11}\ \mathrm{m^3\,kg^{-1}\,s^{-2}}. \quad (615\ \mathrm{ppm})$$

The results of a Heyl-type oscillating torsion balance experiment that was the outgrowth of a University of Virginia program (Rose *et al.*, 1969; Towler *et al.*, 1971) transferred to NBS in 1973 were reported by Luther and Towler (1982, 1984) :

$$G = 6.67259(43)\times10^{-11}\ \mathrm{m^3\,kg^{-1}\,s^{-2}} \quad (64\ \mathrm{ppm}) \tag{19}$$

where the uncertainty is composed of a 40 ppm statistical component and a 50 ppm nonstatistical component.

---

* An expression that is pure numerology is

$$28(\hbar c/m_{\mathrm{p}}^2)\, e^{2/3\alpha} = 6.672312(27)\times10^{-11}\ \mathrm{m^3\,kg^{-1}\,s^{-2}}.$$

Other measurements of G have been reported by Facy and Pontikis (1970, 1971), Pontikis (1972), Sagitov *et al.* (1979) and Karagioz *et al.* (1976, 1981). Both Pontikis' and Karagioz' data are internally inconsistent, indicating the presence of systematic effects that had not been adequately evaluated. Only the Sagitov data appear to be statistically valid. This experiment is similar in principle to the Heyl torsion balance and yields

$$G = 6.674\,49(81)\times10^{-11}\,\mathrm{m^3\,kg^{-1}\,s^{-2}}\ (121\,\mathrm{ppm})$$

where the uncertainty is the statistical standard deviation of the mean of 20 values. No information is given on the evaluation of the systematic uncertainties, so that it is impossible to properly combine these measurements with those of Luther and Towler.

The Luther and Towler result, eq. (19), to the exclusion of all other data, was adopted as the 1986 recommended value, but in view of the difficulties in properly evaluating the systematic uncertainties when attempting to make precise measurements of such a weak and all-pervasive force, the assigned uncertainty was arbitrarily doubled; the recommended value is thus 128 ppm in the output tables.

3. DATA ANALYSIS ALGORITHMS

The standard least-squares approach to the analysis of the fundamental constants is based on minimizing a weighted sum of squares of the residual errors in the experimental data by choosing an appropriate set of values for the unknown adjustable parameters. In brief, each experimental result represents a relationship that exists among a set of physical constants. The totality of the observational data is in general a set of repeated or overlapping constraints on our knowledge of these constants. Assuming a gaussian distribution of errors, this is equivalent to a maximization of the likelihood of the parameter choice. This interpretation, however, includes an implicit assumption about the *a priori* expectation of parameter values. An alternative description of the algorithm emphasizes the *linearity* rather than the *quadratic* aspects and is distribution-free. This is the Gauss-Markov theorem that states that if the relationship between input data and the unknown parameters can be linearized, then the least-squares algorithm produces that linear combination (or weighted average) of the data that gives the minimum variance unbiased estimate for the parameters.

Least squares is thus a generalization to the multivariable case of the simple process of taking a weighted mean of a set of observations of a single quantity. In the same sense that a vector in three-dimensional space can be described by components in a coordinate system chosen in many different ways, the actual variables that one selects as the unknowns in a least-squares analysis are arbitrary. The only essential requirement is that the selected quantities form a complete and independent set in terms of which all of the experimental observations may be expressed.

Each observational input datum is expressed as an algebraic relationship among the auxiliary constants and the unknowns. Completeness of the set of unknowns means that there are enough unknowns to express all of the experimental data; independence means that the observations can be expressed in terms of the unknowns in essentially only one way. (The auxiliary constants in these expressions need not be unique, since auxiliary constants do not contribute any statistical uncertainty to the expressions and may be expressed in a number of equivalent forms.) Independence also implies that no unknown can be expressed as a combination of the other unknowns.

It is convenient and laborsaving, but not necessary (now that the inversion of a 50×50 matrix is no longer the practical impossibility that it was 40 years ago), that the observational equations be statistically independent; if not, the statistical weights assigned to the observations must be expressed in terms of a weight matrix with off-diagonal elements. This weight matrix is the inverse of the variance matrix of the observational equations. If the observational equations are independent, the variance matrix is diagonal and the weights become essentially a set of scalar quantities. The weight matrix can be considered to be the metric tensor of a space that describes the statistical properties of the data.

In analyzing and evaluating the input data, we have made use not only of the usual least-squares algorithm but also an algorithm suggested by Tuninskii and Kholin (1975) as well as the extended least-squares algorithms described by Cohen (1980, 1984) and a modification of the original VNIIM algorithm suggested by Taylor (1982).

In the standard least-squares procedure, the weight w_i assigned to each experimental datum is $1/\sigma_i^2$, where σ_i^2 is the total variance. The new algorithms may all be categorized as procedures that recognize that the variances (and hence the weights) needed in the least-squares analysis are not known exactly, but are only available as *a priori* estimates given by s_i^2 with a statistical weight, ν_i. They each use the consistency of the data to provide additional, *a posteriori* information for improving these estimates.

The Tuninskii algorithm modifies each *a priori* variance s_i^2 by a factor R_i determined by requiring that the 'cost function' for the modification,

$$C = \sum_i (R_i - 1)^2 \tag{20}$$

be a minimum under the constraint that χ^2 calculated with weights determined from the modified variances be equal to its expectation value ν, the number of degrees of freedom in the least-squares analysis.

This proceedure may be criticized on two counts: (i) a modification of the variance of any datum by a given factor is assigned the same cost, independent of any further information that may be available concerning the reliability of the *a priori* estimate of that variance; (ii) the cost function remains finite for $R_i \to 0$ and, as a result, if the cost function is identified with a maximum likelihood estimator for the *a priori* variance, the resultant probability distribution is not normalizable.

Some of the modifications of the cost function considered by Taylor in a preliminary review of the data (Taylor, 1982) avoid the second objection (but not the first); as representative of those algorithms we shall use the cost function $C' = \Sigma (R_i - 1)^2/R_i$, or, equivalently,

$$C' = \sum_i (R_i + 1/R_i). \tag{21}$$

An algorithm that answers both objection (i) and objection (ii) weights each cost component by the effective number of degrees of freedom of that observational datum. Hence in place of eq. (21) one introduces

$$C' = \sum_i \nu_i (R_i + 1/R_i) \tag{22}$$

which reduces the extent of the modification of the initial uncertainty assignments for those data whose uncertainties are more precisely defined by giving such modifications higher cost, and allows an uncertainty that is not well defined (ν_i small) to undergo greater modification.

We have also used two extended least-squares algorithms that are designed to take into account the *a priori* knowledge one has regarding the precision of the *a priori* estimates of the data uncertainty assignments:

a.) The first algorithm (ELS1) uses the residual $r_i = y_i - \widehat{Y}_i$, where $\widehat{Y}_i$ is the adjusted value in the least-squares fit, in addition to the *a priori* standard deviation s_i to evaluate a best estimate of the variance σ_i^2 and writes $\widehat{\sigma_i^2} = a_i s_i^2 + b_i r_i^2$. The values of a_i and b_i are determined by requiring that this expression be an unbiased minimum-variance estimator of σ_i^2. The expectation value of the variance of the residual is equal to $1/w_i - t_{ii}$, where $1/w_i$ is the input variance and t_{ii} is the output variance (Cohen, 1953). The self consistent estimate of the variance of the input datum is the mean of the *a priori* estimate s_i^2 with weight ν_i and the *a posteriori* estimate $r_i^2/(1 - w_i t_{ii})$ with weight 1, or

$$\frac{1}{w_i} = \frac{\nu_i s_i^2 + r_i^2/(1 - w_i t_{ii})}{\nu_i + 1}. \tag{23}$$

b.) If there are no identifiably discrepant data it is appropriate to use all of the data to evaluate the statistically most efficient estimates for the weight of each input. Generalizing the equation given above, one considers the more complete sum, $\widehat{\sigma_i^2} = a_i s_i^2 + \Sigma b_{ij} r_i r_j$, and from this obtains the surprisingly simple result (ELS2) that the best estimate is a mean of the *a priori* value s_i^2 with weight ν_i and the *a posteriori* estimate $(\chi^2/\nu)/w_i$ with weight ν_i; hence one may write

$$\frac{1}{w_i} = \frac{\nu\left(\chi^2/\nu w_i\right) + \nu_i s_i^2}{\nu + \nu_i}$$

or

$$w_i = [1 - (\chi^2 - \nu)/\nu_i]/s_i^2. \tag{24}$$

Equation (24) is statistically stronger than eq. (23) in the ratio $(\nu_i + \nu)/(\nu_i + 1)$. However, eq. (24) may not always have a solution with positive weights; for the weight w_i to be positive it is necessary that $\nu + \nu_i$ be greater than χ^2; if there is more than one discrepant item in the data set, it may be impossible for the algorithm to reduce the weights of the discrepant data sufficiently to yield a sufficiently small value of χ^2. This lack of convergence is an indication that the data are discrepant in the sense that the observed residues are larger than can be reasonably ascribed to random fluctuations in the data; since this is a contradiction of the assumptions underlying eq. (24), it indicates that the equation should not be used for that data set.

The fact that ELS2 does not yield a valid solution in all cases is a drawback to its use. A modification to the algorithm can be introduced to circumvent this; instead of the *a priori* estimate $(\chi^2/\nu)/w_i$ that leads to eq. (24), we modify the *a priori* estimate to $(\chi_o^2/\nu)s_i^2$ (which is the simplest form of the Birge ratio assumption). This algorithm is not applied iteratively (with the value of χ^2 evaluated using the modified weights); here the appropriate value of χ^2 is evaluated with the *a priori* values s_i^2, and is therefore denoted as χ_o^2. The weights are then given by:

$$w_i = [(\nu + \nu_i)/(\chi_o^2 + \nu_i)]/s_i^2. \tag{25}$$

Equation (25) represents a generalization of the Birge ratio; it replaces the uniform scaling of the data using a fixed parameter R_B by a nonuniform scaling with a set of factors lying between 1 (if the precision index is high, $\nu_i \gg 1$) and R_B (if the precision index is low, $\nu_i \to 0$). The resultant value of χ^2, calculated with the modified weights, will not be equal to ν but will lie between χ^2 and ν.

4. DATA ANALYSIS

4.1. Preliminary Review

The consistency of the data can be evaluated in part by considering relationships among data of different types. The most obvious of these is Ohm's law,

$$V = A\Omega.$$

From eq. (4.3) and eq. (4.1), we find

$$\begin{aligned} A_{BI85} &= 0.999\,992\,14(58)\ \mathrm{V}/0.999\,998\,467(69)\ \Omega \\ &= 1\ \mathrm{A} - 6.32(59)\ \mu\mathrm{A}, \end{aligned} \tag{26}$$

which shows a discrepancy of (8.4 ± 2.6) ppm with respect to the direct determination of the ampere, eq. (10). Equation (26) represents an improvement in precision of the determination by a factor of 4.3, or a larger statistical weight by a factor of 19.

Another indirect evaluation of the ampere comes from a comparison of the results of the low-field and the high-field γ_p' determinations. Since the two methods determine the same physical quantity, one may write

$$\gamma_p' = \gamma_{lo}\ \mathrm{s}^{-1}/T_{BI85} = \gamma_{hi}\ A_{BI85}\ \mathrm{s/kg} \tag{27}$$

where γ_{lo} and γ_{hi} are dimensionless quantities. The laboratory units are given by

$$A_{BI85} = K_A\,\mathrm{A}, \qquad T_{BI85} = K_A\,\mathrm{T}, \tag{28}$$

since the magnetic field is determined from the geometry and the measured current in the coils. The tesla is, by definition, $\mathrm{T} = \mathrm{kg\,s^{-2}\,A^{-1}}$ and hence

$$K_A^2 = \gamma_{lo}/\gamma_{hi}. \tag{29}$$

The data from those laboratories that have carried out both high-field and low-field determinations (including the NBS Faraday as a high-field determination of the gyromagnetic ratio) yield the results listed in Table 3.

The mean of these four determinations is $-7.54(41)\,\mu\mathrm{A}$ with the uncertainty being the *a priori* value. The data are discordant; they yield $\chi^2 = 30.0$ with 3 degrees of freedom. The Birge ratio is $R_B = 3.16$ and $P_{\chi^2}(30.0|3) < 10^{-5}$. However, the NIM and ASMW measurements are not precise enough to carry significant weight in comparison with the NBS and NPL measurements, and at least one of those two purportedly high precision values must be in error since they disagree by 3.3 times the standard deviation of that difference. The probability of a discrepancy as large as this, with the standard deviation being based on 13.2 effective degrees of freedom, is approximately 0.25%.

An indirect value of N_A may be found from

$$N_A = \tfrac{1}{2} R_H F(2e/h). \tag{30}$$

The units of this expression are $\Omega\ \mathrm{C\ s^{-1}\ V^{-1}\ mol^{-1}}$; as long as the coulomb or ampere standard utilized in the measurement of the Faraday is consistent with the resistance standard in terms of which R_H is expressed and the volt defined by the assigned value of $2e/h$, eq. (30) is independent of the actual values of those laboratory standards. Then, with $2e/h = 483\,594.0\ \mathrm{GHz}/V_{76-BI}$, expressing F in terms of A_{BI85}s and using the mean value of the quantized Hall resistance given in eq. (12), eq. (30) gives $N_A = 6.022\,1433(80)\times10^{23}/\mathrm{mol}$, a value that is 2.3 ± 1.9 ppm larger than the value qiven in section 4 based on the PTB lattice spacing and the NBS molar volume determinations.

Another relationship in which the values of the electrical units cancel is

$$\gamma_p' = \frac{(\mu_p'/\mu_B)}{2\mu_o R_\infty}(2e/h)\alpha^3 R_H \tag{31}$$

if γ_p' is a low-field determination (and hence with units $\mathrm{s^{-1}}/T_{LAB}$) and if $2e/h$ is expressed in LAB volts and R_H in LAB ohms (all corrected to a common time). With α^{-1} from item (*10.1*) and the quantized Hall resistance from eq. (12), eq. (31) gives

$$\gamma_p' = 26\,751.3617(55)\times10^4\ \mathrm{s^{-1}}/T_{BI85},$$

a result that is more precise than any of the direct measurements and that differs from the most precise of those data (item *5.4*) by $-0.38(31)$ ppm.

A value for the fine-structure constant with a precision comparable to that obtained from the anomalous electron moment can be deduced from the quantized Hall resistance combined with the value of the maintained ohm given in eq. (9). This leads to $\alpha^{-1} = 137.035\,9943(127)$, a figure that is in remarkable agreement with the direct measurement, 137.035 9942(89). That this nearly exact agreement can only be considered to be a happy coincidence is indicated by the more recent Hall coefficient measurements such as the Australian value quoted previously. An independent value for α may also be found from the Los Alamos/Yale muonium hyperfine spectrum measurements (items *11.1* and *12.1*). These data give $\alpha^{-1} = 137.036\,003(26)$, which agrees within its uncertainty with item *10.1*, but with only 1/9 the weight.

One can calculate a value of the muon moment from the muonium hyperfine splitting, $\Delta\nu_{\mathrm{Mhfs}}$, given a value for the fine-structure constant, α. This value of μ_μ/μ_p is statistically independent of item (*11.1*), even though it comes from the same measurements, because the primary error contribution is the uncertainty in the theoretical expression for the hyperfine interval. Using the value 137.035 9942 from the electron anomalous moment implies $\mu_\mu/\mu_\mathrm{p} = 3.183\,345\,68(61)$, a value that is consistent with the direct determinations.

There are still other relationships that may be written by combining those given above, but in general they add no essential new information to that which is already available in the relationships given here. The purpose of this survey is only to indicate the extent to which the data are in agreement or disagreement. The full treatment of the data requires a complete multivariate analysis.

4.2. Multivariate Analysis of the Data

The total number of observational equations in this analysis is $N = 38$, but because of the redundancy of the measurements there are only 12 distinct types. These can be expressed in terms of $M = 5$ adjustable unknowns which, in turn, may be used to calculate all other constants of interest. The twelve types of observational equations are given in Table 4. In this table the quantity E is an abbreviation for 483 594.0 GHz/V; this allows us to write, from eq. (6),

$$2e/h = E/K_\mathrm{V}. \tag{32}$$

The five adjustable unknowns are taken to be α, K_Ω, K_V, d_{220} and μ_μ/μ_p. (For convenience in numerical output, the calculations are actually performed with the variable α^{-1}.) In Table 4, the quantities to the left of the centered dot are auxiliary constants; those to the right are the stochastic variables of the adjustment.

It is clear that some of the data will have very little statistical weight in any least-squares analysis, and there is ample evidence, even from the cursory review, that others should be removed from the set of retained data because of discrepancies. An adjustment of all 38 data by standard least squares gives $\chi^2 = 324.9$ with $\nu = 33$, or a Birge ratio $R_\mathrm{B} = 3.14$. Since the expectation value of χ^2 is 33 ± 8.1 and the expectation value of the Birge ratio is 1 ± 0.12, the discrepancy is clearly evident.

The *pro forma* probability that χ^2 would by chance be this large is $P_{\chi^2}(324.9|33) \approx 10^{-29}$.

A simple listing of the differences between two sets of adjusted values can give a distorted image of the essential difference because it cannot show the correlations that exist among them. In order to obtain a proper representation of these differences, the weight matrix (or, equivalently, the variance matrix) of the data may be interpreted as the metric tensor of the space spanned by the variables of the least-squares adjustment. In that space, several quantities in the list may correspond to directions that are almost parallel. The variance matrix defines the orthogonal directions in the space and the metric distance of the adjusted point from a fixed point (*i. e.*, the distance measured in units of the standard deviation in the direction of the displacement). This distance can be expressed either in the space of the input data of the least-squares adjustment or in the space of the output data (the adjusted values) and is given by

$$d^2 = \sum (x_r - x_r^\circ) w_{rs} (x_s - x_s^\circ) = \sum w_i (y_i - y_i^\circ)^2, \tag{33}$$

where w_{rs} is the weight matrix of the solution, y_i is the stochastic input datum, and y_i° is its value evaluated at the fixed point defined by the origin values $\{x_r^\circ\}$.

Once the two highly discrepant items *5.2* and *7.1* are deleted from consideration, and item *10.2* is dropped because it is based on an incompletely developed theory (the dominant uncertainty comes from the uncertainties in convergence of the calculated energy levels of the two-electron system) the result is a data set with 35 observations and 30 degrees of freedom. The standard least-squares adjustment gives $\chi^2 = 106.6$ with $R_B = 1.89$.

All of the algorithms considered here (including standard least squares with the uncertainties re-expressed using 'external error') yield results that are quite similar, and none of these results differ by more than one standard deviation from the set that has been selected as the 'best', and which will form the basis for the 1986 recommended values.

The largest remaining discrepancy is produced by item *5.5*, the Studentsov *et al.* low-field γ_p' determination. It contributes by itself almost half of the total χ^2 (52.6) and its deviation from the value of γ_p' deduced from the remainder of the data is 7.4 times the standard deviation of that difference.

The data also contain several items that contribute very little to the determination of the adjusted values. None of the direct ampere determinations (items 2.1-2.6) contributes more than 0.5% (and all six, less than 1.3%) of the weight of the adjusted value of the ampere, a value defined more precisely from the ratio of the low-field to the high-field γ_p measurements, or from ohm and volt measurements. The ohm, in turn, is only partially defined by the calculable capacitor data; equally as much weight is attached to the 'indirect' value of the ohm resulting from the combination of the quantized Hall resistance and the implied value of the fine-structure constant – both the direct value of item *10.1* and the indirectly deduced value from such relationships as eq. (29).

The low-weight data - items *2.1-2.6*, item *3.1*, items *5.1*, *5.3* and *5.6*, and item *6.1* --- none of which contribute more than 1/15 as much as another direct or indirect determination of the same quantity, may be omitted from the data set without greatly affecting the results. In addition to these eleven deletions, the relatively low weight and somewhat discrepant ASMW high-field γ_p' determination (item *6.4*) is also eliminated from the set.

5. 1986 RECOMMENDED VALUES OF THE FUNDAMENTAL PHYSICAL CONSTANTS

The remaining 22 data items (58% of the original inventory) become the basis for the 1986 Recommended Values. The standard least-squares procedure gives $\chi^2 = 17.09$ for these data, with 17 degrees of freedom. Since the value of χ^2 is only slightly greater than ν, there is very little difference among the results produced using different algorithms except for ELS1. In this algorithm the weight of each input datum is adjusted based on its own deviation from the consensus value; hence the total reduction of χ^2 is larger than using algorithm ELS2, and χ^2 is reduced to an adjusted value 15.2.

Because algorithm ELS2 appears to be best grounded statistically, we use it as the basis for the recommended set, but all the other algorithms considered in the analysis give almost indistinguishably different values. The recommended values are therefore almost totally independent of the actual algorithm used. The value of χ^2 for the adjustment is 17.01 with 17 degrees of freedom.

A short summary of the recommended values is given in Table 5. Table 6 presents a set of related values, such as the quantities V_{BI-76}, Ω_{BI85} and d_{220} that are a necessary part of the data of the adjustment, but that cannot be considered as fundamental in the same sense as the quantities of Table 5.

The equations relating the quantities in Tables 5 and 6 to the variables of the least-squares adjustment and to the auxiliary constants of Table 2 and eq. (32) are fairly direct. The x-ray data in Table 6 are based on the measurements of Deslattes and Henins (1973) and of Kessler, Deslattes and Henins (1979) who determined the ratios of the wavelengths to the d_{220} spacing in Si. More complete Tables are given in the CODATA report (Cohen and Taylor, 1986; 1987a,b).

There are significant changes in the recommended values of the physical constants from 1973 to 1986. The auxiliary constants m_p/m_e and R_∞ show improvements by factors of 19 and 63, respectively. The overall precision of the 1986 adjustment is approximately a factor of 10 better than that of the 1973 adjustment and the precision of α has been increased by a factor of 18. It will be interesting to see if this rate of improvement can be sustained for another thirteen years; if so, the turn of the 21*st* century should be an exciting period for metrology!

This analysis did not make a distinction between the WQED data and the QED data, as was done in 1973 and 1969. If those measurements that depend on QED information, item (*10.1*), the electron anomalous moment and item (*12.1*), the muonium hyperfine-structure, are deleted, the remaining

20 items (15 degrees of freedom) have $\chi^2 = 16.53$ for standard least squares, which is reduced to 15.24 with algorithm ELS2. The distance of the ELS2 solution from the recommended set is $d = 0.29$. The WQED value of α is 1/137.035 9846(94). This differs from the recommended value by 0.036 ± 0.059 ppm. The WQED value of K_V differs from the recommended value by 0.01 ± 1.03 ppm. Thus there is no clear basis for any distinction between QED and WQED data.

The validity of the theory of the quantum Hall effect (QHE) can also be investigated. If the QHE data are deleted from the recommended set, the solution is displaced by a distance $d = 0.038$ and the value of α^{-1} is 137.035 9884(79); the difference from the recommended value is -0.009 ± 0.071 ppm. A value of α^{-1} from the Hall resistance data and the ohm determinations, 137.035 9943(127), was mentioned in Sect. 4. This differs by only 0.043 ± 0.085 ppm from the value above, where the uncertainty is evaluated taking into account the correlations from the direct ohm (Type 1) determinations. Thus, based on the presently available observational data, there is no evidence of any discrepancy in the QHE theory at the current levels of precision.

The discrepancies between the 'as-maintained' electrical standards and the true SI values of the ohm and the volt indicated in the 1986 adjustment imply that the international standards should be revised. This is more important now than it was in 1969 when the international intercomparisons by BIPM produced the present standards. Both the increased demands of technology for accuracy and the existence of instruments capable of better than part per million resolution in the measurements have stimulated the BIPM Consultative Committee on Electricity to establish a new international standard value for the Josephson frequency-voltage quotient and to establish an international resistance standard based on the quantized Hall effect. It is intended that these new standards will be promulgated in October 1988 and will become effective on January 1, 1990. These standards will, in essence, define two quantities, a Josephson frequency f_J = 483 597.xx GHz and a Klitzing number n_K = 25 812.8xxx, where the final digits are yet to be determined. An ideal Josephson device irradiated at frequency f would produce a potential difference of (f/f_J) V across the junction within a specified uncertainty. The quantized Hall resistance of a MOSFET or a heterojunction at the nth plateau will be taken to be $(n_K/n)\ \Omega$ (again within specified uncertainty). These quantities will provide universal, time-independent realizations of electrical units in terms of atomic standards.

6. REFERENCES

Amin, S. R., C. D. Caldwell and W. Lichten, 1981. Phys. Rev. Lett. 47, 1234

Amin, S. R., C. D. Caldwell and W. Lichten, 1984. in *Precision Measurement and Fundamental Constants II*, B. N. Taylor and W. D. Phillips, eds., Natl. Bur. Stand. (US), Spec. Publ. 617 (U. S. Govt. Printing Office,

Washington, DC, 1984), p. 117

Bailey, J., *et al.*, 1979. Nucl. Phys. B **150**, 1

Barrow, J. D., 1983. Quarterly Journal Roy. Astron. Soc. **24**, 24

Becker, P., *et al.*, 1981. Phys. Rev. Lett. **46**, 1540

Becker, P., P. Seyfried and H. Siegert, 1982. Z. Phys. B **48**, 17

Becker, P., and H. Siegert, 1984. in *Precision Measurement and Fundamental Constants II*, B. N. Taylor and W. D. Phillips, eds., Natl. Bur. Stand. (US), Spec. Publ. 617 (U. S. Govt. Printing Office, Washington, DC, 1984), p. 317

Bender, D., and W. Schlesok, 1974. Metrologia **10**, 1

BIPM, 1983. Comptes Rendus des séances de la 17ᵉ CGPM, 1983 (BIPM, Sèvres, France)

Birge, R. T., 1929. The Physical Review Supplement (Reviews of Modern Physics) **1**, 1

Blevin, W. R., and W. J. Brown, 1971. Metrologia **7**, 15

Bliek, L., *et al.*, 1985. IEEE Trans. Instrum. Meas. **IM-34**, 304; and private communication

Bodwin, G. T., D. R. Yennie and M. A. Gregorio, 1985. Rev. Mod. Phys. **57**(3), Pt. 1, 723

Bower, V. E., and R. S. Davis, 1980. J. Res. Natl. Bur. Stand. **85**, 175

Bowman, H. A., R. M. Schoonover, and C. L. Carroll, 1975. *Reevaluation of the densities of the four NBS silicon crystal standards*, Natl. Bur. Stand. Report NBSIR 75-768

Brown, L. S., G. Gabrielse, K. Helmerson and J. Tan, 1985. Phys. Rev. **32**, 3204

Cage, M. E., R. F. Dziuba and B. F. Field, 1985. IEEE Trans. Instrum. Meas. **IM-34**, 301; and private communication

Chiao, W. C., R. M. Liu and P. Shen, 1980. IEEE Trans. Instrum. Meas. **IM-29**, 238

Close, F. E., and H. Osborne, 1971. Phys. Lett. **34B**, 400

Clothier, W. K., 1965. Metrologia **1**, 181

CODATA, 1973. *Recommended consistent values of the fundamental physical constants, 1973*, CODATA Bulletin No. 11 (International Council of Scientific Unions --- Committee on Data for Science and Technology, 51 Blvd de Montmorency, 75016 Paris, France, Dec. 1973)

Cohen, E. R., 1953. Rev. Mod. Phys. **25**, 706

Cohen, E. R., 1980. in *Metrology and Fundamental Constants*, Proceedings of the International School of Physics 'Enrico Fermi', Course LXVIII, A. Ferro-Milone, P. Giacomo and S. Leschiutta, eds. (North-Holland, Amsterdam), p. 581

Cohen, E. R., 1984. in *Precision Measurement and Fundamental Constants II*, B. N. Taylor and W. D. Phillips, eds., Natl. Bur. Stand. (US), Spec. Publ. 617 (U. S. Govt. Printing Office, Washington, DC, 1984), p. 391

Cohen, E. R., K. M. Crowe and J. W. M. DuMond, 1957. *Fundamental Constants of Physics* (Interscience Publishers, New York)

Cohen, E. R., and B. N. Taylor, 1973. J. Phys. Chem. Ref. Data **2**, 663

Cohen, E. R., and B. N. Taylor, 1986. *The 1986 Adjustment of the Fundamental Physical Constants*, CODATA Bulletin No. 63 (International Council of Scientific Unions --- Committee on Data for Science and Technology, 51 Blvd de Montmorency, 75016 Paris, France, Nov. 1986)

Cohen, E. R., and B. N. Taylor, 1987a. J. Res. Natl. Bur. Stand. **92**, 85

Cohen, E. R., and B. N. Taylor, 1987b. Rev. Mod. Phys. (to be published)

Colclough, A. R., 1979. Acustica **42**, 28

Colclough, A. R., T. J. Quinn and T. R. D. Chandler, 1979. Proc. R. Soc. London, Ser. A: **368**, 125

Colclough, A. R., 1984. in *Precision Measurement and Fundamental Constants II*, B. N. Taylor and W. D. Phillips, eds., Natl. Bur. Stand. (US), Spec. Publ. 617 (U. S. Govt. Printing Office, Washington, DC, 1984), p. 263

Colclough, A. R., 1984. private communication

Cutkosky, R. D., 1974. IEEE Trans. Instrum. Meas. **IM-23**, 305

Dahake, S. L., *et al.*, 1983. IEEE Trans. Instrum. Meas. **IM-32**, 5; private communication

deBièvre, P., M. Gallet, N. E. Holden and I. L. Barnes, 1984. J. Phys. Chem. Ref. Data **13**, 809

Delahaye, F., *et al.*, 1986. Metrologia **22**, 103

Deslattes, R. D., *et al.*, 1974. Phys. Rev. Lett. **33**, 463

Deslattes, R. D., *et al.*, 1976. Phys. Rev. Lett. **36**, 898

Deslattes, R. D., 1980a. in *Metrology and Fundamental Constants*, Proceedings of the International School of Physics 'Enrico Fermi', Course LXVIII, A. Ferro-Milone, P. Giacomo and S. Leschiutta, eds. (North-Holland, Amsterdam), p. 38

Deslattes, R. D., 1980b. Ann. Rev. Phys. Chem. **31**, 435

Deslattes, R. D., and A. Henins, 1973. Phys. Rev. Lett. **31**, 972

Deslattes, R. D., and A. Henins, 1984. private communication

Driscoll, R. L., and R. D. Cutkosky, 1958. J. Res. Natl. Bur. Stand. **60**, 297

Driscoll, R. L., and P. T. Olsen, 1968. Document CCE/68-7, submitted to CCE, CIPM, 12^e session (October 1968)

van Dyck, R. S., jr, 1985. private communication

van Dyck, R. S., jr, F. L. Moore, D. L. Farnum and P. B. Schwinberg, 1986. Bull. Am. Phys. Soc. **31**, 244 van Dyck, R. S., jr, P. B. Schwinberg and H. G. Dehmelt, **1984**. in *Atomic Physics 9*, R. S. van Dyck, jr and E. N. Fortson, eds. (World Scientific Publishing Co., Singapore), p. 53

Elnékavé, N., and A. Fau, 1984. in *Precision Measurement and Fundamental Constants II*, B. N. Taylor and W. D. Phillips, eds., Natl. Bur. Stand. (US), Spec. Publ. 617 (U. S. Govt. Printing Office, Washington, DC, 1984), p. 465; Document CCE/83-2, submitted to CCE, CIPM, 16e session, March, 1983; private communication (1984)

Erickson, G. W., 1977. J. Phys. Chem. Ref. Data **6**, 831

Erickson, G. W., 1983. private communication

Facy, L., and C. Pontikis, 1970. Comptes Rendus Acad. Sci., Ser. B: **270**, 15

Facy, L., and C. Pontikis, 1971. Comptes Rendus Acad. Sci., Ser. B: **272**, 1397

Faller, J. E., 1967. Science **158**, 60

Faustov, R., 1970. Phys. Lett. **33B**, 422

Frieze, W., E. A. Hinds, V. W. Hughes and F. M. J. Pichanick, 1981. Phys. Rev. A, **24**, 279

Giacomo, P., 1981. Metrologia **17**, 69

Goldsmith, J. E., E. W. Weber and T. W. Hänsch, 1979. Phys. Rev. Lett. **41**, 1525

Gorbatsevich, S. V., 1973. private communication

Gräff, G., H. Kalinowsky and J. Traut, 1980. Z. Physik **A297**, 35

Grotch, H., and R. Hegstrom, 1971. Phys. Rev. A **4**, 59

Hammond, J. A., and J. E. Faller, 1967. IEEE J. Quantum Electr. **QE-3**, 597

Hänsch, T. W., *et al.*, 1974. Phys. Rev. Lett. **32**, 1336

Hara, K., H. Nakamura, T. Sakai and N. Koizumi, 1968. unpublished report

Hartland, A., G. J. Davies and D. R. Wood, 1985. IEEE Trans. Instrum. Meas. **IM-34**, 309

Heyl, P. R., 1930. J. Res. Nat. Bur. Stand. **5**, 1243

Heyl, P. R., and P. Chrzanowski, 1942. J. Res. Nat. Bur. Stand. **29**, 1

Holden, N. E., and R. L. Martin, 1984. Pure Appl. Chem. **56**, 653

Hudson, R. P., 1984. (editor), Metrologia **19**, 163

Igarashi, T., Y. Koizumi and M. Kanno, 1968. IEEE Trans. Instrum. Meas. **IM-17**, 226

Igarashi, T., H. Nishinaka and K. Murakami, 1978. Document CCE/78-12, submitted to CCE, CIPM, 15[e] session

Igarashi, T., 1983. private communication

Igarashi, T., 1984. private communication

Jones, R. G., and B. P. Kibble, 1985. IEEE Trans. Instrum. Meas. **IM-34**, 181

Karagioz, O. V., *et al.*, 1976. Izv. Acad. Nauk SSSR, Fiz. Zemli **12**, 106; [Engl. transl.: Izv. Acad. Sci. USSR, Phys. Solid Earth **12**, 351 (1976)]

Karagioz, O. V., A. A. Silin and V. P. Izmaylov, 1981. Izv. Acad. Nauk SSSR, Fiz. Zemli **17**, 92; [Engl. transl.: Izv. Acad. Sci. USSR, Phys. Solid Earth **17**, 66 (1976)]

Kessler, jr, E. G., R. D. Deslattes and A. Henins, 1979. Phys. Rev. A **19**, 215

Kibble, B. P., and G. J. Hunt, 1979. Metrologia **15**, 5 (1979)

Kinoshita T., and W. B. Lindquist, 1981a. Phys. Rev. Lett. **47**, 1573

Kinoshita T., and W. B. Lindquist, 1981b. *Eighth order magnetic moment of the electron, IV*, Report CLNS--81/509, and *V*, Report CLNS--81/510, Cornell University (Newman Laboratory of Nuclear Studies, Sept. 1981)

Kinoshita T., and W. B. Lindquist, 1983. Phys. Rev. D **27**, 867, 877, 886

Klempt, E., *et al.*, 1982. Phys. Rev. D **25**, 652

von Klitzing, K., 1986. Rev. Mod. Phys. **58**, 519

von Klitzing, K., G. Dorda and M. Pepper, 1980. Phys. Rev. Lett. **45**, 494

Kponou, A., *et al.*, 1981. Phys. Rev. A, **24**, 264

Lampard, D. G., 1957. Proc. Inst. Elec. Eng. (London) Part C **104**, 2701

Lewis, M. L., and P. H. Serafino, 1978. Phys. Rev. A, **18**, 867

Lewis, S. A., F. M. J. Pichanick and V. W. Hughes, 1970. Phys. Rev. A, **2**, 86

Lichten, W., 1985. private communication

Luther, G., and W. R. Towler, 1982. Phys. Rev. Lett. **48**, 121; also in *Precision Measurement and Fundamental Constants II*, B. N. Taylor and W. D. Phillips, eds., Natl. Bur. Stand. (US), Spec. Publ. 617 (U. S. Govt. Printing Office, Washington, DC, 1984), p. 573

Mamyrin, B. A., N. N. Aruyev and S. A. Alekseenko, 1983. Zh. Eksp. Teor. Fiz. **84**, 1978; [Engl. transl.: Sov. Phys.-JETP **57**, 1152 (1983)]

Mariam, F. G., 1981. *High precision measurement of the muonium ground-state hyperfine interval and the muon magnetic moment* (Ph. D. thesis, Yale University, 1981)

Mariam, F. G., *et al.*, 1982. Phys. Rev. Lett. **49**, 993

Melton, C. E., W. Massey and B. N. Abels, 1971. Z. Naturforsch. **26a**, 1241

Morelli, C., *et al.*, 1974. *The International Gravity Standardization Net 1971 (IGSN71)*, Publ. Spéc. No. 4, (Bur. central de l'assoc. int. de géodesie, Paris)

Nier, A. O., 1950. Phys. Rev. **77**, 789

van der Pauw, L. J., 1958. Philips Res. Repts. **13**, 1

Petley, B. W., and R. W. Donaldson, 1984. Metrologia **20**, 81

Petley, B. W., and K. Morris, 1974. J. Phys. A **7**, 167

Petley, B. W., and K. Morris, 1979. Nature (London), **279**, 141

Petley, B. W., K. Morris and R. E. Shawyer, 1980. J. Phys. B **13**, 3099

Phillips, W. D., W. E. Cooke and D. Kleppner, 1977. Metrologia **13**, 179

Pontikis, C., 1972. Comptes Rendus Acad. Sci., Ser. B: **274**, 437

Powell, L. J., T. J. Murphy and J. W. Gramlich, 1982. J. Res. Nat. Bur. Stand. **87**, 9

Quinn, T. J., A. R. Colclough and T. R. D. Chandler, 1976. Philos. Trans. R. Soc. London, Ser. A: **283**, 367

Quinn, T. J., and J. E. Martin, 1985. Philos. Trans. R. Soc. London, Ser. A: **316**, 85

Rose, R. D., *et al.*, 1969. Phys. Rev. Lett. **23**, 655

Sagitov, M. U., *et al.*, 1979. Dokl. Acad. Nauk SSSR **245**, 567; [Engl. transl.: Sov. Phys.-Dokl. (Earth Sci.) **245**, 20 (1981)]

Sapirstein, J. R., E. A. Terray and D. R. Yennie, 1984. Phys. Rev. D **29**, 2290

Schlesok W., and J. Forkert, 1985. IEEE Trans. Instrum. Meas. **IM-34**, 173

Seyfried, P., 1984. in *Precision Measurement and Fundamental Constants II*, B. N. Taylor and W. D. Phillips, eds., Natl. Bur. Stand. (US), Spec. Publ. 617 (U. S. Govt. Printing Office, Washington, DC, 1984), p. 313

Siegert, H., and P. Becker, 1984. in *Precision Measurement and Fundamental Constants II*, B. N. Taylor and W. D. Phillips, eds., Natl. Bur. Stand. (US), Spec. Publ. 617 (U. S. Govt. Printing Office, Washington, DC, 1984), p. 321

Siegert, H., P. Becker and P. Seyfried, 1984. Z. Phys. B **56**, 273

Sloggett, G. J., *et al.*, 1984. in *Precision Measurement and Fundamental Constants II*, B. N. Taylor and W. D. Phillips, eds., Natl. Bur. Stand. (US), Spec. Publ. 617 (U. S. Govt. Printing Office, Washington, DC, 1984), p. 469

Sloggett, G. J., *et al.*, 1985. IEEE Trans. Instrum. Meas. **IM-34**, 187; (and private communication)

Sloggett, G. J., W. K. Clothier and B. W. Ricketts, 1986. Phys. Rev. Lett. **57**, 3237

Studentsov, N. V., V. N. Khorev and V. Ya. Shifrin, 1981. Izmeritel. Tekh. **1981**(6) p. 56; [Engl. transl.: Meas. Tech. **24**, 501 (1981)]

Tarbeyev, Yu. V., 1984. in *Precision Measurement and Fundamental Constants II*, B. N. Taylor and W. D. Phillips, eds., Natl. Bur. Stand. (US), Spec. Publ. 617 (U. S. Govt. Printing Office, Washington, DC, 1984), p. 483

B. N. Taylor, 1985. J. Res. Natl. Bur. Stand. **90**, 91 (1985)

Taylor, B. N., 1982. *Numerical comparisons of several algorithms for treating inconsistent data in a least-squares adjustment of the fundamental constants*, Natl. Bur. Stand. Report NBSIR 81-2426

Thompson, A. M., 1968. Metrologia **4**, 1; Procès Verbaux des séances CIPM **51**, p. 57, 72e session, 1983 (BIPM, Sèvres, France, 1983); *ibid.* **53**, p. 83, 74e session, 1985

Thompson, A. M., and D. G. Lampard, 1956. Nature **177**, 888

Towler, W. R., *et al.*, 1971. in *Precision Measurement and Fundamental Constants*, D. N. Langenberg and B. N. Taylor, eds., Natl. Bur. Stand. (US), Spec. Publ. 343 (U. S. Govt. Printing Office, Washington, DC), p. 485

Tuninskii, V. S., and S. V. Kholin, 1975. *Concerning changes in the methods for adjusting the physical constants*, Internal Report, Mendeleev Research Institute of Metrology (VNIIM), Leningrad; Metrologiya **1975**(8), 3 (1975) [in Russian]

Vigoureux, P., 1965. Metrologia **1**, 3

Vigoureux, P., and N. Dupuy, 1973. Report Qu-26 (NPL Quantum Metrology Division)

Vigoureux, P., and N. Dupuy, 1980. Report DES-59 (NPL Quantum Metrology Division)

Wada, T., K. Shida, H. Nishinaka and T. Igarashi, 1985. IEEE Trans. Instrum. Meas. **IM-34**, 306

Wang Z. X., 1984. in *Precision Measurement and Fundamental Constants II*, B. N. Taylor and W. D. Phillips, eds., Natl. Bur. Stand. (US), Spec. Publ. 617 (U. S. Govt. Printing Office, Washington, DC, 1984), p. 505

Wapstra, A. H., and G. Audi, 1985. Nucl. Phys. **A 432**, 1

van der Wel, W., K. J. P. M. Harmans, R. Kaarls and J. E. Mooij, 1985. IEEE Trans. Instrum. Meas. **IM-34**, 314

Williams, E. R., and P. T. Olsen, 1979. Phys. Rev. Lett. **42**, 1575

Winkler, P. F., D. Kleppner, T. Myint and F. G. Walther, 1972. Phys. Rev. **A 5**, 83

Yagola, G. K., V. I. Zingerman and V. N. Sepetyi, 1962. Izmeritel. Tekh. **1962**(5), p. 24; [Engl. transl.: Meas. Tech. **1962**, 387 (1962) (1966)]

Yagola, G. K., V. I. Zingerman and V. N. Sepetyi, 1966. Izmeritel. Tekh. **1966**(7), p. 44; [Engl. transl.: Meas. Tech. **1966**, 914 (1966)]

Zhao Ping, W. Lichten, H. P. Layer and J. C. Bergquist, 1986. Phys. Rev. **A 34**, 5138

Zhao Ping, W. Lichten, H. P. Layer and J. C. Bergquist, 1987. Phys. Rev. Lett. **58**, 1293

VARIABILITY OF THE PHYSICAL CONSTANTS

E. Richard Cohen
Rockwell International Science Center
Thousand Oaks, California 91360
USA

ABSTRACT. The accepted numerical values of the physical constants have shown disturbing variations over the period during which presumably precise measurements have been made. The reasons for this have been varied, but in most instances the explanation has been experimental error or incomplete theoretical understanding of the underlying theory. The variations observed over the past forty years in the values of the atomic constants are reviewed and the reasons for some of the variations are presented. On the other hand, Dirac's Large Number Hypothesis (LNH) introduces the question of the variability of the physical 'constants' on the cosmic time scale. Because it is often possible to determine small relative variations in a physical quantity with great sensitivity, there have been some very stringent limits placed on the magnitude of possible long term variations in the values of the constants. A summary of the arguments and of the experimental basis for the conclusions is presented.

1. INTRODUCTION

It is necessary to understand what is meant by the words 'variability' and 'constant' before it is possible to give a meaning to the contradictory title of this paper. A basic assumption of physics is that the physical quantities that are called 'constants' are indeed unchanging in time and, furthermore, are the same at all points in the universe. In a sense, however, one can only consider the constancy of dimensionless quantities. When a dimensional quantity is found to change, one must question whether it is the quantity itself that is changing or the units with which it is measured. To be able to speak about change, it is necessary to compare the quantity that is being tested with another quantity of the same physical nature that is known to be constant and which is then the unit in terms of which the physical quantity is measured. All measurements, therefore, are really reduced to determining the value of dimensionless ratios.

A very simple example of the variabilty of a unit of measurement is given by the history of the measurement of time. The day was originally divided into twenty-four hours, twelve hours of daylight and twelve hours of darkness. Thus the length of the hour was different during the day from what it was at night, and the length of the hour changed

V. De Sabbata and V. N. Melnikov (eds.),
Gravitational Measurements, Fundamental Metrology and Constants, 91–105.

during the year. This change was only measurable when clocks were developed with sufficient accuracy and stability to be able to detect the variability. (An inherent human time-keeping perception may exist, but its existence and capabilities can be quantified only in comparison with an independent, reliable, external clock. Until such a check exists, one is dealing more with psychology than physics.) These clocks must be based on principles of time-keeping that are independent of the rotation of the earth. Burning candles or water dripping through a small hole from a large reservoir provided such an independent basis. With the introduction of reliable mechanical time-keeping devices, it was then possible to divided the day from noon to noon into twenty-four hours of equal length. The solar day as so defined is not constant thoughout the year. In simpler times, this posed no great problem, but as the needs of technology became more demanding, it was necessary to replace this definition of the unit of time by one based on the mean length of the day averaged over the entire year. However, the rotation of the earth is nonuniform because of changes in its moment of inertia, and tying the definition of the second to the day is obviously unnecessary, since the year was the actual unit of observation, and in 1956 the definition of the second was modified so that it was based solely on the period of revolution of the earth in its orbit. This is, however, a simplification of the actual definition of astronomical time. In the precise definitions, one must in essence go back to Newton's concept of time as 'flowing uniformly'. Inertial time is constructed in such a way as to remove mass-independent apparent forces from the equations of motion and to provide the best fit to the observed positions of the solar system with respect to the 'fixed' stars. The equations of motion of the earth-moon system do not lead to a closed cyclic orbit; the year is only an approximation, defined with reference to an elliptical orbit that best approximates the motion of the earth at any instant. In practice astronomical time is an experimentally observed coordinate variable of celestial dynamics. Since the length of the astronomical year is not constant, the definition is actually in terms of the total theory of the solar system. The ephemeris second is a scaling of the time-like variable in the accepted theory of the equations of motion of the earth. The remaining metrological task is to coordinate and synchronize astronomical inertial time with the time kept by atomic clocks.

This emphasizes two aspects of the question of the variability of the units of measurement. One is the capability to measure the quantity in question by more than one process, the other is the existence of an underlying theory that can 'explain' the variability of the unit in terms that are able to reduce the definition to a more 'fundamental' basis.

A different example is provided by the definition of the unit of mass. The kilogram is defined as the mass of the prototype Kilogram kept in Sèvres. There is no underlying theory supporting this definition other than the conservation of mass. It is an open question whether or not the mass of the Kilogram could change: can it ever be other than 1 kg? Physical processes -- corrosion and adsorption of gases on the surface of the platinum-iridium cylinder -- can modify the amount of mass represented by this prototype, but except for careful preservation, and occasional careful, but not too vigorous, cleaning there is no way to observe a change in the mass of the Kilogram other than by comparisons with secondary copies. A common systematic change in mass of the primary and all secondary copies is undetectable. Only a theory of the rates of corrosion and of adsorption of gases under the maintained conditions of storage and treatment provides the assurance that the kilogram is adequately stable.

A different kind of problem in discussing a change in a physical quantity is associated with the definition of the metre: The metre is the length of the path traveled by light in vacuum during a time interval of 1/299 792 458 of a second. With this definition, is it possible for the speed of light to change? In the purely metrological sense the answer is obviously "No"; any change in c would produce only a change in the definition of the metre, which would, in itself, be undetectable. The proper resolution, however, comes from looking at the underlying physics. This definition of the metre is not, and is not intended to be, an operational definition --- one that can actually be carried out in the laboratory. It is, instead, only conceptual and relies on a subsidiary set of physical realizations (the accepted wavelengths of a chosen set of absorption-stabilized laser transitions). In essence, the present definition of the metre can be expressed as a numerical specification of the relative scales of space-like and time-like coordinates. The metre may equally be defined in terms of the flat-space metric:

$$ds^2 = g_{ij}\,dx^j dx^i \tag{1.1}$$

$$g = \mathrm{diag}\,(c^2, -1, -1, -1) \qquad c = 299\,792\,458\ \mathrm{m/s}. \tag{1.2}$$

Although, as a result of this definition, the quantity c cannot change, it is still possible to ask if the speed of light can change. To answer this question, it is necessary to have a theory of the propagation of electromagnetic waves. With such a theory one might postulate a (possibly time dependent) photon mass or some other operationally observable mechanism that could lead to detectible changes in the speed of light.

2. VARIATIONS IN THE VALUES OF THE FUNDAMENTAL CONSTANTS

Whatever may be the situation with regard to possible variations of the values of the quantities that we normally refer to as 'physical constants', it is all too painfully obvious that the recommended numerical values change from adjustment to adjustment. The history of the changes in what is considered to be the 'best values' is shown in Tables 1 and 2, which cover the changes over the last forty years. These tables list changes in the values of six quantities. A longer list could easily be produced but this would not give any additional independent information nor would it necessarily give a correct impression. The list might have dozens of quantities but no more than four or five would be statistically independent. Changes that have occurred in several quantities are usually traceable to a single change that influences them all. The adjustment of the constants defines a point in an abstract space, and the best that can be done in describing the difference between two adjustments is to look at the distance between the two points measured as a mutiple of its standard deviation. Only in this way is it possible to make meaningful statements about the statistical probability of the validity of a given set of numbers.

However, even this procedure is deceptive, because the primary differences in the results from one recommendation to the next involve changes in the theoretical understanding of the relationships that exist among the data or a better understanding of the 'systematic' errors in the measurements, perhaps even to a greater extent than the improvements in the precision and accuracy of the measurements. The data of Table 1 are an excellent example of this situation.

In the 1947 adjustment (DuMond and Cohen, 1948) the Lamb Shift had just recently been identified. Only the first-order terms (second order in perturbation) of Quantum Electrodynamics had been calculated, and the electron magnetic moment was known only as

$$\mu_e/\mu_B = 1 + \alpha/2\pi + \cdots. \tag{2}$$

By the time of the 1950 adjustment (DuMond and Cohen, 1950, 1951, 1952) QED had advanced to an (incorrect) estimate of the next term in this expansion (Karplus and Kroll, 1950):

$$\mu_e/\mu_B = 1 + \alpha/2\pi - 3(\alpha/\pi)^2 + \cdots. \tag{3}$$

In the 1952 adjustment (DuMond and Cohen, 1953) the second order coefficient was given more precisely as -2.973, but the difference between the two was insignificant (< 0.15 ppm) compared to the available experimental uncertainty. The correct coefficient, $-.328\ldots$, did not become available until 1957 (Petermann, 1957, 1958; Sommerfield, 1957).

A striking example of the interconnectedness of the measurements of the physical constants is the change in the inferred value of the Avogadro constant brought about by the discovery of the Lamb Shift. This constant is determined by measuring the dimensions of a unit cell (or the volume per molecule) and the density of a crystal to find the mass of a single molecule. The dimensions of the crystal were measured in terms of X-ray wavelengths, and the most accurate method for calibrating these was to compare them to the wavelengths of the hydrogen-like spectra of highly ionized atoms. The Lyman series for hydrogen lies in the ultra-violet, but for Be IV to O VIII the Lyman lines are in the X-ray region, from 1 nm to 10 nm. Tyrén (1940) used such lines to calibrate the wavelengths of X-rays observed in high order at grazing incidence. Since Tyrén relied on the accuracy of the Bohr-Sommerfeld theory of hydrogenic spectra, the inferred values of these lines were in error by 25 to 100 ppm because of the effect of the Lamb Shift, and this introduces a 75 to 300 ppm correction to the deduced value of the Avogadro constant.

In the 1986 adjustment the fine-structure constant was determined without any input from atomic fine-structure or hyperfine-structure spectroscopy measurements. In 1963, however, both deuterium fine-structure and hydrogen hyperfine-structure were important, but gave contradictory values for the fine-structure constant α. Values of α^{-1}

deduced from fine-structure measurements were 25 to 30 ppm higer than those derived from the hyperfine-structure measurement. The fine-structure measurements claimed a precision of approximately 5 ppm; the hydrogen hyperfine separation, measured in a hydrogen maser was limited by theoretical uncertainties rather than experimental error. Corrections to the simple Fermi theory arising from the effects of proton size and proton recoil were at the time unknown, but pessimistic estimates of their magnitude ran as large as 20 ppm. The discrepancy between the two measurements is independent of the value of the fine structure constant. The ratio of the hyperfine structure interval to the $S_{\frac{1}{2}} - P_{\frac{1}{2}}$ interval in hydrogen is

$$\frac{\Delta\nu_{\mathrm{hfs}}}{\Delta\nu_{\mathrm{fs}}} = \frac{256}{3}\frac{\mu_{\mathrm{p}}}{\mu_{\mathrm{B}}}\frac{\mu_{\mathrm{e}}}{\mu_{\mathrm{B}}}\frac{[1 + (\frac{3}{2} - \frac{5}{2} + \ln 2)\alpha^2 - X(m_e/m_{\mathrm{p}})\alpha]}{\left[2\frac{\mu_{\mathrm{e}}}{\mu_{\mathrm{B}}}\frac{m_{\mathrm{p}} + m_{\mathrm{e}}}{m_{\mathrm{p}}} - 1 + \frac{5}{8}\alpha^2 - 2\frac{\alpha^3}{\pi}(\ln 137)\right]}, \tag{4}$$

where the quantity X represents the effect of nuclear structure and recoil. The uncertainty in this expression due to the uncertainty in the value of α is less than 0.01 ppm, so that the disagreement could not be resolved except by rejecting one or the other of the two measurements. It is now known that the correction term X is no larger than 1 or 2 ppm and that the precision of the fine-structure measurements was optimistic. Even though the measurements of the hyperfine interval itself were accurate to 0.01 ppm (and would therefore yield a value of α with a precision of 0.005 ppm if the theory of the interval were sufficiently well-known to make full use of this experimental accuracy), on the basis of the information available in 1963 the only valid choice was to reject the hydrogen hyperfine interval data. In support of this choice were preliminary measurements of muonium hyperfine structure that gave $\alpha^{-1} = 137.0392(15)$ (Hughes, 1963) and the measurement of the electron magnetic moment anomaly $a_{\mathrm{e}} = \alpha/2\pi - 0.328(\alpha/\pi)^2 + \cdots$ (Wilkinson and Crane, 1963) that gave $\alpha^{-1} = 137.0381(30)$, where the error is dominated by the uncertainty in the magnitude of the uncalculated third term in the perturbation expansion.

By 1964 the measurement of muonium hfs was producing a much smaller value of α^{-1} : 137.0360(30) (Bailey *et al.*, 1964) and the 1969 adjustment, stimulated by the measurement of the Josephson frequency-voltage quotient (Taylor, Parker and Langenberg, 1969), clearly established the lower value for α^{-1}. But the curse of doing the wrong thing for the right reason continued: The available data on the proton magnetic moment included the very precise work of Mamyrin and Frantsuzov (1965), which dissagreed with the other measurements. Because of the lack of detail in the published material, Taylor *et al.* found it difficult to evaluate the Type-B errors of those measurements and therefore to verify the validity of the assigned uncertainty. Had the Mamyrin and Frantzusov data not been deleted, the 1969 value of the Avogadro constant, for example, would have been reduced by 7.5 ppm and would have been in agreement with the 1986 value within 2 ± 7 ppm.

Significant changes in the values of the constants continue to occur. The recommended value of the Avogadro constant in the 1973 adjustment was $6.022\,045(31)\times10^{23}$ mol^{-1} (Cohen and Taylor, 1973). In the 1986 adjustment, the recommended value was increased to $6.022\,1367(13)\times10^{23}$ mol^{-1} (Cohen and Taylor, 1986). The primary cause of this change (and the cause for most of the major differences in the two sets of values) was the deletion, in 1973, of two Faraday determinations that seemed to be discrepant with the remaining data. In hindsight this 'discrepancy' was not that severe and had those data been retained, the 1963 recommendation would have been 10.8 ppm higher and would have differed from the present recommendation by only 4.4 ppm.

3. SECULAR VARIATIONS IN THE CONSTANTS

The possibility that the physical constants may change with a time scale related to the age of the universe (or the Hubble constant) is a consequence of the Large Number Hypothesis (LNH) (Dirac, 1938). We shall here use a modification of Dirac's formulation because the proton can not now be considered as a truly fundamental particle. We therefore write Dirac's large number as the ratio of the electrical force to the gravitational force between two particles of mass m^*, where m^* is the the mass of some 'fundamental' particle -- perhaps a quark or (more conveniently) a pion. This ratio is $N = e^2/4\pi\epsilon_0 G m^{*2}$. Because the Planck constant is a more 'fundamental' quantity than is the elementary charge (in the sense that one may easily imagine a universe in which different particles have different charges, but not one in which different particles obey quantum mechanics with different values of $\hbar$), we multiply this number by $\alpha^{-1} \approx 137$ and write

$$N_1 = \hbar c/G m^{*2} \approx 7.6\times10^{39}. \tag{5}$$

This number may also be expressed in terms of the Planck mass m_{P}, $N_1 = (m_{\mathrm{P}}/m^*)^2$.

Another large number is the ratio of the scale of the universe to the scale of nuclear size. The radius of a proton is often stated to be equal to half the classical radius of the electron. It is much more appropriate to equate the proton radius to the range of the nuclear force $\hbar/m^* c$; since one has $m^* \approx 2\times 137 m_{\mathrm{e}}$, these two lengths are approximately equal. The size of the universe is characterized by the 'Hubble' constant H_o, which we take to be nominally equal to 100 (km/s)/Mpc or 3.24×10^{-18} s^{-1}; the true value of the Hubble constant is given by $H = hH_o$ with $h \approx 0.55$. The 'radius' is then $c/H_o = 9.25\times10^{25}$ m and

$$N_2 = m^* c^2/H_o \hbar = 6.6\times10^{40}. \tag{6}$$

This may also be described as the ratio of the mass energy of the fundamental particle to the smallest possible quantum (the quantum whose wavelength is $2\pi R$, the 'circumference' of the universe).

The argument that these two large numbers must be related is an invocation of Occam's razor. The possible variability of the constants on a time scale proportional to H_o^{-1} is based on the further observation that

N_2 is proportional to the age of the universe and that N_1 must therefore have the same time dependence if the two numbers are causally related.

However it is not necessary that a functional relationship between N_1 and N_2 must imply a time dependence for N_1. The Hubble constant need not represent the age of the universe but rather a time constant of the universe. Thus H_o may indeed be a constant (setting aside our lack of knowledge of its true size, and the fact that its 'best' numerical value has changed by at least a factor of 10 in the past fifty years); if so, N_1 may also be a constant.

3.1. Measureable Effects

3.1.1. Length.

If the physical constants are changing, the effects would be noticeable in subtle ways that depend on how the detection of such a change would involve different aspects of metrology. The sizes of physical objects (*i.e.*, the length of a metre bar or the dimensions of a resonant cavity) are proportional to the Bohr radius, $a_o = \hbar/m_e c\alpha$. The wavelengths of optical radiations are determined by the Rydberg constant, so that a definition of length in terms of optical or X-ray wavelengths is proportional to $\hbar/m_e c\alpha^2$. On the other hand, the second is defined by the frequency of a hyperfine transition, so that the metre, as presently defined, is itself proportional to $(\mu_p/\mu_N)^{-1}\hbar/m_e c\alpha^4$.

Thus the lattice spacing of a standard silicon crystal, measured by Deslattes' interferometer method, is a comparison of a distance defined by a fixed number of Bohr radii to an optical wavelength, and is therefore proportional to α; measurements to a precision of 1 part in 10^9 could therefore, within an observation time of a few years, detect a change in α with a time scale of the age of the universe.

3.1.2. Time.

A similar condition is associated with the measurement of time. Different realizations of a clock can have different relationships to the fundamental constants. Astronomical or ephemeris time is defined by the mean angular velocity of the earth in its orbit. If the physical 'constants' vary, then the conservation of angular momentum and the invariance of the action integral mean that the mean angular velocity of the earth is given by

$$n \propto G^2 M_E^3 M_S^2 / J^2 \tag{7.1}$$

where J is the constant angular momentum and M_E and M_S are respectively the masses of the earth and the sun.

Atomic (actually molecular) clocks such as the ammonia clock define time in terms of molecular vibration frequencies:

$$\nu_{NH_3} \propto \left(\frac{m_e}{m_p}\right)^{\frac{1}{2}} m_e \alpha^2 c^2/\hbar, \tag{7.2}$$

while atomic clocks of the hydrogen maser or cesium beam type define time in terms of a hyperfine transition frequency:

$$\nu_{hfs} \propto \mu_p \mu_e a_o^{-3} \propto \frac{\mu_p}{\mu_N}\frac{\mu_e}{\mu_B}\frac{m_e^2 m_p^2}{(m_p + m_e)^3}\alpha^4 c^2/\hbar. \tag{7.3}$$

A comparison of an ammonia clock to the Cs clock therefore determines the constancy of $(m_e/mp)^{\frac{1}{2}}\frac{\mu_p}{\mu_N}\alpha^2$.

Still a third type of clock that is dependent on atomic constants is the superconducting-cavity stabilized oscillator (SCSO) (Turneaure and Stein, 1976). The frequency of such a device is inversely proportional to the dimensions of the superconducting cavity, which is ultimately defined by the Bohr radius. The relative rates of a SCSO and a cesium clock is then proportional to $\frac{\mu_p}{\mu_e}\alpha^3$. The null result of a comparison of three SCSO's with four Cs clocks over a period of twelve days provided a one-standard-deviation limit for a relative change in α of 4.1×10^{-12} year^{-1} = 0.040 H_o. This is smaller, by a factor of 10 to 100, than the rate of change implied in a suggestion by George Gamow (1967) that e^2 or α might increase in direct proportion to the age of the universe.

3.2. Nuclear and Geophysical Evidence

The laboratory measurement of the constancy of α is to be compared with passive observations such as that by Dyson (1967) who pointed out that the absence of any observable change in the abundance ratios of radioactive nuclides show that the delicate balance between electrostatic and nuclear forces has been constant over billions of years. Radioactive ^{187}Re is a β emitter, decaying with a half-life of 7×10^{10} years into the isotope ^{187}Os. The decay energy is only 2.4 keV; the difference in the Coulomb binding energy of the two nuclei is 16 MeV. Even a small change in α over the age of the earth (1/4 the age of the universe) would therefore significantly change the decay energy, and hence the lifetime, of ^{187}Re. A smaller value of α in the past would increase the decay rate and mean that no ^{187}Re would still exist; a larger value would have reversed the stability of the two isotopes, and ^{187}Os would have decayed by K-electron capture to Re. Dyson then concludes, after considering and dismissing the possiblity of balancing secular changes in both electrostatic and nuclear forces, that the relative change in α could not have been greater than 0.3×10^{-12} year^{-1}.

Chitre and Pal (1968) used the geological dating of two different radioactivity clocks to refute a possible change in α, comparing the α-decay rate of ^{238}U (0.15×10^{-9} yr^{-1}) and the K-capture rate of ^{40}K (0.54×10^{-9} yr^{-1}). Since the two clocks have greatly differing sensitivity to a change in α, the consistency of the geological ages deduced by them indicates that e^2 or α has not changed by as much as 0.05% in the past 10^9 years.

Another refutation of the possibility that α was smaller at an earlier epoch, is the observation by Peres (1967) that, if so, the valley of stability in the atomic mass surface for heavy elements would have then been closer to the line $N = Z$. As a result, the stable isobar of mass 208 would have been gaseous ^{208}Rn in the early history of the earth and the geological distribution of present-day ^{208}Pb would be uniform. In general, a variation of α and hence of the Coulomb term in nuclear binding energy would imply variations in the chemical history of many presently stable isotopes. This would have led to wide variations in the isotopic composition of the elements, which is, except for the lightest elements, known to be extremely uniform over the earth.

A much more stringent limit on the variation of α has been given by Shlyakhter (1976) from consideration of the consistency of the interpretation of the Oklo Phenomenon. Approximately 1.8×10^9 years ago, a uranium deposit in Gabon, West Africa achieved criticality naturally. From the isotopic composition of of the fission products, particularly Sm and Eu, found in the deposits of the Oklo uranium mine it is possible to determine that thermal neutron capture cross sections were essentially the same then as now. This implies that the resonance energy for neutron capture in Sm has shifted by less than the width of the resonance (0.05 eV) or the thermal energy kT (0.025 eV). Since the Coulomb contribution to the nuclear binding energy is $0.6\,A^2Z^{-1/3}$ MeV $\approx 4\times10^8$ eV, α could not have changed by as much as 1 in 10^{10} in 2×10^9 years. This limit gives

$$|\mathrm{d}\ln\alpha/\mathrm{d}t| < 10^{-17}\,\mathrm{yr}^{-1} \tag{8}$$

From an analysis of the measured wavelengths of O III emission lines in the spectrum of five radio galaxies at red shifts between $z = 0.17$ and $z = 0.26$, Bahcall and Schmidt (1967) find $\alpha(z = 0.2)/\alpha(lab) = 1.001(3)$ (quoted error has been reexpressed as one standard deviation). My own analysis of the data gives even less evidence for a dependence on z: $\mathrm{d}\ln\alpha/\mathrm{d}z = 0.0016(75)$.

Rozenthal (1980) has given arguments about the constancy of the strong force that are similar to Shlyakhter's arguments concerning α. If the strong force had been weaker in the past, the deuteron would not be stable, nucleosynthesis could not proceed and the universe would be entirely hydrogen. If the strong force were slightly stronger, ^{2}He would be stable and the reaction

$$\mathrm{p} + \mathrm{p} \rightarrow {}^2\mathrm{He} + \gamma,$$

which is controlled by the electromagnetic interaction, would proceed much faster than thermonuclear synthesis and the production of heavier elements would have been blocked. If the strong force were significantly increased until the temperature and density of the universe fell below the conditions for nucleosynthesis, there would be no heavy elements. A later decrease to its present value would lead to the decay of ^{2}He and the universe would again be all hydrogen. By quantifying these arguments and similar ones in regard to other marginally stable (or marginally unstable) systems (*e.g.*, ^{8}Be), Rozenthal concludes that the strong force could not have changed by more than 15% after the first half hour of the universe.

3.3. Astronomical Evidence

Since the frequency of optical transitions are proportional to R_∞ and hence to $m_e\alpha^2c^2/\hbar$, while hyperfine transitions are proportional to $m_e\frac{\mu_e\mu_p}{\mu_B^2}\alpha^4c^2/\hbar$, a changing value of α would be a source of non-Doppler redshift and the effect would be twice as large in radio sources as in optical sources. The measurement of the difference between the 21-cm and optical redshifts of galaxies (Ford, Rubin and Roberts, 1971; Roberts, 1972) has been analyzed by Lewis (1975) to give

$$(z_R - z_O)/(1 + z_R) = \mathrm{d}\ln\left(\frac{\mu_p}{\mu_B}\alpha^2\right)/\mathrm{d}z = 0.004(6). \qquad (9)$$

The detection of the ultraviolet fine-structure doublet of Mg II in the optical spectrum by Burbidge *et al.* (1976), and the red-shifted 21-cm H line at the same redshift by Wolfe *et al.* (1976) associated with the radio source AO 0235 + 164 provide data that may be analyzed to give limits on the variability of the fine structure constant and the ratio of the proton and electron magnetic moments separately. The wavelengths of the Mg doublet are quoted by Wolfe *et al.* as 426.05(3) nm and 427.15(3) nm at a radio redshift $z_H = 0.52385(1)$. The laboratory wavelengths are taken to be 279.553 nm and 280.270 nm. These wavelengths give redshifts of 0.52404(11) and 0.52407(11) although Wolfe *et al.* quote a slightly different value, 0.52392(10) for the mean redshift of the doublet.

Comparing the hydrogen hfs redshift to the redshift of the weighted mean Mg line gives

$$\mathrm{d}\ln[\frac{\mu_p}{\mu_B}\alpha^2]/\mathrm{d}z = -0.00014(7); \qquad (10.1)$$

comparing the hydrogen hfs redshift to the Mg fs redshift gives

$$\mathrm{d}\ln(\mu_p/\mu_B)/\mathrm{d}z = 0.007(18); \qquad (10.2)$$

and finally, comparing the fs redshift to the redshift of either Mg line gives

$$\mathrm{d}\ln\alpha/\mathrm{d}z = -0.003(9). \qquad (10.3)$$

Pagel (1977) has analyzed the redshift data on fine-structure doublets associated with radio sources with $2.1 < z < 2.7$ in an attempt to detect a change in m_p/m_e. Since the comparison of metal fs lines to hydrogen hsf involves $\mu_p/\mu_B\alpha^2(1 - m_e/m_p)$, the mass-ratio contributes only a second order effect and the sensitivity is poor. Pagel was able to conclude only that

$$|\mathrm{d}\ln(m_p/m_e)/\mathrm{d}z| = 0 \pm 0.04, \qquad (11)$$

or that m_p/m_e did not change by more than 10% out to $z = 2.7$.

A limit on the variation of hc has been given by Baum and Florentin-Nielsen (1976) from measuremnts of the wavelength and energy of photons from galaxies with redshifts up to $z = 0.19$. They used an interference filter to select a wavelength band at 523 nm with a bandwidth of 36 nm. A special photomultiplier tube with an added control grid permitted the energy distribution of the photoelectrons to be measured. The measurement yielded

$$\mathrm{d}\ln hc/\mathrm{d}z = -0.064 \pm 0.054 \qquad (12)$$

at constant received wavelength (523 nm) and implies that the observed time dependence of hc is not statistical significant, and is at least a factor of ten smaller than than that required by Gamow's conjecture.

Tubbs and Wolfe (1980) have pointed out that the limits on the variability of the fundamental constants in time as observed in several Quasi-Stellar Objects also implies the uniformity of physical laws in space and the uniformity of the values of the constants in regions of space that are outside of each others light-cones and are therefore causally disjoint.

4. CONCLUSIONS

Although our knowledge of the values of the fundamental constants is disturbingly variable there is no indication, either directly from laboratory experiments or indirectly from the interpretion of geological and astronomical evidence that the so-called constants have not, indeed, been constant for the greater part of the history of the universe (since the beginning of nucleosynthesis). Although little can be said based on present experimental or observational evidence concerning the first half hour of creation, the evidence seems to be clear that no variation as strong as linear in time is likely. This however does not entirely rule out possible nonlinear variations. If the values of the constants are correlated with the density of matter or with the mean curvature of space, then the observed values of the constants tell us little about a variation that may have occurred at earlier times but with (possibly) an exponential decay to asymptotic values. This idea has been discussed more fully by Bekenstein (1982) in terms of a characteristic length larger than the Planck length, 1.6×10^{-35} m but smaller than nuclear dimensions, 10^{-15} m, and by Marciano (1984) and Kolb *et al.* (1986) who consider cosmologies in which the values of the fundamental constants depends on the size of the space of the 'extra' dimensions in 10-dimensional superstring models.

Obviously, the determination of the values of the fundamental constants to the highest possible precision will continue to be of importance not only for the purposes of metrology but for wider branches of physics as well, reaching out literally to the farthest edge of the universe.

5. REFERENCES

Bahcall, J. N., and M. Schmidt, 1967. Phys. Rev. Lett. **19**, 1294

Bailey, J., W. Cleland, M. Eckhaus, V. W. Hughes and R. Prepost, 1964. Bull. Amer. Phys. Soc. **II** 9, 81

Baum, W. A., and R. Florentin-Nielsen, 1976. Astrophys. Jour. **209**, 319

Bekenstein, J. D., 1982. Phys. Rev. D **25**, 1527

Burbidge,E. M., R. D. Caldwell, H. E. Smith, L. Liebert and H. Spinrad, 1976. Astrophys. J. Lett. **205**, L117

Chitre,S. M., and Y. Pal, 1968. Phys. Rev. Lett. **20**, 278

Cohen, E. R., and B. N. Taylor, 1973. J. Phys. Chem. Ref. Data **2**, 663

Cohen, E. R., and B. N. Taylor, 1986. *The 1986 Adjustment of the Fundamental Physical Constants*, CODATA Bulletin 63 (Pergamon Press, Headington Hill Hall, Oxford, UK and Maxwell House, Elmsford, New York, USA); *see also* J. Res. Natl. Bur. Stand. **92**, 85 (1987)

Dirac, P. A. M., 1938. Proc. Roy. Soc. (London) **A165**, 199

DuMond, J. W. M., and E. R. Cohen, 1948. Rev. Mod. Phys. **20**, 82

DuMond, J. W. M., and E. R. Cohen, 1950. *Report to the (US) National Research Council, Committee on Constants and Conversion Factors of Physics*, (unpublished, December, 1950)

DuMond, J. W. M., and E. R. Cohen, 1951. Phys. Rev. **82**, 555

DuMond, J. W. M., and E. R. Cohen, 1952. American Scientist **40**, 447

DuMond, J. W. M. and E. R. Cohen, 1953. Rev. Mod. Phys. **25**, 691

Dyson, F. J., 1967. Phys. Rev. Lett. **19**, 1291

Ford, W. K., V. C. Rubin and M. S. Roberts, 1971. Astron. J. **76**, 22

Gamow, G., 1967. Phys. Rev. Lett. **19**, 759

Hughes, V. W., 1963. Bull. Amer. Phys. Soc. **II 8**, 33

Karplus, R., and N. M. Kroll, 1950. Phys. Rev. **77**, 536

Kolb, E. W., M. J. Perry and T. P. Walker, 1986. Phys. Rev. D **33**, 869

Lewis, B. M., 1975. Mem. Roy. Astr. Soc. **78**, 75

Mamyrin, B. A., and A. A. Frantsuzov, 1965. Zh. Eksp. i Theor. Fiz. **48**, 416. [Soviet Physics--JETP **21**, 274 (1965)]

Marciano, W. J., 1984. Phys. Rev. Lett. **52**, 489

Pagel, B. E. J., 1977. Proc. Roy. Soc. (London) **A310**, 245

Peres, A., 1967. Phys. Rev. Lett. **19**, 1293

Petermann, A., 1957. Nuclear Physics **3**, 689; Helv. Phys. Acta **30**, 407

Petermann, A., 1958. Nuclear Physics **5**, 677

Roberts, M. S., 1972. *External Galaxies and Quasi-stellar Objects, IAU Symposium No. 44*, D. S. Evans, editor. (Dordrecht, Reidel) p. 195

Rozenthal, I. L., 1980. Usp. Fiz. Nauk **131**, 239. [Soviet Physics--Uspeki **23**(6), 296 (1980)]

Shlyakhter, A. I., 1976. Nature **264**, 340

Sommerfield, C. M., 1957. Phys. Rev. **107**, 328

Taylor, B. N., W. H. Parker and D. N. Langenberg, 1969. Rev. Mod. Phys. **41**, 375

Tubbs, A. D., and A. M. Wolfe, 1980. Astrophys. Journ. **236**, L105

Turneaure, J. P., and S. R. Stein, 1976. *Atomic Masses and Fundamental Constants-5.* edited by J. H. Sanders and A. H. Wapstra. Plenum Press, New York and London. p. 636

Tyrén, F., 1940. dissertation, Uppsala (unpublished); Zeit. f. Physik **109**, 722 (1938)

Wilkinson, D. T., and H. R. Crane, 1963. Phys. Rev. **130**, 852

Wolfe, A. M., R. L. Brown and M. S. Roberts, 1976. Phys. Rev. Lett. **37**, 179

Table 1. Variations in the Values of the Fundamental Constants
Values of some representative constants over the past 40 years.

	$c/(\mathrm{m/s})$	$N_A/(10^{23}/\mathrm{mol})$ ‡	$e/10^{-19}$ C	α^{-1}	R_∞/m^{-1}	m_p/m_e
1947	299 776 000 (6 000)	6.0232 (6)	1.601 99 (24)	137.021 (10)	10 973 730 (8)	1836.58 (30)
1950	299 790 200 (900)	6.023 52 (11)	1.601 864 (23)	137.0429 (9)	10 973 732.3 ±1.0	1836.139 (34)
1952	299 792 900 (800)	6.022 80 (36)	1.602 066 (70)	137.0377 (16)	10 973 730.9 ±1.2	1836.13 (4)
1955	299 793 000 (300)	6.022 94 (16)	1.602 06 (3)	137.0373 (6)	10 973 730.9 ±1.2	1836.12 (2)
1963	299 792 500 (100)	6.022 52 (9)	1.602 10 (2)	137.0388 (6)	10 973 731 (1)	1836.096 (10)
1969	299 792 500 (100)	6.022 169 (40)	1.602 1917 (70)	137.036 02 (21)	10 973 731.2 ±1.1	1836.109 (11)
1973	299 792 458 ±1.2	6.022 045 (31)	1.602 1892 (46)	137.036 04 (11)	10 973 731.77 (83)	1836.151 52 70)
1986	299 792 458	6.022 1367 (36)	1.602 177 33 (49)	137.035 9895 (61)	10 973 731.534 (13)	1836.152 701 (37)

‡ based consistently on ^{12}C.

Table 2. Changes in Constants.
Standard Deviations from the 1986 values.

	c	e	N_A ‡	α^{-1}	R_∞	m_p/m_e
1947	−2.7	−0.8	+1.77	−1.5	−0.2	+1.4
1950	−2.5	−13.6	+12.76	+7.7	+0.8	−0.4
1955	−1.8	−3.9	+5.15	+2.2	−0.5	−1.6
1963	+0.4	−3.4	+4.12	+4.5	−0.5	−5.7
1969	+0.4	+2.1	+0.81	+0.2	−0.3	−4.0
1973	0	+2.6	−2.96	+0.5	+0.3	−1.7

‡ based consistently on ^{12}C.

THE MEASUREMENT PROBLEM OF THE GENERAL MATTER FIELD THEORY AS REQUIRED BY THE COPENHAGEN SCHOOL

Bidyut Kumar Datta & Renuka Datta *
S.N. Bose School For Advanced Study and Research In Mathematics & Mathematical Physics
AE 374, Salt Lake City
Calcutta - 700 064
India

ABSTRACT. In the famous Einstein-Bohr debates, a fundamental requirement of the Copenhagen school was the resolution of the problem of the 'measurement', when described in special cases by macro-variables and of the 'measured', when described by micro-variables. This could not be accomplished by Einstein in his formulation of the general theory of relativity, as it does not exhibit the inertial manifestations of matter. In accordance with Bohr's view of quantum mechanics, these variables are introduced necessarily, asymmetrically, the 'observer' component obeying the rules of classical physics and the 'observed' component obeying the rules of quantum physics. In the general matter field theory initiated by M. Sachs, the variables of the 'observer' and of the 'observed' are introduced symmetrically as component fields of a closed material system that obey the same rules - the rules of the general matter field theory. The introduction of these variables in this relativistic field theory is done in a fully covariant manner, with a unified field approach - the general relativity and matter fields - that explicitly exhibits the inertial manifestations of matter. So, one of the problems that arose in the famous Einstein - Bohr debates is resolved and a fundamental requirement of the Copenhagen school is fulfilled.

1. GENERAL MATTER FIELD THEORY

At the outset we like to give an overview of the general matter field theory as initiated and developed by M. Sachs (1982). The incorporation of the generalized Mach principle with Einstein's formulation of the general theory of relativity via continuous

* Department of Mathematics
Maulana Azad College
Rafi Ahmed Kidwai Road
Calcutta - 700 013, India

V. De Sabbata and V. N. Melnikov (eds.),
Gravitational Measurements, Fundamental Metrology and Constants, 107–114.

field concept leads to the general matter field theory that exhibits the gravitational and electromagnetic force manifestations together with the inertial manifestations of matter. Faraday's field theory that deals with potential field of force and its action on an ambiguously defined 'test particle' is generalized in a consistent manner. Generalized field theory requires that the 'test particle' should be included in the whole system which must be continuous and necessarily closed. A consistent field theory can be defined for a continuous and closed system in which a part playing the role of a 'test particle' that 'looks at' the environment approaches a sufficiently weakly coupled state, in the asymptotic limit, in which it appears, as a first approximation, to be disconnected from the environment (remaining part of the system).

Another leading concept that is essential for the formulation of the general matter field theory is the principle of relativity which states that all laws of nature must be independent of the frames of reference in which they may be represented. The general covariance of the forms of the laws of nature imposed by the principle of relativity is, in fact, an objectivity of the laws of nature - a law about laws. The principle of relativity, according to A. Einstein (ref. L. de Broglie, 1979), asserts that "the laws of nature have the greatest imaginable logical simplicity". The major revolutionary aspect of the general theory of relativity is obtained from the special theory of relativity by a different philosophical approach to matter by considering (i) abstract realism in place of operationalism and positivism and (ii) continuous field concept as fundamental for the nature of matter in place of atomism.

Einstein was a great advocate of Mach principle and firmly believed that the correct theory of gravitation should be based on Mach principle. He had the conviction that his formulation of the general theory of relativity had incorporated Mach principle. In an ingenious way Kurt Gödel (1949) demonstrated that a 'non-Machian' solution can be constructed within the framework of general relativity. The fact that Gödel's solution follows from general relativity implies that the theory does not incorporate Mach principle. Later I. Ozsvath and E. Schücking (1962) constructed 'non-Machian' universes having no closed time-like lines in the framework of general relativity.

Mach principle leads to the conclusion that the inertial mass of any matter, microscopic, macroscopic or astronomical, is not an intrinsic property of a 'thing'. Rather, the inertia is rooted in the dynamical coupling between the quantity of matter whose motion is studied and all of the matter of a physical system which is necessarily closed. Now, the incorporation of Mach principle with the general theory of relativity implies that the inertial manifestations of matter should be derivable from field properties of space-time. But if one incorporates inertia of matter with all other manifestations of interacting matter in the unified field approach one must have to generalize the Machian view-point which removes all remnants of atomism from the theoretical description of matter. Thus, the generalized Mach principle implies that inertia, electromagnetic properties of elementary matter, nuclear moments, etc. should all be related to the dynamical coupling of this matter to the rest of a closed system - entire content of the universe. Thus, the incorporation of the generalized Mach principle with the general theory of relativity reveals that

(i) the theoretical description of matter contains no seeds of atomism,
(ii) in an exact sense, no matter is free in the universe, i.e., the universe is closed,
(iii) the set of field equations that describes the domain of elementary particles in the microscopic phenomena (as in high energy physics) must also describe the laboratory and astronomical domains of phenomena.

So, the physics of the domain of elementary particles is completely incorporated with the physics of the domain of astronomy and cosmology.

2. MODIFIED RIEMANNIAN GEOMETRY

In Riemannian space-time $g_{\mu\nu}=g_{\nu\mu}$. So, we are to determine 10 independent metric field components $g_{\mu\nu}$ at each point x of space-time. But 16-parameter Einstein group implies that there should be 16 independent relations to yield solutions at each point x of space-time. Thus, it seems that Riemannian space is not quite adequate for the full play of Einstein group of general relativity. Rather, the symmetry group of general relativity for its full logical exploitation should have a generalized Riemannian space as its arena that entails 16 independent metric field components. This can, in fact, be obtained by fully exploiting the logic of space-time as imposed by the theory of general relativity. The theory of general relativity endowed the language of space-time with logic in two parts, algebra and geometry. The algebraic part of the logic of space-time yields (i) the form of the irreducible representations of the underlying Lie group of relativity theory, which obey the algebra of quaternions and (ii) 2 - component spinor variables as the basis functions of these representations. The geometric logic yields a curved coordinate system into which the 2-component spinor fields must be mapped. Of course, the space in this case should be characterised by a generalized Riemannian metric.

In order to determine the 16-component metric field underlying the Riemannian space-time we should seek clues in the principle of relativity per se. For this purpose, we must think both algebraically and geometrically in attacking the problem. A. Einstein and W. Mayer (1932) demonstrated that the most primitive representations of the symmetry group underlying the relativity theory obey the algebra of quaternious whose basis functions are 2-component spinor variables. We fully exploit these algebraic features of the Poincaré group of the special theory of relativity.

If we globally extend from linear transformations of the Poincaré group to nonlinear transformations of the Einstein group, the above algebraic properties of the representations of the Poincaré group must prevail, whereas a change must occur in the geometry of space-time, i.e., a change from the geometry of Euclidean axiomatic system to that of Riemannian system. In other words, the algebraic aspect of the logic of space-time remains unaltered, while its geometric aspect undergoes a change when one proceeds to general relativity from special relativity.

Taking account of the above view points, one must obtain the irreducible representations of the Einstein group by retaining the quaternion algebra as its form. For this purpose the invariant geometric element of the space-time point may be expressed as

$$d\varphi = kq^{\mu}(x)dx_{\mu}\,, \quad k = \text{constant}, \tag{2.1}$$

where $q^{\mu}(x)$ are a set of 4 quaternion field variables that transform in space-time as a contravariant 4-vector under the Einstein group. The quantity $d\varphi$ thus defined, though a geometrical invariant in space-time, is a quaternion and as such can be expressed as a second rank spinor of the type $\tilde{\chi}^{*}\otimes\chi$. So, at each point of space-time $d\varphi$ depends on the space-time coordinates x as well as on the 'spin coordinates'.

Now our main problem is to construct a function that remains invariant with respect to 'spin coordinates' as well as in space-time coordinates, such that there exists a 'real number' field variables corresponding to the quaternion field variables. For this purpose we define the conjugated quaternion as

$$d\tilde{\varphi} = k\tilde{q}^{\nu}(x)\, dx_{\nu}\,. \tag{2.2}$$

Now, we make the following identification (real number field variables $\Longleftrightarrow$ quaternion field variables) :

$$ds^2 \Longleftrightarrow \sigma_o g^{\mu\nu} dx_{\mu} dx_{\nu} = d\varphi\, d\tilde{\varphi} = -1/2(q^{\mu}\tilde{q}^{\nu} + q^{\nu}\tilde{q}^{\mu})\, dx_{\mu} dx_{\nu}, \tag{2.3}$$

where σ_o is a 2-dimensional unit matrix. The symmetrization occurs due to the commutability of the real numbers dx_{μ} and dx_{ν}. The factor (-1/2) is taken in anticipation of the normalization of the quaternion field variables. Then we have

$$\sigma_o g^{\mu\nu} \Longleftrightarrow -1/2(q^{\mu}\tilde{q}^{\nu} + q^{\nu}\tilde{q}^{\mu}). \tag{2.4}$$

Now, we generalize the quaternion

$$Q = \sigma_{\mu} x^{\mu} = \sigma_o x^o - \sigma_1 x^1 - \sigma_2 x^2 - \sigma_3 x^3 \tag{2.5}$$

to the quaternion field

$$q^{\mu}(x) = \sigma_{\nu}({}_{\nu}v^{\mu}(x)) = \begin{pmatrix} ({}_{o}v^{\mu} - {}_{3}v^{\mu}) & -({}_{1}v^{\mu} - i\ {}_{2}v^{\mu}) \\ -({}_{1}v^{\mu} + i\ {}_{2}v^{\mu}) & ({}_{o}v^{\mu} + {}_{3}v^{\mu}) \end{pmatrix}, \tag{2.6}$$

where σ^{k} $(k = 1,2,3)$ are Pauli's spin matrices. Thus, $q^{\mu}(x)$ is expressed as the general form of a 2-dimensional Hermitian matrix in which each element is a 4-vector. The field ${}_{\nu}v^{\mu}$, called a 'tetrad' or a 'vierbein' field, is chosen such that

$$ {}_{o}v^{\mu}\,{}_{o}v_{\mu} = 1, \quad {}_{k}v^{\mu}\,{}_{k}v_{\mu} = -1, \ (k = 1,2,3). \tag{2.7}$$

Thus, $q^{\mu}(x)$ is a set of 4 quaternions, one for each 4-vector component. Furthermore, q^{μ} transforms as (i) a contravariant 4-vector in space-time and (ii) a second rank spinor of the type $\tilde{\chi}^{*}\otimes\chi$ in 'spinor-space'.

We define the conjugate of $q^{\mu}(x)$ by

$$\tilde{q}^{\mu}(x) = q^{\mu}(-\sigma_{o}, \sigma_{k}) = \tilde{\sigma}_{\nu}({}_{\nu}v^{\mu}(x)) = \begin{pmatrix} (-{}_{o}v^{\mu} - {}_{3}v^{\mu}) & -({}_{1}v^{\mu} - i\ {}_{2}v^{\mu}) \\ -({}_{1}v^{\mu} + i\ {}_{2}v^{\mu}) & (-{}_{o}v^{\mu} + {}_{3}v^{\mu}) \end{pmatrix}. \tag{2.8}$$

It is evident that $q^{\mu}\tilde{q}^{\nu}$ transforms as a second-rank contravariant tensor in space-time. So, this fulfils the requirement of the invariant structure of the left-hand member of the equation (2.3). In order to obtain the proper local limit of the metric field, we require that

$$ {}_{\nu}v^{\mu}(x) \xrightarrow{\text{loc.lim.}} \pm\,\delta_{\nu}^{\ \mu}, \tag{2.9}$$

where

$$\delta_{o}^{\ o} = \delta_{oo} = 1, \quad \delta_{k}^{\ k} = -\delta_{kk} = -1. \tag{2.10}$$

This gives

$$(q^{o}, q^{1}, q^{2}, q^{3}) \xrightarrow{\text{loc.lim.}} (\sigma_{o}, \sigma_{1}, \sigma_{2}, \sigma_{3}), \tag{2.11a}$$

$$(\tilde{q}^{o}, \tilde{q}^{1}, \tilde{q}^{2}, \tilde{q}^{3}) \xrightarrow{\text{loc.lim.}} (-\sigma_{o}, \sigma_{1}, \sigma_{2}, \sigma_{3}). \tag{2.11b}$$

Then the correspondence (2.4) yields that

$$g^{\mu\nu}(x) \xrightarrow{\text{loc.lim.}} \begin{pmatrix} 1 & 0 & 0 & 0 \\ 0 & -1 & 0 & 0 \\ 0 & 0 & -1 & 0 \\ 0 & 0 & 0 & -1 \end{pmatrix} \tag{2.12}$$

as required by the special relativity.

Furthermore, for the quaternion formalism, it is easily seen that

$$q^{\mu}q_{\mu} = g_{\mu\nu}q^{\mu}q^{\nu} = -46_{o} = \text{invariant.} \tag{2.13}$$

3. THE MEASUREMENT PROBLEM IN FIELD CONCEPT

We assume that there are objective elements underlying the real world and that space and time, according to the theory of general relativity, simply play a subjective role in the representation of the laws of nature. "According to J. Stachel (ref. P.G. Bergmann, 1987), it took Einstein several years to realize fully that in GR the coordinates are not physical quantities but merely convenient labels for identifying world points (non-uniquely). In this respect GR differs profoundly from special relativity." Now, if one applies the axiomatic basis of relativity theory to its logical extreme, one must arrive at the conclusion that in contrast with the quantum mechanics, bits of matter cannot be the fundamental building blocks with which one can construct a theory of matter. Rather, the 'relation' of some kind forms the basis for a theory of matter in this context. As the fundamental starting point one can plausibly take the 'observer - observed' relation, where the 'observer' is considered in regard to its interaction relation with the 'observed', and vice-versa in accordance with the concept of the generalized field theory as discussed in section 1. 'Observer' refers to an intrinsic component of a fundamental interaction. So, logically the general theory of relativity needs a theoretical structure which makes no distinction between the part of the system identified as 'observer' and the part identified as 'observed'. This symmetry requirement of the theory of relativity implies that the variables that describe the 'observer' and those that describe the 'observed' (microscopic, macroscopic or astronomical) must have the same covariant form. Thus, in this relativistic field theory the introduction of both types of variables is done in a fully covariant manner, with the help of a unified field that explicitly exhibits the inertial manifestations of matter together with the gravitational and other manifestations of interacting matter. In Bohr's view of quantum mechanics, these variables are introduced, necessarily, asymmetrically, the 'observer' component obeying the rules of classical physics, and the 'observed' component obeying the rules of quantum physics. So, one of the problems that arose in the famous Einstein - Bohr debates is resolved and a fundamental requirement of the Copenhagen school is fulfilled.

4. DISCUSSION

In the general theory of relativity by applying the 'observer-observed' symmetry relation, one is in a good position to investigate this interaction relation in the asymptotic region that approaches a sufficiently weakly coupled state in which it appears, as a first approximation, to be uncoupled to the environment. The assumption of this elementarity of the interaction relation - the 'elementary interaction' - implies that there exists no uncoupled state, i.e., there can never be an actual separation between two parts, 'observer' and 'observed'. So, by utilizing the axiomatic basis of the theory of general relativity to its logical extreme, one can plausibly conclude that the 'elementary interaction' is the only relation that prevails as the unbreakable entity in this theory, called the general matter field theory. Thus, the 'elementary interaction', i.e., 'observer-observed' relation is an objective entity, for it is independent of the interchange of the 'observer' component (the subject) and the 'observed' component (the object) of the relation.

In the general matter field theory global features of the whole universe appear in the quantitative description of the elementary particle physics. The dependence of the inertial mass of an interacting elementary particle on the curvature of space-time is one such feature. So, one can predict the behaviour of matter that interacts over the laboratory and astronomical domains by analyzing the discoveries of the behaviour of the matter in the domain of elementary particles. Though any physical problem, in fact, deals with an extremely small portion of the Universe corresponding to extremely high energy-momentum transfer between matter and matter in the microscopic domain, it does not pose any problem, for the portion concerned is a part of the continuous and closed field that does extend into the astronomical domain.

ACKNOWLEDGEMENT

In preparing this lecture-note we have freely consulted the following book and owe our indebtedness to the author.

M. Sachs, **General Relativity and matter**, D. Reidel Publishing Co., Dordrecht, Holland (1982).

REFERENCES

Bergmann, P.G. (1987): 'Observables in general relativity', abstract of the lecture delivered at the Xth course of the International School of Cosmology & Gravitation, Erice (Italy), May 2-12, 1987.

De Broglie, L. (1979) : **Annales de la Foundation Louis de Broglie 4, 56.**

Einstein, A. and Mayer, W. (1932) : **Preuss. Akad. Wiss. Phys. Math. Klass Sitz.**, p. 522.

Gödel, Kurt (1949) : **Rev. Mod. Phys. 21**, 447.

Ozsvath, I. and Schücking, E. (1962) : **Nature 193**, 1168.

Sachs, M. (1982) : **General Relativity and Matter**, D. Reidel Publishing Co. (Dordrecht).

ON THE RELATIONS BETWEEN FUNDAMENTAL CONSTANTS

Venzo de Sabbata
Department of Physics of University
Via Paradiso, 12
I-44100 Ferrara, Italy
and I.N.F.N., Sezione di Bologna, Italy

ABSTRACT. We like to consider two astrophysical empirical laws, try to give some theoretical basis for them and to show that the fine structure constant α is a typical dimensionless parameter not only of atomic physics but also of gravitational physics.

1. INTRODUCTION

The astrophysical laws for which we are concerned, are:

a) the relation between intrinsic angular momentum S and the magnetic moment U of stars [1],[2],[3], that may be written as

$$S = qU \tag{1}$$

where q is a universal constant

b) the relation between total angular momentum J and the mass M of astronomical system [4] that can be written in the form

$$J = pM^2 \tag{2}$$

where p is a universal constant.

Now, following the analysis of Wesson [5] we know that particle physics involves three constants with dimensions $[e] = M^{1/2} L^{3/2} T^{-1}$, $[h] = ML^2 T^{-1}$, and $[c] = LT^{-1}$. These form a complete set in the sense that only one dimensionless number $[\alpha] = 0$ can be formed from them. On the other hand gravitation is developed in general involving two constants that is $[G] = M^{-1} L^3 T^{-2}$ and c. These two constants are not sufficient to form a set because from them we cannot form any dimensionless constant: we need at least one more dimensional constant (provided it have different dimensionality).

A new dimensional constant can be introduced, for instance, just looking at the empirical law (2): that is p which has the dimension of $[p] = M^{-1} L^2 T^{-1}$. With this new constant, also in the case of gravitation we will have a set of three constants, namely (G,p,c), which form a complete set in the sense that they are combinable to form one dimen-

V. De Sabbata and V. N. Melnikov (eds.),
Gravitational Measurements, Fundamental Metrology and Constants, 115–129.

sionless constant $\beta = G/pc$. If we choose (1) and instead of p we introduce the new constant q which has the dimension of $L^{-1/2} M^{1/2}$ we may form with the set (G,q,c) an other complete set with the dimensionless constant $\gamma = c^2 / q^2 G$ [6], [7], [8].

If the aim is to incorporate a new dimensional constant, we need a gravity theory wider than general relativity, and looking at the empirical law (2) where rotation is involved, one is led to consider the Einstein-Cartan theory. We will see in fact that it is possible to deduce the law (1) just introducing torsion into the gravitational equations and we will see that α is not only the dimensionless constant for particle physics, but also for gravitation. Before to deduce theoretically the Blackett law, we desire to say some words about torsion.

2. MAXWELL'S EQUATIONS IN A RIEMANN-CARTAN SPACE

If in the general relativity theory we introduce torsion that, according to Trautman, is the simplest and more natural modification of the general relativity, we can give a geometrical description of matter also at a microscopical level: in fact general relativity is concerned with macroscopical bodies, but if we go down to a microscopical level, we find that matter is formed by elementary particles which are characterized not only by a mass but also by a spin and this latter is measured in units of $\hbar$. So if we wish to extend general relativity to include microphysics, we must take into account that matter is described by mass and spin density: mass is related to curvature by means of the energy-momentum tensor, so that in the framework of a generalization of the theory, spin should be related through the spin density tensor to some other geometrical property of the space-time according to the spirit of a geometric theory of gravity.

We know that this requirement is satisfied by the E.C.S.K. (Einstein, Cartan, Sciama, Kibble) theory which provides a relation between torsion and spin. In this way also the constant $\hbar$ is involved in gravitation physics. Moreover if we introduce a coupling between torsion and electromagnetic potential [9] , the fine structure constant enter in the photon-torsion interaction lagrangian.

In fact in this latter case we may introduce a phenomenological photon-torsion interaction [10]

$$L_I = \alpha G \varepsilon^{\mu\nu\alpha\beta} A_\mu F_{\nu\alpha} \partial_\beta \varphi \qquad (3)$$

where α is the fine structure constant, G the gravitational constant $\varepsilon^{\mu\nu\alpha\beta}$ the totally antisymmetric tensor, A_μ the electromagnetic potential, $F_{\mu\nu} = \partial_\mu A_\nu - \partial_\nu A_\mu$ and φ is the axial-vector part of the torsion tensor determined by the pseudoscalar potential (see also [11],[12],[13],[14],[15],[16]). We can notice that this potential has been introduced through an explicit computation as a result of the minimal interaction of the torsion background with the fermion-antifermion pairs produced by photon according to the vacuum polarization effect.

Until recently people believe that the electromagnetic field cannot be coupled to torsion in order to preserve local gauge invariance: one cannot applied the minimal coupling principle

$$\eta_{ik} \rightarrow g_{ik} \quad , \qquad \partial_k \rightarrow \nabla_k \tag{4}$$

that is one cannot replace the flat with curve metric and the derivative with covariant derivative in the presence of torsion to the special relativistic photon lagrangian without breaking gauge invariance. Several attempts has been made [17], [18], but they was successful however only at the cost of imposing arbitrary geometrical constraints upon torsion [17] , or introducing a modified definition of gauge transformation [18] .

Therefore these attempts are beyond the Einstein-Cartan theory so that in the context of Einstein-Cartan theory it was

(*) "photons neither produce nor feel torsion"

But really we like to emphasize that this statement (*) can be assumed valid as long as the electromagnetic field is treated as a classical not quantized field.

When a quantum point of view is adopted, however, the proposition (*) is valid only in the first approximation since, in general, according to a quantum description of the electromagnetic field, in a space with torsion one must always expect an interaction between the photon and the torsionic background.

In fact a photon can disintegrate into a virtual electron-positron pair for a certain fraction of the time (the "vacuum polarization" effect). Since a massive fermion is minimally coupled to torsion, the pair interacts with the torsionic background and this interaction induces an "indirect" photon-torsion coupling which preserves the gauge invariance. We shall motivate the introduction of this non minimal coupling, that we call semi-minimal as it is minimal at microscopical quantum level (the interaction of the $e^+ e^-$ pairs with torsionic background), through an explicit perturbative computation.

Without enter in the details of the procedure, we notice only that the virtual pairs production induces in vacuum a current density proportional to field intensity, and the Maxwell equations, written in momentum space are so modified:

$$(p^2 \delta^i_k - p^i p_k) A^k(p) = 4\pi \left[J^i(p) + \pi^{ik}(p) A_k(p) \right] \tag{5}$$

where π^{ik} is the polarization tensor which is given (through the usual Feynman rules for the photon self energy diagram) (see Fig 1)

$$\pi_{ik}(p) = \frac{ie^2}{(2\pi)^4} \int d^4q \, \mathrm{Tr} \left[S_F(q) \gamma_i S_F(q-p) \gamma_k \right] \tag{6}$$

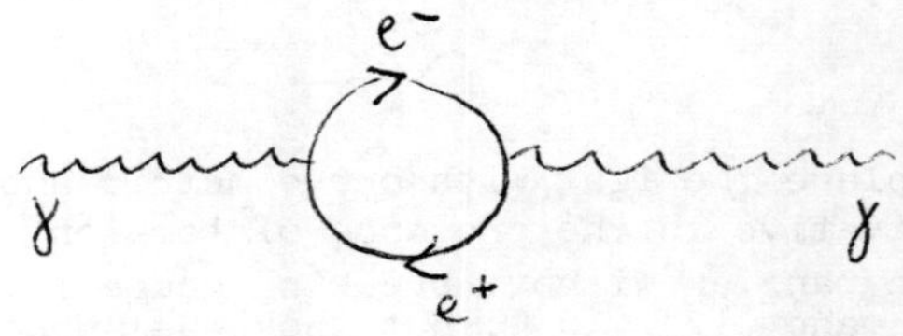

Fig 1

where $S_F(p)$ is the electron propagator in momentum space and in vacuum is

$$S_F^{(o)}(p) = -i/(i\gamma_k p^k + m) \tag{7}$$

and the vacuum polarization tensor $\Pi^{(o)}_{ik}(p)$ produces only a charge renormalization effect ($\Pi^{(o)}_{ik}(p) = (p^2)(p^2\eta_{ik} - p_i p_k)$).

Following this procedure, writing the Dirac equation in U_4, the propagator in the presence of torsion and developing the modified propagator to the first order in the spin-torsion coupling constant (that we take equal to G), we find that the Maxwell equations in the presence of torsion are modified as follows

$$\begin{cases} \partial_{[i}F_{jk]} = 0 \\ \partial_k F^{ik} = 4\pi J^i + \dfrac{2\alpha}{3\pi}\eta^{iklj}F_{kl}Q_j \end{cases} \tag{8}$$

being

$$Q^l = \frac{1}{16}\eta^{ijkl}K_{ijk} \tag{9}$$

(where K_{ijk} is the usual contorsion tensor).

Now all these things that allow us to introduce the interaction between photon and torsion through the virtual pairs production, that is through the vacuum polarization effect, can be translate in the following, so to say, macroscopical empirical procedure replacing the minimal coupling principle.

1) Apply the minimal coupling principle only to the field tensor, and not to the potential vector; then $F_{ik} = 2\partial_{[i}A_{k]}$ and $\partial_{[i}F_{jk]} = 0$

still hold, but the other equation becomes

$$\nabla_k F^{ik} \equiv F^{ik}_{\ |k} + K_{kl}^{\ \ i} F^{kl} + K_{kl}^{\ \ k} F^{li} = 4\pi J^i \tag{10}$$

where a bar denotes covariant derivative involving the Christoffel symbols

2) Use for the coupling only the totally antisymmetric part of the torsion, that is:

$$K_{ijk} = -\frac{2}{3}\eta_{ijkl} Q^l \tag{11}$$

3) Replace the spin-torsion coupling constant G by $G' = \alpha G/\pi$

By applying these rules, we obtain the following electromagnetic equations:

$$\begin{cases} \partial_{[i} F_{jk]} = 0 \\ \frac{1}{\sqrt{-g}} \partial_k \left(\sqrt{-g}\, F^{ik}\right) = 4\pi J^i + \frac{2\alpha}{3\pi} \eta^{iklj} F_{kl} Q_j \end{cases} \tag{12}$$

which in the case of a flat metric are reduced exactly to (8).

In this case the rules 1), 2), 3) constitute the new procedure which we call "semi-minimal" because it is minimal at a microscopical level, i.e. when considering the interaction of the e^- e^+ pairs with torsion, even if it is non minimal macroscopic photon-torsion interaction.

The things that so far we have said are valid in the Einstein-Cartan theory, but we can also introduce a propagating torsion, imposing that the axial-vector part of the torsion tensor (that is (9)) is determined by the pseudoscalar potential φ according to $Q_\beta = \partial_\beta \varphi$ so that we have the Lagrangian (3) for the photon-torsion interaction.

3. DIRAC LARGE NUMBER HYPOTHESIS

As regards the Planck constant $\hbar$, it is involved in gravitation physics when a torsion-spin interaction is considered. At this purpose, I like to show an example which represents an extension of the Dirac large-numbers hypothesis [19] .

First of all two words on the strong gravity. In fact it has been pointed out [20] [21] , that the strong gravity theory [22] provides the physical basis for the large-numbers hypothesis of Dirac. Now in 1938 Dirac was impressed by some large dimensionless numbers constructed using the fundamental parameters of microphysics (such as the Planck constant $\hbar$, the mass of the nucleon etc.) and the cosmology (such as the Hubble constant, the mass of the Universe etc.) combined with the gravitational constant G and the velocity of light c. These large numbers are of the order of D^n where $D^n \sim 10^{40}$. For example the ratio between the present age the Universe ($\sim H^{-1}$) and the time need by light to cross and elementary particle ($\sim r/c$ where $r \sim \hbar/mc$) is given by

$$\frac{mc^2}{\hbar H} \sim D \tag{13}$$

where m is a typical hadronic mass (which represents also the ratio between the radius of the Universe and the characteristic intrinsic size of an elementary particle). The ratio between the Compton wave-length of the proton and its gravitational radius is

$$\frac{\hbar c}{2Gm^2} \sim D \tag{14}$$

The famous ratio between electric to the gravitational force between protons and electrons:

$$\frac{e^2}{GMm} \sim D \tag{15}$$

The ratio of the present age of the Universe to the atomic unit of time e^2/mc^3

$$\frac{tmc^3}{e^2} = \frac{mc^3}{e^2 H} \sim D \tag{16}$$

Moreover the ratio between the mass M contained inside the Hubble radius ($M \sim Rc^2/G$ which we indicate as the mass of Universe M_U) and the mass of the proton is

$$\frac{M_U}{m} \sim \frac{c^3}{GHm} \sim D^2 \tag{17}$$

and so on.

Now Dirac observes that the ratio (13) is not a constant, but is growing linearly in time because the age of the universe is growing: if we believe that the equality of all these ratios is not a numerical coincidence, characteristic only of our present epoch, we are led then to conclude that when a dimensionless number is of the order of D^n, it should be variable in time as t^n. This is the so-called large numbers hypothesis of Dirac that Dirac expresses with these words:

> "Any two dimensionless large numbers that will be formed in nature, are connected by a simple mathematical relation whose coefficients are of the order of unity. If one finds by some elementary considerations that some of these numbers vary with time, then they must do this in a way that the relation between them does not change".

An immediate consequence of this assumption is that in the system of units in which the light velocity, the Planck constant and the proton mass are constants, the Newton constant of gravity must decrease in time like $G \propto t^{-1}$ as we can see from (14) and (15). To reconcile these effects with the standard Einstein theory which requires a constant value of G, one may assume that general relativity is valid in a system of units different from that based on the atomic constants. In that case we must distinguish the proper time interval ds_E of the Einstein theory and the interval ds_A based on the laboratory measurements of times and distances such as the wave-length of the spectral lines or the frequency of monochromatic radiation. The other time interval ds_E is supposed to correspond to phenomena governed by the gravitational interaction, such as planetary motion. In this way in the Einstein units G_E is constant while in the atomic units $G_A \propto t^{-1}$ and, considering the motion of a planet around the sun, if the mass is constant then $r_A \propto t^{-1}$ that is the radius of the orbit decreases in time.

4. STRONG GRAVITY

But the point we wish to stress here is that the strong gravity theory may provide a physical interpretation of this large numbers coincidences. Suppose in fact that the universe and a hadron are two physical systems governed by similar interval laws, differing only for a scale factor which transforms the weak gravitational field into the strong gravity field, and assume that the strong gravity potential of a hadron,

$$\Phi(h) \sim G_f\, m/rc^2 \tag{18}$$

for $r \sim \hbar/mc$, is of the same order of magnitude as the gravitational potentialof the universe

$$\Phi(U) \sim G M_U/Rc^2 \tag{19}$$

where $R = c/H$; It follows that

$$\frac{G_f}{G} \sim \frac{M_U}{m}\frac{r}{R} \sim D \tag{20}$$

in agreement with the value suggested for the strong gravity coupling constant.

If the Newton coupling constant varies in time as deduced from Dirac large number hypothesis that is as $G \propto t^{-1}$ then from (20) we have that at the epoch $t_0 = (HD)^{-1} = \hbar/mc^2 = 10^{-23}$ sec, it was $G=G_f$ (notice that t_0 is the beginning of hadronic era) so that the strength of ordinary gravity and of strong gravity coincided.

One may speculate then that the strong gravitational interactions have been frozen with their primordial strength, but confined inside the hadronic matter, while the other interaction becoming weaker and weaker in time, have reached the present value $G = 10^{-40}\, G_f$. It is natural then to identify the atomic proper time interval of Dirac, ds_A, with that of strong gravity, which is effective at a nuclear level only, and the Einstein interval, ds_E, with that of ordinary macroscopic gra vity. In this sense the strong gravity theory provide a physical basis for the large numbers hypothesis.

5. SOME COSMOLOGICAL DEDUCTION

We can also notice that starting from the analogy between the gravitational interaction at a cosmological and at a hadronic level, and given the ratio of the two coupling constants G/G_f , one can deduce the value of the Hubble radius, or of the critical mass of the universe directly from the values of the corresponding microphysical quantities. In fact from (17) and (20) one has

$$\frac{R}{r} \sim D \tag{21}$$

Moreover if we consider also the torsion, that is if we introduce spin and torsion into the strong gravity field theory, then the gravitational potential of a nucleon is modified by the spin contributions. In particular at a distance $r \sim \hbar/mc$ we can write with suitable approximation an effective potential $\Phi(h) = \Phi_0(h) + \Phi_1(h)$:

$$\Phi_0(h) \sim G_f \frac{m^2}{\hbar c} \tag{22}$$

$$\Phi_1(h) \sim G_f^2 \frac{\sigma^2 r^2}{c^6}$$

where $\Phi_1(h)$ is the contribution of the spin being $\sigma \sim \hbar/r^3$ the hadronic spin density.

Since the assumption $\Phi_0(h) \sim \Phi_0(U)$ (see (18),(19),(20)) provides the large numbers correlations between microphysics and cosmology, for the case of mass, radius and coupling strength (see (20)), we suppose also $\Phi_1(h) \sim \Phi_1(U)$. Then (22) gives

$$G_f \sigma r \sim G \Sigma_U R \tag{23}$$

where $\Sigma_U \sim S_U/R$ is the density of intrinsic angular momentum of the Universe. From (20) and (21) we have

$$\sigma \sim \Sigma_U \tag{24}$$

which means that the angular momentum density of the universe and of elementary an particle are the same, or, in other words

$$S_U = \sigma R^3 \sim \left(\frac{R}{r}\right)^3 \hbar \sim \hbar D^3 \tag{25}$$

A relation identical to (25) has been obtained also in ref. [23] on the ground of different assumptions however. It should be noted that interpreting this angular momentum as due to a rigid rotation of the universe as a whole, that is putting $\Sigma_U = M_U \omega / R$ one obtains that the angular velocity is

$$\omega \sim \frac{R \Sigma_U}{M_U} \sim \frac{R}{M_U} \frac{\hbar}{r^3} \sim \frac{c}{R} \sim H \qquad (26)$$

and is perhaps remarkable that an angular velocity of the same order of magnitude is predicted also by Gôdel's [24] rotation model of universe

$$\omega_G = 2\sqrt{2\pi G \rho} \sim \left(\frac{G M_U}{R^3}\right)^{1/2} \sim \frac{c}{R} \sim H \qquad (27)$$

which is an angular velocity compatible with the measured isotropy of the cosmic microwave background: in fact such a value is in agreement even with the more drastic upper limit ω_0 for the rotation of the universe imposed by measurements of the sun peculiar velocity through the cosmic microwave background, that is $\omega \leq \omega_0 \sim 10^{-18}$ sec.

Therefore in the framework of strong gravity with torsion one is led to construct the following large numbers [25] :

$$\frac{G_f}{G} = \frac{R}{r} \sim D \quad (10^{-40}) \text{ related to coupling strengths}$$

$$\frac{M_U}{m} \sim \left(\frac{R}{r}\right)^2 \sim D^2 \quad (10^{-80}) \text{ related to mass} \qquad (28)$$

$$\frac{S_U}{\hbar} \sim \left(\frac{R}{r}\right)^3 \sim D^3 \quad (10^{-120}) \text{ related to spin}$$

Also Ivanenko [26],[27] , find evidence for the relation (25) on the basis of the recent discovery of the rotation of the Metagalactic system of radio-sources as due to geometrical rotation of the Universe.

6. THE THEORETICAL BASIS FOR THE BLACKETT RELATION AND THE $J=PM^2$ RELATION.

But now return back to the two astrophysical laws (because this is the main topic of the speech):

$$S = qU \qquad (1)$$

$$J = pM^2 \qquad (2)$$

that is the relation (1) between intrinsic angular momentum S and the magnetic moment U of stars and the relation (2) between the total angular momentum J and the mass M of astronomical systems. From the experimental data of ref. [3] which report the maximum and minimum values of the measured magnetic fields we can obtain for q the mean value $q = 1.3 \cdot 10^{15}\ g^{1/2} cm^{-1/2}$ while the data of ref. [5] gives for p the value $p = 7.9 \cdot 10^{-16}\ g^{-1} cm^{2} sec^{-1}$. The remarkable result is that both relations seems to indicate that the fine-structure constant α is the typical dimensionless parameter not only of atomic physics but also of gravitational physics [6], [7], [8].

As is well known, Blackett [1] on the basis of some observed ratio of the magnetic-dipole moment over the angular mimentum of some planets and stars, established the empirical relation

$$U = \beta \frac{\sqrt{G}}{c} S \tag{29}$$

where β is a numerical coefficient of the order of unity. The other empirical relation (2) which relates the total angular momentum J and the mass M of the astronomical systems was extensively investigated by Brosche [4] and by Wesson [5].

At first sight these two empirical laws seem to be completely unrelated but we will show that a close connection can be established between these two laws and since to the relation (1) can be given a theoretical basis with the introduction of the spin-torsion interaction and some considerations concerning the early universe, also the relation(2) acquires, indirectly, a cosmological interpretation. First of all, in view of a possible unification of all the fundamental interactions, Wesson [5] argued that a new dimensional constant is to be introduced in gravitational physics, so that its combination with the Newton coupling constant G and the light velocity c can form a dimensionless parameter characteristic of the gravitational interaction.

The new universal constant, according to Wesson, is p defined by the mass-angular momentum relation (2), and the dimensionless number we can obtain is

$$\beta = G/cp \tag{30}$$

Wesson assumed that β is typical parameter of gravitational physics, like the fine structure constant $\alpha = e^2/hc$ is typical of microphysics. Now here we wish to point out that also the empirical law (1) provides a new universal constant q and starting with q we can form another dimensionless number [6]:

$$\gamma = c^2 / q^2 G \tag{31}$$

which we could choose, instead of β, as the dimensionless constant characteristic of the gravitational interaction.

The remarkable result we wish to stress is that, from the experimental values of p and q one gets

$$\beta \sim 0.28 \cdot 10^{-2}, \qquad \gamma \sim 0.79 \cdot 10^{-2} \tag{32}$$

and then $\beta \sim \gamma$ up to an order of magnitude.

If one does not believe that this numerical coincidence is accidental, then this result seems to suggest that the two set of constants (G,c,p) and (G,c,q) are equivalent in the sense that both define the same dimensionless parameter $\beta = \gamma$ characteristic of gravitation.

It follows that the two constants p and q are not independent, since equating the two definitions (30) and (31) we obtain the relation

$$q^2/p = c^3/G^2 \tag{33}$$

which establishes a connection between the two empirical laws, and suggest the possibility they have a commom origin.

Therefore once that a theoretical explanation is provided for one of these two laws, and one of the two constants p or q is fixed, then from (33) another fundamental constant can be deduced or, equivalently, the existence of another law of nature can be predicted.

Now, as previously said, we have in fact a cosmological explanation of the first law (1) S = qU, if we introduce a spin-torsion interaction during the early stages of our universe. The Blackett law appears as a consequence of a resonant spin alignement during the early stage of the universe.

We do not go in the detail of calculation that one can find in the references [28], [29], and will give only an outline.

Before to do that, we can observe that Blackett was led to the relation S = qU observing that we know a fundamental relation between magnetic moment and spin angular momentum for microscopic charged particle: the Bohr magneton $e/2m_e c$; this is enormously greater respect to 10^{-15} given for macroscopic bodies. This fact is due to the enormous charge density of electron. If on the contrary we substitute the gravitational mass $G^{1/2}m$ at the place of the electric charge, where is just the gravitational mass $G^{1/2}m$ responsible for producing magnetic field of a rotating body, then we would expect the ratio of the magnetic moment to the angular momentum to be smaller then the Bohr magneton by the ratio $G^{1/2}m_e/e$ and in this way we are led to the value of $\sim 10^{-15}$ $cm^{1/2}\,g^{-1/2}$, which is the value seen above for some celestial body.

Now we will see that, as said, the relation (1) has a cosmological significance since it can be deduced in a very simple way as a natural consequence of the initial conditions of the universe provided that torsion is introduced into the gravitational equations according to the Einstein-Cartan theory.

It is in fact well known that the spin-torsion interaction Lagrangian in a Riemann-Cartan space is

$$\mathcal{L} = -\frac{1}{2} S^{kij} K_{ijk} \tag{34}$$

where S_{ijk} is the canonical spin tensor and K_{ijk} is the contorsion tensor related to the torsion tensor Q_{ijk} by

$$K_{ijk} = -Q_{ijk} - Q_{kij} + Q_{jki} \tag{35}$$

being $Q_{ij}{}^{k} = \Gamma^{k}_{[ij]}$ the antisymmetrical part of the affine connection. In the case of Dirac particle (the spin is a totally antisymmetric tensor) we have

$$S^{ijk} = \eta^{ijkl} S_l \tag{36}$$

(where η^{ijkl} is the completely antisymmetric symbol), and the interaction Lagrangian contains only the axial-vector part of the torsion tensor:

$$\mathcal{L} = - S_k Q^k \tag{37}$$

(where $Q_k = \frac{i}{2} \eta_{kijl} Q^{ijl}$) and going to the particle rest system ($S^k \equiv (0, \vec{S})$) the interaction energy between spin $\vec{S}$ and torsion $\vec{Q}$ is

$$E = -\mathcal{L} = -\vec{S} \cdot \vec{Q} \tag{38}$$

We can observe that this formula is formally analogous to the interaction energy of a magnetic moment $\vec{\mu}$ in a constant magnetic field $\vec{H}$, namely

$$E = -\vec{\mu} \cdot \vec{H} \tag{39}$$

so that we can understand, without go through calculation (see ref.[28], [29]) that torsion induces spin alignement as well as a magnetic field and if to each spin is associated a magnetic moment, torsion can also produce a magnetic field inside the matter. Applying these considerations to the early universe we find that to each portion V of hadronic matter (V is the volume of the portion) containing a total angular momentum density J=S/V and then a torsion $Q = 4\pi G_f J/c^2$, it was associated a magnetic field (produced by the nucleons orientation)

$$B \sim \frac{8\pi}{3c} (2\alpha G_f)^{1/2} J \tag{40}$$

(α the fine structure constant and G_f strong gravity constant) that is a dipole magnetic moment

$$U \sim (\alpha G_f)^{1/2} \frac{S}{c} \tag{41}$$

At this point we can observe that the same relation between angular and magnetic moment holds also in the case of a particle. In fact

$$\frac{\mu}{\hbar} \sim \frac{e}{mc} \sim (\alpha G_f)^{1/2} / c \tag{42}$$

This result fits nicely with Dirac's two metric hypothesis which we have already mentioned that is the atomic metric, operative at a microscopic level and the usual macroscopic metric connected with gravitation. These two metrics coincided at the beginning of the hadron era, but the atomic metric has not changed in time, while the macroscopic metric developed with the universe (and, as seen before, this provides a basis for understanding of the large-number hypothesis [19] , [20] , [21]).

Therefore the ratio U/S for a particle was frozen to the value (42) it had at that epoch ($t_g \sim 10^{-23}$ sec when, according to Dirac large number hypothesis for which the gravitational constant decreases in time as $G \propto t^{-1}$, the gravitational constant coincides with the strong gravity constant G_f), while the magnetic field of a celestial body has developed in time up to its present value

$$U \sim (\alpha G)^{1/2} S/c \tag{43}$$

This explain also why Blackett's experiment [30] which measured the magnetic field of a rapidly rotating body, provided a null result. In fact, according to our interpretation [31] , this law (43) holds only for astronomical bodies (and not for all rotating matter), because it is a consequence of a very hot and collapsed initial state of the Universe.

7. CONCLUSION

So we find the blackett relation

$$S = qU \tag{1}$$

with

$$q \sim \frac{c}{(\alpha G)^{1/2}} \tag{44}$$

which is in remarkable agreement with astrophysical data (that is with $q = 1.3 \cdot 10^{15}\ g^{1/2}\ cm^{-1/2}$). Using now (44) and the relation (33)

$$q^2/p = \frac{c^3}{G^2} \tag{33}$$

connecting the two astophysical laws, we can deduce that the ratio of the angular momentum over squared mass ($J=pM^2$) is a universal constant p given by

$$p \sim G/\alpha c \sim 3\cdot 10^{-16}\ g^{-1} cm^2 sec^{-1} \tag{45}$$

again in agreement with experimental data.

But in our opinion the most remarkable result of this investigation [6] , [7] , [8] , is that inserting into the definitions (30) and (31) of the dimensionless numbers β and γ the value of p and q obtained both from our theoretical predictions (44) and (45) we obtain

$$\beta \sim \alpha \qquad\qquad \gamma \sim \alpha \tag{46}$$

where $\alpha = 0.73 \cdot 10^{-2}$ is the fine structure constant.

This means that the two empirical, and seemingly unrelated, astrophysical laws (1) and (2) not only seem to define the same constant $\beta = \gamma$ but also suggest that the dimensionless parameters of gravitation and microphysics are the same $\alpha = \beta = \gamma$. Therefore, according to Wesson arguments (see ref. [5]) they may represent a promising clue to the unification of gravitation with the other fundamental interactions: the dimensionless coupling constant for particle physics is the same as the dimensionless coupling constant for gravitation. Instead of two sets having three constants each (e,h,c) and (G,p,c,) (or (G,q,c)) we have four distinct constants namely e,h,c, and G.

In any case we think that torsion (and perhaps strong gravity) play an important role at this respect and that even independently from this relevance towards a unified theory, the deep connection and the general validity of these two astrophysical laws is worthy of further theoretical investigations.

venzo de sabbata

Erice 2-12 may 1987

REFERENCES

[1] P.M.S.Blackett - Nature (London) 159,658(1947

[2] D.Ahluwalia and T.Y.Wu - Lett.Nuovo Cimento 23,406(1978)

[3] S.P.Sirag - Nature(London) 278,358(1979)

[4] P.Brosche - in "Spin, Torsion, Rotation and Supergravity" ed. by Peter G.Bergmann and Venzo de Sabbata, Plenum N.Y. pg. 375-382 (1980)

[5] P.S.Wesson - Phys.Rev. D23,1730 (1981)

[6] V.de Sabbata and M. Gasperini - Lett.Nuovo Cimento 38,93(1983)

[7] V.de Sabbata and M. Gasperini - Proc. of Einstein Int.Found. 1,55 (1983)

[8] V.de Sabbata and M. Gasperini - Acta Cosmological (Poland) 12,79 (1983)

[9] V.de Sabbata and M. Gasperini - Phys.Lett. 77A,300 (1980)

[10] V.de Sabbata and M. Gasperini - "Introduction to Gravitation"World Sci.Singapore 1985,pg.268

[11] V.de Sabbata and M. Gasperini - Lett. Nuovo Cimento 30.193 (1981)

[12] V.de Sabbata and M. Gasperini - Lett. Nuovo Cimento 28,229 (1980)

[13] V.de Sabbata and M. Gasperini - Phys.Lett. 83A,115 (1981)

[14] V.de Sabbata and M. Gasperini - Phys.Rev.D23,2116(1981)

[15] V.de Sabbata and M. Gasperini - Nukleonika (Poland) 25,1373(1980)

[16] V.de Sabbata and M. Gasperini - Lett. Nuovo Cimento 30,363(1981)

[17] M.Novello - Phys.Lett. 59A,105(1976)

[18] S.Hojman,M.Rosenbaum,M.P.Ryan and L.C.Shepley - Phys.Rev.D17,3141 (1978)

[19] P.A.M.Dirac - Proc.Roy.Soc.165A,199(1938)

[20] V.de Sabbata and P.Rizzati - Lett.Nuovo Cimento 20,525(1977)

[21] C.Sivaram and K.Sinha - Phys.Lett.60B,181(1976)

[22] C.Isham,A.Salam and J.Strathdee - Phys.Rev. D3,867 (1971)

[23] R.M.Muradyan - preprint E2-9804 (Dubna 1977)

[24] K.Gôdel - Rev.Mod.Phys. 21,447 (1949)

[25] V.de Sabbata and M.Gasperini - Lett.Nuovo Cimento 25,489 (1979)
[26] D.Ivanenko - Astron. Circulae A.Sci. USSR, Moscow N.1254,April 1983
[27] D.Ivanenko - in "Proc.Eddington Centenary Symposium Vol.I ed. by Venzo de Sabbata and T.M.Karade, World Sci. Singapore 1984,pg.170
[28] V.de Sabbata and M.Gasperini - Lett.Nuovo Cimento 27,133(1980)
[29] V.de Sabbata and M.Gasperini - in "Origin of Galaxies" ed. by Venzo de Sabbata, World Sci. Singapore 1981,pg.181
[30] P.M.S. Blackett - Phil.Trans.Roy.Soc.London Ser.A245,309(1952)
[31] V.de Sabbata and M.Gasperini - Lett.Nuovo Cimento 27,449(1980)

QUANTUM ELECTRODYNAMICS AND FUNDAMENTAL CONSTANTS

R. N. Faustov
Centre of Fundamental Metrology
USSR State Committee for Standards
9, Leninsky Prospect
117049, Moscow
USSR

ABSTRACT. The present status of precision tests of quantum electrodynamics (QED) is summarized. Precision atomic physics tests of QED are brought up to date; included is a discussion of the fine structure of positronium and hydrogen, the decay rate of positronium and ground-state hyperfine splitting in muonium and positronium. The most recent results for the anomalous magnetic moment of the electron are presented. The implications of QED for the fundamental physical constants - especially for the fine-structure constant and the Rydberg constant - are considered in connection with the Josephson and quantum Hall physics.

1. INTRODUCTION

Quantum electrodynamics (QED) is the only quantum field theory model producing reliable quantitative predictions of high precision. Due to this property QED serves as the main object for studing the applicability of fundamental principles of relativistic quantum theory and a pattern for constructing some other models, the so-called gauge theories of strong, weak and electromagnetic interactions of elementary particles (e.g. quantum chromodynamics, the Salam-Weinberg model etc.). By now practically no doubts remain as to the validity of QED itself.

Research performed recently in the realm of low-energy phenomena (atomic spectra, leptonic g-factors etc.) has significantly enhanced precision of experimental data and theoretical calculations [1,7]. Further advance in this direction encounters difficulties both of technical and principle nature. Calculations of higher orders of the perturbation theory in the fine-structure constant become less and less feasible (even if the most modern computers are envolved). Further the substantial manifestation of strong interaction effects (nucleon and nucleous structure, hadro-

V. De Sabbata and V. N. Melnikov (eds.),
Gravitational Measurements, Fundamental Metrology and Constants, 131–142.

nic vacuum polarization etc.) and weak interaction effects (exchange of intermediate vector bosons) reveals itself in studying the Lamb shift in hydrogen and muonic helium ion, hydrogen hyperfine structure muon g-factor etc [1]. At present the accurate calculation of the contribution of these effects is not accessible. On the other hand the bounds on the accuracy within modern experimental methods exist.

High-energy tests of QED, namely the electron-positron colliding beam experiments permit penetration at very small distances. However these tests also has attained limits, where weak interactions (the contribution of weak neutral currents) play a significant role [1].

The comparison of the predictions of QED and the relativistic bound-state theory with the results of precision measurements provides an important source of data, which permit to improve the accuracy of values of the fundamental constants (specifically the fine-structure constant, the Rydberg constant, masses and magnetic moments of the electron, proton, muon and so on). The most accurate data of such kind can be obtained from the investigation of electromagnetic interactions of leptons either with external fields or inside the simplest hydrogenic atoms (positronium, muonium).

2. MUONIUM HYPERFINE SPLITTING

QED and the relativistic bound-state theory yield the following expression [1,2,4,5,7] for the hyperfine splitting of the ground-state energy level of the muonium ($e^-\mu^+$) atom:

$$\nu_{Mu}^{th} = E_F \left[1 + a_e + \frac{3}{2}(Z\alpha)^2 + \varepsilon + \frac{\delta_\mu}{1 + a_\mu} \right], \qquad (1)$$

where the lowest order Fermi splitting

$$E_F = \frac{16}{3}\alpha^2 R_\infty \frac{\mu_\mu}{\mu_B}\left(1 + \frac{m_e}{m_\mu}\right)^{-3}, \qquad \alpha = \frac{e^2}{4\pi\hbar c},$$

$$\frac{\mu_\mu}{\mu_B} = \frac{m_e}{m_\mu}(1 + a_\mu) = \frac{\mu_\mu}{\mu_p}\frac{\mu_p}{\mu_B}, \qquad \mu_B = \frac{e\hbar}{2m_e c},$$

$$\frac{\mu_\mu}{\mu_p} = 3.18334547(47), \qquad \frac{\mu_p}{\mu_B} = 1.521032202(15)\times 10^{-3},$$

m_e and m_μ are the electron and muon masses, μ_μ and μ_p are the muon and proton magnetic moments, μ_B is the Bohr magneton, a_e and a_μ are the anomalous magnetic moments of the electron and muon in terms of the corresponding magnetons. The term $(3/2)(Z\alpha)^2$ is the Breit relativistic correction for the electron. The quantity ε determines the contribution of purely radiative corrections under the condition of the muon being static (i.e. $m_e/m_\mu = 0$):

$$\varepsilon = \alpha\,(Z\alpha)(\ln 2 - \tfrac{5}{2}) - \frac{8\alpha}{3\pi}(Z\alpha)^2 \ln\alpha\left[\ln\alpha - \ln 4 + \frac{281}{480}\right] + \frac{\alpha}{\pi}(Z\alpha)^2\, 15.38(29) + \frac{\alpha^2}{\pi}(Z\alpha)D, \qquad Z=1, \qquad (2)$$

where the coefficient D has not yet been calculated. The last term in the square brackets in eq. (1) characterizes relativistic and radiative corrections due to the muon recoil (finite mass):

$$\delta_\mu = -\frac{3}{\pi}(Z\alpha)\frac{m_e m_\mu}{m_\mu^2 - m_e^2}\ln\frac{m_\mu}{m_e} + \frac{(Z\alpha)^2\, m_e m_\mu}{(m_\mu + m_e)^2}\left[2\ln\alpha^{-1} - 8\ln 2 + 3\frac{11}{18}\right] - \frac{\alpha}{\pi^2}(Z\alpha)\frac{m_e}{m_\mu}\left[2\ln^2\frac{m_\mu}{m_e} - \frac{13}{12}\ln\frac{m_\mu}{m_e} - 20.05(27)\right] - \frac{\alpha^2}{\pi^3}(Z\alpha)\frac{m_e}{m_\mu}\left[\frac{4}{3}\ln^3\frac{m_\mu}{m_e} - \frac{35}{18}\ln^2\frac{m_\mu}{m_e}\right]. \qquad (3)$$

The calculation of ν_{Mu} is facilitated by the fact that besides the fine-structure constant $\alpha \cong 1/137$, there is a second small parameter in the form of the electron to muon mass ratio $m_e/m_\mu \cong 1/200$. Using the known values of the fundamental constants we obtain the theoretical prediction

$$\nu_{Mu}^{th} = 4\ 463\ 303.4(1.5)\ \text{kHz} \qquad (0.34\ \text{ppm}), \qquad (4)$$

where the theoretical uncertainty (1.5 kHz) combines the errors from the measured values of the ratio μ_e/μ_μ (0.7 kHz) and fine-structure constant (0.8 kHz) with the estimate for the contribution ($\sim$1 kHz) from the uncalculated coefficient D. The experiment gives [7]

$$\nu_{Mu}^{exp} = 4\ 463\ 302.88(16)\ \text{kHz} \qquad (0.036\ \text{ppm}) \qquad (5)$$

in good agreement with the theoretical result:

$$\nu_{Mu}^{th} - \nu_{Mu}^{exp} = (0.5 \pm 1.5) \quad \text{kHz}.$$

3. HYPERFINE SPLITTING IN MUONIC HELIUM ATOM

If in the ordinary helium atom ^4He one electron is replaced by a negative muon then one obtains an atomic system ($^4He^{++}\mu^- e^-$) in which the hyperfine splitting is closely related to the hyperfine splitting in muonium. Indeed as the muon mass is about 200 times larger than the electron mass and the charge of the helium nucleus (α particle) Z=2, so the Bohr radius of a muon in helium is 400 times smaller than the Bohr radius of an electron in hydrogen. Thus the bound system ($^4He^{++}\mu^-$) in the ground state can be considered in the first approximation as a small pseudonucleus with the effective charge Z=1, mass $M=m_\alpha + m_\mu$ and the magnetic moment equal to that of a muon, since the spin and the magnetic moment of a helium nucleus are zero. Hence the hyperfine splitting in muonic helium is caused by the interaction of the electron and muon magnetic moments. The ground-state hyperfine interval in muonic He is equal in value, but opposite in sign to that in muonium.

More accurate calculations with regard for the pseudonucleus structure, the Coulomb interaction of the electron with the muon and helium nucleus, relativistic and radiative corrections give the value [6]

$$\nu_{\mu He}^{th} = 4\ 464.8(0.5)\ \text{MHz} \qquad (6)$$

in good agreement with the experimental result [15]

$$\nu_{\mu He}^{exp} = 4\ 465.004(29)\ \text{MHz} \qquad (6.5\ \text{ppm}) \ . \qquad (7)$$

The uncertainty in the theoretical value (6) arises from uncalculated radiative and relativistic corrections of higher order in the fine-structure constant.

4. MUONIUM LAMB SHIFT

Recently the Lamb shift in muonium [9,10] was measured for the first time. Presently there are two experimental results

$$S^{exp}_{Mu} = \begin{cases} 1070^{+12}_{-15} \pm 2 \text{ MHz} & \text{ref. [9]} \\ 1054 \pm 22 \text{ MHz} & \text{ref. [10]} \end{cases} \tag{8}$$

which should be compared with the theoretical value [10]

$$S^{th}_{Mu} = 1047.6(0.3) \text{ MHz}. \tag{9}$$

5. POSITRONIUM HYPERFINE SPLITTING

The study of the hydrogenic positronium atom ($e^- e^+$) is of special interest, as the nucleus recoil efects manifest themselves here to the greatest extent due to the equality of the electron and positron masses, and also owing to the existence of specific annihilation effects in the electron-positron system. At the same time the absence of the small parameter of the mass ratio brings about significant computational complications. The theory leads to the following expression [1,2,7] for the positronium hyperfine interval in the ground state

$$\nu^{th}_{Ps} = \alpha^2 R_\infty \Big[\frac{7}{6} - \frac{\alpha}{\pi}(\ln 2 + \frac{16}{9}) + \\ + \frac{5}{12}\alpha^2 \ln\alpha^{-1} + \alpha^2 K \Big] . \tag{10}$$

By now only some contributions to the coefficient K have been calculated, which make in total 0.232. As a result using the known values of the fundamental constants we obtain

$$\nu^{th}_{Ps} = (203402.51 \pm 10) \text{ MHz}. \tag{11}$$

The most precise experimental value [14] is

$$\nu^{exp}_{Ps} = 203389.10(.74) \text{ MHz} \qquad (3.6 \text{ ppm}) \tag{12}$$

6. POSITRONIUM AND HYDROGEN FINE STRUCTURE

The $1^3S_1 - 2^3S_1$ interval in the orthopositronium energy spectrum has been measured with high precision by Doppler-free two-photon laser spectroscopy [13]. The following value was found [13] for the 1S - 2S transition frequency:

$$f^{exp} = 1\ 233\ 607\ 185(15)\ \text{MHz} = \left[\frac{3}{8}\ R_\infty - 83545(15)\right]\ \text{MHz} \qquad (0.012\ \text{ppm}). \qquad (13)$$

Taking into account the relativistic and radiative corrections, including the Lamb shift, up to the terms of order $\alpha^3 R_\infty$ we obtain the theoretical value [7,13]

$$f^{th} = \left[\frac{3}{8}\ R_\infty - 83533(10)\right] \text{MHz} = 1233607197(11)\ \text{MHz} \qquad (14)$$

in good agreement with the experiment.

Recently the similar measurement has been carried out for atomic hydrogen in two independent experiments [16,17]. The following values are obtained for the 1S-2S transition frequency

$$f_H^{exp} = \begin{cases} 2\ 466\ 061\ 395.6(4.8)\ \text{MHz} & \text{ref. [16]} \\ 2\ 466\ 061\ 397\ (25)\ \text{MHz} & \text{ref. [17]} \end{cases} . \qquad (15)$$

In confronting these values with the theoretical predictions the quantity of primary interest is the 1S Lamb shift, which cannot be measured directly. Assuming that all other contributions including the 2S Lamb shift are known with required accuracy from the theory, we find the experimental values of the ground-state Lamb shift

$$S_H^{exp}\ (1S) = \begin{cases} 8184.8(5.4)\ \text{MHz} & \text{ref. [16]} \\ 8182\ (25)\ \text{MHz} & \text{ref. [17]} \end{cases} . \qquad (16)$$

The theory predicts

$$S_H^{th}\ (1S) = \begin{cases} 8173.25(8)\ \text{MHz} & \text{ref. [20]} \\ 8173.01(8)\ \text{MHz} & \text{ref. [21]} \end{cases} . \qquad (17)$$

The overall agreement of the theory with experiment is rather poor. The further improvement of experimental accuracy is foreseen.

7. POSITRONIUM DECAY RATE

Rather high accuracy (200 ppm) is attained now in measuring

the triplet decay rate λ_T of orthopositronium in the ground (1^3S_1) state [12]. Some recent experimental results are

$$\lambda_T^{exp} \text{ (in } \mu s^{-1}) = 7.0516(13); 7.051(5); 7.045(6); 7.056(7). \tag{18}$$

Bound-state QED and perturbation theory yield the following expression [7] for the dominating three-photon decay rate (in μs^{-1})

$$\lambda_3^{th} = \frac{2(\pi^2-9)}{9\pi\hbar} m_e \alpha^6 c^2 \left[1 - A\frac{\alpha}{\pi} - \frac{1}{3}\alpha^2 \ln \alpha^{-1} - B\alpha^2\right] = \\ = 7.03830(7), \tag{19}$$

$$A = \begin{cases} 10.266(11) \\ 10.282(3) \end{cases} .$$

The coefficient B is uncalculated as yet. Thus at present the most accurate experimental result [12] exceeds the theoretical prediction by 10 standard deviations. In order to resolve this discrepancy the coefficient B would have to be about 300.

8. ANOMALOUS MAGNETIC MOMENT OF THE ELECTRON

A highly accurate value of the anomalous magnetic moment (g-factor) of the electron has been obtained by studying the motion of a single electron in the electromagnetic Penning trap (geonium). The latest value of the electron anomaly in units of the Bohr magneton is [8]

$$a_e^{exp} = \frac{1}{2}(g_e - 2) = 1\,159\,652\,193(10) \times 10^{-12} \quad (0.0086 \text{ ppm}) \tag{20}$$

The uncertainty assigned to this value except the experimental error (4×10^{-12}) takes account of the theoretically estimated corrections due to the finite size of the trap.

The calculation of contributions from QED diagrams up to the fourth order in fine-structure constant yields the theoretical value [7].

$$a_e^{th} = \frac{\alpha}{2\pi} + c_2\left(\frac{\alpha}{\pi}\right)^2 + c_3\left(\frac{\alpha}{\pi}\right)^3 + c_4\left(\frac{\alpha}{\pi}\right)^4 + \ldots + \delta a =$$

$$= \begin{cases} 1\,159\,652\,306(111) \times 10^{-12} & \text{ref. [7]} \\ 1\,159\,652\,650(127) \times 10^{-12} & \text{ref. [22]}, \end{cases} \tag{21}$$

$$C_2 = -0.328\ 478\ 966; \quad C_3 = \begin{cases} 1.1765(13) & \text{ref. [7]} \\ 1.204(5) & \text{ref. [22]} \end{cases}$$

$$C_4 = -0.8(1.4); \quad \delta a\,(\text{hadrons} + \tau\ \text{lepton}) = 4.4\times10^{-12}\ .$$

The differing values of C_3 originate from dissimilar numerical contributions of the light-by-light scattering to the sixth-order anomaly. The uncertainty in the theoretical value of a_e combines the error from α with the error in numerical calculation of C_3 and C_4. The difference between the theoretical and experimental values is

$$a_e^{th} - a_e^{exp} = \begin{cases} 113(111)(10)\times10^{-12} & \text{ref. [7]} \\ 457(127)(10)\times10^{-12} & \text{ref. [22]} \end{cases} .$$

Before the completion of more accurate (preferably analytical) calculations of C_3 and C_4 it is premature to draw definite conclusions about the comparison of the theory and experiment.

9. FINE-STRUCTURE CONSTANT

So far in numerical evaluation of the theoretical predictions we have been using the value of the fine-structure constant found the Josephson effect measurements [21,24]

$$\alpha_J^{-1} = 137.035981(12) \qquad (0.088\ \text{ppm}) \quad . \qquad (22)$$

However it is possible to invert the procedure of confronting the theory and experiment. Assuming the validity of QED we can extract a value of the fine-structure constant from the combination of theory and results of precision measurements. In so doing we get from the data on hyperfine splitting in muonium

$$\alpha_{Mu}^{-1} = 137.036\ 003(26) \qquad (0.19\ \text{ppm}) \qquad (23)$$

and from the data on electron (g - 2) with the upper value of C_3 in eq. (21)

$$\alpha_{g-2}^{-1} = 137.0359942(89) \qquad (0.065\ \text{ppm}) \quad . \qquad (24)$$

Both these values lie somewhat higher than the Josephson one. However it should be kept in mind that the Josephson value of α is obtained from the formula [24] which contains some other fundamental constants, specifically the proton gyromagnetic ratio γ_p. The latter introduces the bulk of the uncertainty in α . Some recent measurements of γ_p disagree with each other.

The value of the fine-structure constant or rather the ratio e^2/h is directly available from the measurements of the quantum Hall effect:

$$\alpha_{QH}^{-1} = \begin{cases} 137.035\,994(13) & (0.095 \text{ ppm}) \quad \text{ref. [24]} \\ 137.036\,012(11) & (0.080 \text{ ppm}) \quad \text{ref. [23]} \end{cases} \tag{25}$$

It should be mentioned however that the accuracy of the formula for the quantized Hall resistance is not established rigorously owing to the incompleteness of the theory.

The recommended (1986) value of α is [24]

$$\alpha^{-1} = 137.0359895(61) \qquad (0.045 \text{ ppm}) . \tag{26}$$

10. RYDBERG CONSTANT

Lately in several experiments [18,19] the Rydberg constant R_∞ has been determined by measuring the wavelengths of the Balmer transitions in hydrogen and deuterium with utmost precision. The values obtained agree quite well with each other

$$R_\infty = 10\,973\,731 + \begin{cases} 0.569(6) \text{ m}^{-1} & \text{ref. [18]} \\ 0.569(7) \text{ m}^{-1} & \text{ref. [19]} \\ 0.573(3) \text{ m}^{-1} & \text{ref. [19]} \end{cases} , \tag{27}$$

but disagree with the recommended (1986) value [24]

$$R_\infty = 10973731.534(13) \text{ m}^{-1} \quad (0.0012 \text{ ppm}) . \tag{28}$$

In addition fine-structure splittings and isotope shifts have been measured in good agreement with QED predictions [20] (see the Table).

The measurements if the 1S-2S interval mentioned above in Section 6 can also be used to infer the Rydberg constant

$$R_\infty = 10973731 + \begin{cases} 0.492(21) \text{ m}^{-1} & \text{ref. [16]} \\ 0.50(11) \text{ m}^{-1} & \text{ref. [17]} \end{cases} \tag{29}$$

The upper value strongly disagrees with all other values of R_∞ quoted in this Section.

TABLE. Fine-structure splittings and isotope shift.

		Theory (MHz)	Experiment (MHz)
Fine structure			
$3P_{1/2} - 3P_{3/2}$	H	3250.1	3250.3(4)
	D	3251.0	3250.9(4)
$4P_{1/2} - 4P_{3/2}$	H	1371.13	1370.9(3)
	D	1371.50	1371.8(3)
Isotope shift H - D			
$2S_{1/2} - 3P_{1/2}$		124261.1	124261.8(3)
$2S_{1/2} - 3P_{3/2}$		124262.0	124262.4(3)
$2S_{1/2} - 4P_{1/2}$		167752.8	167752.4(3)
$2S_{1/2} - 4P_{3/2}$		167753.2	167753.3(3)

REFERENCES

1. R.N. Faustov, Izmeritel´naya Tekhnika No.7, p.22 (1984), English translation Measurement Techniques 27, 595 (1984)

2. N.A. Bojkova, Yu.N. Tyukhtyaev, and R.N. Faustov, in Proceedings of the International Seminar on High Energy Physics and Quantum Field Theory vol.I, p.116, IHEP, Protvino, 1983, (in Russian).
N.E. Nyun´ko, Yu.N. Tyukhtyaev, and R.N. Faustov, ibid. p.104, (in Russian).

3. A.P.Martynenko and R.N.Faustov, Yadernaya Fizika 45, 770 (1987).

4. V.V. Starshenko and R.N. Faustov, JINR Rapid Communications No.7-85, p.39, Dubna, 1985.
M.I. Eides and V.A. Shelyuto, Phys.Lett. B146, 241 (1984)

5. M.I. Eides, S.G. Karshenboim, and V.A. shelyuto, Phys. Lett. B177, 425 (1986).

6. M.Ya.Amusia, M.Yu. Kuchiev, and V.L. Yakhontov, PTI preprint 883, Leningrad, 1983.

7. T. Kinoshita and J.Sapirstein, in Atomic Physics 9, p.38, World Scientific, Singapore, 1984.

8. R.S. Van Dyck, Jr. et al., ibid., p.53.

9. C.J. Oram, ibid., p.75.

10. A. Badertscher et al., ibid., p.83.

11. F.G. Mariam et al., Phys.Rev.Lett. 49, 993 (1982).

12. C.I. Westbrook et al., Phys.Rev.Lett. 58, 1328 (1987).

13. S. Chu et al., Phys.Rev.Lett., 52, 1689 (1984).

14. M.W. Ritter et al., Phys.Rev. A30, 1331 (1984).

15. C.J. Gardner et al., Phys.Rev.Lett. 48, 1168 (1982).

16. E.A. Hildum et al., Phys.Rev.Lett. 56, 576 (1986).

17. J.R. Barr et al., Phys.Rev.Lett. 56, 580 (1986).

18. F. Biraben et al., in Proceedings of CLEO´86-IQEC´86, San Francisko, California, 1986.

19. Ping Zhao et al., Phys.Rev. A34, 5138 (1986); Phys.Rev. Lett. 58, 1293 (1987).

20. G.W. Erickson, J.Phys.Chem.Ref.Data 6, 831 (1977).

21. J. Sapirstein, Phys.Rev.Lett. 47, 1723 (1981); in AIP Conference Proceedings No.136, p.100, N.-Y., 1985.

22. M.A. Samuel, Phys.Rev.Lett. 57, 3133 (1986).

23. G.J. Sloggett et al., Phys.Rev.Lett. 57, 3237 (1986).

24. E.R.Cohen and B.N.Taylor, ´The 1986 Adjustment of the Fundamental Physical Constants´, CODATA Bulletin No.63, Nov. 1986.
B.N. Taylor, J.Res.Natl.Bur.Stand.U.S. 90, 91 (1985).

SEARCHING FOR THE SOURCE OF THE FIFTH FORCE

Carrick Talmadge and Ephraim Fischbach
Physics Department
Purdue University
West Lafayette, Indiana 47907 USA

ABSTRACT. We discuss the formalism for studying the constraints on a possible "fifth force" implied by existing data. It is shown that recent results from composition-dependent experiments are consistent with a new coupling, in which the fundamental charge is predominantly *either* nuclear isospin *or* baryon number. We also show, however, that neither of these solutions by itself can account for the geophysical results, but that a multicomponent model of the fifth force with two terms may be capable of accounting for the existing data.

I. INTRODUCTION

A large number of experiments are presently underway to search for a new intermediate-range force, usually referred to as the "fifth force".[1] (A list of the currently known experiments is given in the Appendix.) Results from those of Thieberger,[2] Stubbs, *et al.*,[3] Adelberger, *et al.*,[4] Niebauer, *et al.*,[5] and most recently by Boynton, *et al.*,[6] have already appeared, and others are expected in the near future. These data, when combined with the original results of Eötvös, Pekár, and Fekete (EPF),[7] and with the geophysical analyses of Stacey, *et al.*,[8] and Hsui,[9] comprise a sufficiently large body of information from which we can hope to find answers to some of the questions which presently confront us. Evidently our first question must be whether the "fifth force" in fact exists, and if so to what attribute(s) of

V. De Sabbata and V. N. Melnikov (eds.),
Gravitational Measurements, Fundamental Metrology and Constants, 143–179.

matter does it couple. These questions are intimately related to each other, since the evidence for the fifth force would be much more compelling if there existed a simple phenomenological picture which could account for all of the available data.

The recent paper by Boynton, *et al.*,[6] represents a potentially significant step forward in our understanding of the phenomena related to the fifth force. In the first place, their experiment, which is the most sensitive to date, sees a signal compatible with a new intermediate-range composition-dependent force, albeit at a level below that expected from a coupling to baryon number or hypercharge. Secondly, their accompanying phenomenological analysis demonstrates that the recent experimental results of Refs. 2-6 may be compatible with one another, if the dominant coupling of the postulated fifth force is to nuclear isospin, $I_z = N - Z$, rather than to baryon number $B = N + Z$, as had originally been proposed.[1] Here N and Z denote the numbers of protons and neutrons which, along with the electron lepton number L and the mass M, are the only known macroscopic quantum numbers that the coupling can depend on. The analysis of Boynton, *et al.*,[6] leaves open a number of questions, however, most notably whether a dominant coupling to I_z would be compatible with the geophysical data,[8,9] or with laboratory determinations of the Newtonian constant of gravitation G.[10] In addition there is the question of the extent to which such a picture would be compatible with the original EPF data. These issues were not addressed by Boynton, *et al.*, because they involve considerations which go beyond the immediate details of their model. Specifically, both the geophysical data and the laboratory determinations of G are sensitive to possible composition-independent forces, which the experiments of Refs. 2-6 are not. Likewise the comparison of their model to the EPF experiment is complicated by our still incomplete knowledge of the local environment of that experiment.

The object of the present paper is to examine in a systematic way the constraints implied by all the existing data, in an effort to establish whether a simple phenomenological description of these data exists. In this work, which is an extension of our previous efforts to formulate a multicomponent model of the fifth force,[11] we demonstrate the following:

1) The recent results from the *composition-dependent* experiments in Refs. 2-6 can be explained in terms of a single new Yukawa coupling, in which the fundamental charge is predominantly *either* I_z *or* B. The existence of the "I_z-solution" was first pointed out by Boynton, *et al.*,[6] and the "B-solution" is presented here for the first time.

2) When results from the *composition-independent* experiments are included, par-

ticularly the geophysical analysis of Stacey, *et al.*,[8] then a single coupling whose strength is proportional to either I_z or B *cannot* account all of the data.

3) A generalized multicomponent model with couplings to *both* I_z *and* B may be capable of accommodating all the existing data.
4) Specific experimental tests, which we discuss below, can discriminate among the various scenarios.

II. DESCRIPTION OF THE EXPERIMENTAL CONSTRAINTS

In this section we review the formalism through which the results of a given experiment are used to constrain the parameters which describe the fifth force in a particular model. The simplest model for a new intermediate-range force is one in which the Newtonian gravitational potential $V_N(r)$ for two point masses i and j is modified by the introduction of an additional term $\Delta V(r)$ so that

$$V(r) = V_N(r) + \Delta V(r) = -\frac{G_\infty m_i m_j}{r}\left(1 + \alpha e^{-r/\lambda}\right). \tag{2.1}$$

Here G_∞ is the Newtonian gravitational constant, m_i and m_j are the masses of the particles, r is the distance between them, λ is the range of the new interaction, and α is its strength relative to gravity. If $\Delta V(r)$ is assumed to arise from a purely gravitational coupling, then α and λ must be fixed constants (*i.e.*, the same for all i, j) so as to preserve the "universal" nature of the gravitational interaction.[12] As we observe in Ref. 1, however, $\Delta V(r)$ could also arise from an intermediate-range coupling of matter to a new (non-gravitational) field whose source was baryon number B. In this case α would no longer be a universal constant, but would depend on i and j:

$$\begin{aligned} \alpha \longrightarrow \alpha_{ij} &= -\left(\frac{B_i}{\mu_i}\right)\left(\frac{B_j}{\mu_j}\right)\xi, \\ \xi = \frac{f^2}{G_\infty m_{\mathrm{H}}^2}; &\qquad \mu_{i,j} = m_{i,j}/m_{\mathrm{H}}. \end{aligned} \tag{2.2}$$

Here f is the strength of the coupling to baryon number (or hypercharge), and $m_{\mathrm{H}} = m({}_1\mathrm{H}^1)$. In an experiment which did not specifically look for composition dependent effects, α_{ij} would seem to be approximately constant, since $B/\mu \cong 1$ for all materials. However, the object of the Eötvös experiment, and the recent experiments in Refs. 2-6, is to look for precisely the kinds of effects which would be expected if α_{ij} were not constant, and hence such experiments allow us to focus uniquely on non-gravitational intermediate-range forces.

The generalization of (2.1) represented by (2.2) can be extended further in two significant ways. To start with we can introduce charges other than B as the sources of the new fields. In addition we can let the fields coupling to these charges have different masses, so as to produce forces with correspondingly different ranges. In such a model the acceleration $\vec{a}_i$ of a point mass μ_j in the presence of a point mass μ_i is given by[11,12]

$$\vec{a}_j = \vec{g}_N \left\{ 1 + \sum_{k=1}^{N_\alpha} \alpha_k (1 + r/\lambda_k) e^{-r/\lambda_k} - \sum_{\ell=1}^{N_\xi} \xi_\ell q_{\ell i} q_{\ell j} (1 + r/\lambda_\ell) e^{-r/\lambda_\ell} \right\}. \qquad (2.3)$$

Here α_k (not to be confused with α in (2.2)) and ξ_ℓ are a set of positive constants which respectively describe the universal and non-universal (*i.e.*, composition-dependent) contributions, and $\vec{g}_N$ is the Newtonian acceleration. $q_{\ell i}$ and $q_{\ell j}$ are the normalized charges, which for couplings to baryon number and lepton number are given respectively by

$$q_\ell = \frac{B}{\mu} \equiv q_B; \qquad q_\ell = \frac{L}{\mu} \equiv q_L. \qquad (2.4)$$

As we will discuss shortly, the number of charges and their dependences on B and L is one of the key experimental and theoretical questions confronting us at present. Returning to (2.3) we note that the composition-independent terms proportional to α_k are intrinsically positive, while the composition-dependent terms proportional to ξ_ℓ enter with an overall negative sign. This distinction between the two sets of terms reflects inferences drawn from simple theories: The exchange of scalar ($J = 0$) or tensor ($J = 2$) fields typically leads to an attractive force whose source is mass-energy, and which is thus composition-independent. By contrast, the exchange of vector ($J = 1$) fields leads to composition-dependent forces, which can be either attractive or repulsive depending on i and j. One of the important constraints on phenomenological descriptions of the fifth force comes from the geophysical evidence that the non-Newtonian contribution is *repulsive* for at least some macroscopic test objects. We know of no simple theory in which such a repulsive interaction can arise from a coupling to mass-energy, and hence the only remaining attributes of bulk matter to which a new field can couple are N, Z, and L. But a coupling to any combination of these charges will lead to an anomalous composition-dependent acceleration of the type seen in the EPF data. Thus the geophysical and EPF data are mutually consistent, and point to a new coupling whose source is some linear combination of N, Z, and L.

The problem is now to find the appropriate combination. It should be noted at the outset, that it is not necessarily the case that the magnitude of the fifth force that a source exerts on a test mass is determined by the global values of the respective charges N, Z, and L. In particular, since electrons are free to move in response to an external field, the fifth force may be partially "shielded" in analogy to an electromagnetic field.[11] If this were the case, the forces objects exert on one another would be determined by detailed dynamical considerations which would require knowing more than N, Z, and L. The possibility of shielding arising from a coupling to L has been discussed previously,[11,13] along with specific experimental tests to check for this possibility. However, in the absence of any compelling evidence for such a phenomenon, we will adopt the traditional view that the strength of the fifth force depends only on N, Z, and L. For ordinary neutral bulk matter, $Z = L$, and hence we can express the vector charge Q_1 which is the source of the fifth force in the form

$$Q_1 = f(c_B B + c_L L), \tag{2.5}$$

where f is a fundamental constant (analogous to the electric charge e), and c_B and c_L are dimensionless constants. The EPF data suggest that the dominant contribution in that experiment comes from B, with $c_L/c_B = (4 \pm 4) \times 10^{-3}$. If this were always the case, we could simply define $c_B \equiv 1$ and write

$$Q_1/f = B + c_L L; \qquad c_L \ll 1. \tag{2.6}$$

However, the recent work of Boynton, *et al.*,[6] indicates that in order to achieve consistency among the experiments in Refs. 2-6, it may be necessary to take the dominant charge to be $Q_1/f = I_z = N - Z = B - 2L$. As we show below, consistency among these experiments can also be achieved if Q_1/f is dominantly B, as originally suggested.[1] This solution requires that λ be sufficiently large ($\lambda \gtrsim 200\,\text{m}$), and also that the contribution from L although small, not be entirely negligible. It follows that the combination of Boynton, *et al.*, and the present work suggest that the most natural charges to describe the fifth force are $B = N + Z$ and $I_z = N - Z = B - 2L$. We thus choose to parametrize Q_1/f in the form

$$Q_1/f = \cos\theta_5 B + \sin\theta_5 I_z; \qquad -\frac{\pi}{2} \leq \theta_5 \leq +\frac{\pi}{2}, \tag{2.7}$$

which allows for the relative admixture of B and I_z to have any value from $-\infty$ to $+\infty$. Eq. (2.7) can be compared to other parametrizations that have been used previously:

$$Q_1/f = \beta B + (1-\beta) I_z, \qquad \text{[Boynton, et al., Ref. 6]}, \tag{2.8a}$$

$$Q_1/f = \cos\theta_5 B + \sin\theta_5 L, \qquad \text{[Adelberger, et al., Ref. 4]}, \tag{2.8b}$$

$$Q_1/f = \cos\theta N + \sin\theta Z, \qquad \text{[Glashow, Ref. 14]}, \tag{2.8c}$$

We see that our parametrization is similar to that of Boynton, *et al.*, with

$$\tan\theta_5 = (1-\beta)/\beta. \tag{2.9}$$

However, the parametrization of Boynton, *et al.*, leads to the unphysical limit that $Q_1/f \to \infty$ as the coupling approaches a pure L or Z coupling. Since the parametrization in Eq. (2.7) does not suffer from this normalization problem, we adopt this in lieu of that of Boynton, *et al.*

Returning to (2.3) we can use (2.7) to write the normalized charge q_ℓ in the form

$$q_\ell = q_\ell(\theta_{5\ell}) = \cos\theta_{5\ell}\frac{B}{\mu} + \sin\theta_{5\ell}\frac{I_z}{\mu}, \tag{2.10}$$

where the notation $\theta_{5\ell}$ allows for the possibility that each of the composition-dependent components has not only a characteristic range λ_ℓ, but a characteristic mixing angle $\theta_{5\ell}$ as well. From (2.3) the acceleration difference $\Delta a_{jj'}$ between two test masses j and j' in the presence of a point source is given by

$$\Delta\vec{a}_{jj'} = \vec{g}_N \left\{ -\sum_{\ell=1}^{N_\xi} \xi_\ell\, q_\ell(\theta_{5\ell})\, \left(\Delta q_\ell(\theta_{5\ell})\right)_{jj'} \left[(1+r/\lambda_\ell)e^{-r/\lambda_\ell}\right] \right\}, \tag{2.11}$$

$$[\Delta q_\ell(\theta_{5\ell})]_{jj'} = q_{\ell j}(\theta_{5\ell}) - q_{\ell j'}(\theta_{5\ell}).$$

For an extended source Eq. (2.11) can be suitably generalized to take account of the fact that $\Delta\vec{a}_{jj'}$ and $\vec{g}_N$ are not necessarily parallel. It is evident from (2.11) that whatever the source is, the component $\Delta a_{jj'}$ in the direction being detected has the form

$$\Delta a_{jj'} \equiv \Delta a = \sum_{\ell=1}^{N_\xi} \xi_\ell q_s(\theta_{5\ell}) \Delta q_d(\theta_{5\ell}) \mathcal{F}(\lambda_\ell;\theta_{5\ell}), \tag{2.12}$$

where q_s denotes the source charge, $\Delta q_d \equiv (\Delta q)_{jj'}$ is the difference in charges of the detector, and where $\mathcal{F}(\lambda_\ell;\theta_{5\ell})$ is the appropriate generalization of the quantity in [ ] in Eq. (2.11). $\mathcal{F}(\lambda_\ell;\theta_{5\ell})$ is implicitly a function of $\theta_{5\ell}$ since the source of the fifth force field need not be chemically homogeneous. We can illustrate the notation in (2.12) by considering as an example the experiment of Boynton, *et al.*, which measures the acceleration difference Δa between Be and Al in a direction perpendicular to the face of an (almost) vertical cliff. Assuming that the experiment takes place at the base of the cliff, a distance x from the face, then for a single component theory ($N_\xi = 1$) in (2.12)

$$\Delta a = \xi q_s(\theta_5)\Delta q_d(\theta_5)\mathcal{F}(\lambda;\theta_5), \tag{2.13a}$$

$$\mathcal{F}(\lambda;\theta_5) = \pi G_\infty \bar{\rho}\lambda e^{-x/\lambda}, \tag{2.13b}$$

where $\bar{\rho} \cong 2700\,\mathrm{kg\,m^{-3}}$ is the average local density of the cliff (assumed to be infinitely high).

Since the result of any experiment is a determination of Δa for a given source and pair of test masses, Eq. (2.13) implies a constraint among the unknown theoretical parameters ξ, θ_5, and λ, where we assume for the time being that $N_\xi = 1$. This constraint can be exhibited in the form of a curve or pair of curves showing the allowed values of ξ as a function of θ_5 or λ for each experiment, and we can then ask whether the existing experiments point to a common set of values for these parameters. As we have already noted, Boynton, *et al.*,[6] have found a set of parameters which can account for at least part of the existing data, and we have found another. In what follows we discuss in general terms how such solutions come about.

Returning to Eq. (2.13a) we write

$$\xi q_s(\theta_5)\Delta q_d(\theta_5) = \frac{\Delta a}{\mathcal{F}(\lambda;\theta_5)} \equiv k(\lambda;\theta_5). \tag{2.14}$$

The dimensionless quantity k is a measure of the intrinsic sensitivity of an experiment: Two experiments carried out at the same site will have the same $\mathcal{F}$, and hence the one which can detect the smaller acceleration difference Δa will correspond to the smaller value of k. Given an experimentally-determined value of k, one can derive via (2.14) a constraint on ξ in the form

$$\xi = \frac{k(\lambda;\theta_5)}{q_s(\theta_5)\Delta q_d(\theta_5)}. \tag{2.15}$$

Eq. (2.15) is of fundamental importance since it directly exhibits the constraint implied by the data on the parameter ξ, which if it were nonzero would mean that there was in fact a fifth force. We see from (2.14) or (2.15) that at those values of θ_5 where either $q_s(\theta_5)$ and/or $\Delta q_d(\theta_5)$ vanishes, an experiment loses its sensitivity to the presence of a fifth force, since even an arbitrarily large value of ξ would then be compatible with the measured k. The two poles in θ_5 where ξ blows up are shown in Fig. 2 of Adelberger, *et al.*,[4] for the experiments of Refs. 2-4. It follows that from the preceding discussion that even though an experiment has high intrinsic sensitivity (small k), it can end up setting relatively weak constraints on ξ, if it happens to choose a source for which $q_s \cong 0$, and/or a detector for which $\Delta q_d \cong 0$. Although it might appear at first sight that the vanishing of q_s or Δq_d for a given experiment would be accidental, this is not necessarily the case. As was noted by Talmadge, *et al.*,[11] and by Adelberger, *et al.*,[4] for values of $\sin\theta_5$ near unity in (2.7), the coupling is predominantly to I_z, and for most terrestrial sources $\langle I_z \rangle \approx 0$. This is true in particular for SiO_2 and $CaCO_3$, which are among the most common constituents of rocks. Hence at different sites q_s may differ from zero only by virtue of relatively small quantities of other minerals present in the rock. It follows that q_s is quite sensitive to the presence of substances (including water) with $\langle I_z \rangle \neq 0$, and that as a consequence q_s can vary considerably from one location to another. One of the significant contributions of Boynton, *et al.*,[6] was to explore throughly the possibility of a coupling to I_z by evaluating $q_{Is} = (I_z/\mu)_s$ for the different sites at which the recent composition-dependent tests had been carried out. Their results, which are shown in the appropriate column of Table I, made it clear that the failure of the experiments of Stubbs, *et al.*,[3] and Adelberger, *et al.*,[4] to see an effect could be attributed to the near vanishing of q_I at the University of Washington Nuclear Physics Laboratory (NPL), where both of these experiments were carried out.

In contrast to the solution of Boynton, *et al.*, in which compatibility among the composition-dependent experiments is largely a consequence of the vanishing of $q_s(\theta_5)$ if $\sin\theta_5 \approx 1$, the second solution that we have found depends on the near-vanishing of $\Delta q_d(\theta_5)$ for various pairs of materials, which can happen if $\sin\theta_5 \approx 0$. This solution also depends on the relatively small strength of the source at NPL as λ becomes appreciably larger than 100 m. The vanishing of Δq_d may again appear to be coincidental, and in fact the detailed implementation of this scenario requires that θ_5 fall into a relatively narrow range, as we discuss in detail below. However, for $\sin\theta_5 \approx 0$, $\Delta q_d(\theta_5)$ nearly vanishes for many pairs of materials as we see from Fig. 1.

We have calculated $\Delta q_d(\theta_5)$ for the $92 \times 91/2 = 4,186$ pairs that can be formed using the first 92 elements of the Periodic Table, and found the value of θ_5 for which each individual Δq_d vanishes. Shown in Fig. 1 is a histogram giving the number of pairs which vanish in a given interval of θ_5 as a function of θ_5. It is evident from

Table I: Experimental parameters for the recent tests for a new composition-dependent force. Values for $q_{Is} = (I_z/\mu)_s$ are those values calculated by Boynton.[15]

Experiment	q_{Bs}	q_{Is}	Δq_B	Δq_I	$10^{11}\Delta a$ $(\mathrm{m\,s^{-2}})$
Adelberger, *et al.*, [Al-Be]	1.008	+0.000 ±0.003	+0.00205	−0.0745	+0.44 ±1.17
Boynton, *et al.* [Al-Be]	1.008	+0.007 ±0.001	+0.00205	−0.0745	−2.53 ±0.71
Niebauer, *et al.* [U-Cu]	1.008	+0.000 ±0.003	−0.00134	+0.1395	+130 ±500
Stubbs, *et al.* [Cu-Be]	1.008	+0.000 ±0.003	+0.00249	−0.0228	+0.93 ±1.03
Thieberger [Cu-H_2O]	1.008	+0.012 ±0.001	+0.00172	+0.2007	+85 ±28

the figure that many pairs vanish near $\sin\theta_5 \cong 0$, although one must hasten to add that the difference between $\sin\theta_5 = 0$ and $\sin\theta_5 \cong 0$ can be quite important, for reasons to be discussed below. It is easy to see why Δq_d vanishes for many pairs near $\sin\theta_5 = 0$. From the definition of Δq_d we have

$$\Delta q_d = 0 = \cos\theta_5 \Delta\left(\frac{B}{\mu}\right) + \sin\theta_5 \Delta\left(\frac{I_z}{\mu}\right) \Longrightarrow \tan\theta_5 = -\frac{\Delta(B/\mu)}{\Delta(I_z/\mu)}. \tag{2.16}$$

For most materials $\Delta(I_z/\mu) \gg \Delta(B/\mu)$, from which it follows that $|\sin\theta_5| \cong |\tan\theta_5| \ll 1$. As a side comment we observe that the envelope of the histogram has an approximately Gaussian or Lorentzian shape. If approximated as a Gaussian, it has a maximum at $\theta_5 = 0.320°$, and a full-width at half maximum of $\sigma_\theta = 0.707°$.

Returning to Table I we see that the recent composition-dependent experiments quoted in Refs. 2-6 differ considerably not only in the magnitudes of the acceleration differences, but also in the magnitudes and signs of the normalized charges $q_B = B/\mu$ and $q_I = I_z/\mu$, for the respective sources and detectors. The implication of the preceding discussion is that these differences are related: For appropriate values of θ_5 either $q_s(\theta_5)$ or $\Delta q_d(\theta_5)$ can depend sensitively on the source or detector, and in either case large variations in the measured acceleration difference can result. In the next section we present the specific details of these two scenarios.

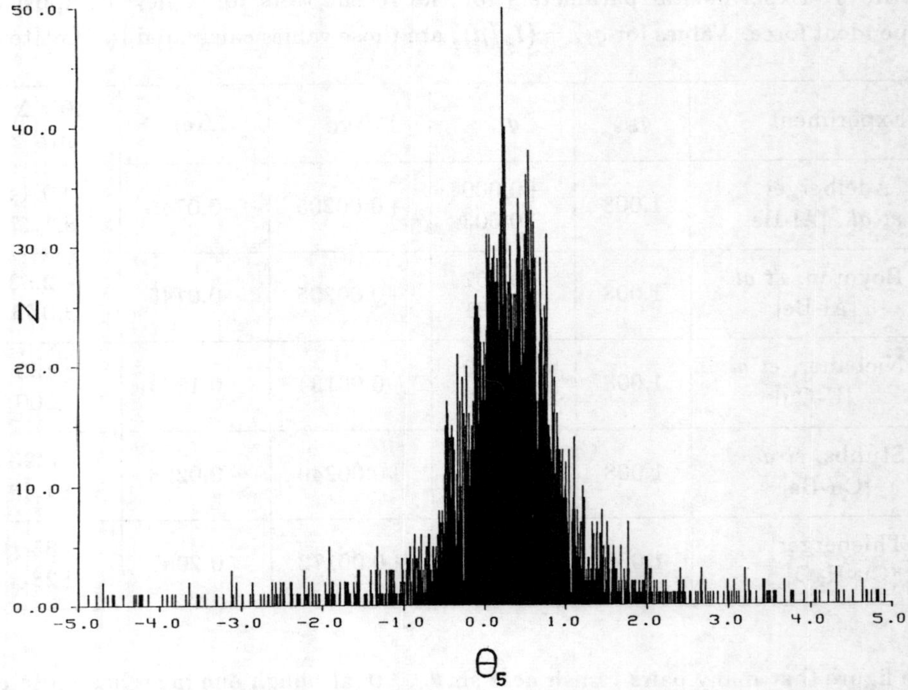

Figure 1: Histogram of values of θ_5 (in degrees) at which $\Delta q_d(\theta_5) = 0$ for all of the natural elements.

III. INPUT DATA

a) Discussion of Composition-Dependent Tests

We are interested in the question of whether the available experimental data are consistent with the existence of a single non-Newtonian coupling characterized by a strength ξ, range λ, and mixing angle θ_5 in Eq. (2.7). Such a coupling may exist if a set of values for these parameters can be found which satisfy Eq. (2.15), when the appropriate experimental quantities are inserted. We discuss in this section exactly what data we use as input.

Returning to Eq. (2.15) we write this constraint equation in a more explicit form, so as to exhibit the errors on the various experimentally determined quantities:

$$\xi = \frac{k \pm n\delta k}{\Delta q_d(\theta_5)[\cos\theta_5 q_{Bs} + \sin\theta_5 (q_{Is} \pm \delta q_{Is})]}, \tag{3.1}$$

where $n = 1, 2, \ldots$ leads to a 1σ, 2σ, ... bound on ξ. The error on $k = \Delta a/\mathcal{F}$ is dominated by the error in Δa, and since this is Gaussian we can take δk to be Gaussian as well. By contrast, the error δq_{Is} on q_{Is} represents an estimate of the maximum likely uncertainty in q_{Is}, and is not Gaussian. We note in passing that since $(B/\mu)_s = q_{Bs} \cong 1$ for all substances, any uncertainty in the chemical composition of the source produces a negligible error δq_{Bs} in q_{Bs}. To handle the uncertainty in q_{Is} we write

$$q_{Is} \pm \delta q_{Is} \equiv q_{Is} + \kappa \delta q_{Is}, \qquad -1 \leq \kappa \leq +1, \tag{3.2}$$

so that ξ in (3.1) is now a function of κ. The constraint curves are obtained by finding the maximum and minimum values of ξ for each θ_5 (assuming λ is fixed), as κ and δk range over their allowed values. Evidently the κ which extremizes ξ is that for which the expression in [ ] in (3.1) vanishes, namely

$$\kappa(\theta_5) = \frac{-q_s(\theta_5)}{\sin\theta_5 \delta q_{Is}} \equiv \bar{\kappa}. \tag{3.3}$$

Whenever the denominator in (3.1) vanishes ξ becomes infinite, and there is no bound from experiment. Since this is a concern only if $\bar{\kappa}$ falls in the physically allowed range $-1 \leq \bar{\kappa} \leq +1$, we can deal with this possibility by writing:

$$\kappa(\theta_5) = \begin{cases} -1 & \text{if } \bar{\kappa}(\theta_5) < -1; \\ \bar{\kappa}(\theta_5) & \text{if } -1 \leq \bar{\kappa}(\theta_5) \leq +1; \\ +1 & \text{if } +1 < \bar{\kappa}(\theta_5). \end{cases} \tag{3.4}$$

This prescription maximizes the possible values of ξ. Stated another way, the values of κ which give the upper and lower bounds on ξ that we are searching for are either $+1$ or -1, except if $\bar{\kappa}$ happens to lie in the range -1 to $+1$. In this case $\bar{\kappa}$ gives the extremum of ξ, but since this extremum is now infinite, there is no bound under these circumstances on ξ.

It now remains to evaluate $k = \Delta a/\mathcal{F}$ for each of the relevant experiments. In analyzing the composition-dependent tests in Refs. 2-6, we can simply use the quoted results for Δa, combined with values of $\mathcal{F}$ obtained by knowing the local topography. The composition-independent tests will be handled separately as we discuss below. The input data to the calculation of k is the actual acceleration difference Δa quoted by the various experiments, and since this differs from the minimum detectable acceleration difference $(\Delta a)_{\min}$ which characterizes the intrinsic sensitivity of the apparatus, we list the latter separately in Table II below.

Table II: Comparison of the sensitivities of the various composition-dependent tests.

Experiment	Δa_{min} ($m\,s^{-2}$)
Adelberger, *et al.*	1.2×10^{-11}
Boynton, *et al.*	7.1×10^{-12}
Eötvös, Pekár, and Fekete	1.3×10^{-11}
Niebauer, McHugh, and Faller	5.0×10^{-9}
Stubbs, *et al.*	1.0×10^{-11}
Thieberger	3.6×10^{-11}

In the case of Thieberger[2] we list the sensitivity inferred from his measurement of the velocity of the Cu ball. This is because the error he quotes for the H_2O-Cu acceleration difference [$\Delta a(H_2O\text{-}Cu) = (8.5 \pm 1.3) \times 10^{-10}\,m\,s^{-2}$] is predominantly due to the absolute calibration of his apparatus, which has no direct bearing on how small a signal he can measure. We note from Table II that the original EPF experiment is in fact comparable in sensitivity to the best modern experiments. Note also that the claimed sensitivity of the 1935 repetition of the Eötvös experiment by Renner[16] [$\Delta a = 3.4 \times 10^{-12}\,m\,s^{-2}$] would make it substantially more sensitive than the best of the modern composition-dependence tests. We have noted elsewhere[1] that the Renner results are internally inconsistent, for which reason we have not incorporated his data into our various analyses.

We return now and give the values of k for each of the relevant composition-dependent experiments. For present purposes the EPF experiment is not being included, because of lingering uncertainties in the local topography at the time of the experiment, which make it somewhat difficult to calculate $\mathcal{F}$.

i) Evaluation of k for the Thieberger Experiment. For Δa we use

$$\Delta a(H_2O\text{-}Cu) = 8.5(1 \pm 0.33) \times 10^{-10}\,m\,s^{-2}. \tag{3.5}$$

The central value is taken directly from Thieberger's paper, but for the error we use the 33% error quoted for the final result [see caption to his Fig. 3], which includes additional uncertainties due to the determination of $\mathcal{F}$, and due to the minimum velocity measured at his BNL site. We can obtain an estimate of $\mathcal{F}$ by assuming that Thieberger's experiment takes place at the top of an infinite homogeneous vertical cliff, so that we can use Eq. (2.13b) with $\bar{\rho} = 2750\,kg\,m^{-3}$, and $x/\lambda \cong 0$. We then find

$$\mathcal{F}(\lambda = 100\,\mathrm{m}; \theta_5) \approx 5.8 \times 10^{-5}\,\mathrm{m\,s^{-2}}. \tag{3.6}$$

A more careful integration over the cliff taking into account its finite height gives

$$\mathcal{F}(\lambda = 100\,\mathrm{m}; \theta_5) \cong 5.1 \times 10^{-5}\,\mathrm{m\,s^{-2}}. \tag{3.7}$$

In either case, the assumption that the cliff is chemically homogeneous means that $\mathcal{F}$ does not in fact depend on θ_5. Combining (3.5) and (3.7) we find

$$k(\lambda = 100\,\mathrm{m}) \cong (+1.7 \pm 0.6) \times 10^{-5}. \tag{3.8}$$

Alternatively we can evaluate k directly from the left-hand side of Eq. (2.14), using Thieberger's quoted value,

$$\xi\lambda = (1.2 \pm 0.4)\,\mathrm{m}, \tag{3.9}$$

which implies that at $\lambda = 100\,\mathrm{m}$, $\xi = (1.2 \pm 0.4) \times 10^{-2}$. Since $k(\lambda; \theta_5)$ is independent of θ_5 for the presumed homogeneous cliff, we can evaluate k for any θ_5 we choose, in particular for $\theta_5 = 0$ which corresponds to a pure B coupling. Using $q_{Bs} \cong 1$ and $\Delta q_{Bd} = -1.72 \times 10^{-3}$, we find

$$k(\lambda = 100\,\mathrm{m}) \cong (+2.1 \pm 0.7) \times 10^{-5}, \tag{3.10}$$

which is in reasonable agreement with the value found in (3.8). Presumably the latter result is more reliable, since it is directly derivable from Thieberger's quoted results, and this is the value we will use in the present analysis.

ii) Evaluation of k for the NPL Experiments. In the NPL experiments of Stubbs, *et al.*,[3] and Adelberger, *et al.*,[4] the acceleration difference Δa is related to the maximum shift $\Delta\theta_{\max}$ in the equilibrium orientation of the suspended test masses, via the equations

$$2m\Delta a(L/2) = \tau_{\max} = K\Delta\theta_{\max}. \tag{3.11}$$

Here $\tau_{\max}$ is the maximum torque, $L = 3.90\,\mathrm{cm}$ is the length of one of the sides of the square from whose corners the test masses are suspended,and $m = 10.04\,\mathrm{g}$ is the mass of each of the 4 masses (*i.e.*, 2 each of Be and Cu for Stubbs, *et al.*). K is the torsion constant of the fibre, which is given in terms of the moment of inertia I and period $T(\cong 420\,\mathrm{s})$ by

$$K = \frac{4\pi^2 I}{T^2} = \frac{8\pi^2 m L^2}{T^2} \cong 6.84 \times 10^{-9}\,\mathrm{N-m}. \tag{3.12}$$

For their two sets of test masses, the quoted values for $\Delta\theta_{\max}$ are

$$\begin{aligned} \Delta\theta_{\max}(\text{Cu-Be}) &= (+0.53 \pm 0.59)\,\mu\text{rad}, \qquad [\text{Stubbs, } et\ al., \text{ Ref. 3}], \\ \Delta\theta_{\max}(\text{Al-Be}) &= (+0.25 \pm 0.67)\,\mu\text{rad}, \qquad [\text{Adelberger, } et\ al., \text{ Ref. 4}], \end{aligned} \tag{3.13}$$

and hence by combining Eqs. (3.11)-(3.13) we find

$$\begin{aligned} \Delta a(\text{Cu-Be}) &= (+0.93 \pm 1.03) \times 10^{-11}\,\text{m}\,\text{s}^{-2}, \\ \Delta a(\text{Al-Be}) &= (+0.44 \pm 1.17) \times 10^{-11}\,\text{m}\,\text{s}^{-2}. \end{aligned} \tag{3.14}$$

Alternatively, we can express the results in Eq. (3.14) in the form of 2σ limits:

$$\begin{aligned} \Delta a(\text{Cu-Be}) &\lesssim 3.0 \times 10^{-11}\,\text{m}\,\text{s}^{-2}, \\ \Delta a(\text{Al-Be}) &\lesssim 2.8 \times 10^{-11}\,\text{m}\,\text{s}^{-2}. \end{aligned} \tag{3.15}$$

We can estimate $\mathcal{F}(\lambda = 100\,\text{m}; \theta_5)$ by approximating the NPL site as a homogeneous hill, which is tilted from the horizontal by an angle δ such that $\sin\delta \approx 0.15$ for $\lambda = 100\,\text{m}$. Since the NPL is located roughly midway up the hill, the analog of Eq. (2.13b) becomes,

$$\mathcal{F}(\lambda;\theta_5) \cong 2\pi G_\infty \bar{\rho}\lambda \sin\delta \cong 1.3 \times 10^{-5}\,\text{m}\,\text{s}^{-2}, \tag{3.16}$$

where we have used $\bar{\rho} \cong 2100\,\text{kg}\,\text{m}^{-3}$, and $\lambda = 100\,\text{m}$. Combining (3.15) and (3.16) we find,

$$\begin{aligned} |k(\text{Cu-Be})| &\lesssim 2.3 \times 10^{-6}, \\ |k(\text{Al-Be})| &\lesssim 2.1 \times 10^{-6}. \end{aligned} \tag{3.17}$$

iii) Evaluation of k for Boynton, *et al.* In the experiment of Boynton, *et al.*,[6] the fractional change in the period $T = 2\pi/\omega_0$ of his ring, which is half Al and half Be, is given by

$$\begin{aligned} \left(\frac{\Delta T}{T}\right)_{\max} &= \frac{\xi q_5(\theta_5)\Delta q_d(\theta_5)\mathcal{F}(\lambda;\theta_5)}{\bar{\ell}\omega_0^2} = \frac{\Delta a}{\bar{\ell}\omega_0^2}, \\ \bar{\ell} &= \frac{3\pi}{4}\left[\frac{R_2^4 - R_1^4}{R_2^3 - R_1^3}\right]\frac{M_T}{M_A}. \end{aligned} \tag{3.18}$$

Here R_1 and R_2 are the inner and outer radii of the ring, and $M_A(M_T)$ is the active (total) mass of the ring. For the Be-Al ring,

$$\bar{\ell} \cong \frac{3\pi}{4}\left[\frac{(4.45\,\text{cm})^4 - (3.97\,\text{cm})^4}{(4.45\,\text{cm})^3 - (3.97\,\text{cm})^3}\right] 1.0 \cong 13.25\,\text{cm}. \tag{3.19}$$

From Ref. 6 we have $T = 975\,\mathrm{s}$, so that $\omega_0 = 0.00644\,\mathrm{s}^{-1}$, and

$$\left(\frac{\Delta T}{T}\right)_{\max} = (-4.6 \pm 1.3) \times 10^{-6}. \tag{3.20}$$

Combining Eqs. (3.18)-(3.20) we then find,

$$\Delta a \cong (2.53 \pm 0.71) \times 10^{-11}\,\mathrm{m\,s}^{-2}. \tag{3.21}$$

We have evaluated $\mathcal{F}$ numerically for the site near Mt. Index where the experiment of Boynton, *et al.*, was carried out, and we find

$$\mathcal{F}(\lambda = 100\,\mathrm{m}; \theta_5) \cong 5.32 \times 10^{-5}\,\mathrm{m\,s}^{-2}, \tag{3.22}$$

and hence

$$k(\lambda = 100\,\mathrm{m}) \cong (-4.8 \pm 1.3) \times 10^{-7}. \tag{3.23}$$

iv) Evaluation of k for Niebauer, *et al.* For the Galileo-type experiment of Niebauer, *et al.*,[5] Δa is the quantity directly measured, so that we can use their quoted value as is:

$$\Delta a(\text{U-Cu}) = (+0.13 \pm 0.50) \times 10^{-8}\,\mathrm{m\,s}^{-2}, \tag{3.24a}$$

$$\Delta a(\text{U-Cu}) \lesssim 1.13 \times 10^{-8}\,\mathrm{m\,s}^{-2} \qquad [2\sigma\ \text{limit}]. \tag{3.24b}$$

We can estimate $\mathcal{F}$ by assuming that their experiment is carried out in an environment which is approximately flat over a distance of order $\lambda \cong 100\,\mathrm{m}$. Then, with $\bar{\rho} \cong 2750\,\mathrm{kg\,m}^{-3}$,

$$\mathcal{F}(\lambda = 100\,\mathrm{m}; \theta_5) \cong 2\pi G_\infty \bar{\rho} \lambda \cong 1.15 \times 10^{-4}\,\mathrm{m\,s}^{-2}, \tag{3.25}$$

so that

$$|k(\lambda = 100\,\mathrm{m})| \lesssim 9.8 \times 10^{-5}. \tag{3.26}$$

b) Discussion of Composition-Independent Tests

The preceding discussion dealt with the data from composition-dependent tests for the fifth force. As we will see in the next section, the constraints on ξ and θ_5 that are implied by these experiments are mutually compatible, and generate two sets of allowed solutions for ξ and θ_5. However, the effects of the putative fifth force can also show up at useful sensitivity levels in several other systems which include *i*) the comparison of the geophysical and laboratory values of the Newtonian constant of gravitation G, *ii*) the possible composition-dependence of the laboratory values of G, and *iii*) the Kreuzer experiment.[17] Since one of the motivations for the fifth force is an apparent discrepancy between the geophysical and laboratory values of G, it is interesting to ask whether the magnitude of this discrepancy can be accounted for by the same phenomenological picture which describes the composition-dependent data we have just discussed.

Returning to Eq. (2.3) we note that for a laboratory experiment carried out over a distance scale $r \ll \lambda_k$, $\vec{a}_j$ becomes

$$\begin{aligned} \vec{a}_j &= \vec{g}_N \left\{ 1 + \sum_{k=1}^{N_\alpha} \alpha_k - \sum_{\ell=1}^{N_\xi} \xi_\ell q_{\ell i} q_{\ell j} \right\} , \\ \vec{g}_N &= -\hat{r} \frac{G_\infty M_\oplus}{R_\oplus^2} . \end{aligned} \tag{3.27}$$

Since the expression in { } is a constant, the effect of the terms in { } is to modify the Newtonian gravitational constant such that

$$G_\infty \longrightarrow G_\infty \left\{ 1 + \sum_{k=1}^{N_\alpha} \alpha_k - \sum_{\ell=1}^{N_\xi} \xi_\ell q_{\ell i} q_{\ell j} \right\} \equiv G_0 . \tag{3.28}$$

Hence the laboratory value of the Newtonian constant G_0 differs from the value G_∞ that would be obtained in the absence of the fifth force, and this difference can be detected in at least two ways: To start with G_0 and G_∞ can be directly compared by measuring G_∞ over distance scales where the fifth force does not contribute for one reason or another. The geophysical measurements of Stacey and collaborators,[8] and Hsui[9] accomplish this by measuring gravity at sufficiently great depths below the surface of the Earth that the fifth force contributions cancel by spherical symmetry. The second way of detecting a difference between G_0 and G_∞ is to look for variations among individual determinations of G_0 as a function of

Table III: Determination of laboratory G for various experiments [*cf.* Refs. 10, 18-23]. See de Boer[24] and Gillies[25] for an excellent review of the various laboratory G determinations.

Experiment	m_1	m_2	$10^{11}G$ ($\mathrm{N\,m^2\,kg^{-2}}$)
Luther and Towler (1982)	W	W	6.6726(5)
Sagitov (1979)	Steel	Cu	6.67449(81)
Karagioz (1976)	Steel	Cu	6.668(2)
Renner (1973)	Hg	Cu	6.670(8)
Facy and Pontikis (1972)	Brass	Bronze	6.67122(61)
"	Brass	Pb	6.67126(66)
"	Brass	Ag	6.67162(51)
"	Brass	Hg	6.67157(51)
Rose, *et al.* (1969, 1976)	W	Al	6.6699(14)
Heyl (1930, 1942)	Steel	Pt	6.673(3)

Table IV: Numerical values for B/μ and I_z/μ for the materials used in Table III.

Material	B/μ	I_z/μ
Al	1.008515	0.037352
Fe	1.008994	0.070598
Cu	1.008951	0.089077
Brass	1.008949	0.087434
Bronze	1.008850	0.112666
Ag	1.008713	0.130462
W	1.008093	0.196753
Pt	1.008009	0.202082
Hg	1.007979	0.204118
Pb	1.007942	0.210314

the composition of the different test masses. We begin by discussing the latter experiments.

Table III summarizes the most recent high-precision determinations of G_0. We see immediately that any attempt to use these results to search for a systematic composition-dependence must confront the problem that the data are not mutually consistent within the quoted error bars. In lieu of a more rigorous approach, we have simply multipled all the errors by a common factor of 1.55, which is roughly the value needed to achieve statistical consistency. The hypothesis of a coupling to a particular charge q_ℓ in (2.10) can then be tested by fitting the data to Eq. (3.28) for the presumed charge. For these purposes, any composition-independent contribution from α_k can be absorbed into G_∞, and hence for a coupling to a single composition-dependent term we can write

$$G_O = (G_O)_{12} = G_\infty \left[1 - \xi_\ell q_{\ell 1} q_{\ell 2}\right], \tag{3.29}$$

where 1 and 2 (which may be the same material) denote the two substances whose gravitational attraction is being measured. Two particular cases are of special interest, namely $q_\ell = q_B = B/\mu$ and $q_\ell = q_I = I_z/\mu$. Fitting to the data of Table III using the values for q_B and q_I in Table IV, we find for a pure B coupling

$$\begin{aligned} G_O &= G_\infty \left[1 - \xi_B q_{B1} q_{B2}\right], \\ \xi_B &= +0.017 {}^{+0.075}_{-0.089}, \\ G_\infty &= (6.73 \pm 0.57) \times 10^{-11}\,\mathrm{N\,m^2\,kg^{-2}}. \end{aligned} \tag{3.30}$$

At the 2σ level the results in (3.30) lead to the constraint

$$-0.16 \lesssim \xi_B \lesssim +0.17. \tag{3.31}$$

Similarly for a pure I_z coupling we have

$$\begin{aligned} G_O &= G_\infty \left[1 - \xi_I q_{I1} q_{I2}\right], \\ \xi_I &= -0.0035 \pm 0.0048, \\ G_\infty &= (6.714 \pm 0.0007) \times 10^{-11}\,\mathrm{N\,m^2\,kg^{-2}}, \end{aligned} \tag{3.32}$$

which at the 2σ level gives the constraint

$$-0.013 \lesssim \xi_I \lesssim +0.006. \tag{3.33}$$

The limits in (3.33) are probably unrealistically stringent. For example, if we retain only the highest precision results, which are those of Luther and Towler,[10] Sagitov,[23] and Facy and Pontikis,[20] then we obtain

$$\begin{aligned} G_\infty &= (6.717 \pm 0.0007) \times 10^{-11}\,\mathrm{N\,m^2\,kg^{-2}}, \\ \xi_I &= +0.0021 \pm 0.0049, \\ -0.012 \lesssim \xi_I &\lesssim +0.008, \qquad [2\sigma \text{ limit}]. \end{aligned} \tag{3.34}$$

The maximum variation in ξ_I that we found by omitting various sets of data points was

$$-0.030 \lesssim \xi_I \lesssim +0.028. \tag{3.35}$$

As we discuss in Sec. IV below, the limits in (3.31)-(3.35) are consistent with the values of ξ_B and ξ_I that emerge from the two solutions we find for compatibility among the results for composition-dependent tests. It should be noted that although negative values of ξ_B and ξ_I are formally allowed by the fits, they should be regarded as unphysical since $\xi_{B,I} < 0$ would correspond to an *attractive* force between like materials, such as the W masses in the experiment of Luther and Towler.[10]

We conclude the discussion of the laboratory data for G_0 by giving the results for the general charge q_ℓ in (2.10). To find the best values of G_∞ for a given value of $\theta_{5\ell}$, we minimize the function

$$\chi^2(\xi_\ell, G_\infty) = \sum_{i=1}^{N} \left\{ \frac{(G_0)_i - G_\infty[1 - \xi_\ell q_{\ell 1} q_{\ell 2}]_i}{K\delta(G_0)_i} \right\}^2 \tag{3.36}$$

with respect to ξ_ℓ and G_∞. Here $(G_0)_i \pm \delta(G_0)_i$, $i = 1, \ldots, N$, is the value of G_0 determined in any of the experiments in Table III, and $K = 1.55$ is the scale factor we introduced earlier. The "best" values of ξ_ℓ and G_∞ ($\bar{\xi}_\ell$ and $\bar{G}_\infty$) produce a minimum χ_0^2 in χ^2, but if we want the range of values of ξ_ℓ allowed at the n-standard deviation ($n\sigma$) level, this is given by the quantities $\xi_{\ell\pm}$, which satisfy,

$$\chi^2(\xi_{\ell\pm}, \bar{G}_\infty) = \chi_0^2 + n^2. \tag{3.37}$$

In order to take proper account of the systematic errors in the various measurements, the values of $\xi_{\ell\pm}$ shown in Fig. 6 below are the maximum values obtained upon dropping various combinations of the data points.

In the preceding discussion we have focussed on the relatively small composition-dependent variations that would be expected among the different determinations of G_0 due to the fifth force. Another signal for the fifth force would be a difference between a suitable average value of G_0, and G_∞ as determined by the geophysical data.[8,9] We use for G_∞ the value determined by Stacey[8]

$$G_\infty = (6.720 \pm 0.002 \pm 0.024) \times 10^{-11}\,\mathrm{N\,m^2\,kg^{-2}}, \tag{3.38}$$

where the first error is the 1σ statistical uncertainty, and the second is the maximum admitted systematic uncertainty arising from possible errors in the density measurements. For G_0 we take the values from Table III, and obtain a best fit for $\xi_{\ell\pm}$ using the χ^2 defined in Eq. (3.36), when G_∞ is constrained to be that value given in Eq. (3.38). Systematic uncertainties inherent in the G_0 determinations are again characterized by taking the maximum value $\xi_{\ell+}$ and the minimum value $\xi_{\ell-}$ obtained by dropping various combinations of the G_0 data points. As an example of what one expects to find, we consider the case of a comparison between the result in Eq. (3.38) and the value of G_0 given by the recent CODATA determination[26]

$$G_0 = (6.67259 \pm 0.00085) \times 10^{-11}\,\mathrm{N\,m^2\,kg^{-2}}, \tag{3.39}$$

which is the result of Luther and Towler,[10] except that the error has been inflated. Assuming as before the existence of a single composition-dependent coupling in (2.11), G_0 for the tungsten-tungsten (W-W) measurement and G_∞ are related by

$$G_0 = G_\infty\left[1 - \xi[q_W(\theta_5)]^2\right], \tag{3.40}$$

from which we find (at the 2σ level)

$$\xi_\pm(\theta_5) = \frac{1}{[q_W(\theta_5)]^2}\cdot\frac{(G_\infty - G_0)}{G_\infty} = \frac{+0.0071 \pm 0.0039}{[q_W(\theta_5)]^2}. \tag{3.41}$$

We note that $\xi_\pm$ in this constraint curve has a single pole corresponding to the vanishing of $q_W(\theta_5)$. When we use the more general prescription discussed above to obtain a limit for $\xi_\pm$ from all of the modern G_0 determinations, we find that while there is a region of reduced sensitivity corresponding to the vanishing of the various $q_{\ell i}(\theta_5)$ near $\theta_5 = \pm\pi/2$, no singularities actually occur.

We next obtain an additional constraint on possible composition-dependent effects by considering the Kreuzer experiment.[17] This experiment involves the measurement of the torque induced in a torsion balance, whose pendant was composed of Al, by a teflon cylinder $[(C_2F_4)_xH_2]$ in motion through a mixture of dibromomethane $[CH_2Br_2]$ and trichloroethylene $[C_2HCl_3]$ which was 74% Br by weight. The temperature of the liquid solution was adjusted such that the teflon cylinder and the solution had the same mass density, which ensured that the pendant was insensitive to ordinary gravitational effects. For this combination of materials, Kreuzer concluded that:

$$(G_{\text{Al-Teflon}} - G_{\text{Al-sol'n}})/G = (+1.2 \pm 4.4) \times 10^{-5}. \tag{3.42}$$

A later reanalysis by Gilvarry and Muller[27] made the claim that the limit determined by Kreuzer from his data was better than the data allowed for. However, more recent analyses by Morrison and Hill[28] and by Ni[29] have supported Kreuzer, by noting that Gilvarry and Muller assumed an unphysically large non-linear contribution to the thermal coefficient of expansion of the solution. In what follows we will thus use the result in (3.42) quoted by Kreuzer. For the materials he used, we find $q_{B,\text{teflon}} = 1.007889$, $q_{I,\text{teflon}} = 0.04053$, $q_{B,\text{sol'n}} = 1.008659$, $q_{I,\text{sol'n}} = 0.09082$, $q_{B,\text{Al}} = 1.008515$, and $q_{I,\text{Al}} = 0.03735$. Again the cases of immediate interest are for $q_\ell = q_B$ and $q_\ell = q_I$. For these couplings, we find at the 2σ limit

$$\begin{aligned} -0.042 \lesssim \xi_I \lesssim +0.055, \\ -0.098 \lesssim \xi_B \lesssim +0.129. \end{aligned} \tag{3.43}$$

Comparing these results to Eqs. (3.31) and (3.35), we see that the Kreuzer experiment provides a tighter constraint for a pure baryon number coupling than do the laboratory determinations of G, whereas for a pure isospin coupling the G determinations are able to set a more stringent limit. This is primarily due to the vanishing of $q_{I,\text{Al}}$ in the vicinity of $\theta_5 = \pm 90°$. As can be seen from Fig. 6 in the next section, however, the Kreuzer experiment is able to provide a more stringent limit on values of ξ_ℓ, except in the neighborhood of $q_{I,\text{Al}} \to 0$ or $\Delta q_{I,\text{teflon}-\text{sol'n}} \to 0$.

Another useful constraint would arise from a sufficiently precise determination of G_0 in the Splityard Creek Lake experiment.[30] This is because one of the sources (water) in this experiment has the opposite sign of q_I ($q_I = -0.112$) relative to the sources in the other G_0 determinations, and thus we would expect the effect due to a coupling to I_z (such as in the model of Boynton, *et al.*) to enter with an opposite sign relative to the other G_0 determinations.

IV. DISCUSSION OF SINGLE-COMPONENT SOLUTIONS

We present in this section our results for the constraints on $\xi = \xi(\theta_5)$ for various values of λ, using the input from Sec. III. Fig. 2 shows the 2σ constraint curves for $\lambda = 100\,\text{m}$ implied by the various composition-dependent experiments, for the full range of θ_5, $-\pi/2 \leq \theta_5 \leq +\pi/2$. For the experiments which have seen positive results [Thieberger (t), and Boynton, *et al.* (b)] the allowed values of ξ and θ_5 lie between the curves with the corresponding labels. For the remaining experiments [Stubbs, *et al.* (s), Adelberger, *et al.* (a), and Niebauer, *et al.* (n)] which set upper limits on the magnitude of a possible effect, the values of ξ, θ_5 allowed at the 2σ level fall below the corresponding curves. We see from these figures that the curves

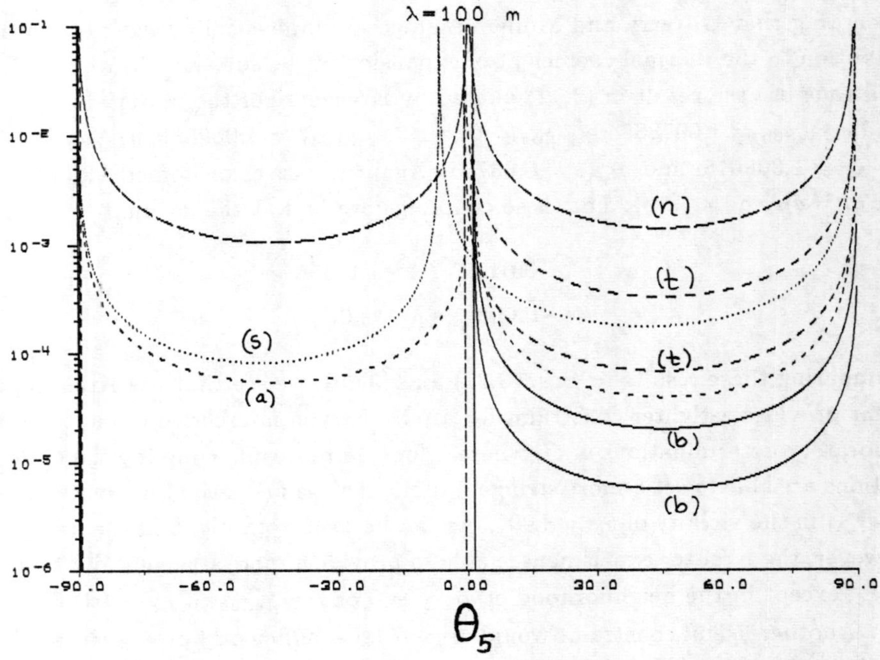

Figure 2: Constraints on $\xi = \xi(\theta_5)$ for $\lambda = 100\,\mathrm{m}$ from various composition-dependent experiments, as discussed in the text. θ_5 is in degrees.

overlap near $\theta_5 = -90°$, and converge (but do not overlap) near $\theta_5 = 0°$. [An apparent overlap region near $\theta_5 = +90°$ is merely a continuation of the $\theta_5 = -90°$ solution, since each constraint curve will be periodic in π.] Figs. 3 and 4 show these two regions in greater detail. We see from Fig. 3 that for values of ξ and θ_5 in the shaded region, the constraints implied by all the composition-dependent experiments are simultaneously satisfied. This is the solution found by Boynton, *et al.*,[6] which corresponds to a coupling predominantly to isospin. By way of contrast, the detailed view of the region near $\theta_5 = 0°$ shown in Fig. 4 indicates that there is no solution near $\theta_5 = 0°$, for $\lambda = 100\,\mathrm{m}$. This is in large measure due to the stringent limit set for Be-Cu by Stubbs, *et al.*,[3] especially when compared against the magnitude of the effect seen by Thieberger.[2] We see from Fig. 5, however, that at $\lambda = 1000\,\mathrm{m}$ a solution near $\theta_5 = 0°$ does exist, as shown by the darkened region in the figure. This solution comes about by virtue of the fact that the sensitivity of Thieberger's experiment increases more rapidly with increasing λ than does that of

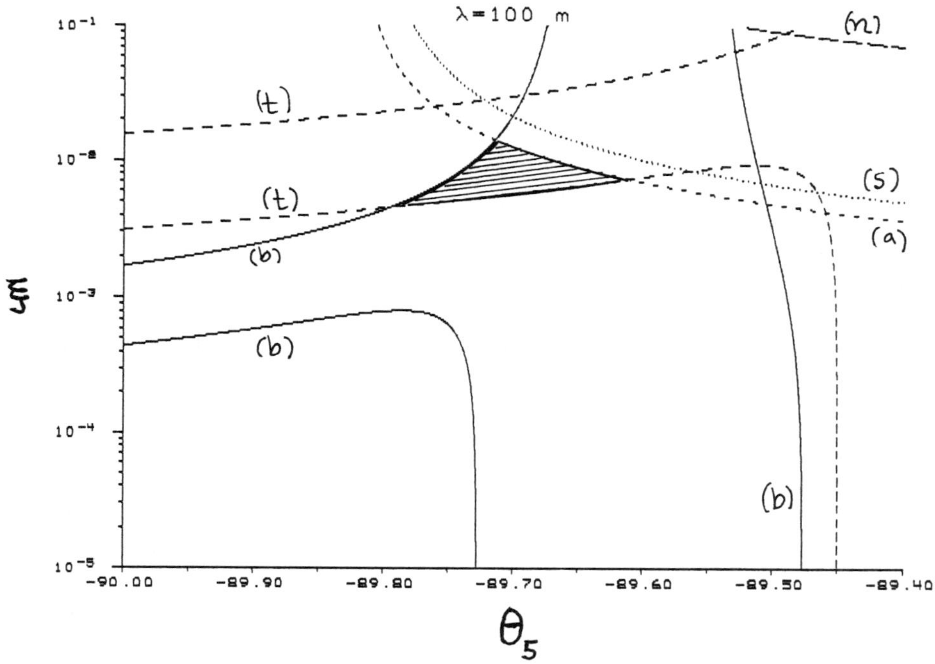

Figure 3: Blow-up of the region near $\theta_5 = -90°$ from Fig. 2, demonstrating the solution of Boynton, *et al.*, near a pure I_z coupling, as indicated by the shaded region.

Stubbs, *et al.* In fact such a "second" solution (*i.e.*, near $\theta_5 = 0°$) becomes possible for values of λ as small as $\sim 200 - 300$ m. We note in passing that the "first" solution of Boynton, *et al.*, exists for $\lambda \gtrsim 30$ m up to at least $\lambda \cong 1000$ m. We also see from Fig. 5 that ξ is a rapidly varying function of θ_5 in the vicinity of $\theta_5 = 0°$, so that although a solution is possible for $\theta \cong 2.5°$, none exists for $\theta = 0°$, which would correspond to a pure B coupling. This is another reflection of the conclusion originally drawn by Stubbs, *et al.*, that their data are not consistent with a pure B coupling, at the level expected from Thieberger and Stacey.

Fig. 6 exhibits the constraints implied by the composition-independent experiments [Kreuzer (k), G_0-determinations (g), and mine/borehole determinations of G_0 (m)]. For the G_∞ determination we have used Stacey's[8] data, as discussed in the previous section, which are supported by the recent results of Hsui.[9] As before the limits implied by the Stacey data, which indicate a positive effect, are shown by

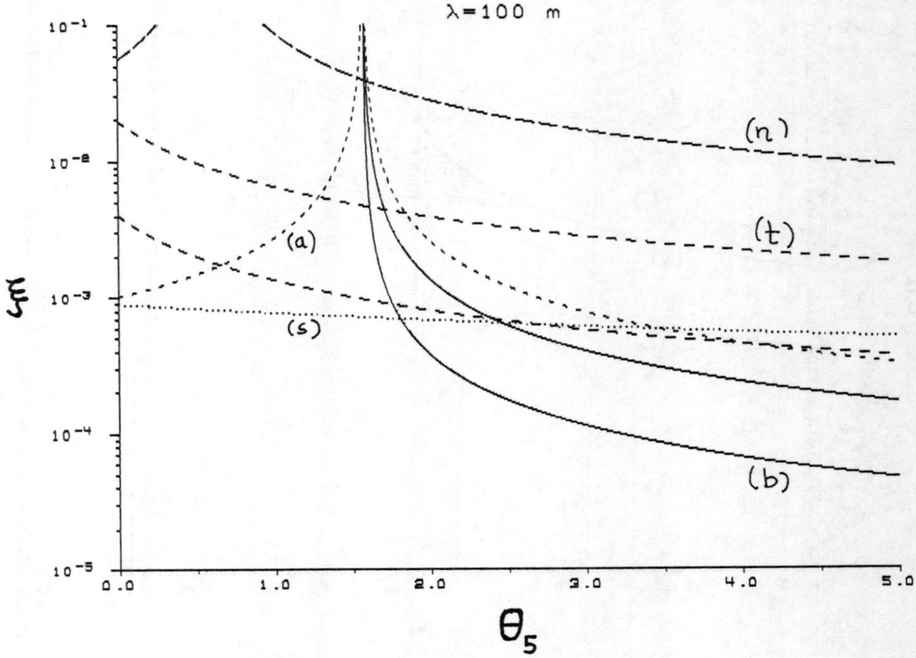

Figure 4: Blow-up of the region near $\theta_5 = 0°$ from Fig. 2, showing that in fact no solution exists for a pure B coupling at $\lambda = 100\,\text{m}$.

the two curves labeled (m) in the figure, whereas the other curves correspond to the 2σ limits discussed in the text. Taken by themselves, these data are consistent with the two sets of solutions shown by the shaded regions in Fig. 6. However, when we superimpose on these solutions the two solutions found earlier from the composition-dependent measurements (shown by the vertical bars near 0° and −90°), then no common overlap region exists. In qualitative terms, the primary difficulty is in reconciling the relatively large ($\mathcal{O}(1\%)$) discrepancy between G_∞ and G_0 implied by the Stacey data, with the relatively small values of ξ_B or ξ_I that emerge from the composition-dependent experiments.

It is worth noting that since the composition-independent tests can involve contributions from the α_k in (3.28), which the composition-dependent tests do not, a direct comparison between these may be more complicated than implied by the single-component model that we have been assuming. This was the reason why the Stacey data were not included in the analysis of Boynton, *et al.* We point

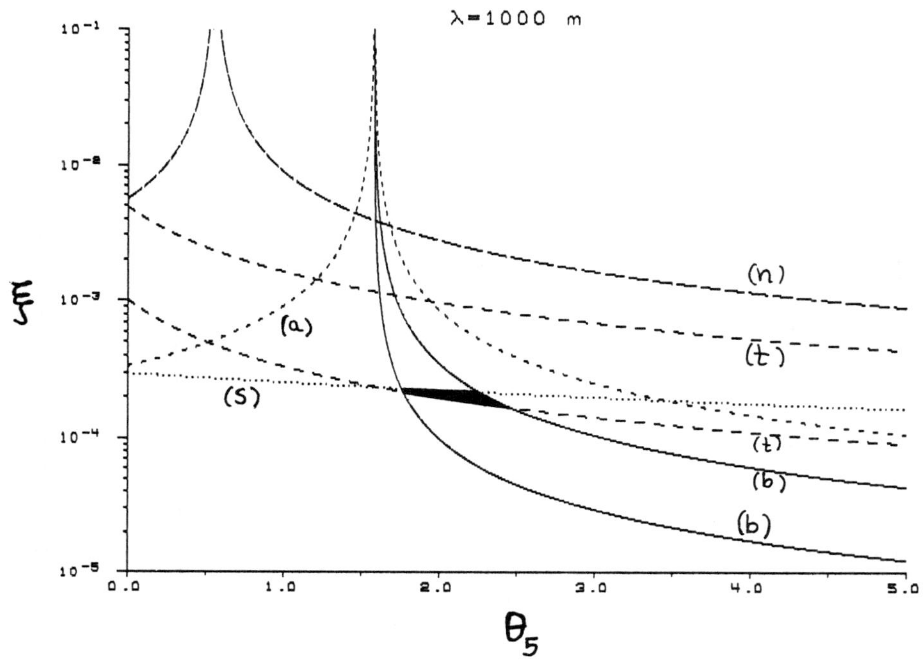

Figure 5: Blow-up of the region near $\theta_5 = 0°$ from Fig. 2 at $\lambda = 1000\,\mathrm{m}$, showing the solution for θ_5 near a pure B coupling. The solution is indicated by the dark region.

out, however, that simply including an additional composition-independent term cannot by itself bring all the data into agreement, and in fact only serves to make matters worse. Returning to (3.28) we consider as an example the comparison of the G_0 experiment of Luther and Towler[10] (which used two W masses), and Stacey's[8] determination of G_∞, for a model with a single α and a single ξ_ℓ. From Eq. (3.28) we have

$$G_0 = G_\infty \left[1 + \alpha - \xi_\ell q_{\ell \mathrm{W}}^2\right] = G_\infty \left[1 - (\xi_\ell q_{\ell \mathrm{W}}^2 - \alpha)\right]. \tag{4.1}$$

Since G_0 and G_∞ can be determined independently, as can ξ_ℓ (from composition-dependent experiments), it is possible in principle to establish whether or not $\alpha \neq 0$. However, since $\alpha > 0$ in all known theories, the effect of $\alpha \neq 0$ would be to further reduce the correction term on the right-hand-side of (4.1), which from Fig. 6 is already too small to account for the difference between G_0 and G_∞. Notice that

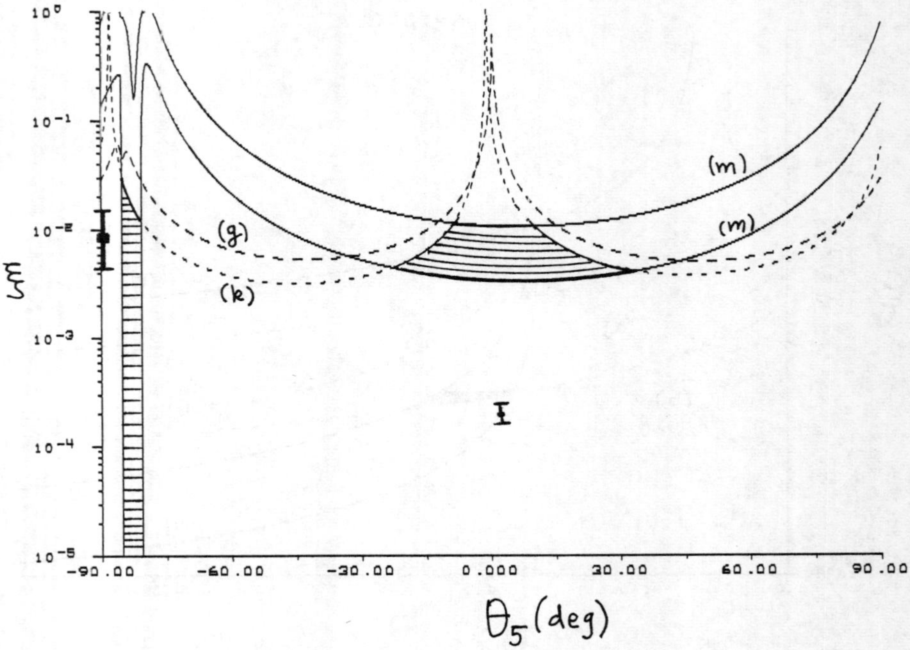

Figure 6: Constraints implied by various composition-independent experiments, as discussed in the text. Curve (k) is that resulting from Kreuzer's experiment,[17] curve (g) is from the laboratory determinations of G,[10,18–23] and curve (m) is from the mine data of Stacey, *et al*,[8] which is supported by the recent analysis of Hsui.[9] The allowed values for constraint (m) are indicated by the shaded regions.

this argument depends on the assumption that the α_k are intrinsically positive, which is equivalent to assuming that the source of the composition-independent terms is a quantum number like M which always gives an *attractive* force. As noted previously we do not know of a consistent theory where M could be the source of a *repulsive* force between two macroscopic objects, although an imaginative suggestion along these lines has been put forward recently by Peccei, Solà, and Wetterich.[31]

It should be noted that one can directly test whether the "I_z-solution" of Boynton, *et al.*, or the "B-solution" that we have presented here, or perhaps some other alternative is correct. Each of these makes quite specific predictions for the results to be expected from the ongoing experiments, when different materials are used. Boynton, *et al.*,[6] have already noted that for their "I_z-solution", the signal expected in their current experiment comparing polyethylene $[(CH_2)_x]$ and Cu should be larger by a factor of ~ 3, and have opposite sign, compared to what they measured for Be-Al. Using the values

$$\begin{aligned} \left(\frac{B}{\mu}\right)_{CH_2} &= 1.00670; \quad \left(\frac{I_z}{\mu}\right)_{CH_2} = -0.1429, \\ \left(\frac{B}{\mu}\right)_{Cu} &= 1.00895; \quad \left(\frac{I_z}{\mu}\right)_{Cu} = +0.0891, \end{aligned} \tag{4.2}$$

we find that for our "B-solution" the effect for $(CH_2)_x$-Cu should have opposite sign, and should be larger than that for Be-Al by a factor of $\sim (6-17)$ depending on the precise value of θ_5. The fact that there is a large range for the predicted value reflects the extreme sensitivity of the "B-solution" to θ_5, which we have noted earlier. In any case we see that these two solutions can easily be distinguished from each other, and likewise from other alternatives as well.

We conclude this discussion by noting that the "I_z solution" of Boynton, *et al.*, does not satisfactorily account for the EPF data, as we see from Fig. 7, line I. However, this figure also demonstrates that if the Pt data are excluded, then an adequate fit is obtained for the remaining data, as shown by line II. We also note that the Pt data taken alone also fall along a straight line (III), albeit with a nonzero intercept. It is also worth noting that the EPF data are also incompatible with the "B solution" even though it contains only a small admixture of I_z $(\mathcal{O}(4\%))$.

V. DISCUSSION OF MULTICOMPONENT SOLUTIONS

As was shown in the previous section, a simple single-component model with a coupling predominantly to either B or I_z, while compatible with the modern composition-dependent experiments, cannot explain either the EPF data,[7] or the mine/borehole results.[8,9] This naturally leads to the question of whether models with several independent components could accommodate the existing data. We now examine the requirements that a multicomponent model must satisfy in order to achieve this goal.

i) Since the effective source for the EPF experiment is at least 15 – 20 m from their point of observation, this would imply that there must be a longer-range component which couples to B, which we characterize by a coupling strength ξ_B and a range λ_B. We take $\lambda_B \gtrsim 200$ m as suggested by Fig. 5. Such a coupling is allowed by the composition-dependent experiments, since the NPL

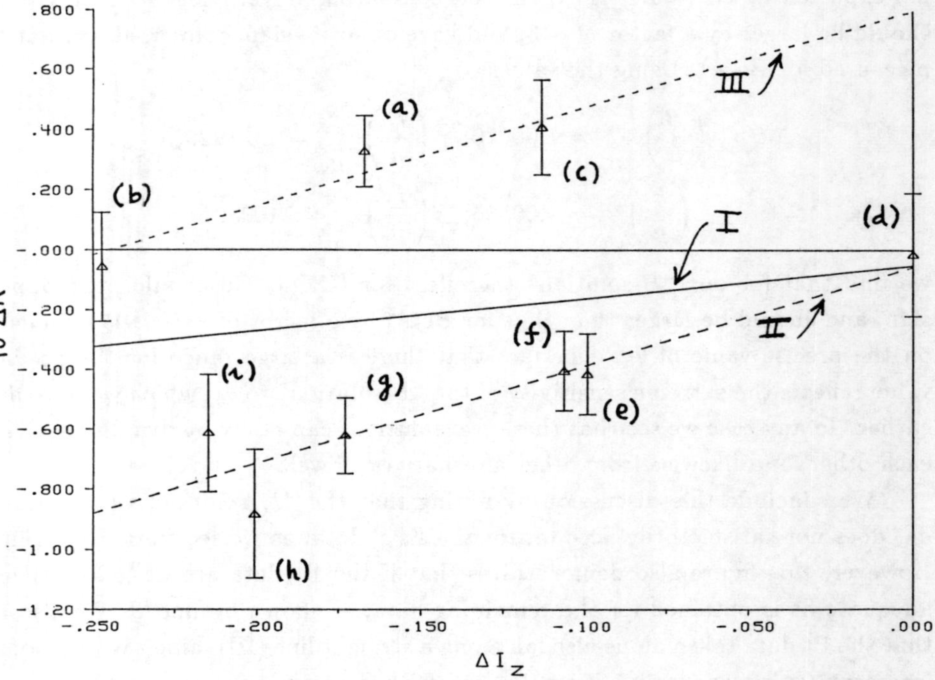

Figure 7: Results of a fit of the EPF data to I_z. Here, line I represents a fit to all of the EPF data, line II a fit in which the data involving Pt are excluded, and line III a fit to only the data involving Pt. The data points are represented by (a) magnalium-Pt, (b) snakewood-Pt, (c) Cu-Pt, (d) Ag-Fe-SO_4, (e) asbestos-Cu, (f) $CuSO_4 \cdot 5H_2O$-Cu, (g) $CuSO_4$ (sol'n)-Cu, (h) H_2O-Cu, and (i) tallow-Cu.

experiments are intrinsically less sensitive to longer-range components than the experiments in Refs. 2 and 6, due to the relatively small mass distribution at the NPL site.

ii) There must in addition be a shorter-range component which couples to a generalized $q(\theta_5)$, whose range should be less than about 20 m, in order to reproduce the composition-dependence implied by the experiment of Boynton, *et al.*[6] We characterize such a coupling by a strength ξ_Q and a range $\lambda_Q \lesssim 20\,\text{m}$.

iii) We will require that ξ_B and ξ_Q be positive in accordance with the discussion in Sec. II. As we will see below, this will set stringent limits on possible models.

iv) The short-range limit for the multicomponent model should be compatible with the mine data. As we see in Fig. 6, such a requirement immediately limits the effective short-range coupling to $-20^\circ \lesssim \theta_5 \lesssim +30^\circ$, and additionally sets the limit $2 \times 10^{-3} \lesssim \xi_Q \lesssim 1.5 \times 10^{-2}$.

These criteria, when combined, set severe limits on possible models which one can construct, and in practice are quite difficult to meet, if one requires that each constraint be satisfied at the 2σ level. We first give a brief formulation of the two component model under consideration. Following Eq. (2.12) we consider couplings to B and to a general charge $q(\theta_5)$,

$$\Delta a = q_{Bs}\Delta q_B \xi_B \mathcal{F}(\lambda_B) + q_s(\theta_5)\Delta q_d(\theta_5)\xi_Q \mathcal{F}(\lambda_Q), \qquad (5.1)$$

where $\lambda_B \gtrsim 200\,\text{m}$ and $\lambda_Q \lesssim 20\,\text{m}$. It is not difficult to see how one could formulate such a model which would fit *i)*-*iii)* above. Firstly, by setting $\lambda_B \gtrsim 200\,\text{m}$ we have essentially removed the constraints implied by the NPL experiments on a coupling to B, due to their decreased sensitivity to longer-ranged forces. Secondly, since q_{Is} vanishes for the NPL experiments, we have similarly removed the constraints that follow for a coupling to I_z. We now have two experiments remaining, Boynton, *et al.*, and Thieberger, while at the same time we have two degrees of freedom left in our model, represented by the parameters ξ_B and $\xi_Q = \xi_I$. We can "use up" these degrees of freedom in satisfying the remaining constraints, and in fact still have some freedom left over in the specification of λ_B and $\lambda_Q = \lambda_I$, which we can then use to "tune" the overall magnitude of ξ_B and ξ_I. As an example, we have the following solution for $\lambda_B = 1000\,\text{m}$, and $\lambda_I = 10\,\text{m}$ and $5\,\text{m}$ respectively:

$$\begin{aligned} \xi_B &= (+1.48 \pm 1.10) \times 10^{-4}, \\ \xi_I &= (+4.9 \pm 2.6) \times 10^{-2}, \\ \lambda_B &= 1000\,\text{m}, \quad \lambda_I = 10\,\text{m}. \end{aligned} \qquad (5.2)$$

and

$$\begin{aligned} \xi_B &= (+1.48 \pm 0.92) \times 10^{-4}, \\ \xi_I &= +0.15 \pm 0.07, \\ \lambda_B &= 1000\,\text{m}, \quad \lambda_I = 5\,\text{m}. \end{aligned} \qquad (5.3)$$

Both of these multicomponent solutions can accommodate the EPF results (at least qualitatively), and the second solution nominally satisfies Stacey's mine result, although it is incompatible with both the Kreuzer experiment and the limit inferred from the G_0 determinations.

We next examine a solution corresponding to choosing the charge $q(\theta_5)$ in (5.1) to be nearly pure B. Such a solution will exist, since it will look approximately like a single-component solution near $\theta_5 = 0°$. However, it will not be possible in general to satisfy requirement *iv*) if the 2σ criterion on each constraint is enforced. It appears that the closest one can come to meeting all of the various conditions is the following set of parameters:

$$\begin{aligned} \xi_B &= (1.5 \pm 0.5) \times 10^{-4}, \\ \xi_Q &= (+1.1 \pm 0.9) \times 10^{-3}, \\ \theta_5 &= 14°, \\ \lambda_B &= 1000\,\mathrm{m}, \quad \lambda_Q = 15\,\mathrm{m}. \end{aligned} \tag{5.4}$$

This set of parameters satisfies the constraints from Boynton and Niebauer, but only satisfies the other three constraint curves within 2.3σ. Additionally, this solution is (barely) consistent with the Stacey data. Because of the choice of θ_5, it is also consistent with the Kreuzer and G_0-determination constraints, as we see from Fig. 6.

We see from the above discussion that a two-component model of the type discussed is able to account for the available data, although somewhat marginally. The principal hurdle in this scenario comes from the results of the mine analysis of Stacey, *et al.*,[8] which appear to imply an effect much larger than is suggested by any of the other experiments, with the possible exception of Thieberger. The geophysical constraints could in principle be satisfied if one introduces a *repulsive* field coupling only to mass, in addition to the other two components discussed above. It appears, however, to be very difficult in practice to construct a sensible quantum field theory which could produce such an effect.

Although it appears a simple matter to bring the EPF results into *qualitative* agreement with the other results — that is, to produce a model which couples to baryon number only at large distances — it is difficult at present to ascertain whether the EPF data are in *quantitative* agreement with the other results. The reason for this is that the additional degrees of freedom provided by adding more than one distance scale, make it is uncertain whether one could eventually know the matter distribution at the EPF site in sufficient detail to make a definitive statement. The matter distribution could be a particularly difficult problem if one discusses a short-range coupling to isospin, for which the local water table would probably dominate, or a very long-range coupling for which underground geological formations would likely produce the dominant effect. For this reason, we have chosen for the present to regard the EPF data only as a possible clue as to what the composition-dependence should look like at large distances.

It is natural to ask what sort of model would be required to bring all the experimental results into more satisfactory agreement. In the next section we outline some possible avenues that can be explored.

VI. SUMMARY AND CONCLUSIONS

We discuss in this section additional strategies for constructing models which may be capable of more adequately describing the existing data.

1) Enlarge the models to include additional Yukawa components. This is always a possible, if somewhat unappealing, alternative since eventually the number of parameters in such models becomes comparable to the available data. In such a circumstance we would simply be reparametrizing the data, and not learning anything from them. In particular, such models tend to have limited predictive value, which means that they are also of limited use.

2) Reject some of the data. We have consistently taken the view that all of the experiments and analyses which have been published to date appear to be carefully done. However, the geophysical analyses of Stacey, *et al.*,[8] and Hsui[9] are subject to uncertainties which the recent "laboratory" experiments in Refs. 2-6 are not. An example, cited by Stacey, *et al.*, would be a bias in the gravity gradient due to a mass irregularity located at an appropriate depth below the surface of the Earth. Whether or not this proves to be a real problem will hopefully become clearer in the near future, when results become available from the Los Alamos/Scripps determination of G in a Greenland borehole. Since the comparison of the geophysical and laboratory values of G is the only means presently available to fix the strength of the fifth force (independent of λ), any significant bias in the geophysical data could substantially change the present analysis. Another candidate for being excluded is, of course, the original EPF experiment or perhaps the (earlier) EPF data using Pt as the comparison mass. As we have discussed elsewhere,[32,33] if the Pt data are ignored, the remaining points correlate with B, I_z, $1/\rho$ (where ρ is the density), and presumably with other quantities which vary monotonically with atomic number Z. By noting that thet EPF data can in this way support a correlation to I_z (see Fig. 7), we can bring the EPF experiment at least quantitatively in line with the phenomenological analysis of Boynton, *et al.* However, dropping the Pt data, for which there is no compelling reason, would be done on the assumption that they are somehow in error. But we would be then faced with

the question of why this error would generate spurious results which happen to exactly simulate a correlation of $\Delta\kappa$ and $\Delta(B/\mu)$.

3) Introduce unconventional couplings. The present analysis assumes that a source and a test mass can couple only through the global quantum numbers N, Z, L, and M. It is possible of course that the fifth force couples to some other quantum numbers, such as total quark number, or to some hitherto unknown quantum numbers. Moreover, it is also assumed that the couplings of N, Z, L, and M behave in the "usual" way, for example that they respect the usual symmetries P, C, T, CP, ... *etc.* This is an important assumption to test, since our experience with the strong, electromagnetic, and weak interactions indicates that the weaker the interaction the fewer symmetries it obeys. In fact this issue has been discussed by us previously[34] in connection with the decays $K^0_{L,S} \to \pi^0\gamma_Y$. Another example of our reliance on conventional couplings is the assumption that a repulsive force between like objects (as the geophysical data suggest the fifth force to be), can only arise from the exchange of a C-odd vector field. This has the consequence that a repulsive force necessarily leads to an Eötvös-type anomaly, as we have observed previously.[11]

ACKNOWLEDGMENTS

We are deeply indebted to Paul Boynton for numerous valuable conversations, and to Sam Aronson for critically reading this manuscript.This work was supported in part by the United States Department of Energy.

APPENDIX: ONGOING FIFTH FORCE EXPERIMENTS

Laboratory Eötvös experiments

- Cowsik, Tandon, Krishnan, *et al.* [Tata Institute, Bombay]
- C. W. F. Everitt and P. Worden [Stanford]
- J. Faller and P. Keyser [JILA, Colorado]
- O. V. Karagioz [Moscow State University]
- R. Newman and P. Nelson [U. California, Irvine]

Eötvös experiments near a mountain

- E. G. Adelberger, J. H. Gundlach, B. R. Heckel,K. D. McMurry, F. J. Raab, W.

F. Rogers, C. W. Stubbs, H. E. Swanson, and R. Watanabe [U. Washington, Seattle]

- P. Boynton, D. Crosby, P. Ekstrom, and A. Szumilo [U. Washington, Seattle]
- V. Fitch, M. V. Isaila, and M. A. Palmer [Princeton]
- R. Krotkov, H. Sakai, and J. Shafer [U. Massachusetts, Amherst]
- R. Newman and P. Nelson [U. California, Irvine]
- F. D. Stacey, *et al.* [Queensland, Australia]

Repetitions of the Galileo experiment

- V. Cavasinni, E. Iacopini, E. Polacco, and G. Stefanini [CERN/Pisa]
- J. Faller and T. Niebauer [JILA, Colorado]
- K. Kuroda [National Research Laboratory of Metrology, Japan]
- S. Richter, W. T. Burton, Jr., D. C. Hyland, M. D. Koontz, and W. R. Palmer [Harris Corp., Florida]
- A. Sakuma [BIPM, Paris]
- Tsubokawa, *et al.* [International Latitude Observatory, Japan]

Floating Ball Differential Accelerometer Experiments

- P. G. Bizzeti and A. M. Sona [U. Firenze, Italy]
- R. Davisson [U. Washington, Seattle]
- P. Thieberger [Brookhaven]

Searches for Non-Newtonian Gravity

- M. E. Ander, F. N. Spiess, G. E. Backus, A. D. Chave, J. Hildebrand, and M. A. Zumberge [Los Alamos/Scripps] — Measurement of G using a borehole in Greenland
- A. Hsui [U. of Illinois, Urbana] — Measurement of G using boreholes in Michigan and Texas
- C. Jekeli [Air Force Geophysics Laboratory] — Measuring G using high towers
- D. Long [Eastern Washington University] — Repetition of his 1976 laboratory experiment
- G. I. Moore, F. D. Stacey, G. J. Tuck, and B. D. Goodwin [Queensland, Australia] — Measurement of G on geophysical scales using a lake
- P. Kasameyer, N. Burkhard, and D. Carlson [Lawrence Livermore National Laboratory], Testing the inverse square law using towers.
- Y. Ogawa, *et al.* [U. of Tokyo] — Dynamical test of the inverse-square law
- H. J. Paik, H. A. Chan, and M. V. Moody [Maryland] — Measurements of the Laplacian of the gravitational field over various distances scales

- J. Price [Stanford] — Test of Newtonian gravity over millimeter ranges
- S. K. Runcorn [Newcastle Upon Tyne, England] – Test of Newtonian gravity using a lake.

Other experiments

- E. Amaldi, R. Bizzarri, A. Degasperis, G. Muratori, G. V. Pallotino, G. Pizzella, F. Ricci, and C. Rubbia [U. Rome and CERN] — An Eötvös experiment using a gravity-wave detector and external rotating masses
- N. Beverini, *et al.* [Pisa, LANL, Rice, Texas A&M, Genoa, Kent State, Case Western Reserve, CERN, NASA/Ames] — Comparison of p and $\bar{p}$ masses at LEAR in a vertical drift tube.
- G. Shoemaker [California State, Sacramento] — Dynamic beam balance comparison of two masses.
- C. C. Speake and T.J. Quinn [BIPM, Paris] — Beam balance comparison of two masses.

REFERENCES

1. E. Fischbach, D. Sudarsky, A. Szafer, C. Talmadge, and S. H. Aronson, Phys. Rev. Lett. **56**, 3 (1986); **56** 1427 (E) (1986); E. Fischbach, D. Sudarsky, A. Szafer, C. Talmadge, and S. H. Aronson, *Proceedings, 2nd Conference on Intersections between Particle and Nuclear Physics*, (Lake Louise, Canada, May 26-31, 1986), edited by D. Geesaman, AIP Conference Proceedings #150, p. 1102.
2. P. Thieberger, Phys. Rev. Lett. **58**, 1066 (1987).
3. C. W. Stubbs, E. G. Adelberger, F. J. Raab, J. H. Gundlach, B. R. Heckel, K. D. McMurry, H. E. Swanson, and R. Watanabe, Phys. Rev. Lett. **58**, 1070 (1987).
4. E. Adelberger, *et al.*, Phys. Rev. Lett. **59**, 849 (1987).
5. T. M. Niebauer, M. P. McHugh, and J. E. Faller, Phys. Rev. Lett. **59**, 609 (1987).
6. P. E. Boynton, D. Crosby, P. Ekstrom, and A. Szumilo, Phys. Rev. Lett. **59**, 1385 (1987).
7. R. v. Eötvös, D. Pekár, and E. Fekete, Ann. Phys. (Leipzig) **68**, 11 (1922); R. v. Eötvös, D. Pekár, and E. Fekete, *Roland Eötvös Gesammelte Arbeiten*, edited

by P. Selényi (Akadémiai Kiado, Budapest, 1953) pp. 307-372; the latter paper has been translated into English by J. Achzehnter, M. Bickeböller, K. Bräuer, P. Buck, E. Fischbach, G. Lübeck, and C. Talmadge, University of Washington preprint 40048-13-N6; for an earlier translation see Annales Universitatis Scientiarum Budapestinensis de Rolando Eötvös nominate, Sectio geologica, volume **7**, (1963) pp. 111-165.

8. F. D. Stacey, G. J. Tuck, S. C. Holding, A. R. Maher, and D. Morris, Phys. Rev. D**23**, 1683 (1981); F. D. Stacey and G. J. Tuck, Nature **292**, 230 (1981); F. D. Stacey, in *Science Underground*, (Los Alamos 1982) edited by M. M. Nieto, *et al.*, AIP Conference Proceedings #96 (American Institute of Physics, New York, 1983), p. 285; F. D. Stacey, Sci. Prog. Oxf. **69**, 1 (1984); S. C. Holding and G. J. Tuck, Nature **307**, 714 (1984); S. C. Holding, F. D. Stacey, and G. J. Tuck, Phys. Rev. D**33**, 3487 (1986); F. D. Stacey, G. J. Tuck, G. I. Moore, S. C. Holding, B. D. Goodwin, and R. Zhou, Rev. Mod. Phys. **59**, 157 (1987).

9. A. T. Hsui, Science **237**, 881 (1987).

10. G. G. Luther and W. R. Towler, Phys. Rev. Lett. **48**, 121 (1982).

11. C. Talmadge, E. Fischbach, and S. H. Aronson, in *Searches for New and Exotic Phenomena*, Proceedings of the VIth Moriond Workshop, Les Arcs, France, 25-31 January, 1987, edited by J. Tran Thanh Van (Editions Frontières, Gif Sur Yvette, France, 1987, to be published).

12. E. Fischbach, D. Sudarsky, A. Szafer, C. Talmadge, and S. H. Aronson, in *Proceedings of the XXIII International Conference on High Energy Physics*, Berkeley, 16-23 July, 1986, edited by S. C. Loken, (World Scientific, Singapore, 1987), p. 1021.

13. E. Fischbach, H. T. Kloor, C. Talmadge, S. H. Aronson, and G. T. Gillies, submitted to Phys. Rev. Lett.

14. S. L. Glashow, in *Proceedings of the 1986 Moriond Workshop on Neutrino Masses*, Tignes, France, January, 1986, edited by J. Tran Thanh Van (Editions Frontières, Gif Sur Yvette, France, 1986).

15. P. Boynton, private communication.

16. J. Renner, Matematikai és Természettudományi Értesitő **53**, 542 (1935).

17. L. B. Kreuzer, Phys. Rev. 169, 1007 (1968).

18. P. R. Heyl, J. Res. Natl. Bur. Stand. **5**, 1243 (1930); P. R. Heyl and P. Chrzanowski, J. Res. Natl. Bur. Stand. **29**, 1 (1942).

19. R. D. Rose, H. M. Parker, R. A. Lowry, A. R. Kulthau, and J. W. Beams, Phys. Rev. Lett. **23**, 655 (1969); R. A. Lowry, W. R. Towler, H. M. Parker, A. R. Kulthau, and J. W. Beams, in *Atomic Masses and Fundamental Constants, vol. 4*, edited by J. H. Sanders and A. H. Wapstra (Plenum Press, New York, 1972), p. 521; G. G. Luther, W. R. Towler, R. D. Deslattes, R. A. Lowry, and J. W. Beams, in *Atomic Masses and Fundamental Constants, vol. 5*, edited by J. H. Sanders and A. H. Wapstra (Plenum Press, New York, 1976), p. 592.

20. L. Facy and C. Pontikis, C. R. Hebd. Seances Acad. Sci. Ser. B: **273**, 437 (1972).

21. Ya. Renner, in *Determination of Gravity Constant and Measurement of Certain Fine Gravity Effects*, edited by Yu. D. Boulanger and M. U. Sagitov (Nauka Press, Moscow, 1973). [English translation: NASA-TT-F-15722, 1974.]

22. O. V. Karagioz, V. P. Ismaylov, N. I. Agafonov, E. G. Kocheryan, and Yu. A. Tarakanov, Izv. Akad. Nauk SSSR Fiz. Zemli **5**, 106 (1976). [English Translation: Phys. Solid Earth **5**, 351 (1976).]

23. M. U. Sagitov, V. K. Milyukov, Ye. A. Monakhov, V. S. Nazarenko, and Kh. G. Tadzhidinov, Dokl. Akad. Nauk SSSR **245**, 567 (1979). [English translation: Dokl. Akad. Nauk SSSR, Earth Sciences **245**, 20 (1981).]

24. H. de Boer, in *Precision Measurements and Fundamental Constants II*, edited by B. N. Taylor and W. D. Phillips, Nat. Bur. Stand. (U.S.),Spec. Pub. 617 (1984), 561.

25. G. T. Gillies, Metrologia **24** (suppl.), 1 (1987).

26. E. R. Cohen and B. N. Taylor, J. Res. Natl. Bur. Stand. **92**, 85 (1987); E. R. Cohen and B. N. Taylor, Physics Today **40**, No. 8, BG 11 (1987); E. R. Cohen, in *International School of Cosmology and Gravitation: Gravitation Measurements, Fundamental Metrology and Constants, Erice, Sicily*, edited by V. de Sabatta, (D. Reidel, to be published).

27. J. J. Gilvarry and P. M. Muller, Phys. Rev. Lett. **28**, 1665 (1972).

28. D. Morrison and H. A. Hill, Phys. Rev. D**8**, 2731 (1973).

29. W.-T. Ni, Astrophys. J. **181**, 939 (1973).

30. F. D. Stacey, private communication. See also F. D. Stacey, G. J. Tuck, G. I. Moore, S. C. Holding, B. D. Goodwin, and R. Zhou, Rev. Mod. Phys. **59**, 157 (1987).

31. R. D. Peccei, J. Solà, and C. Wetterich, DESY preprint 87-026; see also J. Ellis, N. C. Tsamis, and M. Voloshin, Phys. Lett. **194B**, 291 (1987).

32. C. Talmadge, *Reanalysis of the Eötvös Experiment*, Ph. D. thesis Purdue University, 1987 (unpublished).

33. E. Fischbach, D. Sudarsky, A. Szafer, C. Talmadge, and S. H. Aronson, submitted to Annals of Physics.

34. S. H. Aronson, H. Y. Cheng, E. Fischbach, and W. Haxton, Phys. Rev. Lett. **56**, 1342 (1986); **56**, 2334 (E) (1986).

LORENTZ NONINVARIANCE AND THE UNIVERSALITY OF FREE FALL IN QUASI-RIEMANNIAN GRAVITY

M. Gasperini
Dipartimento di Fisica Teorica dell'Università
Corso M.D'Azeglio 46,
10125 Torino, Italy
and I.N.F.N., Sezione di Torino, Italy

ABSTRACT. The possibility of deviations from geodesical motion is discussed in the hypothesis that the Lorentz gauge symmetry of general relativity is broken. In particular it is stressed that, in the limit of a weak static field, an eventual correction to the Newtonian acceleration is independent from the distance of the source, and should affect then the experimental tests of the equivalence principle both in the terrestrial and in the solar gravitational field.

1. INTRODUCTION

In a recent paper [1] it has been stressed that a deviation from Lorentz invariance implies, in general, a breakdown of the so-called weak equivalence principle, i.e. of the universality of free fall. In particular the consequences of the model of Nielsen and Picek [2,3], which takes into account an eventual breaking of Lorentz symmetry in the context of weak interactions, have been carefully discussed [1].

The object of this paper is to discuss the possibility of non-geodesical motion in the hypothesis that Lorentz invariance is not an exact local gauge symmetry for the gravitational interactions.

Following a phenomenological approach, and using in particular a quasi-riemannian model [4] as an effective theory to describe gravity in the case of broken Lorentz symmetry, we find that in this context the universality of free fall is not necessarily violated, because it is always possible to restore the geodesic character of the motion by a suitable choice of the parameters governing the breaking of the SO(3,1) symmetry in the matter part of the Lagrangian.

Without such a fine tuning, however, the motion is not geodesic. In that case it would be possible, in principle, to arrange a composition dependent acceleration of two macroscopic masses to the gravitational source, by assuming that the gravitational interactions of baryons, unlike those of leptons, are not locally Lorentz invariant.

The obtained deviation from geodesic motion, however, in the static limit and in the weak field approximation is independent from the distance of the source, and it is then severely constrained, as we shall

V. De Sabbata and V. N. Melnikov (eds.),
Gravitational Measurements, Fundamental Metrology and Constants, 181–190.

see, by the results of the Dicke-Braginskii experiments [5,6] performed in the gravitational field of the sun.

Therefore a violation of Lorentz symmetry could not explain, at least in the context of the model we are considering here, an eventual anomalous result for the accelerations of various materials to the earth obtained with Eötvös-type experiments [7,8] .

2. EQUATIONS OF MOTION IN A QUASI-RIEMANNIAN THEORY OF GRAVITY

The possibility that the gravitational interactions be described by a theory which is covariant under general coordinate transformations, but not locally Lorentz invariant, was originally suggested by Weinberg [5] in the context of higher dimensional unified theories, and the general structure of these models, called quasi-riemannian, has been recently investigated in refs.[9,10,11] .

In this paper we consider a quasi-riemannian theory in four dimensions, as an effective geometric model to discuss the consequences of an eventual violation of Lorentz symmetry in the context of the gravitational interaction. Assuming in particular that the SO(3,1) gauge symmetry of gravity is broken, but local rotational invariance is preserved (like in the model of Nielsen and Picek [2,3] and in the case discussed by Zee [12]), we are led to a quasi-riemannian theory with tangent space group G_T = SO(3).

As shown in ref.[9] , the action for this theory can be conveniently formulated in terms of the components of the usual anholonomic basis $V^a = V^a_\mu dx^\mu$ and of the original SO(3,1) connection $\Omega^{ab} = \Omega_\mu{}^{ab} dx^\mu$, which is supposed to be torsionless

$$dV^a + \Omega^a{}_b \wedge V^b = 0 \tag{1}$$

(conventions: Latin letters denote tangent space indices, Greek letters are holonomic world indices; however a,b,c,d... run from 1 to 4, while i,j,k,... from 1 to 3).

In the case G_T = SO(3), the Lorentz connection Ω may be decomposed into the SO(3) connection $\omega^{ij} \equiv \Omega^{ij}$ and the SO(3) vector-valued one-form $\varpi^i \equiv \Omega^{i4}$, and the basis V^a into V^i and V^4 , transforming respectively as an SO(3) vector and scalar. Following then the standard procedure developed in ref.[9], the most general second-order gravitational action, with local SO(3) symmetry, can be easily constructed in terms of ω^{ij} , ϖ^i , V^i and V^4 , and in four dimensions it depends (besides the usual Newton constant G) on four additional parameters $k_1 \dots k_4$ (plus an eventual cosmological constant term which for simplicity here is neglected).

In order to display clearly the differences between the quasi-riemannian model and the standard Einstein theory, it is convenient to express the generalized gravitational action S_g explicitly in terms of the original variables V^a and Ω^{ab} . We have then [13]

$$S_g = \frac{1}{2\chi}\int [\, R^{ab}(\Omega) \wedge {}^*V^{ab} + k_1 R^{4b}(\Omega) \wedge {}^*V_{4b} +$$

$$+ k_2\, \Omega^{a4} \wedge {}^*\Omega_{a4} + k_3\, \Omega^a{}_4 \wedge V^4 \wedge {}^*(\Omega_{a4} \wedge V^4) +$$

$$+ k_4\, \Omega_{a4} \wedge V^a \wedge {}^*(\Omega_b{}^4 \wedge V^b)\,] \tag{2}$$

where the first term is the usual Einstein action (the SO(3,1) scalar curvature), $\chi = 8\pi G/c^4$ is the Newton coupling constant, and the k coefficients are dimensionless. Finally, an asterisk denotes the Hodge duality operation (here we follow the same conventions as in ref.[9]) and we have used the definition $V^{ab} \equiv V^a \wedge V^b$. It is immediate to verify that this action, constructed without using the coordinate basis in the world manifold, is automatically a scalar under general coordinate transformations, but not with respect to local Lorentz rotations.

Now we consider the matter part of the action, S_m . With the choice G_T = SO(3) as the local gauge group, the dynamical energy-momentum tensor T_{ab} , which represents the sources in the conventional Lorentz invariant Einstein theory, can be decomposed into the SO(3) tensor T_{ij}, the SO(3) vector T_{i4}, and the SO(3) scalar T_{44} (as the theory is constructed with a torsionless connection, the enrgy-momentum tensor we are considering is symmetric, $T_{ab} = T_{ba}$, and corresponds to unpolarized matter).

If the local Lorentz symmetry is broken also for the source terms contained in S_m, the components T_{ij}, T_{i4} and T_{44} contribute then with different strength to the field equations, and by varying S_m with respect to V^a one obtains in general the following expression

$$\delta_V S_m = -\int \theta_a \wedge \delta V^a \tag{3}$$

where [13]

$$\theta_a \equiv \theta_{ab}\, {}^*V^b = T_{ab}\, {}^*V^b + a_1 T_{44}\, {}^*V^4 \delta_a{}^4 +$$

$$+ a_2 T_{a4}\, {}^*V^4 + a_3 T_{4b}\, {}^*V^b \delta_a{}^4 \tag{4}$$

and the a_i are independent dimensionless parameters governing the breaking of the local SO(3,1) symmetry in the matter Lagrangian (the symmetry is restored in the limit $a_1=a_2=a_3=0$). The Lorentz nonivariance of the matter Lagrangian is represented then by the modified source θ_{ab} .

Imposing eq.(1) as a constraint, and varying the total action $S_g + S_m$ with respect to V^a (see ref.[13]), one obtains the generalized field equations of the quasi-riemannian theory, which can be written in components as follows (we set for simplicity χ = 1)

$$G_\alpha{}^\beta = T_\alpha{}^\beta + \tau_\alpha{}^\beta + H_\alpha{}^\beta \tag{5}$$

Here $G_\alpha{}^\beta$ is the usual symmetric Einstein tensor (constructed in terms of the Christoffel symbols) following from the variation of the Einstein scalar curvature, $H_\alpha{}^\beta$ (whose explicit expression is not needed here) represents the contribution of the variation of all the Lorentz breaking terms in the gravitational action S_g , and finally $T_\alpha{}^\beta + \tau_\alpha{}^\beta$ is the generalized source due to a breaking of Lorentz symmetry in S_m. From eq.(4) we have explicitly

$$\tau_\alpha{}^\beta = \Theta_\alpha{}^\beta - T_\alpha{}^\beta = a_1 T_4{}^4 V_\alpha{}^4 V^\beta{}_4 + a_2 T_\alpha{}^4 V^\beta{}_4 + a_3 V_\alpha{}^4 T^\beta{}_4 \tag{6}$$

Note that the index 4, appearing in the coordinate expression of the field equations, is always to be understood as a tangent space index, i.e.

$$T_4{}^4 = T_\mu{}^\nu V^\mu{}_4 V_\nu{}^4 , \qquad T_\alpha{}^4 = T_\alpha{}^\mu V_\mu{}^4 \tag{7}$$

and so on. The field equations (5) are then generally covariant, as expected, but locally only SO(3) invariant, as the left-hand side of eq. (5) is a world tensor (in the indices α , β) but not a Lorentz scalar, as can be seen explicitly from eqs.(6,7).

In general relativity, the equations of motion for a test body in a given external field can be obtained by integrating the conservation law $T_\alpha{}^\beta{}_{;\beta} = 0$ (a semicolon denotes the usual Riemann-Christoffel covariant derivative) which follows from the field equations $G_\alpha{}^\beta = T_\alpha{}^\beta$ and the contracted Bianchi identity $G_\alpha{}^\beta{}_{;\beta} = 0$ (Papapetrou's method [14]). In first approximation, for a structureless "pole-particle" , one gets then the geodesic equation.

In the case of the quasi-riemannian theory we are considering, the Bianchi identity for $G_\alpha{}^\beta$ implies in general, through equation (5)

$$(T_\alpha{}^\beta + \tau_\alpha{}^\beta + H_\alpha{}^\beta)_{;\beta} = 0 \tag{8}$$

In the absence of sources the field equations imply $H_\alpha{}^\beta{}_{;\beta} = 0$. In this case a test body in a given external field can be consistently represented by a covariantly conserved energy-momentum tensor ($T_\alpha{}^\beta{}_{;\beta} = 0$), provided that the breaking of Lorentz symmetry in the matter Lagrangian is absent (or neglegible).

In the case of non-vacuum solutions of the generalized field equations, on the contrary, the motion of matter in general is not geodesic, even if the matter Lagrangian is locally Lorentz invariant (i.e. $\tau_\alpha{}^\beta = 0$). In fact the requirement $T_\alpha{}^\beta{}_{;\beta} = 0$ can be satisfied in this case only if $H_\alpha{}^\beta{}_{;\beta} = 0$. But this last condition is no longer a consequence of the fiel equations, it is to be imposed as four additional constraints to be satisfied by the four parameters $k_1 \ldots k_4$, and in general only the trivial solution $k_1 = k_2 = k_3 = k_4 = 0$ could be allowed, reducing then the theory to general relativity.

It is always possible however to restore the geodesical motion, if we allow for a violation of local Lorentz symmetry also in the matter Lagrangian ($\gamma_\alpha{}^\beta \neq 0$). In fact the condition $T_\alpha{}^\beta{}_{;\beta} = 0$ implies then $(\gamma_\alpha{}^\beta + H_\alpha{}^\beta)_{;\beta} = 0$, and this system of four equations for the seven unknown parameters $\{k_i, k_4, a_i\}$ has in general infinite nontrivial solutions. It is then possible, in principle, to related the parameters characterizing the breaking of local SO(3,1) symmetry of the source to those of the pure gravity part of the action, so as to achieve the geodesic motion of matter.

Therefore the quasi-riemannian geometric model, describing a gravitational violation of local Lorentz symmetry, does not necessarily imply a breakdown of the universality of free fall, unlike other model of Lorentz noninvariance [1] . Demanding explicitly that the weak equivalence principle be satisfied could be also a useful requirement to reduce the number of free parameters of a quasi-riemannian theory. An example of quasi-riemannian cosmological model, in which the motion of matter is geodesic, and the field equations are depending only on two arbitrary parameters (and which generalizes a previous Lorentz noninvariant solution [15]) has been discussed in ref. [13] .

To conclude this section, I will deduce explicitly the generalized equations of motion for a test body in an external field (which is solution of the vacuum field equations), in the general case of broken local SO(3,1) symmetry also in the matter Lagrangian.

As $H_\alpha{}^\beta{}_{;\beta} = 0$ in this case, the conservation equation for the quasi-riemannian theory is, from eqs.(6,8),

$$T^{\mu\nu}{}_{;\nu} + a_1 (T^{44} V^\mu{}_4 V^\nu{}_4)_{;\nu} + a_2 (T^{\mu 4} V^\nu{}_4)_{;\nu} + a_3 (T^{\nu 4} V^\mu{}_4)_{;\nu} = 0 \tag{9}$$

and can be written explicitly as

$$\partial_\nu (\sqrt{-g}\, T^{\mu\nu}) + \sqrt{-g}\, \Gamma_{\nu\alpha}{}^\mu T^{\alpha\nu} + a_1 [\partial_\nu (\sqrt{-g}\, T^{44} V^\mu{}_4 V^\nu{}_4) + \sqrt{-g}\, \Gamma_{44}{}^\mu T^{44}] + a_2 [\partial_\nu (\sqrt{-g}\, T^{\mu 4} V^\nu{}_4) + \sqrt{-g}\, \Gamma_{4\alpha}{}^\mu T^{\alpha 4}] + a_3 [\partial_\nu (\sqrt{-g}\, T^{\nu 4} V^\mu{}_4) + \sqrt{-g}\, \Gamma_{\nu 4}{}^\mu T^{\nu 4}] = 0 \tag{10}$$

where Γ is the Christoffel symbol, and $g = \det(g_{\mu\nu})$.

In order to obtain the equation governing the response of the test particle to the external field, we integrate this equation over the spacelike hypersurface Σ intersecting the world line of the particle at t = const. By applying the Gauss theorem, expanding the gravitational field in power series around the coordinates x^μ of the center of mass according to the Papapetrou procedure [14], and neglecting the coupling to internal momenta for a structureless pole-particle [14] , we obtain finally

$$\frac{d}{dx'^4}\int_\Sigma d^3x'\sqrt{-g}\,T^{\mu 4}(x') + \Gamma_{\nu\alpha}{}^{\mu}(x)\int_\Sigma d^3x'\sqrt{-g}\,T^{\alpha\nu}(x') +$$

$$+ a_1 \frac{d}{dx'^4}\left[V^{\mu}{}_4(x)\int_\Sigma d^3x'\sqrt{-g}\,T^{44}(x')\right] + a_1\Gamma_{44}{}^{\mu}(x)\int_\Sigma d^3x'\sqrt{-g}\,T^{44}(x') +$$

$$+ a_2\left[\frac{d}{dx'^4}\int_\Sigma d^3x'\sqrt{-g}\,T^{\mu 4}(x') + \Gamma_{4\alpha}{}^{\mu}(x)\int_\Sigma d^3x'\sqrt{-g}\,T^{\alpha 4}(x')\right] +$$

$$+ a_3\left[\frac{d}{dx'^4}V^{\mu}_4(x)\int_\Sigma d^3x'\sqrt{-g}\,T^{44}(x') + \Gamma_{\nu 4}{}^{\mu}(x)\int_\Sigma d^3x'\sqrt{-g}\,T^{\nu 4}(x')\right] \qquad (11)$$

Now we multiply by dx^4/ds and, in the limiting case in which radius of the world tube appraches zero, $x'^{\mu} \to x^{\mu}$, we put

$$m\,\dot{x}^{\mu}\dot{x}^{\nu} = \dot{x}^4\int_\Sigma d^3x'\sqrt{-g}\,T^{\mu\nu}(x') \qquad (12)$$

where a dot denotes differentiation with respect to the proper time s along the particle world line. We obtain then

$$(1+a_2)\ddot{x}^{\mu} + \Gamma_{\nu\alpha}{}^{\mu}\dot{x}^{\alpha}\dot{x}^{\nu} + (a_1+a_3)\frac{d}{ds}\left(V^{\mu}{}_4\dot{x}^4\right) +$$

$$+ a_1\Gamma_{44}{}^{\mu}\dot{x}^4\dot{x}^4 + (a_2+a_3)\Gamma_{\nu 4}{}^{\mu}\dot{x}^{\nu}\dot{x}^4 \qquad (13)$$

Remembering that $\dot{x}^4 = \dot{x}^{\nu}V_{\nu}{}^4$, the acceleration (13) can be written explicitly as a function of the velocity, of the derivatives of the metric and of the vierbein field, in the following form (which is useful for explicit computations)

$$(1+a_2)\ddot{x}^{\mu} + \Gamma_{\nu\alpha}{}^{\mu}\dot{x}^{\alpha}\dot{x}^{\nu} + (a_1+a_3)\left(\ddot{x}^{\nu}V_{\nu}{}^4V^{\mu}{}_4 +\right.$$

$$\left.+ \dot{x}^{\beta}\dot{x}^{\nu}V_{\nu}{}^4\partial_{\beta}V^{\mu}{}_4 + \dot{x}^{\beta}\dot{x}^{\nu}V^{\mu}{}_4\partial_{\beta}V_{\nu}{}^4\right) +$$

$$+ a_1\Gamma_{\nu\lambda}{}^{\mu}V^{\nu}{}_4V^{\lambda}{}_4V_{\alpha}{}^4V_{\beta}{}^4\dot{x}^{\alpha}\dot{x}^{\beta} +$$

$$+ (a_2+a_3)\Gamma_{\nu\beta}{}^{\mu}V^{\beta}{}_4V_{\alpha}{}^4\dot{x}^{\nu}\dot{x}^{\alpha} = 0 \qquad (14)$$

If the gravitational interaction violates local Lorentz invariance but preserves SO(3) symmetry, according to the quasi-riemannian model discussed here, this is the generalized equation of motion determining the response of a test particle to a given exyernal field. Note however that, if the matter Lagrangian is locally SO(3,1) invariant, then $a_1 = a_2 = a_3 = 0$, and the equation of motion reduces to the usual geodesic equation, even if the pure gravity part of the action violates the local Lorentz symmetry.

3. VIOLATION OF LORENTZ SYMMETRY AND THE RESULTS OF EÖTVÖS-TYPE EXPERIMENTS

The possibility of a composition dependent acceleration of various materials to the Earth, recently pointed out through a careful reanalysis of the Eötvös experiment [8] , and suggested also by gephysical data [16], could be interpreted as the consequence of an intermediate range "fifth force" coupling to baryon number [8,17,18] .

Looking for other possible sources of anomalous accelerations, one of the possibilities is a gravitational breaking of the local Lorentz symmetry, since this breaking may imply non-geodesical motion, as shown in the previous section.

In this context it is even possible that the violation of the weak equivalence principle depend on the baryon content of a test body, provided one assumes that only the paths of baryons are not geodesics, that is Lorentz invariance is not the local gauge symmetry for the gravitational interactions of baryons, unlike the leptonic gravitational interactions.

This hypothesis of course is completely "ad hoc". It should be noted, however, that the possibility of different gravitational interactions for leptons and baryons is not a new idea in the scientific literature, remember for example the strong gravity theory and the tensor-meson dominance hypothesis [19] (such two-tensor theory may simulate Lorentz non-invariant effects in weak interactions [20] and, interestingly enough, a two tensor theory with a massive component was also considered as a possible explanation of the anomalous energy dependence of the K°-$\overline{K}^\circ$ parameters [21] ; however a tensor field cannot give the repulsive interaction which seems to be suggested by the geophysical data[16-18] , and moreover to reproduce an intermediate-range force the massive graviton coupled to baryons should be very light, contrarily to the original f-gravity theory [19]).

If we suppose that the gravitational interactions of baryons are described by the quasi-riemannian theory considered in the previous section, we have in general two consequences: the gravitational field produced by a source of given mass deviates from that predicted by general relativity, according to the baryon content of the source, and test bodies with different internal baryonic structure fall differently in a given external field.

Assuming that in a macroscopic context both these effects produces very small corrections to the standard theory, that is $k_a \ll 1$, $a_i \ll 1$, in order to estimate, to first order in the parameters describing the breaking of Lorentz symmetry, the difference in the gravitational acceleration of a baryon and a lepton, for example on the Earth surface, we can use the equation of motion (14) inserting, in first approximation, the metric obtained solving the field equations with all the k coefficients equal to zero, i.e. the solution of the general relativistic field equations.

In the case of a static and spherically symmetric source of mass M we have then the Schwarzschild field which, in polar coordinates x^μ =

$(r, \vartheta, \varphi, t)$ is given by

$$g_{\mu\nu} = \mathrm{diag}\,(e^{\lambda}, r^2, r^2 \sin^2\vartheta, e^{\nu}) \tag{15}$$

where

$$e^{\nu} = e^{-\lambda} = 1 - \frac{2GM}{r} \tag{16}$$

Considering a radial motion ($\dot{\vartheta} = \dot{\varphi} = 0$) in this metric, from eq.(14) we obtain

$$(1 + a_1 + a_2 + a_3)\,\ddot{t} + (1 + \tfrac{1}{2}a_2 + \tfrac{1}{2}a_3)\,\nu' \dot{r}\,\dot{t} = 0 \tag{17}$$

$$(1 + a_2)\,\ddot{r} + \frac{\lambda'}{2}\dot{r}^2 + \frac{\nu'}{2} e^{\nu-\lambda}\,\dot{t}^2\,(1 + a_1 + a_2 + a_3) = 0 \tag{18}$$

(a prime denotes the derivative with respect to r).

They can be easily integrated (see for example ref.[22]), but we are interested here only in the static limit, in which $\dot{r} = 0$, $\dot{t} = 1$, and in the weak field approximation, in which terms higher than linear in GM/r^2 are neglected; in this case the contribution of the Lorentz non-invariant terms produces only a constant renormalization of the Newton coupling, as the radial equation reduces to

$$\ddot{r} = -\frac{GM}{r^2}\left(1 + \frac{a_1 + a_3}{1 + a_2}\right) \tag{19}$$

In this approximation baryons are coupled then to the Newton gravitational field with a strength different from that of leptons, and the gravitational force acting on a macroscopic mass m_1, containing B_1 baryons of mass m_H, is given by

$$m_1 \ddot{r}_1 = -\frac{GM}{r^2} m_1 \left[1 + \left(\frac{a_1 + a_3}{1 + a_2}\right) \frac{m_H}{m_1} B_1\right] \tag{20}$$

Comparing the acceleration of two test masses, m_1 and m_2, in the Earth gravitational field, one gets

$$\frac{\Delta a}{g} = \frac{a_1 + a_3}{1 + a_2}\,\Delta\left(\frac{B}{\mu}\right) \tag{21}$$

where $\Delta a = \ddot{r}_1 - \ddot{r}_2$, $g = -GM/r^2$ is the local acceleration of gravity, $\Delta(B/\mu) = B_1/\mu_1 - B_2/\mu_2$, and $\mu_{1,2} = m_{1,2}/m_H$ denotes the masses of the test bodies in units of barionic mass.

A violation of Lorentz symmetry in the gravitational interactions of baryons leads then to a composition dependent acceleration. An interpretation, based on eq.(21), of the nongeodesic effects that seem to appear in the Eötvös experiment (according to the reanalysis of ref.[8]) would imply [8,17,18] $(a_1 + a_3)/(1 + a_2) \sim 10^{-6}$, as $\Delta(B/\mu) \sim 10^{-3}$. This geometric model of Lorentz nonivariance predicts however a constant shift in the gravitational coupling strength of baryons, so that the same effect should be observed also at different distances from the source. Since this is not the case, as implied by the results of the experiments of Dicke [5] and Braginskii [6] that, performed in the solar gravitational field, give $\Delta a/g \lesssim 10^{-12}$, this interpretation of the Eötvös data is to be ruled out, unless, in some more sofisticated model of spontaneous breaking of the Lorentz gauge symmetry, the phenomenological parameters a_i become position dependent variables, so that the coefficient of $\Delta(B/\mu)$ decreases with the distance from the source.

4. CONCLUSION

To discuss the consequences of an eventual violation of local Lorentz symmetry in the context of the gravitational interactions, we have assumed in this paper a model of gravity with broken SO(3,1) symmetry, based on a four-dimensional quasi-riemannian theory with SO(3) as the tangent space group. Considering the equations of motion for a test body in this theory, it has been shown that in general the motion is not geodesic; with a suitable choice of the parameters characterizing the strength of the Lorentz breaking terms in the matter Lagrangian, however, it is possible in principle to restore the covariant conservation of the energy-momentum tensor, and to obtain the geodesic equation of motion.

Therefore a gravitational interaction without local Lorentz symmetry does not necessarily implies a breakdown of the principle of the universality of free fall. This one of the two main points stressed in this paper.

The other result obtained with this phenomenological investigation is that, in the hypothesis that the gravitational interactions of baryons are not locally Lorentz invariant, it would be possible to obtain a non-vanishing, composition dependent, difference in the gravitational acceleration of two test bodies. The gravitational Lorentz noninvariance produces however, in the Newtonian approximation, a position independent gravitational coupling, which is strongly constrained by the results of Dicke [5] and Braginskii [6]. It seems therefore that a constant violation of the universality of free fall, of the strenght required to explain an eventual anomalous result of the Eötvös experiment, is to be ruled out.

Nevertheless, it is perhaps worth stressing that a gravitational Lorentz noninvariance of baryons may lead to a composition dependent effective coupling constant, as this violation could be a possible interesting explanation if an anomalous acceleration would be discovered also in the solar gravitational field, since in that case the effect could not be ascribed to the intermediate range interaction of refs.[8,17,18].

ACKNOWLEDGEMENTS

It is a pleasure to thank V.De Sabbata for his invitation to the 1987 Erice School on Cosmology and Gravitation, and for many interesting conversations. I am also very grateful to C.W.Kim for reading a preliminary draft of this paper and for his very useful comments.

REFERENCES

1. E.Fischbach et al., Phys.Rev.D32,154(1985)
2. H.B.Nielsen and I.Picek, Phys.Lett.114B,141(1982)
3. H.B.Nielsen and I.Picek, Nucl.Phys.B211, 269(1983)
4. S.Weinberg, Phys.Lett.138B,47(1984)
5. P.G.Roll, R.Krotkov and R.H.Dicke, Ann.Phys.26,442(1964)
6. V.B.Braginskii and V.I.Panov, Zh.Eksp.Teor.Fiz.61,873(1971)
7. R.V.Eötvös, D.Pekàr and E.Fekete, Ann.Phys. 68,11(1922)
8. E.Fischbach et al, Phys.Rev.Lett.56,3(1986)
9. S.P.de Alvis and S.Randjbar-Daemi, Phys.Rev.D32,1354(1985)
10. K.S.Wiswanathan and B.Wong, Phys.Rev.D32,3108(1985)
11. M.Gasperini, Phys.Rev.D33, 3594(1986)
12. A.Zee, Phys.Rev.D25,1864(1982)
13. M.Gasperini, Class.Quantum Grav.4,485(1987)
14. A.Papapetrou, Proc.Roy.Soc.209A,248(1951)
15. M.Gasperini, Phys.Lett.163B,84(1985)
16. F.D.Stacey anf G.J.Tuck, Nature 292,230(1981)
17. E.Fischbach et al, Proc. of the XXIII Int.Conference on High Energy Physics (Berkeley, 1986)
18. see also the lecture of E.Fischbach in this volume
19. C.J.Isham, A.Salam and J.Strathdee, Phys.Rev.D3,867(1971)
20. M.Gasperini, Phys.Lett.141B,369(1984)
21. S.H.Aronson et al, Phys.Rev.D28,495(1983)
22. M.Gasperini and A.Tartaglia, Mod.Phys.Lett.A (1987) (in press)

STATUS OF THE NEWTONIAN GRAVITATIONAL CONSTANT

G. T. Gillies
Department of Physics
University of Virginia
Charlottesville, Virginia 22901
USA

ABSTRACT. During the past 20 years, there has been renewed interest in Newtonian gravity and, in particular, on the constancy of the Newtonian gravitational constant, G, in space and time. Curiosity alone has provided the classical motivation for studies of this very weak force. However, the push towards a unified field theory is now beginning to drive the development of new ways of measuring G precisely. Reviewed here is the status of our knowledge of the absolute value of G, searches for couplings of gravity to other forces, and measurements of the dependency of G on various physical quantities and parameters.

1. INTRODUCTION

It is not unusual for an undergraduate Physics program to include a measurement of the Newtonian gravitational constant, G, as part of the laboratory curriculum. Most students in such a program may not realize it but their measurements of G, which can often be made to 5% or so, are within a factor of 50 to 100 of the most precise absolute measurements of G ever made. This situation is unique in Physics, since all other fundamental physical constants (excluding the Stefan-Boltzmann constant) are known with uncertainties typically no larger than a few parts per million (ppm). The extreme difficulties encountered in measuring G arise from the several systematic effects which, in all experiments to date, have acted to mask the isolated gravitational force [1]. Relative measurements of G have been undertaken in efforts to uncover couplings between gravity and the other forces of nature, and to search for dependencies of G on some particular quantity or parameter of interest. Such experiments do not normally call for high precision absolute metrology but instead demand very sensitive and stable signal detection schemes.

From Newton's time up until that of Einstein, the most compelling reason for measuring the constant of proportionality of the inverse square law of gravitation arose from curiosity about the mean density of the earth. These two quantities are related by the expression

V. De Sabbata and V. N. Melnikov (eds.),
Gravitational Measurements, Fundamental Metrology and Constants, 191–214.

$$G\,\delta = \frac{3\,g}{4\rho R} \tag{1}$$

where ρ and R are the mean density and radius of the earth, and g is the gravitational acceleration at its surface. Clearly, then, a measurement of G makes it possible to determine the mass of the earth (in kilograms) with an uncertainty commensurate with that of the G measurement itself. In fact the geocentric gravitational constant, ie, the product GM, is known to an uncertainty of about 10^{-8} [2]. If, therefore, one could conceive an independent method of determining the mass of the earth with high precision, the new reduced level of uncertainty could be translated directly into a more precise determination of G. Unfortunately, no such method is yet available.

Around the turn of this century, interest in absolute measurements of G waned somewhat, but a great deal of interest arose in searches for variations in G and couplings of gravity to the electromagnetic force and the newly discovered phenomenon of radioactivity. This interest has continued to the present and, in fact, grown substantially. Hand-in-hand with it has come renewed interest in determining the absolute value of G, spurred in part by modern technological advances. Moreover, the push towards the achievement of a totally unified description of the forces of nature (including gravity) has also done much to stimulate new work in this area. It is difficult to see how the scale factor(s) of such a theory could be constructed without including G. The difficulty, then, is that the precision of the determinations of the new scale factors would be limited by the uncertainties in our knowledge of G. Given the extreme slowness with which advances in the determination of G have taken place over the last two centuries, one can predict that progress towards the verification and ultimate acceptance of such a unified theory might stall. In principle, this problem can be avoided by establishing a fixed value for G as has been done for the speed of light, c. The problem then becomes one of arriving at an acceptable theoretical prediction for G in the same way that c is predicted by electromagnetism. At present, no such theoretical prediction for G is available, although there are a few efforts in progress towards this goal; see [1] for a review. From the historical perspective, there is no accepted experimental evidence to indicate that G is indeed related to any of the other fundamental constants. Therefore, not many have been motivated to work on this problem. On the other hand, a unified field theory may lead to a self-consistent prediction for the value of G. Nevertheless, it will still be necessary to experimentally verify such a prediction at an acceptable precision level before we risk closing the issue by assigning a self-consistent but physically inaccurate value to G.

Finally, the recent suggestions of the existence of forces weaker than gravity [3] and the anomalously high values of G as measured geophysically [4] provide additional motivation for further high precision studies of the scale dependence of G.

It is clear, then, that an unambiguous 1 ppm measurement of G (not to mention continued searches for variations in G) would be most welcome if no other reason than to end the tenacity with which G has yielded its digits!

In what follows, the status of the various measurements of G are reviewed. We begin by considering what is known about the absolute value of G, then go on to examine the searches for couplings of gravity to other forces and measurements of variations of G. Other reviews on this topic are also available [5,6,7].

2. THE ABSOLUTE VALUE OF G

2.1. History

Although there were pre-Newtonian theoretical and experimental investigations of masses and the ponderomotive forces acting on them, the history of G, or equivalently δ, begins in the post-Newtonian era with the geophysical studies of Bouguer and de la Condamine. They measured the period shift of a pendulum and the deviation of a plumb-line from vertical as produced by the attraction of the mountains Corilleras and Chimborazo in Ecuador. Others used this technique to make similar measurements, but there is no agreement at all among the different results. Pendulum gravimeters were also used in mines early on, and the name of Airy is most often associated with the development of this technique. The principle was to measure g as a function of depth in mine, and from the measured vertical gravity gradient, one can then extract the value of G. This method has been used recently by Stacey and colleagues, but more will be said later about their work. The results of all these early measurements are summarized beautifully by A. S. MacKenzie [8] and E. D. Preston [9], and a modern interpretation of geophysical measurements of G is given by Stacey, et al [4].

Laboratory measurements of G began with the torsion balance as developed for this purpose by Cavendish [10]. He measured the deflection of the suspended beam resulting from the attractive action of nearby masses. Others repeated his work and also used the torsion balance in different, more metrologically sound modes. For instance, Reich [11] realized that a torsion pendulum's oscillation frequency shifted when the attracting masses were removed. This made measurements of G dynamic rather than static in nature, and was to become the preferred technique for virtually all later experimentation. Kunz [12] conceived the idea of driving a torsion pendulum at its resonance frequency via externally oscillating masses. While an elegant technique in principle, it can lead to very large amplitude oscillations of the balance beam. Since the oscillation period becomes a strongly nonlinear function of the amplitude at large amplitudes, great care must be taken when using this method. Other workers employed vertical pendulums and beam balances to make measurements of G; for a review of these techniques, see [6]. Listings of the results of the different determinations of G carried out prior to 1900 have been reported in several places, most notably by Poynting [13].

2.2 Modern Laboratory Measurements

Prior to 1986, the official CODATA evaluation of G was based largely on the work of Heyl and Chrzanowski [14]. There was relatively little work done on the absolute value of G during the 1940's and 1950's. In the early 1960's, J. W. Beams and colleagues of the University of Virginia for the first time introduced the concept of feedback into torsion balance measurements of G [15]. They did this by placing the torsion balance on a turntable that could be driven with a constant acceleration equal to that of the balance beam acted on by the large attracting masses. This experiment had as its goal the elimination of uncertainties due to surrounding gravity gradients via the averaging effect of the turntable's rotation. These workers also attempted to incorporate magnetic suspension of the test mass into the experiment in order to eliminate the large source of drift otherwise unavoidable in torsion fiber based measurements. These improvements and others were potentially useful but the experiment failed to produce the expected 1 ppm accuracy [16] and in the end, the fedback rotational mode was abandoned for operation in the dynamic torsion pendulum mode (with a tungsten torsion fiber). By this time, the apparatus had been taken to the U.S. National Bureau of Standards in Washington, D.C., and it formed the basis for the experiment by Luther and Towler which to date claims the smallest measurement uncertainty of any determination of G.

During this time, there was similar work elsewhere on the determination of G. At the Quantum Physics Division of the U.S. National Bureau of Standards in Boulder, Colorado, Dr. J. E. Faller and his student W. A. Koldewyn were using magnetic suspension to replace the torsion fiber, but encountered difficulties in stabilizing the restoring torque constant of the suspension. They used large, thick-walled hollow cylinders as the attracting masses, a clever innovation which was intended to ease the absolute metrology requirements of their experiment. A "gravitational doughnut" of this type produces a gravitational force that vanishes at its center of mass and at infinity. In fact, there is a region near the face of the cylinder where the force reaches a maximum and around which the force varies slowly. By operating the torsion balance in this region, where the resultant field changes very little with respect to distance, the requirements on the exactness of determination of the distance between the center of masses of the attracted and attracting bodies could be relaxed somewhat. Nevertheless, mass inhomogeneities in the material of the cylinders limit the level of improvement that can be obtained [17].

An experiment was designed for use in the Grotta Gigante in Trieste, Italy which uses another unique arrangement of masses. Professors A. H. Cook and A. Marussi (deceased) have constructed a very large scale torsion balance in which for the first time the gravitational forces in a G experiment will dominate the disturbance forces. Since it is virtually impossible to make high accuracy solid spheres on a scale of 1 m, they decided to assemble together cylindrical structures which were arranged such that the lower order, even zonal harmonics vanished. Large scale cylinders can be machined with relatively high accuracy, and with sufficient care, one can produce a cylindrically symmetric body that has the field of a point source, ie, of a sphere. In their experiment, the large masses are on the order of 500 kg and the small masses are on the order of 10 kg. No result for G has been produced to date by this experiment.

Since 1974, eleven redeterminations of G have been published [7], and three of them have claimed precisions of 10^{-4}. The results of these high precision experiments are shown in Table I.

TABLE I

Authors	Year	Technique	Result $(G \pm \Delta G) \times 10^{-11}$ $m^3kg^{-1}s^{-2}$
Facy, Pontikis [18]	1972	Resonant Pendulum	6.6714 ± 0.0006
Sagitov et al. [19]	1979	Torsion Pendulum	6.6745 ± 0.0008
Luther, Towler [20]	1982	Torsion Pendulum	6.6726 ± 0.0005
CODATA [21]	1986	-	6.67259 ± 0.00085

Table I. The three most precise laboratory determination of G. Also shown is the 1986 CODATA value.

Cohen and Taylor [21] discuss the experimental values in detail. The most interesting thing to note is that all these values exclude each other within the limits of the errors quoted. If we weight each of these three results equally, then it is clear that we do not know the value of G with an uncertainty of 10^{-4} as is otherwise suggested by the individual measurements. The experiment by Luther and Towler is generally thought (in the western world) to have been the most carefully done. There is relatively little documentation covering it, though. The work of Sagitov et al. is largely unknown in the west, due sadly to a typographical error in the original paper announcing their result: The first digit to the right of the decimal point in their value of G was dropped, making it appear to be largely uninteresting when in fact it was among the best pieces of work done to date. The elegant experiment by Facy and Pontikis involved a resonant torsion pendulum, but the result for G as obtained by this technique has never been independently verified.

2.3. Modern Geophysical Measurements

As mentioned above, Stacey et al. [4] have made determinations of the absolute value of G by gravimetry in mines. They have also produced a preliminary result for G by weighing masses with a beam balance situated in a variable depth hydroelectric reservoir. Their interest in geophysical determinations of G was stimulated when they noticed that all the available gravity surveys of mines produced values of G about 1% larger than the laboratory values. They undertook a series of their own measurements in two mines in Australia, and have arrived at the values of G shown in Table II. Their beam balance result is cited for comparison.

TABLE II

Location	Method	Result $(G \pm \Delta G) \times 10^{-11}$ $m^3kg^{-1}s^{-2}$
Hilton	Gravimetry	6.720 ± 0.024
Mount Isa	Gravimetry	$6.704^{+0.089}_{-0.025}$
Splityard Creek	Beam Balance	6.668 ± 0.040

Table II. Results of the geophysical and beam balance determinations of G by Stacey et al. [4].

The consistently higher geophysical values of G coupled with the care with which Stacey et al. have taken their data provided part of the reason why Fischbach et al. [3] suggested that there is a force weaker than gravity which acts over geophysically appropriate distance scales (ie, about 200 m). One can take an alternative approach to this interpretation and say that perhaps there is some as yet undiscovered systematic error which affects all of the torsion balance experiments or a different error which affects all of the geophysical experiments. Since G has seldom been measured by any technique other than these two it is not unreasonable to suppose that such a situation might exist. On the other hand, a comparison of the geophysical results with the laboratory results is instructive. If we take the high error limit of Sagitov et al. and the low error limit of Stacey et al.'s Mount Isa experiment, we find that they miss overlapping by no more than the difference between the means of the highest and the lowest of the laboratory values. Perhaps the results due to these two different measurement techniques are not that disparate after all.

2.4 Measurements in Progress

The comments of the last section highlight the importance of finding independent methods of measuring G. This of course, has been a long standing goal of both metrologists and physicists for several years now, although at present there are just a few active programs in progress.

The most promising experiment is that led by Prof. H. de Boer of the Physikalisch-Technische Bundesanstalt (PTB) in Braunschweig, West Germany. His group has built a torsion balance which operates in the deflection mode. It has several unique features which deserve mention. They have resurrected an idea due originally to Burgess [22] wherein the load carried by the balance beam (ie, the attracted masses) is greatly increased by virtue of incorporating a fluid suspension which buoyantly lifts virtually all of the mass. A delicate torsion fiber then provides the residual support and restoring force. The angular displacement of the beam is measured by a quadrant electrometer and the absolute intermass spacings are monitored continuously by optical interferometry. The entire apparatus is installed on a servo controlled pneumatic suspension and it has produced preliminary results [23].

The Cavendish Laboratory in England has (appropriately enough!) been the site of some recent measurements of G, as well. C. Speake built a beam balance with electrostatic feedback that has been employed to make a preliminary determination of the gravitational constant [24]. Additionally, Prof. A. J. F. Metherell and colleagues have been investigating the question of the optimum shape of the masses to be used in determinations of G. Their results are now in print [25,26]. Personnel from the Bureau International des Poids et Mesures (BIPM) in France have been collaborating with the Cavendish for several years on these projects. At present, a refined version of a servo controlled, flexure strip balance [27] and a vertical pendulum have been built by Quinn et al. (the latter instrument was begun at the Cavendish) and are being prepared for use at the BIPM in measurements of and searches for variations in G [28,29].

There are several torsion balance experiments being conducted in the United States, but mostly for searches for a fifth force or tests of the weak equivalence principle. With respect to the measurement of G, the group at the University of Virginia has built and is now operating a superconducting levitation system which at present suspends only a mirror, but is designed to levitate a small dumbbell. By using superconducting levitation, rather than active, room temperature magnetic suspension, they hope to circumvent many of the stability problems that have arisen in the previous attempts to replace the torsion fiber by a magnetic field [17].

2.5. Future Work

The tremendous growth of mankind's technical capabilities in the last twenty years is now making the exploitation of outer space a reality.

From the gravitational physicist's point of view, this is a welcome opportunity to do experiments with test masses in an environment free from the large terrestrial gravity gradients. Of course, the gravity gradients due to the spacecraft itself, and the residual field of the earth will still place limits on the precision of G measurements made in orbit, but quite possibly at a level below which the inhomogeneities of the materials themselves become a problem. There have been several space-based measurements of G proposed, and citations to them can be found in section 1 of the bibliography on G assembled by Gillies [7].

It will take several years for any such experiment to be prepared, tested, approved and flown. In the meantime, progress on conceiving new earth based measurements of the absolute value of G must continue. There are two proposed experiments which have received some attention recently. The first is by an East German group [30] who suggest that the mechanical oscillations of a superconducting ring levitated between attracting masses will lead to an improved value of G. The second is a possible experiment described by Opat [31] in which molecular interferometry would be employed to measure gravitational effects and perhaps G. This latter experiment could be most interesting since it opens the door to incorporating the high precision techniques of atomic physics into gravitational experiments.

Speake and Gillies [1] have reviewed the nature of the various problems which pose precision limits on the standard torsion balance measurements of G, and Stegna and Sagitov [32] have examined several of the modern era experiments in great detail.

3. SEARCHES FOR COUPLINGS OF GRAVITY TO OTHER FORCES

3.1 Gravito-electromagnetic couplings

The various similarities between gravitation and magnetism did not escape the early natural philosophers and scientists. Gilbert, for instance, concluded in 1600 that gravity was the result of the natural magnetism of the earth, and that the force of gravity must diminish with distance away from the earth in much the same way as the attraction between bar magnets falls off with distance [33]. Of course, the gravitational force has always been observed to have only one polarity, ie, it is always attractive; this is not the case for either electricity or magnetism. This intrinsic difference is no less mysterious today than it was in Gilbert's, or for that matter, Aristotle's time.

The asymmetry in the forces does not stop there. The strength of the electromagnetic force between two fundamental particles (say, between two protons) is roughly 10^{40} times the gravitational force between them. This, and the fact that all masses interact with all other masses gravitationally, makes G so difficult to measure: It is virtually impossible to isolate and measure the gravitational force in the same way that one can isolate and measure electromagnetic forces. These difficulties are compounded when one is searching for a coupling between gravity and the electromagnetic forces. The principle question becomes one of insuring that the effects of the two forces are sufficiently separated in the experiment to prevent one from masking the other. At first glance, this might seem an insurmountable problem at even modest levels of precision, but in fact, interesting limits have been placed on the size of such couplings.

L. A. Bauer of the Carnegie Institution of Washington, D.C., made measurements of permanent magnetic materials in both the magnetized and unmagnetized states. He concluded that the magnetic moment of the material somehow altered the weight of the magnet [34]. M. G. Lloyd repeated Bauer's experiments and concluded that there was no such coupling at the level of 5×10^{-12} gauss^{-1} [35]. Richardson and Simons studied an electrified beam balance in the hopes of seeing if the high voltage side of the balance weighed more or less than the grounded side. They found a null result bounded at 1.2×10^{-7} V^{-1} [36]. F. E. Nipher spent several years trying to establish the existence of a gravitoelectric coupling, and claimed to see evidence for it [37] in rapid energy release events (explosions), but there has never been reported any independent substantiation of his work.

A somewhat related topic is the question of gravitoelectric induction of the type proposed by Faraday and sought by Blackett. Woodward has proposed a modern experiment to investigate this point, and his paper reviews the literature on this subject [38].

3.2. Gravitonuclear Couplings

The discovery of radioactivity gave physicists much food for thought at the turn of the 20th century. It took Rutherford's patient investigations of the nuclear nature of matter and Fermi's postulation of the neutrino to make it clear that new forces were at work. Nevertheless, many researchers were already at work during this time on the question of the role of gravity, if any, in these interesting new phenomenon. Much of the interest in this area was stirred by R. Geigel [39] who claimed to have experimentally uncovered an absorption of gravitational energy in radioactive substances. A flurry of activity surrounded this announcement, and shortly thereafter, it was shown by W. Kaufmann [40] that Geigel's find was in fact due only to a thermal gradient in his balance. Interest in this area had been established, though, and many eminent researchers spent much time building elaborate torsion and beam balances to study the question. They included Sagnac, Thomson, Southerns, Zeeman, Rutherford and Compton. For a listing of the citations to their work, see section 8 of reference [7].

In more recent times, interest has again arisen in this area, due to an effect observed by D. H. Whitaker of Los Alamos National Laboratory [41]. He reports that a sample of radioactive ^{198}Au exhibits a seven standard deviation shift in its nuclear decay rate when subjected to a centrifugal (pseudogravitational) field of about 200,000 g. His result is not widely known, so it has never received independent study.

3.3. Gravitational Shielding, Absorption, and Permeability

In discussing questions of the coupling of gravity to other forces, one should also consider the possibility of non-Newtonian self-couplings of the gravitational field of a given mass to other mass distributions. From a purely empirical standpoint, such an effect might exhibit itself in the form of either a reduction or augmentation of the gravitational force on the test mass, ie, either a shielding or absorption effect, respectively. In either case, the effect can be parameterized as a type of gravitational permeability, in analogy to magnetic permeability. Another way of looking at this would be to assume a dependence of G on material, perhaps of a nature not inconsistent with a violation of weak equivalence. A shielding effect, then, would indicate that the screening material has a "diagravimagnetic" or antigravity behavior, while an absorption effect would then correspond to a "paragravimagnetic" or supragravity behavior. Weber has explored the way in which these effects arise in general relativity due to the nonlinearity of the field equations. Forward has done the same and produced detailed comparisons of these effects with their electromagnetic analogs. The estimated values of the general relativistic phenomenon, based on the calculations of these authors, are far too small to ever be observable.

The earliest experiments in this field involved the use of Cavendish balances with screens of different density materials placed between the balance beam and the attracting masses. The intent of the experimenters was to see if measurements of G were affected by the presence of intervening screens. The classic tests of this type are summarized in Table III, wherein the results are expressed in terms of a bound on the variation of the gravitational constant, $\Delta G/G$.

The results shown in Table III are consistent with no effect, but this has not been the case with all experiments designed to study shielding and absorption. Q. Majorana [45] built a very sensitive beam balance to search for the absorption of the gravitational force acting on a suspended lead sphere with and without screens of mercury and lead around it. In both cases, he found a positive result for what he called his "quenching factor," a universal constant which governed the size of the screening effects he predicted. His results are shown in Table IV. The apparatus used by Majorana in his first experiment, involving a mercury screen around the lead test mass, is shown in Figure 1. As seen there, the mercury could be pumped into and out of the screening chamber through a hose ("M" in Figure 1). In his second apparatus, not shown, a lead brick housing was made to fit around the test mass.

At about this time, a German theorist, K. F. Bottlinger had developed a parameterized theory of gravitational absorption in which the equivalent of Majorana's quenching factor was included. According to Bottlinger, if we call q a gravitational "ray" of some initial intensity q_o which passes through material of density ρ, and thickness L, then λ, the absorption coefficient with units of $cm^2/gram$,

TABLE III

Author	Date	Screening Masses	Result, $\Delta G/G$
Austin and Thwing [42]	1897	Lead, Mercury, Zinc, Iron, Water, Alcohol and Glycerine	< 0.002
Laager [43]	1904	Lead, Aluminum, Copper, Iron, Zinc, and Paraffin	< 0.001
Erismann [44]	1911	Mercury, Water and Paraffin Oil (all liquids)	< 0.001

Table III. Results of the classic searches for gravitational permeability using a Cavendish balance apparatus. The data of Kleiner [49] are excluded from this table, since it is not clear from his publication what the limit on $\Delta G/G$ would be.

is defined by the equation

$$q = q_o \, e^{-\int \lambda \rho dL} \approx q_o \, (1 - \lambda \int \rho dL) \tag{2}$$

An expression of this type [46] has served as the standard parameterization for studies of such effects from Majorana's time to the present. In what follows, we shall adhere to this interpretation.

Braginsky, Rudenko and Rukman [47] did a dynamic version of Majorana's experiment in which a steel rotor was spun between two brass cubes fixed on a balanced yoke made of aluminum. The blades of the rotor were intended to shield the field of the earth from the upper brass cube as they rotated through the gap in the yoke. The lower brass cube served to cancel the Newtonian attraction of the blade on the upper cube. The motion of this balance was detected synchronously at the gravitational field chopping frequency (twice the rotation speed of the two blade steel rotor). They found a null result for the absorption coefficient at a level about four times smaller than that of Majorana. Later, Braginsky and Martynov [48] used an evacuated, electro-statically sensed Cavendish balance to detect the motion of an external mass screened by a segmented brass cylinder. The result was null for a value of the absorption coefficient a full order of magnitude less than that of Majorana. Note also, however, that in Braginsky's second experiment the earth was not used as the source of the gravitational force acting on the test mass. This may be important, as we shall see later. The results of the Soviet workers are also displayed in Table IV.

TABLE IV

Author	Date	Technique	Screening Material	Result, λ cm^2/gram
Majorana	1920	Beam Balance	Mercury	6.7×10^{-12}
Majorana	1922	Beam Balance	Lead	2.8×10^{-12}
Braginsky, et al.	1962	Synchronously sensed beam balance	Steel	$< 7.0 \times 10^{-13}$
Braginsky, et al.	1966	Cavendish balance	Brass	$< 2.8 \times 10^{-13}$

Table IV. A comparison of the gravitational screening experiments of Majorana with those of Braginsky and colleagues.

The smallest limits on λ have been established by gravimeter and horizontal pendulum experiments during solar eclipses, with the moon serving as the screening mass. Several groups have participated in this kind of exercise, and the appropriate publications are indexed in section 3 of reference [7]. The most carefully done of the dozen or so such experiments appears to be that of Slichter, Caputo and Hager [50]. They used a LaCoste-Romberg gravimeter to search for gravity variations before, during and after the total solar eclipse of February 15, 1961. Power spectrum analyses of their data indicate that $\lambda < 8.3 \times 10^{-16}$ cm^2/gram.

Such high precision limits on the size of λ have led Groten to conclude that "Gravity absorption should no longer be considered in theory in the way we did twenty years ago. Today such considerations seem too simplified [51]." He is most correct in stating this. Bottlinger's and Majorana's assumption that λ is a universal constant and is the same for all material everywhere in space (ie, that it has no composition or scale dependence) has been proven incorrect, except for the case (consistent with general relativity) where λ is uninterestingly small. On the other hand, Majorana was known to be a very careful and competent experimentalist. Could it be true that λ is not a universal constant but rather is a quantity with a certain composition and range dependence? That is, could λ be measured to be vanishingly small in certain classes of experiments, yet be measured to have a finite value in others (for instance, in the work of Majorana)?

3.4. Couplings of Gravity to Forces Weaker than Gravity

The suggestion by Fischbach et al. [3] that a new, weak, short range force exists has caused considerable controversy. From a reanalysis of the Eötvös data, they concluded that the new force couples to baryon number with a range of about 200 m. This conjecture would explain the (apparently) anomalously high value of G determined geophysically by Stacey et al. [4]. Much subsequent experimental activity has been undertaken in response to this suggestion and the results are mixed at this point (see the article by Fischbach in these Proceedings for a review of the experimental situation). One thing is clear so far: If all the experimental results are free of errors and if they are to be reconciled, the original concept of a coupling of this new force to baryon number alone must be abandoned or significantly altered. It may be that, as discussed above, the experiment of Majorana and an assumption of a new short range coupling to lepton number hold the critical clue [52]. All of the absorption experiments during solar eclipses involve the moon whose orbital radius places it beyond the interaction distance of the proposed new short range force. On the other hand, the Cavendish balance experiments involve intermass spacings which are too small, leaving only the first experiment of Braginsky and colleagues as an independent check on Majorana's result, at least in terms of the range of the new force. Any further intercomparison of these experiments then becomes a question of understanding the nature of the coupling mechanism of the new force. This line of reasoning is now being followed [52] and promises to be interesting.

4. VARIATIONS IN G

4.1. The Temporal Constancy of G and of Masses

The question of a time variation of G has a long history, with the first attempt to put such an effect on a physical basis being Dirac's Large Numbers Hypothesis (see the article by Hellings in these Proceedings). Various cosmological models developed in the early 1960's also called for a time varying G. These included the Hoyle-Narlikar theory, the Brans-Dicke theory, and others. All of them predicted rather similar time rates of change of G, typically on the order of the Hubble age of the universe, ie, $(dG/dt)/G \approx 10^{-11}$ per year.

As theoretical interest in the existence of such an effect grew, laboratory experiments aimed at testing the theories were conceived. The earliest attempts, at MIT and Princeton, used fedback gravimeters to measure $(dg/dt)/g$, in the hopes of deriving $(dG/dt)/G$ from the data. The results are reviewed by Gillies in section 11 of [7]. Although much care went into these efforts, cosmologically interesting levels of precision were never attained. From the experimental standpoint, interest then began to shift to the possible use of astronomical observations. Several workers studied the spin down rates of the earth and ancient eclipse records in attempts to establish upper limits on $(dG/dt)/G$. Moreover, the Apollo missions made it possible to place retroreflectors on the moon's surface, thereby permitting high accuracy lunar laser ranging. This makes it possible to put a limit on time varying gravity by monitoring the tidal acceleration of the moon. While useful, the technique has not yet achieved the level of precision gained by other techniques. Independent of the ranging efforts, the orbital velocity of the moon has also been observed and compared against time as kept by an atomic clock. These experiments have led to the only non-null result to date: $(dG/dt)/G = (-6.2 \pm 2.2) \times 10^{-11}$ per year [53], a value which is not in agreement with any other measurement. The highest precision results have been produced by the group at the Jet Propulsion Laboratory and by the workers at the Harvard-Smithsonian Astrophysical Center (most formerly at MIT) who have used radar ranging to the Viking landers on the surface of Mars. In both cases, the results are consistent with a null value for G-dot (see the articles by Hellings and Reasonberg in these Proceedings). This technique shows promise of placing even tighter limits on the size of such an effect if additional data can be taken during the course of an upcoming Soviet mission to land a spacecraft on the surface of Phobos, a moon of Mars.

The question is not closed, however [54]. The quantity measured by the radar ranging method is $\{d(GM)/dt\}/GM$ where M is the mass of the body in question, in this case Mars. If G and M were separately varying, this technique could not distinguish the size of the individual rates. The results gathered to date are consistent with the predictions of general relativity which requires that both G and rest masses remain constant in time. However, there are several other theoretical interpretations possible, among them those due to at least two versions of Dirac's Large Numbers Hypothesis (ie, with multiplicative or additive matter creation). From these alternate points of view, G, rest masses and the planet's orbital period could be covarying in such a way as to lead to a null results consistent with the careful and accurate measurements made by the radar ranging groups. Furthermore, the role of Mach's principle in the interpretation of the astronomical data is unclear. The nature of the balance between the centrifugal and gravitational forces acting on the planet could be such as to produce an intrinsically null result in any astronomical measurement of G-dot.

In considering difficulties of this nature, Dirac foresaw that it would, perhaps, be necessary to measure the stability of the gravitational force in terms of the electrical force, as can only be done in an appropriate laboratory experiment. J. W. Beams realized that a variation of his rotating Cavendish balance experiment to measure the absolute value of G could accomplish this [55], and Ritter and Beams designed an experiment based largely on this concept. Begun in 1973, various aspects of this experiment have been well developed [56], but the experiment as a whole has never been run.

A separate experiment designed to search for spontaneous matter creation was also designed by Ritter at the University of Virginia [57]. He and several students have been developing the room temperature version of it since 1974. In it, the inertial spindown of a magnetically suspended cylinder is monitored. If all other drag torques can be made negligibly small or can be accurately accounted for, then the residual angular deceleration of the rotor might be due to a secular growth in its moment of inertia resulting from matter creation. Much effort has gone into the design of stable, active permanent magnetic suspensions to serve as ultra-low drag bearings for the rotor. One such suspension system is shown in block diagram in Figure 2. It is similar to that developed by Faller and Koldewyn [17], but has some unique features [58]. Figure 3 is a photograph showing an overall view of the experiment resting on its vibration isolation.

4.2. The Spatial Constancy of G

So much has been written in recent years about the possibility of a scale dependence of G (ie, a violation of the inverse square law) that only a few highlights of the story need by given here.

Although MacKenzie's 1894 result seemed to rule out the possibility of such an effect on the scale of 3 to 7 cm, interest in such an effect on a laboratory scale grew when D. Long suggested that a reanalysis of past absolute measurements of G supported a violation of the inverse square law. A dozen different laboratory experiments were initiated to test this claim, and it is now firmly established that at least over distances up to 1 m, G does not depend on intermass spacing. The same conclusion can be drawn at the other end of the distance scale. Astronomical observations have shown that the inverse square law of gravity is indeed obeyed exactly on large distance scales, except in the cases where general relativistic effects become nonnegligible, as in the case of the advance of the perihelion of the orbit of Mercury.

At distances of about 10 m to 10 km, though, the story is different. The only data available at present are those of Stacey et al. [4] and those of a few other gravity surveys which have been made available to Stacey. As mentioned above, these workers find a consistently higher value of G, by about 1% or so, in this range. It is now very important that independent measurements of the gravitational force be made at these distances. The classical Cavendish techniques for doing this are virtually useless over such large distances, with the possible exception of a space-based experiment. Therefore, we must rely on other methods. Fortunately, a new gravity gradiometer being built by Paik, et al. at the University of Maryland [59] shows great promise of being useful at these intermediate distances. Preliminary experiments with it have been completed and development work on it is underway. When this instrument becomes fully operational, it should provide the definitive means of testing Fischbach et al.'s suggestion that there exists a composition dependent force weaker than gravity.

For a thorough listing of references to searches for variations in G with distance, see section 10 of [7].

4.3. The Dependence of G on Temperature

In 1785 Dr. George Fordyce, working in England, published some rather surprising experimental results which claimed to show that certain bodies lost a small but measurable amount of weight on being either heated or cooled. This announcement stimulated Count Rumford to repeat Fordyce's experiments using the best available bullion balance to weigh various masses over a wide range of temperatures. He concluded that the results of Fordyce must have contained a systematic error, for all of the new data indicated that the gravitational force had no temperature dependence, at least at the level of 3.3 parts in 10^8 K^{-1}.

Over a century later, and again in England, Prof. J. H. Poynting set out to improve upon the limit established by Count Rumford. He and his associate, P. Phillips, also used a beam balance from which were suspended masses that could be heated and cooled. They were successful in placing a limit of about 10^{-10} K^{-1} on the size of any temperature dependent weight change. Several others repeated the Fordyce experiment during the early 1900's, but always with a null result.

In 1916, though, Prof. P. E. Shaw looked at the matter from a different point of view. He reasoned that it would be particularly interesting to examine the temperature dependence of an isolated gravitational interaction as is produced in a Cavendish balance. In the work of Rumford, Poynting and the others, the source of the Newtonian attraction was the earth. Since the size of the attracted body (hung on the balance) is truly minuscule by comparison with the earth, Shaw suggested that, ". . . the earth may have a preponderating influence, whereby change of its temperature alone would affect the mutual attraction [60]." Therefore, he set about building a Boys-type deflection mode torsion balance and measured the intermass attraction as a function of the temperature of the attracting masses. He concluded that G was indeed affected by temperature. His result took the form $G = G_o(1 + \alpha T)$ where $\alpha = 10^{-5}$ K^{-1}. This result caused great excitement and there was much discussion of it in the contemporary literature. Shaw was a careful man and repeated the measurement under several different experimental conditions. In the end, he found that an unaccounted for thermal gradient was disturbing the balance beam. When the apparatus was improved through more careful thermal shielding, he was able to conclude that $\alpha < 2 \times 10^{-6}$ K^{-1}. No further repetitions of this work have been carried out since then. The citations for the publications covering the work referred to here are available in section 7 of the bibliography on G [7].

4.4. Other Tests

Some other experimental tests of both Newtonian and relativistic theories of gravity can be interpreted as searches for variations in G. For instance, Eotvos- and Kreuzer-type experiments can be looked on as searches for a composition dependence in the gravitational force. Normally speaking, such experiments are intended to place limits on the ratio of passive gravitational to inertial mass, and on the ratio of active to passive gravitational mass, respectively. The results are reviewed in a recent paper by C. Will [61].

At least two workers have looked for gravitational anisotropies in crystalline materials. The motivation has been that since so many properties of crystalline materials depend upon direction (eg., birefringence), then it might be reasonable to assume that the gravitational field of such materials has some preferred direction, too. The typical experiment of this genre consisted of weighing a given sample with its crystalline axis oriented, say, parallel to the earth's field. The experiment was then repeated with the crystal's axis now rotated such that it was perpendicular to the earth's field. Null results were obtained in such experiments by Heyl and by Poynting. Cavendish-type experiments searching for a similar effect have also been done by MacKenzie and by Zeeman, but again, with null results. For the citations to these works, see section 4 of [7].

Searches for an anisotropy of G on a much larger scale have also been undertaken. Vinti looked for a possible effect of the anisotropy of G on celestial orbits. Warburton and Goodkind searched for evidence of a preferred reference frame with a superconducting gravimeter. Williams and Metherell began an experimental search for a gravitational anisotropy using a sensitive torsion balance, but its results were never reported. These experiments are all reminiscent of the Hughes-Drever class of experiment which placed upper limits on the anisotropy of inertial mass. The appropriate references are given in section 14 of [7].

Finally, there is a growing literature on the topics of gravitational experiments with fundamental particles and searches for a dependence of G on the state of quantization of test masses. Early experiments by McReynolds and by Dabbs et al. measured the gravitational acceleration of free neutrons, for instance. Using similar methods, Koester verified the equivalence of gravitational and inertial mass for the neutron. More recently, the University of Rochester group led by A. C. Melissinos has performed experimental tests of Newtonian gravity on fundamental particles moving at relativistic velocities in accelerators. Page and Geilker have reported a Cavendish balance experiment in which the arrangement of the masses was linked to a stochastic radioactive decay process. In their opinion, the results constituted indirect evidence for quantum gravity. Little credence is given the that interpretation by most of the community, however. Anandan has predicted relativistic gravitational effects in charged particle interferometry. He has also predicted that the Newtonian gravitational potential will produce observable effects on a superfluid Josephson interferometer. Stedman has reviewed the whole field of ring interferometric tests of classical and quantum gravity. The citations to all of these articles are available in section 13 of [7].

5. FORTHCOMING STUDIES

Much has been learned about G since the time of Cavendish, but there are still several very fundamental questions which remain unanswered.

For instance, does antimatter fall upwards in a gravitational field? Only one experimental effort to date has addressed this issue, and it was not able to give an unambiguous answer. Recently, however, a collaboration of workers at Los Alamos, CERN, Texas A & M University and elsewhere have conceived a way to monitor antiprotons in the LEAR facility in Geneva. They expect to be able to hold them in captivity long enough to see if G is negative for antimatter [62]. Preliminary work with H^- ions is currently underway at Los Alamos National Laboratory in preparation for on-line work at CERN in the near future.

Another question of interest has to do with the role of spin in gravity. Classical considerations of Poincaré invariance and some grand unification scenarios predict an anomalous spin interaction between gravitationally coupled masses. Two ways of experimentally identifying such an interaction are being studied, one by R. Newman et al. at the University of California, Irvine, and the other by R. C. Ritter et al. at the University of Virginia. Both of the experiments have put preliminary limits on the ratio of the strength of the anomalous interaction to that between the magnetic moments of two electrons.

Newman's experiment uses a cryogenic torsion balance with magnetized, 10 gram split toruses made of $GdNi_5$ and NdNi. One of the split toruses is suspended above a second which is fixed in place inside the dewar. The size of any anomalous interaction is measured by monitoring the angular displacement produced in the suspended, spin polarized sample following reversal of the spin polarization of the fixed sample. The present limit on the size of the anomalous interaction from this experiment is one part in 10^{10} [63].

The experiment at the University of Virginia uses a standard Cavendish balance in the dynamic mode, but with special masses of Dy_6Fe_{23}. In this material, the component of the external magnetic field due to the orbital angular momentum of the electrons is exactly compensated by that due to their spin angular momentum at a certain temperature. However, before the experiment is begun, the test masses are inserted in a region of strong magnetic field to align the spins. Therefore, when kept at their compensation temperature, the masses have all their spins aligned, but have no net external field of their own. Thereafter, great care is taken to keep the interacting components of the experimental arrangement in a region of low magnetic field. This involves careful shielding and the installation of Helmholtz coils. The experiment then consists of monitoring the oscillation period of the balance with the attracting spin-polarized masses parallel and then anti-parallel to those on the suspended balance beam. The present sensitivity is such that even with only partially compensated spin polarized masses the limit on the size of the anomalous interaction is $(0.45 \pm 3.7) \times 10^{-10}$ [64].

In summary, it is safe to predict that G will remain difficult to measure, both in absolute value and in terms of its properties, if any. It is also safe to say, though, that its importance to physicists will most likely grow, and that while the relatively high uncertainty in G remains frustrating to some, it should keep many metrologists and precision measurements experts in employment for years to come!

ACKNOWLEDGEMENTS

It is a pleasure for me to thank the "Ettore Majorana" Centre for

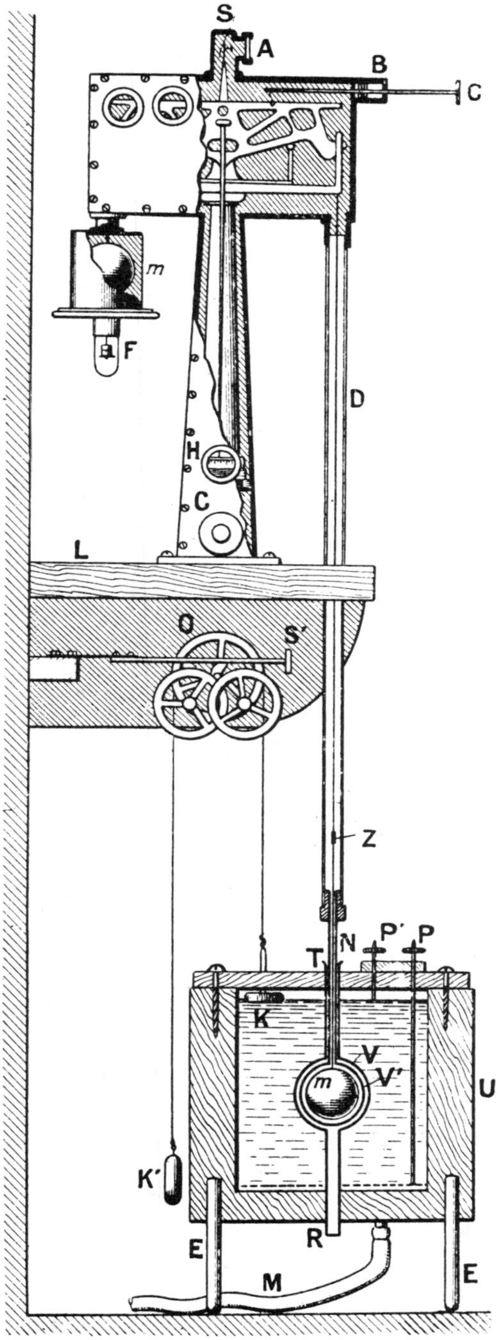

Figure 1. The apparatus used by Q. Majorana to search for an absorption of gravity. A lead sphere, m, was shielded by a screen of mercury in the container, U.

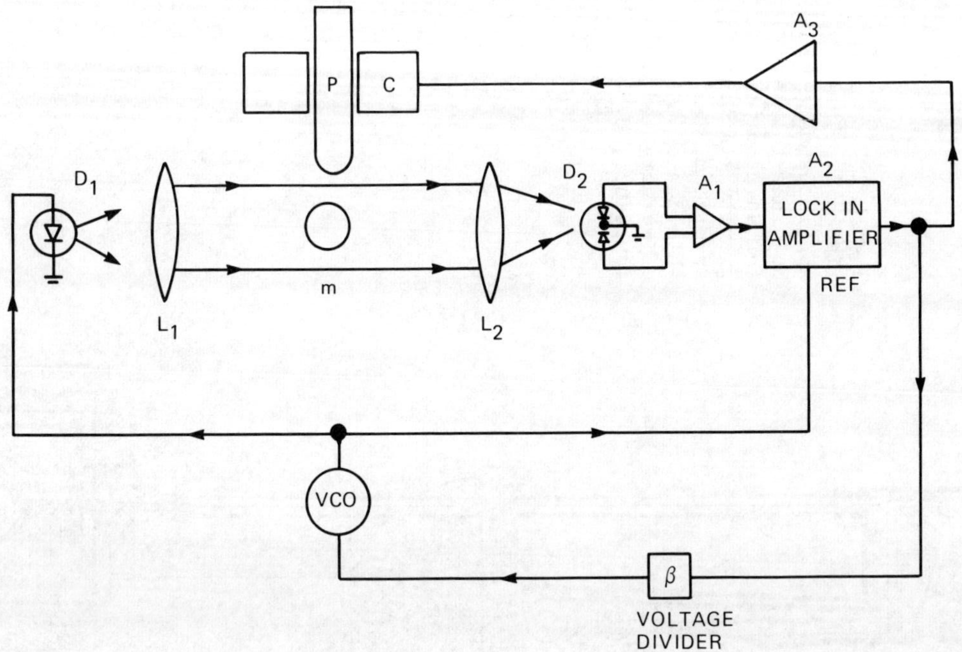

Figure 2. Block diagram of a differentially sensed, lock in amplifier controlled magnetic suspension system. A beam of light passing from the emitter, D_1, to the detector, D_2, monitors the height of the suspended body, m. The feedback circuit, A_3, controls the current in the servo coil, C, such that the mass always remains stably levitated.

Figure 3. Photograph of the University of Virginia matter creation experiment. The suspension system rests on a 500 kg stone which, in turn is vibration isolated by pneumatic springs atop another massive stone. The measurement and control electronics are in an adjacent room.

Scientific Culture for the opportunity to lecture on G during the 10th course of the International School of Cosmology and Gravitation held in Erice, Sicily during May 2 - 12, 1987. I am most grateful to Professors Venzo de Sabbata and Vitaly Melnikov for the invitation to participate, and to the Centre for financial assistance with the travel expenses and for hospitality for myself and my wife. While there, I profitted from many stimulating conversations with Dr's. E. Braun, E. R. Cohen, R. Drever, R. Faustov, E. Fischbach, R. Hellings, Y. Loskutov, R. Reasonberg, N. Rosen, V. Rudenko, V. Shelest and C. Speake. I am also grateful to my friends and colleagues at the Bureau International des Poids et Mesures in Sèvres, France for many interesting conversations and for their hospitality during a visit there following the Erice School. In particular, I acknowledge Dr's. C. Speake, T. Quinn and P. Giacomo. I also thank Prof. R. C. Ritter for helping me to improve this manuscript.

REFERENCES

1. C.C. Speake and G.T. Gillies, Zeit. f. Naturforsch., in press.
2. A.J. Ferrari, et al., J. Geophys. Res. **85**, 3939 (1980).
3. E. Fischbach, et al., Phys. Rev. Lett. **56**, 3 (1986).
4. F.D. Stacey, et al., Rev. Mod. Phys. **59**, 157 (1987).
5. R.C. Ritter, 'Experimental searches for gravitational effects outside of general relativity,' In: Proc. of the 2nd Marcel Grossmann Meeting on General Relativity, edited by R. Ruffini (North Holland, Amsterdam, 1982), 1039 - 1070.
6. H. de Boer, 'Experiments relating to the gravitational constant,' In: Precision Measurement and Fundamental Constants II, edited by B. N. Taylor and W. D. Phillips (U.S. Government Printing Office, Washington, D.C., 1984), 561 - 572.
7. G.T. Gillies, 'The Newtonian Gravitational Constant: An Index of Measurements,' Metrologia **23**(Supplementary Issue), (1986/87), 56 pp.
8. A.S. MacKenzie, The Laws of Gravitation: Memoirs by Sir Isaac Newton, Pierre Bouguer, and Henry Cavendish together with Abstracts of Other Important Memoirs, (American Book Company, New York, 1900), 160 pp.
9. E.D. Preston, Philos. Soc. Wash. **12**, 369 (1895).
10. H. Cavendish, Philos. Trans. R. Soc. London, **88**, 469 (1798).
11. F. Reich, Abh. d. Kön. S. Ges. d. Wissensch. (Leipzig), Math.-Naturwiss. Klasse 1, 385 (1852).
12. J. Kunz, Phys. Rev. **29**, 910 (1927).
13. J.H. Poynting, 'Gravitation,' Encyclopedia Britannica, 11th edition, (1910).
14. P.R. Heyl and P. Chrzanowski, J. Res. Natl. Bur. Stand. **29**, 1 (1942).
15. J.W. Beams, Phys. Today **24**(5), 34 (1971).
16. R.D. Rose, et al., Phys. Rev. Lett. **23**, 655 (1969).

17. W.A. Koldewyn and J.E. Faller, Bull. Am. Phys. Soc. **17**, 472 (1972).
18. C. Pontikis, C. R. Acad. Sci. **274**, 437 (1972).
19. M.U. Sagitov, et al., Dokl. Akad. Nauk SSSR **245**, 567 (1979).
20. G.G. Luther and W.R. Towler, Phys. Rev. Lett. **48**, 121 (1982).
21. E.R. Cohen and B.N. Taylor, CODATA Bulletin **63**, 12 (1986).
22. G.K. Burgess, Phys. Rev. **14**, 247 (1902).
23. H. de Boer, Metrologia, in press.
24. C.C. Speake and G.T. Gillies, Proc. Roy. Soc. London, in press.
25. A.J.F. Metherell, et al., 'Optimizing the shape of the attracting mass in precision measurements of G,' In: Precision Measurements and Fundamental Constants II, edited by B.N. Taylor and W.D. Phillips, (U.S. Government Printing Office, Washington, D.C., 1984), 581 - 586.
26. A.J.F. Metherell and T.J. Quinn, Metrologia **22**, 87 (1986).
27. T.J. Quinn, et al., Metrologia **23**, 87 (1986/87).
28. C.C. Speake and T.J. Quinn, Nature **321**, 567 (1986).
29. C.C. Speake and T.J. Quinn, 'A new method for determining G, the Newtonian constant of gravitation,' BIPM Preprint, (April 17, 1987), 8 pp.
30. U. Bleyer, et al., Gerlands Beitr. Geophys. (Leipzig) **86**, 148 (1977).
31. G.I. Opat, 'Molecular Interferometry - A possible gravitational field measuring technique,' In: Proc. of the 3rd Marcel Grossmann Meeting on General Relativity, edited by Hu Ning, (Science Press, Beijing, 1983), 1491 - 1495.
32. L. Stegna and M.U. Sagitov, The Constant of Gravitation, (Kiado Akadémiai, Lorand Eötvös University, Budapest, 1979).
33. W. Gilbert, De Magnete Magneticisque Corporibus et de Magno Magnete Tellure Physiologia Nova, (London, 1600).
34. L.A. Bauer, Terrestrial Magnetism and Atmospheric Electricity, **13**, 25 (1908).
35. M.G. Lloyd, Terrestrial Magnetism and Atmospheric Electricity **14**, 67 (1909).
36. O.W. Richardson and L. Simons, Philos. Mag. **43**, 138 (1922).
37. F.E. Nipher, Trans. Acad. Sci. St. Louis **23**, 383 (1920).
38. J.F. Woodward, Gen. Rel. Gravit. **12**, 1055 (1980).
39. R. Geigel, Ann. Phys. (Leipzig) **10**, 429 (1903).
40. W. Kaufmann, Ann. Phys. (Leipzig) **10**, 894 (1903).
41. D.H. Whitaker, 'The effect of acceleration on the decay rate of ^{198}Au,' Los Alamos National Laboratory Technical Report No. LA-9388-MS, (June 1982), 7 pp.
42. L.W. Austin and C.B. Thwing, Phys. Rev. **5**, 294 (1987).
43. F. Laager, 'Versuch mit der Drehwage die Abhängigkeit der Gravitation vom Zwischenmedium nachzuweisen,' Dissertation, University of Zürich, (Haller'sche Buchdruckerei, Bern, Switzerland, 1904), 34 pp.
44. T. Erismann, Vierteljahrsch. d. Naturforsch. Gesellsch. Zür. **53**, 157 (1908).
45. Q. Majorana, J. Phys. Radium 1, 314 (1930).

46. K.F. Bottlinger, Sitzungsber. d. Math.-Physik. Klasse d. Kön. Bay. Akad. d. Wissensch. z. München **44**, 223 (1914).
47. V.B. Braginsky, et al., Zh. Eksp. Teor. Fiz. **43**, 51 (July 1962).
48. V.B. Braginsky and V.K. Martynov, Vest. Mosk. Univ., Fiz. **21**(2), 60 (1966).
49. A.Kleiner, Arch. Sci. Phys. Nat. **20**, 420 (1905).
50. L.B. Slichter, et al., J. Geophys. Res. **70**, 1541 (1965).
51. E. Groten, private communication, November 23, 1981.
52. E. Fischbach, private communication, May 9, 1987.
53. T.C. Van Flandern, Astrophys. J. **248**, 813 (1981).
54. R.C. Ritter and G.T. Gillies, Bull. Am. Phys. Soc. **31**, 805 (1986).
55. J.W. Beams, Bull. Am. Phys. Soc. **18**, 267 (1973).
56. R.C. Ritter and J.W. Beams, 'A laboratory measurement of the constancy of G,' In: On the Measurement of Cosmological Variations of the Gravitational Constant, edited by L. Halpern, (Univ. Presses of Florida, Gainesville, Florida, 1978), 29 - 70.
57. R.C. Ritter, et al., Nature **271**, 228 (1978).
58. W.S. Cheung, et al., Prec. Engng. **2**, 183 (1980).
59. H.J. Paik, Phys. Rev. D **19**, 2320 (1979).
60. P.E. Shaw, Philos. Trans. R. Soc. London **216**, 349 (1916).
61. C.M. Will, Theory and Experiment in Gravitational Physics, (Cambridge University Press, Cambridge, 1981).
62. T. Goldman and M.M. Nieto, Phys. Lett. **112B**, 437 (1982).
63. D. Graham and R. Newman, 'Experimental test of the role of intrinsic spin in gravitation,' In: Abstracts of the 11th Int. Conf. on General Relativity and Gravitation, edited by B. Laurent and K. Rosquist, vol. 2, (Int. Soc. for Gen. Rel. and Gravit., Stockholm, 1986), 614.
64. R.C. Ritter, et al., 'Experimental search for a new spin interaction between electrons in bulk matter,' In: Abstracts of the Quangzhau Meeting on Experimental General Relativity and Gravitation, in press.

TIME VARIATION OF THE GRAVITATIONAL CONSTANT

R.W. Hellings
Jet Propulsion Laboratory
Pasadena, California 91109
USA

ABSTRACT. The idea that the Newtonian gravitational constant may vary on cosmic time scales, originally suggested by P.A.M. Dirac, has become a generic prediction of several unified field theories that include gravitation along with the other quantum fields. Experiments looking for the effects of such theories have limited the variation to be less than $10^{-11}\mathrm{yr}^{-1}$. In the next few years, that limit may be reduced as a result of the Soviet-Franco-American dynamics experiment on the Soviet mission to Phobos in 1990.

1. INTRODUCTION AND SUMMARY

It has been fifty years now since Dirac [1] first suggested the idea that the gravitational constant G may be a variable. What had lead Dirac to suggest this strange idea was the fact that a certain cosmic numerical coincidence could be shown to be not so coincidental after all if the gravitational constant were to change at the same rate as the universe evolved, that is at a rate $\dot{G}/G \sim H_0 \sim 5 \times 10^{-11}$ yr^{-1}, where H_0 is the present value of the Hubble constant. If the gravitational constant were to change, then the ratio of a gravitationally-determined time, like the period of a planet, to a non-gravitational time, such as the period of a atomic photon, would change as the universe evolved. The ratios of all gravitational times, of course, would stay the same. The idea of a variable G was thus seen to be roughly equivalent to the idea that there are two dynamical time scales in physics, one for gravitational systems and one for atomic systems. Within the last decade, it has been shown that there would be an observable difference between the effects produced by a cosmic influence in gravitational physics and those produced by a cosmic influence in atomic physics. Five years ago, both effects were searched for via the perturbations they would induce in the orbits of the planets when planetary positions were measured using atomic time standards. Measurements of the range from Earth to Mars using the Viking landers on the surface of Mars succeeded in setting the first theoretically interesting limit on these effects. The results limited a cosmic effect in gravitational physics to $\dot{G}/G < 0.5 \times 10^{-11}$ yr^{-1} and a cosmic effect in atomic physics to $\dot{\xi} < 0.9 \times 10^{-11}$ yr^{-1}. The straightforward interpretation of Dirac's idea was thus shown to be not viable, and it seemed for

V. De Sabbata and V. N. Melnikov (eds.),
Gravitational Measurements, Fundamental Metrology and Constants, 215–224.

awhile that the case might be closed.

Recently, however, the idea of a time variability in the fundamental "constants" has been given new life as a result of theoretical attempts to unify gravitation with the other fundamental interactions of physics into a single unified "Theory Of Everything" (i.e., a "TOE"). It is obvious that a theory that postulates that gravity and electromagnetism are really just two manifestations of the same single force must find a way to explain how gravity came to be so much weaker, particle for particle, than electrodynamics. One natural way to explain this would be to suggest that it was not always so — that gravity was once as strong as electrodynamics but that is has weakened as the universe has evolved. This approach is encouraged by the fact that the ratio of the electrical to gravitational force in an atom is about the same size as the age of the universe in atomic-sized time units. It was, in fact, just this cosmic numerical coincidence that started Dirac off in this direction so long ago. It is a fact that many of the TOE's, especially Kaluza-Klein or supersymmetric string theories which postulate that the universe contains more than three spatial dimensions, generically predict a change in the effective coupling constants of physics, tied to the cosmic evolution of the sizes of the extra dimensions. While none of these theories are at a level of development yet where they are able to make hard predictions of the rate of variation, one may nevertheless expect that the rate of change will be roughly equal to the Hubble expansion rate of the three usual dimensions. Since existing observational limits are very near this same order of magnitude, we are clearly at a critical point where an improvement in the level of sensitivity in the tests for this effect is most important.

A significant improvement in the accuracy of the $\dot{G}/G$ tests may soon be forthcoming as a result of the proposed collaboration between France, the Soviet Union, and the United States in a mission to Phobos. Covariance studies indicate that a few years of tracking of a Phobos lander by the tracking stations of NASA's Deep Space Network may be able to reduce the uncertainty in the size of $\dot{G}/G$ by almost an order of magnitude. This raises the possibility that, in the near future, we may be able to place very strong limits on the size of this effect, or, perhaps, to detect the effect, thereby providing evidence for a quantum-field-theoretical unification of all forces into one.

In the sections to follow, we will first (Section II) discuss Dirac's Large Numbers Hypothesis which was the basis for the early $\dot{G}/G$ prediction and then (Section III) continue with the development of this idea into a formalism defining two types of cosmic effects in local physics. Then, in Section IV, the predictions of the modern unified field theories for variation of the fundamental physical constants will be sketched. Finally, in Sections V and VI, we will discuss the method of testing these theories using classical planetary orbits and will discuss past results and expected future results of the tests.

2. THE LARGE NUMBERS HYPOTHESIS

It is striking that most fundamental dimensionless ratios which occur naturally in physics are either of the order of unity or they are of the order of some simple power or root of 10^{40}. To P.A.M. Dirac this was too much of a coincidence; it had to be a hint of some deeper physical relationship. It is easy to tie many of these numbers to each other in a rational way, but there are two basic ratios whose connection would seem to be a mystery. The first of these is the ratio of electrical to gravitational force in a hydrogen atom:

$$\frac{e^2/r^2}{Gm_e m_p/r^2} = 0.227 \times 10^{40}, \tag{1a}$$

where r is the distance between the proton of mass m_p and the electron of mass m_e, e is the charge on the electron, and G is the gravitational constant. The second ratio is that between the radius of the universe and the classical electron radius, or, equivalently, between the age of the universe and the natural time scale for the electron:

$$\frac{1/H_0}{e^2/m_e c} = 6.57 \times 10^{40} \left(\frac{50}{H_0}\right), \tag{1b}$$

where H_0 is the present value of Hubble's constant in km/sec/kpc.

Dirac postulated, in what has come to be known as the Large Numbers Hypothesis (or LNH), that these numbers were in fact related. If this relation is not to be an accident of the present epoch, then the ratio of the electrical force to the gravitational force must be increasing along with the age of the universe. This means that the dynamics of local gravitational systems and of local non-gravitational systems will differ more and more as time goes by. When a gravitational period, such as the period of a planet, is compared to a non-gravitational period, such as the period of a photon emitted in an atomic transition, the ratio of the two periods must not be constant but must change as the universe evolves.

3. CLASSICAL REALIZATIONS OF THE LNH

As pointed out in recent papers [2], there are at least two conceptually different ways in which an incommensurability between orbital periods and atomic times might be produced. One way [2,3] is to modify gravitational physics by postulating additional cosmic fields which affect the form of the physical metric tensor (i.e., the geometry measured by atomic rods and clocks). The Brans-Dicke theory [3] is typical of such a theory. In these theories, the drift in the orbital periods results from a time-dependent renormalization of the gravitational constant due to the slowly changing values of the cosmic fields. One writes the effective gravitational constant as

$$G_{\text{eff}} = G\left[1 + (\dot{G}/G)(t - t_0)\right], \tag{2}$$

where G and $\dot{G}$ are the values of the effective gravitational constant and its time derivative at some $t = t_0$. With this G_{eff} in the metric, there appears a term in the geodesic equation which gives rise to a coordinate acceleration

$$\delta a^i = -\frac{\dot{G}}{G}\left[\frac{GM}{r}\frac{x^i}{r^2}(t - t_0)\right], \tag{3}$$

in addition to the usual Newtonian and relativistic terms.

A second way [2,4] in which an incommensurability between atomic times and gravitational periods might be produced would be for there to exist some cosmic influence which directly affected atomic physics while the field equations of gravitation remained unchanged. In this case, there would be an observable effect of the clock drift on solar system measurements due to the fact that the time associated with each astrometric observation is the time given by an atomic clock. The form of the effect on the coordinate acceleration equations is obtained by first defining a parameter $\dot{\xi}$ which gives the rate at which atomic time (dt) will drift away from the proper time inherent in relativistic dynamics (called Einstein time dt_E)

$$dt_E/dt = 1 + \dot{\xi}(t - t_0). \tag{4}$$

After transforming the geodesic equation to reflect this new definition of the affine parameter, one obtains a non-geodesic equation of motion

$$\frac{d^2x^\mu}{dt^2} + \Gamma^\mu_{\alpha\beta}\frac{dx^\alpha}{dt}\frac{dx^\beta}{dt} = \dot{\xi}\left(g^{\mu 0} - \frac{dx^\mu}{dt}\frac{dx^0}{dt}\right), \tag{5}$$

where the metric is that measured by atomic rods and clocks and $\Gamma^\mu_{\alpha\beta}$ are the Christoffel symbols derived from that metric. There is thus in the coordinate acceleration an additional term springing from the right-hand side of Eq. (5)

$$\delta a^i = \dot{\xi}\left[\frac{GM}{r}\frac{x^i}{r^2}(t - t_0) - \frac{dx^i}{dt}\right]. \tag{6}$$

From a naive application of the standard cosmological models, one expects $\dot{\xi} \sim \dot{G}/G \sim H_0 \sim 5 \times 10^{-11}\mathrm{yr}^{-1}$.

4. $\dot{G}$ AND HIGHER-DIMENSIONAL THEORIES

The success of the SU(2)×U(1) unification of the electromagnetic and weak interactions and numerous hints of symmetries between lepton and hadron physics have sent theorists off on a quest for a further Grand Unified Theory (i.e., a "GUT") of all quantum field theories. So far success has eluded them, and so many different avenues are still being pursued. One of these avenues involves leaping ahead in the unification process to include gravitation along with the other three quantum fields in a completely unified Theory Of Everything (i.e., a "TOE"), and one of the most intriguing methods for this unification is to postulate that the universe contains more than three spatial dimensions, the preferred numbers being nine or ten.

There is almost a feeling of magic that occurs the first time you see Einstein's field equations in $D=4+d$ spacetime dimensions separate into the normal Einstein equations in 4 dimensions plus the field equations of Yang-Mills gauge theory. The subtle connection between General Relativity and gauge theory is the attraction of these higher-dimensional Kaluza-Klein theories (the supersymmetric string theories are an example). One begins with simple geometry in $4+d$

dimensions, and, by explicitly separating the metric tensor into a 4-dimensional part and a compact d-dimensional part, one recovers both the normal Einstein theory of gravitation in 4-dimensions and a 4-dimensional effective gauge theory of elementary particles, formed out of the extra degrees of freedom inherent in the additional d dimensions. The properties of particles and the physics of their interactions are determined by the structure of the compact extra-dimensional space. In particular, the effective coupling constants of the 4-dimensional field theories are related to the circumferences or the volume of the compact space.

One generally begins [5] by writing the D-dimensional metric tensor as

$$g_{MN}(x,y) = \begin{bmatrix} g_{\mu\nu}(x) + \gamma_{mn}(y)\xi^m_\alpha(y)A^\alpha_\mu(x)\xi^n_\beta(y)A^\beta_\nu(x) & -\gamma_{mn}(y)\xi^m_\alpha(y)A^\alpha_\mu(x) \\ -\gamma_{mn}(y)\xi^m_\alpha(y)A^\alpha_\mu(x) & \gamma_{mn}(y) \end{bmatrix} \tag{7}$$

where M and N go from 0 to $3+d$, μ and ν go from 0 to 3, and m and n go from 4 to $d+3$. The x-coordinates are assumed to cover spacetime while the y-coordinates are cyclic and cover the compact d dimensions. The $g_{\mu\nu}$ is the spacetime metric and γ_{mn} is the metric of the compact d-dimensional subspace, assumed to be invariant under a group of isometries with Killing vectors $\xi^m_\alpha(y)$. The quantities $A^\alpha_\mu(x)$ will turn out to be the gauge fields.

When the D-dimensional scalar lagrangian density is written in terms of $g_{\mu\nu}$ and γ_{mn}, the variation of the action integral gives the spacetime Einstein equations and the Yang-Mills field equations in their usual form, provided that the Killing vectors are normalized according to

$$\int d^d y \sqrt{\gamma}\; \xi^m_\alpha \xi^n_\beta \gamma_{mn} = \kappa^2 \delta_{\alpha\beta}, \tag{8}$$

where κ^2 is the bare gravitational constant, related to the effective spacetime gravitational constant G by

$$\kappa^2 = 16\pi G \int d^d y \sqrt{\gamma}. \tag{9}$$

This choice of normalization leads (see Weinberg [5]) to a relation between the circumferences of the compact space along the integral curves of the Killing vectors and the coupling constants of the gauge theory. The bare coupling constants are given by [6]

$$\alpha_i = K_i G / R_{KK}, \tag{10}$$

where R_{KK} is the mean radius of the compact space and K_i are numbers that depend on the compact space topology.

If these theories had been worked out thoroughly enough to give complete field equations and to pick out the relevant cosmological solutions in D-dimensions, then equations (9) and (10) could be used to predict the variation of the effective coupling constants of fundamental physics. Unfortunately, this level of detail in the theoretical development of the theories has not yet been acheived. However, the few rough cuts at the problem which have been made, such as that of Chodas and Detweiler [7], give predictions like $\dot{R}_{KK}/R_{KK} \sim H_0/f$, where f is some number near the number of extra dimensions in the compact space. Many theorists would like to assume that the evolution of the compact dimensions would occur during the earliest stages of the evolution of the universe, so that the coupling constants would be effectively constant at the present epoch. However, the rates are determined by field equations and not by assumption, and solutions to the field equations with this property are very specialized and very difficult to come by (see, for example, Wu & Wang [8]). The general prediction would seem to be a variation of both the gravitational constant and the coupling constants of non-gravitational physics at a rate perhaps a little less than the Hubble rate. This variation may be one of the few testable predictions of these theories, which otherwise are indistinguishable from a non-unified theory at all but the highest energies.

5. TESTING $\dot{G}$ IN THE SOLAR SYSTEM

Solar system astrometric and radiometric data have been searched for the effects of a varying gravitational constant and for the effects of the cosmic variation in atomic clocks which would be produced as a result of a variation in the coupling constants of atomic physics. The data which have proven most sensitive to these effects are the range data from Earth to Mars. Beginning in July 1976 and ending in July 1982, 1136 range measurements between tracking stations of the Deep Space Network and the Viking Landers on Mars were taken with an average interval between measurements of about two weeks. The range measurement itself was accurate to about two meters while uncertainties in the time delay produced by interplanetary plasma, along with unknown delays inherent in the tracking station and spacecraft signal paths, left a calibration uncertainty of about nine meters. In order to fit the data, the Earth's orbit had to be adjusted along with the orbit of Mars. However, the Earth's motion is also constrained by astrometric observations of the other bodies of the solar system. It therefore proved necessary to include all reliable solar system data and all relevant solar system parameters in adjusting the solar system model to fit the Viking data. The data used were:

1. 1136 range measurements to the Viking Landers on Mars (from July 1976 to July 1982),
2. 645 range measurements to the Mariner 9 spacecraft in orbit around Mars (from November 1971 to October 1972),
3. 1305 radar bounce range measurements from the surfaces of Mercury and Venus (from 1964 to 1977),
4. 2954 Lunar laser range measurements (from 1969 to 1980),
5. 44755 optical position measurements of the sun and planets, right ascension and declination (from 1911 to 1979).

In order to determine the values of $\dot{G}/G$ and $\dot{\xi}$ using these data, Eqs. (3) and (6) were numerically integrated to get the perturbations which each parameter would produce in the Earth-Mars range. The signature of each perturbation was then searched for in the data. Other parameters which were solved for simultaneously with $\dot{\xi}$ and $\dot{G}/G$ were:

1. Initial orbital elements of the moon and planets,
2. The masses of Venus, Earth, and the outer planets,
3. The masses of the three asteroids (Ceres, Pallas, and Vesta) which have a large effect on Mars' orbit,
4. The average densities of the remaining asteroids, divided into two groups according to compositional type,
5. Other parameters representing corrections for systematic effects known to exist in individual data sets or parameters unique to each data set.

These parameters were fit to the data in a least squares sense, resulting in estimates and standard deviations for each of the parameters. If all errors in the data sets were random Gaussian-distributed errors, then the formal standard deviation given by the least squares fit would have been the correct estimate of the uncertainty in the values of the parameters. Experience has shown, however, that there typically exist small unmodeled errors in the data sets, so additional care was required to estimate the realistic uncertainty that should be assigned to the values.

The major uncertainty in the determination of $\dot{G}/G$ and $\dot{\xi}$ came from uncertainty in the masses of the asteroids. It was determined that the three largest asteroids (Ceres, Pallas, and Vesta) had large periodic effects in the Earth-Mars range which could be separated, allowing their masses to be determined in the solutions. The 200 largest remaining asteroids in the belt which had fairly well-determined radii [9] were divided into two groups. The C-group consisted of asteroids generally in the outer regions of the belt having compositional type C, F, P, and D. The S-group contained asteroids of type S, M, E, and R which are found in greatest abundance in the inner regions of the belt [10]. The asteroids in each group were treated as point masses with a common density (to be determined in the solutions) and with volumes determined from their measured diameters. This model of the asteroid belt, three individual masses plus two groups with uniform density, was found to be adequate in that: 1) the model fit the data well enough to eliminate visible trends in the residuals, and 2) solving for the masses of additional individual asteroids did not produce significant changes in the estimates of $\dot{G}/G$ or $\dot{\xi}$. The results of this analysis gave [11]

$$\dot{G}/G = (0.2 \pm 0.4) \times 10^{-11}\mathrm{yr}^{-1}, \qquad (11.a)$$

$$\dot{\xi} = (0.1 \pm 0.8) \times 10^{-11}\mathrm{yr}^{-1}. \qquad (11.b)$$

The effects of uncertainties in the observed radii of the asteroids were also assessed. The nominal radius values were taken from the most recent compilation by Tedesco [9]. Any systematic errors in the radii for a given group would have simply scaled the size of their effect, and would thus have been absorbed into a new value for the density. This value would be in error, but it would not have affected the estimates of $\dot{G}/G$ or $\dot{\xi}$. The effects of random errors in the radii were investigated by a Monte Carlo technique of randomly adjusting the radii of the 200 asteroids by amounts consistent with their known uncertainties (about 10% for well-observed asteroids and 50% for asteroids which have not been well-observed). The resulting range of values for $\dot{G}/G$ and $\dot{\xi}$ fell well within the error limits quoted in Eq. (11). In addition, the effect of the 2000+ asteroids which were not included in the two groups was found to be negligible, producing a correction to the modeled asteroid effect of less than 5%.

It may be concluded that the available Solar System astrometrical data severely limit the existence of a cosmic influence on either gravitational ($\dot{G}/G$) or non-gravitational ($\dot{\xi}$) local physics at the level expected from Dirac's Large Numbers Hypothesis.

VI. FUTURE PROSPECTS

In July 1988, the Soviet Union will launch two identical spacecraft in the direction of Phobos, one of the moons of Mars. The spacecraft will arrive in Mars orbit in January 1989. In March or April of 1989 one orbiter will execute a flyby of Phobos. The flyby will be at low-altitude (50 m) and low-velocity (2–5m/sec). The second orbiter will execute a similar flyby about two months later. During each encounter, a small long-lived lander will be placed on the surface of Phobos. These landers will perform investigations on the surface composition and internal structure of Phobos and the dynamics of Phobos and Mars. The total weight of the lander is about 40 kg. The structure consists primarily of a triangular box, 60 cm wide and 12 cm high, on which are mounted three solar panel wings, sensors, and an antenna. After landing, a sun sensor tips the solar panels toward the sun. The box is supported 80 cm above the surface by three legs. It is anchored to the surface by a penetrator fired at touchdown from a plate at the bottom of the legs. The lander is equipped with a French-built sun sensor which will determine the orientation of Phobos. The lander has a target lifetime of one year, but could last much longer.

Doppler and range tracking of the Phobos landers will permit the orbit of Phobos to be determined and, thereby, the range to the center of mass of Mars to be inferred. The data will be reduced by comparing them to a model consisting of an analytic ephemeris for Phobos, developed for this application by researchers at the *Bureau des Longitudes* in Paris, and a JPL numerically-integrated planetary ephemeris. Sun Sensor data will be used to determine the angular amplitude of the forced librations of Phobos. The librations will be part of the model as well. The reduced sun sensor, range, and Doppler data will be added to existing Mars data from the Mariner 9 (1972-73) and Viking (1976-82) missions. All relevant parameters, including the orbital elements of Phobos (along with the secular acceleration), $\dot{\xi}$, $\dot{G}/G$, and the asteroid masses, will be solved for in a least-squares sense.

A preliminary covariance study was performed in which range data to Mars were assumed to have been taken once per week, each range having an uncertainty of 15m. These simulated data were then added to the existing solar system data sets and least squares fits were made of the parameters of the solar system model using combined data. Results were compared with no Phobos data, $1\frac{1}{2}$ year of

Phobos data, and five years of Phobos data. The results are shown below. In the solutions, the masses of ten asteroids were determined simultaneously with $\dot{G}/G$. There is a strong coupling between the effects of $\dot{G}/G$ and the perturbations of the asteroids, and the inclusion of so many asteroids in the fit was to obtain a realistic estimate of how uncertainty in the asteroid masses could corrupt the determination of $\dot{G}/G$.

TABLE I. Covariance study results

Phobos Data in solution	none	$1\frac{1}{2}$ years	5 years
$\dot{G}/G$ uncertainty	8.0	3.0	1.6

note: Units for $\dot{G}/G$ are parts in 10^{12} per year.

As may be seen, even the $1\frac{1}{2}$ years of data provided during the nominal Phobos mission will yield an important improvement in the knowledge of $\dot{G}/G$. This is because the signal from this effect will have built up during the eight years between the end of the Viking data and the assumed beginning of the Phobos data, so that even a few data points would serve to constrain the value of this parameter. If, in addition, the lander continues for as long as five years, then the limits on $\dot{G}/G$ will be a factor of five better than at present and will constrain the higher-dimensional theories at an extremely interesting level.

REFERENCES

1. P.A.M. Dirac, Nature **139** (1937) 323.
2. P.J. Adams, V.M. Canuto, I. Goldman, & R.W. Hellings, Phys. Rev. **D28** (1983).
3. C. Brans & R.H. Dicke, Phys. Rev. **124** (1961) 925. P.G. Bergmann, Int. J. of Th. Phys. **1** (1968) 25. C.M. Will and K. Nordtvedt, Jr., Ap. J. **177** (1972) 757. R.W. Hellings and K. Nordtvedt, Jr., Phys. Rev. **D7** (1973) 3593. N. Rosen, Ann Phys.(N.Y.) **84** (1974) 455.
4. V. Canuto, P.J. Adams, S.H. Hsieh, & E. Tsiang, Phys. Rev. **D16** (1977) 1643. P.J. Adams, Int. J. Th. Phys. **22** (1983) 421. V. Canuto & I. Goldman, Nature **296** (1982) 709. P.A.M. Dirac, Proc. Royal Soc. **A365** (1979) 19.
5. S. Weinberg, Phys. Lett. **125b** 265 (1983).
6. W.J. Marciano, Phys. Rev. Lett. **52** 489 (1984).
7. A. Chodas & S. Detweiler, Phys. Rev. **D21** 2167 (1980).
8. Y.S. Wu & Z. Wang, Phys. Rev. Lett. **57** 1978 (1986).
9. E.F. Tedesco, in *Planetary Rings*, edited by A. Brahic, Proceedings of the I.A.U. Colloquium no. 75 (Reidel, Dordrecht, 1983).

10. C.R. Chapman, J.G. Williams, & W.K. Hartmann, Ann. Rev. Astron. Ap. **16** 33 (1978).
11. R.W. Hellings, *et al.*, Phys. Rev. Lett. **51** 1609 (1983).

THE SUPERNOVA SN 1987A AND THE NEUTRINO MASS

T.H. Ho
Centre of Theoretical Physics, CCAST, World Laboratory
and
Institute of Theoretical Physics, Academia Sinica
P.O. Box 2735, Beijing, Peoples' Republic of China

ABSTRACT. It is shown that the Kamioka neutrino data from the SN 1987A strongly suggests a mass of the neutrino of $m_{\nu_e} = 4.54 \pm 0.78$ eV.

1. MEASURING THE MASS OF THE NEUTRINO IS IMPORTANT FOR THE FOLLOWING REASONS:

   a) it is a fundamental constant;
   b) it is closely related to various unification schemes;
   c) it may help to show if the Universe is open or closed.

2. A SHORT SUMMARY OF THE PRESENT RESULTS OF ν - MASS MEASUREMENT IS SHOWN BELOW:

Mass limit of	Reference
$17 < m_{\nu_e} < 40$ eV (Moscow)	[1]
$0 \leq m_{\nu_e} < 19$ eV (Zurich)	[2]
$0 \leq m_{\nu_e} < 28$ eV (Los Alamos)	[3]
$0 \leq m_{\nu_e} < 32$ eV (Tokyo)	[4]
$0 \leq m_{\nu_e} < 30$ eV (Beijing)	[5]

if one were to take experiments [1] and [2] too seriously, one would get

$$m_{\nu_e} \sim 17 - 18 \text{ eV}.$$

However, since it is an extremely difficult experiment, since also complicated molecular structure corrections have to be taken into account, it is unclear what the mass of the e-neutrino really is!

V. De Sabbata and V. N. Melnikov (eds.),
Gravitational Measurements, Fundamental Metrology and Constants, 225–229.

3. THE NEUTRINO BURST FROM THE SUPERNOVA SN 1987 A

Recently, some neutrino signals from SN 1987 A were recorded by four groups: the Mont-Blanc , Kamioka , IMB and Barksan . The data obtained by Kamioka, IMB and Barksan have a simultaneity within one minute, whilst the Mont-Blanc data occurred 4 to 5 h earlier. It seems that the Mont-Blanc obtained an accidental record. However, the Barksan data up to now are not formally published ,whilst the Mont-Blanc had another evidence from a gravitational wave *), which shows a signal with an amplitude of 135 K, against a background noise of 29 K, about 1.4 s before the first neutrino. It is thus difficult to exclude the reliability of Mont-Blanc events, even though the 135 K event is related to a total gravitational-wave energy, emitted by the supernova, of the order of 2400 solar masses, as calculated with the standard formulae of general relativity

However, if the gravitational-wave signal is a reliable one, then one can easily obtain the information about the neutrino mass, provided the neutrino is emitted at the same time as the gravitational wave, or with a little delay. i.e.,

$$m_{\nu_e} < 7.5 \text{ eV} \times \left(\frac{E_\nu}{10 \text{ MeV}}\right) .$$

Since the energy of the electron or positron of the first neutrino signal is 5.8 MeV, by assuming that this signal is arising from

$$\tilde{\nu}_e + p \rightarrow n + e^+ ,$$

which is the dominant mechanism in the detector, then one has

$$m_{\nu_e} \leq 7.5 \times \left(\frac{(5.8 \pm 1.2) + 1.8}{10}\right) = 5.7 \pm 0.9 \text{ eV}.$$

4. KAMIOKA DATA AND THE NEUTRINO MASS LIMIT

Among all these neutrino data, the Kamioka events are the most interesting, and they can be listed as done in Table I.

---

*) As we heard from Prof. G.Pizzella in this School meeting.

TABLE I
The Kamicka data of the neutrino burst

Event No.	Event time (s)	E_e (MeV)	Electron angle ϑ_e	E_ν (MeV)*)
1	0	20.0 ± 2.9	18 ± 18	21.8 ± 2.0
2	0.107	13.5 ± 3.2	15 ± 27	15.3 ± 3.2
3	0.303	7.5 ± 2.0	108 ± 32	9.4 ± 2.0
4	0.324	9.2 ± 2.7	70 ± 30	11.1 ± 2.7
5	0.507	12.8 ± 2.9	135 ± 23	14.9 ± 2.9
6	0.686	6.3 ± 1.7	68 ± 77	8.1 ± 1.7
7	1.541	35.4 ± 8.0	32 ± 16	37.4 ± 8.0
8	1.728	21.0 ± 4.2	30 ± 18	22.9 ± 3.2
9	1.915	19.8 ± 3.2	38 ± 22	21.7 ± 3.2
10	9.219	8.6 ± 2.7	122 ± 30	10.6 ± 2.7
11	10.433	13.0 ± 2.6	49 ± 26	14.9 ± 2.6
12	12.439	8.9 ± 1.9	91 ± 39	10.8 ± 1.9

From Table I, one can see that there are six events grouped together as a first peak during the neutrino burst. If the time width of this peak is narrow enough (there are many different models of supernova explosion, but in the initial stage, there usually is an initial neutrino peak), then we have a new way to estimate m_{ν_e} . Let the difference of the arrival times of the i-th and the j-th neutrino be $t_i - t_j$, which is known from the experiments. The original time difference of the i-th and j-th events is $t_i^0 - t_j^0 \equiv Y_{ij}$, which is unknown but can be related to $(t_i - t_j)$ as follows:

$$Y_{ij} = (t_i - t_j) - L\, m_\nu^2\, c^4\, (1/E_i^2 - 1/E_j^2)/2c \tag{1}$$

$$= (t_i - t_j) - 0.158\text{x}10^{-2}\quad L/\ell_0\ \left[(10\ \text{MeV}/E_i)^2 - (10\ \text{MeV}/E_j)^2 \right] X$$

in which $X = \left(\frac{m_\nu c^2}{1\,eV}\right)^2$, $\ell_0 = 10^4$ light years, $L/\ell_0 = 16.3$, as $L = (50 \pm 2)$kpc, since this distance is estimated from two types of variable stars, Cephid and RRLyr.

C.W. Kim [11] and then C.R. Ching et al. [12] assume $Y_{ij} \simeq 0$,with which they obtain the correlations of the ν -mass value shown in Tables II and III.

TABLE II
The possible solutions (I) of m_ν

$Y_{ij} = 0$	X $(eV)^2$	m_ν (eV)
Y_{21}	19.25	4.39 ± 1.96
Y_{41}	20.85	4.57 ± 1.51
Y_{61}	20.54	4.53 ± 1.11
Y_{24}	21.73	4.66 ± 2.54
Y_{26}	20.79	4.56 ± 1.38
Y_{46}	20.28	4.50 ± 2.44

TABLE III
The possible solutions (II) of m_ν

$Y_{ij} = 0$	X $(eV)^2$	m_ν (eV)
$Y_{8,10}$	412	20.3 ± 6.48
$Y_{9,10}$	414	20.4 ± 6.67
$Y_{8,9}$	341	$18.5 \pm$ *)

*) Very large error but still $m_{\nu_e} > 6.59$.

From Table II, and including the errors, the average value is:

$$m_{\nu_e} = 4.54 \pm 0.78 \text{ ev},$$

whilst the average from Table III is $m_{\nu_e} = 19.7$ eV, where the error is large, but $m_\nu > 6.59$ eV, so that these two definitely belong to two different possible solutions.

Since the confidence level for the second solution is lower than that for the first one, we would like to recommend the value

$$m_{\nu_e} = (4.54 \pm 0.78)\ \text{eV}$$

If this value is correct and if we assume $m_{\nu_e} \simeq m_{\nu_\mu} \simeq m_{\nu_\tau}$, then

$$0.46 \leq \Omega = \rho_0/\rho_c \leq 0.67,$$

which seems to coincide with the value of Ω , determined from a cluster or supercluster of galaxies mentioned by Prof. Melnikov:

$$0.4 \leq \Omega \leq 0.7 \quad [13]$$

We have applied the same method to the IBM data, but no new results can be obtained since, in the IBM detector, the threshold is

$$E_e > 20\ \text{MeV}.$$

The author would like to thank Prof. A.Zichichi and V.de Sabbata for their hospitality during the School meeting; he is also grateful to Prof. M.Jacob for his hospitality arranging a short visit in CERN.

REFERENCES

[1] S. Boris et al. – Phys.Rev.Lett. 58 (1987) 2019.
[2] M. Fritschi et al. – Phys.Lett. B173 (1986) 485.
[3] J.F.Wilkerson et al. – Phys.Rev.Lett. 58(1987)2023.
[4] H. Kawakami et al. – Phys.Lett. B187 (1987)198.
[5] H.Z. Sun et al. – Talk given in the 1986 Osaka Workshop about neutrino and nuclear β decay.
[6] M.Aglietta et al. – IAU Circular Nr. 4323, and to be published in Europhysics Letters.
[7] K.Hirata et al. – Phys. Rev.Lett. 58(1987)1490.
[8] R.M. Bionta et al. – Phys. Rev.Lett. 58(1987)1494.
[9] A.Pomansky – Proc. XXII Recontres de Moriond, Les Arcs 8–15 March 1987, ed. J.Tran Thanh Van (Editions Frontières, Gif sur Yvette, 1987), Vol.I. (It was heard at this School meeting that the Barksan data may not be so reliable).
[10] G.Pizzella – This School meeting.
[11] C.W.Kim – Private communication, but in Kim's calculation, he put $E_\nu = E_\ell$, so he obtained an intersecting point from Y_{i1} , Y_{i2} and Y_{i4} , with $m_{\nu_e} = 3.6$ eV.
[12] C.R.Ching,T.H.Ho,Z.J.Tao and A.X.Huo – Preprint AS–ITP–87–015.
[13] V.N.Melnikov – This School meeting.

THE FIFTH FORCE EXPERIMENT AT THE TIFR

R.Cowsik, N.Krishnan, P.Saraswat, S.N.Tandon & S.Unnikrishnan
Tata Institute of Fundamental Research
Homi Bhabha Road
Bombay 400 005
India

ABSTRACT. A resonance torsion balance experiment designed to be sensitive to a fifth force coupling to baryons with a strength 0.1% of gravity and with any range greater than a few meters is described. Particular emphasis is laid on those aspects of the design of this experiment which confer immunity from perturbations and errors often encountered in torsion balance experiments.

1. INTRODUCTION

The question of the existence of new forces in nature has been of interest in the modern idiom ever since the speculation of Lee & Yang[1] in 1955 that baryon number conservation corresponded to a new (i.e. till then unrecognised) gauge symmetry. Pursuing an analogy with electrodynamics where the conservation of electric charge is seen to follow from the gauge invariance of the electromagnetic field, they argued that the observed fact of baryon number conservation could be explained as arising from the gauge invariance of a new bosonic field which coupled to baryons. Massless (or low mass) gauge bosons of this field would cause forces between macroscopic bodies detectable in laboratory experiments and distinguishable from the gravitational force because of the lack of proportionality between mass and baryon number taken together with the fact that the mass defect per nucleon is a function of nuclear species.

In the years since 1955 a number of other motivated suggestions for new forces have arisen; recent instances of these are the model of Moody & Wilczek[2] where the axion field gives rise to spin dependent forces (with a range of a few cm.), and the models of Chang, Mohapatra & Nussinov[3] in which the exchange of Goldstone bosons combined with the QCD anomaly leads to long range forces of spin dependent as well as spin independent types. (See also Y. Fujii[4]) The conjecture of Fischbach et.al.,[5] of a fifth force coupling to baryons with a strength as large as 1% of gravity should be seen in this context; however, it has the important difference of being based on experimental data from the past and consequently carries a different flavour. (That these data

V. De Sabbata and V. N. Melnikov (eds.),
Gravitational Measurements, Fundamental Metrology and Constants, 231–239.

do not clearly rule out alternate possibilities is what really makes the whole situation more exciting).

On the experimental front, constraints on theoretical speculations are imposed primarily from three types of measurements, viz.,i) torsion balance experiments, ii) astronomical observations of the orbital parameters of solar system bodies as well as of the binary pulsar and iii) mine gravity measurements. (See fig.1). Of these, the latter two really

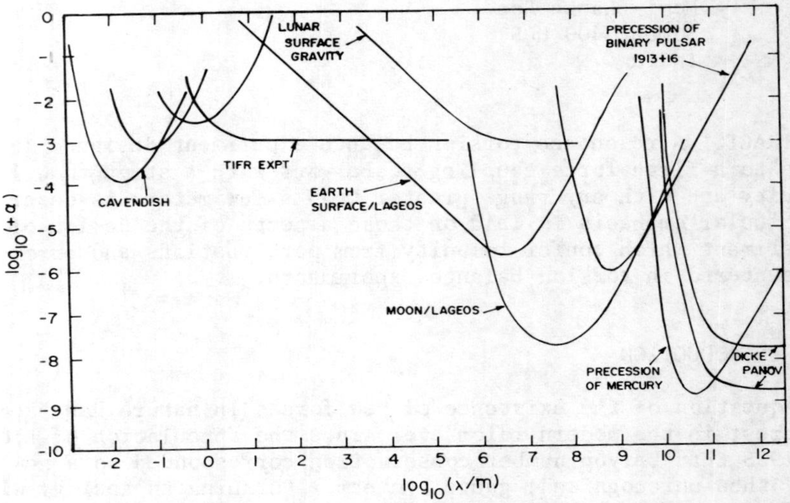

Figure 1. Showing the limiting constraints in parameter space imposed by current experimental evidence (from A. de Rujula[6]); the entire area upwards from a curve is excluded by the particular measurement(s) labelling that curve. The region of this space to be explored in the present experiment is shown by the curve labelled TIFR EXPT.

probe departures from Newtonian gravity through the addition of a Yukawa term:

$$V(1,2) = -\frac{Gm_1m_2}{r}(1 + \alpha e^{-r/\lambda}) \qquad (1)$$

(and a baryon-dependent force mimics this closely owing to the approximate mass proportionality of nucleon number). Of the three classes of experimental method mentioned above, torsion balance experiments provide the most versatile probe, for, depending on the mode in which they are operated, they can place constraints on departures from Newtonian gravity; or on new couplings to e.g., baryon number; or even (as in new experiments planned or in execution at Irvine[7] and at Virginia[8]) on the existence of spin dependent macroscopic couplings. In fact, of the three lines of evidence for the fifth force put forward by Fischbach et.al.,[5] the least uncertain is still that from the classic torsion balance experiments of Eotvos.

2. THE TIFR EXPERIMENT

A torsion balance experiment designed to be sensitive to a fifth force coupling to baryons with a strength 0.1% of gravity and with a range greater than a few metres has been set up at the Tata Institute of Fundamental Research (TIFR). This section will (i) put forward the method underlying the measurement, (ii) describe the set-up of the experiment, and (iii) outline the relative immunity from perturbations and errors that has been built into this experiment.

2.1. The Method

This is a resonance torsion balance experiment. The method adopted is to torsionally suspend a dual ring (its two halves are made of lead and copper) in an environment well shielded from various sources of noise and to move a large 'source' mass around this torsion balance in a circular orbit with a period equal to the natural torsional period of the balance. If the fifth force exists, the two halves of the balance experience different accelerations towards the source mass and the resulting periodic torque on the balance causes its torsional oscillation amplitude to resonantly increase with time. In the early resonance regime the increase in amplitude is linear with time and observation of such a signature will indicate a fifth force; additionally, positioning the source mass in the opposite phase relative to balance oscillation should cause its amplitude to resonantly decrease and this will constitute further proof of the action of the source mass on the balance.

The parameters of the balance and the source mass are such that a fifth force 1% as strong as gravity will cause a resonant amplitude growth rate of 10^{-6} rad./day. The angular position of the torsion balance is recorded by an autocollimating optical lever with an angular resolution in a 1 Hz. bandwidth of 3×10^{-9} rad.

A very important aspect of an experiment of this kind is that it can put a limit on the strength of the fifth force essentially independent of its range (beyond a range of about a metre), unlike experiments which involve local masses of large spatial extent, such as cliffs and hills.

2.2. The Set-up

The salient features of the instrument are shown schematically in Fig.2. The suspension used is tungsten wire of diameter 105 μm, and length 250 cm. The mass of the suspended ring is 1500 g., it has outer and inner radii of 10 and 7 cm., respectively and a thickness of 1 cm. The mother materials of which the ring is constructed have a density homogeneity somewhat better than 1 part in 10^3. For a number of reasons we find it convenient to demand that the first three mass moments of the two halves of the ring be the same and this, together with the demand for identical external appearance is possible despite the

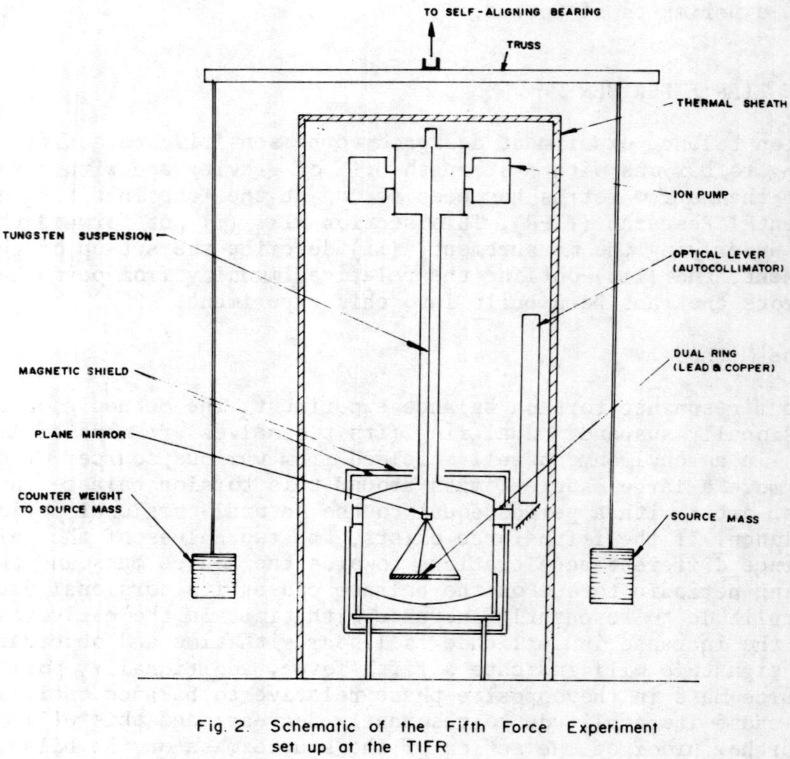

Fig. 2. Schematic of the Fifth Force Experiment set up at the TIFR

density difference between lead and copper simply by machining two grooves in the lead half of the ring (see fig.3): the five free parameters available in the two grooves (the inner and outer groove radii and the common groove depth) are adjusted to suit the three demand conditions. The complete balance ring consists of these semi-rings sandwiched between thin aluminium rings and bonded with thin layers of epoxy resin.

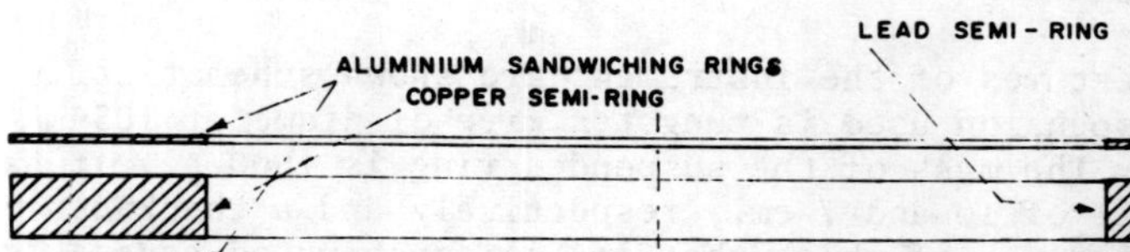

Figure 3. Exploded section of the balance, showing the copper and lead halves together with the sandwiching aluminium rings.

The first level of shielding of this balance consists of the ultra high vacuum chamber in which it is suspended; a pressure of $3x10^{-8}$ Torr is dynamically maintained by two ion pumps (60 l/s) switched on after the initial bakeout (at 100°C for 24 hrs.). The bottom of the chamber is surrounded by a two-stage magnetic shield which acts to cut down the magnetic field in the vicinity of the ring to 10^{-3} Oe and magnetic field gradients to less than 10^{-4} Oe/cm. This entire system, together with the autocollimator is enclosed in a thermal sheath; the measured temperature profile at a point inside the sheath shows the absence of any thermal wave at the period of the balance ($\sim$ 13 min.) with an amplitude greater than a millidegree.

Two 250 kg. source masses hang in the plane of the balance by means of a suspension system consisting of a truss mounted on a self-aligning thrust bearing supported from fixed points on the floor. The truss has two arms at right angles, each arm carrying a 'source' mass, and at the far end of each arm is a 'counterweight' mass to balance the entire truss structure when it is loaded on the bearings. (The counterweight masses are smaller than the source masses but hang on longer lever arms).

Data from the experiment consists of the torsion angle measured by the autocollimator, and a record of the angular position of the source masses as a function of time. Housekeeping data on magnetic field and field gradients and of temperature and its gradients in the volume enclosed by the magnetic shields is also gathered, all of this data being acquired and stored by a dedicated minicomputer.

2.3. Immunization from Noise

The use of ultra-high vacuum is a major means of isolation of the balance from gas related noise and damping torques; apart from reduction of gas drag, ultra-high vacuum acts to reduce the size of radiometric torques that are generated by temperature differentials that might exist across the chamber. Radiometric torques are further reduced because of the circular shape of the ring: to the extent that it is perfectly circular, specular reflection of incident gas molecules and thermal photons causes radial momentum transfer to the ring and hence no torque.

The presence of the magnetic shield eliminates the possibility of spurious torques arising from time varying external magnetic fields coupling to the differing magnetic susceptibilities of the two halves of the ring. The thermal sheath cuts down systematic effects of a thermal nature.

A possible source of systematic error, a tilt of the floor correlated with mass movement may cause energy to be periodically fed into the torsion balance. This is eliminated in the present experiment by the stratagem of supporting the source masses from fixed points on the floor, for movement of the masses does not then give rise to time-varying tilts of the floor.

In torsion balance experiments which involve the movement of the source mass, a major source of systematic error is the coupling of the quadrupole moment components of the balance to gravitational field gradients generated by the source mass thus causing time varying torques on the balance quite unrelated to the existence of a fifth force. This is a particularly heightened inconvenience, for the source mass, being small, must necessarily be close to the balance to generate an observable torque and, being close, sets up large gravity gradients. Our approach to a solution for this problem is two-fold: (i) through appropriate balance mass design and (ii) by symmetrizing the source mass distribution. With reference to Fig.4 the torque generated by the source mass, good to terms involving the quadrupole moment of the balance mass is

$$\tau_Z = \frac{GM}{R^3}(Q_{xx} - Q_{yy})\,\frac{x}{R}\,\frac{y}{R} + \frac{GM}{R^3}\,Q_{xy}\,\frac{y^2 - x^2}{R^2} + \frac{GM}{R^3}\left(Q_{xz}\,\frac{y}{R} - Q_{yz}\,\frac{x}{R}\right) \tag{2}$$

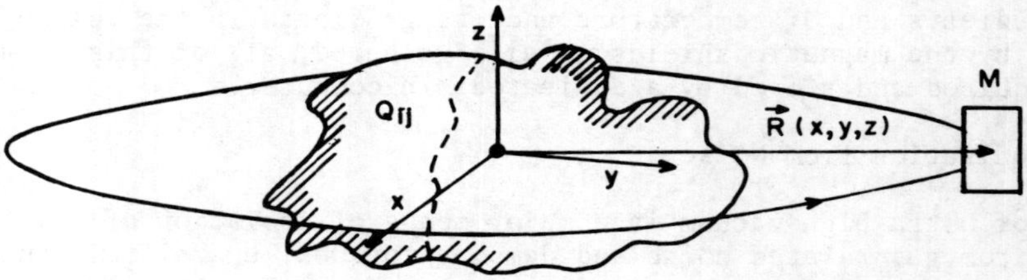

Figure 4. M is the source mass moving in a circular orbit about the balance mass which is characterized by quadrupole moment components Q_{ij}. The torsional suspension is along the z-axis and M moves in a plane perpendicular to it. The origin of the coordinate system is the centroid of the balance mass.

The first two terms in this expression arise from effective ellipticity in the balance mass and are bilinear in x and y while the third term, which is linear in x and y, arises even in a perfectly circular balance which is not perpendicular to the z-axis (torsion axis). When the source mass is moved in a circular orbit in a plane z = constant and with a frequency ω, the first two terms have frequency 2ω while the last term is of frequency ω. It is the last term which is most troublesome because it is of the same frequency as a fifth force torque; the first two terms are rejected heavily first by construction of a highly axially symmetric balance, second by operating with two identical source masses kept 90° apart (see fig.5) and finally because the balance being of very high Q is an excellent filter. To guard against strong couplings

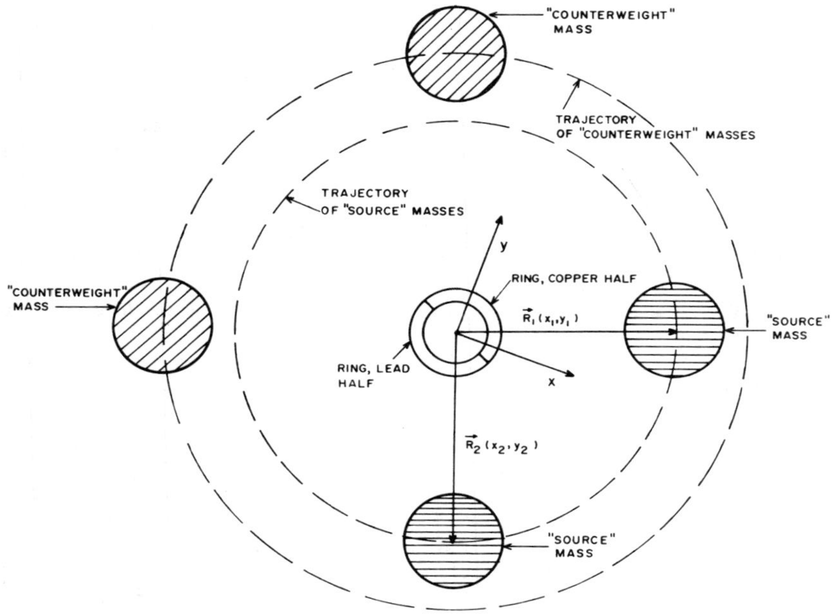

Figure 5. The source and counterweight masses together with the balance in a horizontal section containing the centre of mass of the balance. The magnitude of the fifth force torque is largest at the phase of source mass positioning shown here. Notice that $x_2 = y_1$, $y_2 = -x_1$, a consequence of the 90° separation of the source masses.

through the third term we ensure, first, that during manufacture the ring is as nearly circular & planar as possible and second that during integration, this (nearly) perfect circle is in a plane perpendicular to the torsion axis.

Despite the care being taken to minimize the components Q_{xz} and Q_{yz}, it is important to have independent proof that any observed effect is not due to gravitational gradients. A simple means of doing this stands out on looking at equation (2): the third term has a z and R dependence quite different from the $1/R^2$ dependence of a fifth force term, so by performing runs of the experiment with changed values of z and R one may pinpoint the source of any observed signal.

3. CURRENT STATUS

All essential elements of instrumentation for this experiment have been built, the entire apparatus has been put together and preliminary runs have been performed. Operating principles of the experiment have in the process been smoothed out. However, in these runs, a resonant buildup of the torsional oscillation amplitude of the balance was found, which peaked at the wrong phase of source mass positioning for it to be due to the fifth force; identifying and eliminating this source of coupling (presumably due to the gravitational fields or field gradients of the source masses) is a task ahead. It is unlikely that quadrupole moment components of the ring generate this coupling because of the care taken in its manufacture and in the integration of the balance. On the other hand, the composite system that is torsionally suspended consists of the plane mirror (which reflects the autocollimator beam) and below it, the ring, and this composite could have a relatively large net quadrupole moment.

Drift in the torsional suspension is widely noticed in torsion balance experiments, and this one is no exception. Over the typical time span of a run (10-12 hrs.), the net drift observed in the suspended system is 6×10^{-6} rad. As part of this is certainly due to the autocollimator, an upper limit to the drift in the suspension system is 10^{-5} rad./day, and its drift performance is probably much better. In any case, the observed drift is small enough and smooth enough not to cause any problems in the interpretation of the data.

A matter of concern to us is that the level of human activity in the neighbourhood of the experimental area in Bombay has of late increased dramatically and it appears difficult to obtain any result over the background of vibration induced noise. However, we have for some time been preparing an experiment to test the principle of equivalence (with a sensitivity of 1 part in 10^{14}) and the site for this at Gauribidanur on the Deccan Plateau (seismically one of the quietest places in the world) is ready; we now intend to perform the fifth force experiment there and a shift is in progress. (This site has the advantage of being remote from centres of habitation as well, the nearest hamlet being 5 km. away). We expect to have first results from the experiment by December 1987.

4. ACKNOWLEDGEMENTS

Funds for this experiment have come in part from the DST-sanction No.12(20)/STP-II dated 20.4.1983. We are indebted to our colleagues, C. Viswanadham, U.D. Vaishnav, C.D. Khatu, D.B. Mane, R.L. Mishra, R.A. Parekhji, S. Pereira, G.P. Puthran and M.J. Reddy for engineering support.

Communicated at this School by N. Krishnan.

References

1. T.D. Lee and C.N. Yang, Phys. Rev., 98, 1501 (1955).

2. J.E. Moody and F. Wilczek, Phys. Rev., D30, 130 (1984)

3. D. Chang, R.N. Mohapatra and S. Nussinov, Phys. Rev. Lett., 55, 2835 (1985).

4. Y. Fujii, Ann. Phys., 69, 494 (1972).

5. E. Fischbach, D. Sudarsky, A. Szafer, C. Talmadge and S.H. Aronson, Phys. Rev. Lett., 56, 3 (1986)

6. A. de Rujula, Phys. Lett., B180, 213 (1986).

7. Talk by R.D. Newman, Moriond Conference, Jan. 1987.

8. Talk by G.T. Gillies, this School.

ANALYSIS OF GROUND STATES OF GENERAL RELATIVITY THEORY AND RELATIVISTIC THEORY OF GRAVITATION

A.A.Logunov, Yu.M.Loskutov, M.A.Mestvirishvili
Moscow State University
Physical Faculty
II7234 Moscow
USSR

ABSTRACT. This paper discusses two subjects: I) demerits of the General Relativity Theory (GRT) as a physical theory of gravity and 2) construction of the Relativistic Theory of Gravitation (RTG) possessing all the attributes of field theories.

I. ON INCOMPATIBILITY OF GRT WITH CONSERVATION LAWS

As known, the gravitational field (GF) of GRT, as differed from other physical fields, does not possess the localizability property [1] */ and its energy-momentum is not a tensor quantity in the full sense of the word, i.e. under arbitrary coordinate transformation the components of the energy-momentum pseudotensor $T^{\mu\nu}_{(g)}$ of the gravitational field (EMPTGF) do not obey the tensor law. For example, Shrödinger showed [2] that with some choice of the coordinates of a three-dimensional space, or with their rearithmetization, all the components of the EMPTGF in the exterior of a massive static body can be made zero. Einstein responded to that as follows: "As to Shrödinger's reasonings, their cogency is explained by an analogy with electrodynamics in which the energy stresses and density of any field are nonzero. Still, I cannot find a reason why this should also be the case with gravitational fields. Gravitational fields can be specified without introducing energy stresses and density" (ref. [1], p.627). This and other statements of Einstein [1] testify to his having departed from the representation of a gravitational field as a physical field of the Faraday-Maxwell type not deprived of the localizability property. At the same time, as to be seen

*/ "... for an infinitesimal region, the coordinates can be chosen in such a way, that a gravitational field will be absent there". - Einstein wrote (ref. [1], p. 423).

V. De Sabbata and V. N. Melnikov (eds.),
Gravitational Measurements, Fundamental Metrology and Constants, 241–282.

in what follows, the integral energy-momentum conservation laws do not hold in GRT either. Hilbert [3] was the first to point to the disagreement between this and other physical theories regarding energy conservation: "I maintain, ... that for General Relativity Theory, i.e. in the case of the general invariance of the Hamilton function, there are not at all energy equations corresponding to those in orthogonal-invariant theories. I could even note this fact as a feature inherent in General Relativity Theory". Unfortunately, this remark of Hilbert was not comprehended by his contemporaries.

Interpreting the nonlocalizability of the gravitational field in GRT just as "... an unusual circumstance..." (ref.[4], p.698) the advocates of geometrized approach maintain at a time that it "... does not hinder from defining the total energy and other notions though some reasonable reserves should be made" (ref.[4], p.698). According to many authors as well as those of [4], "... the pseudotensor appearing in GRT does not in itself ruin the theory"*) and "does not lead to an arbitrariness in its observational predictions therefore it can be readily accepted" (ref.[4]p.702).

Let us show now that such statements**) are fallacious.

Let, for simplicity, a static spherically-symmetric body having the active mass M be the source of the gravitational field. The total energy P^0 of the source (or its inertial mass) is defined in GRT as

$$P^0 \equiv \int (-g)\left(T^{00}_M + T^{00}_g\right) dV = \lim_{r\to\infty} \oint h^{00k}\, dS_k . \qquad (I.I)$$

---

*) There are some attempts to justify the arbitrariness inherent in the EMPTGF by pleading that "this is not the first case of using in the theory the quantities whose definition has some arbitrariness left" (ref.[4], p.702). Such a reasoning cannot basically be concurred with because there are no two arbitrariness alike. For instance, the gauge arbitrariness in electrodynamics is allowed so far as it does not affect the physical quantities to be defined. As to the arbitrariness in the EMPTGF, it is the one affecting them.

**) The authors of [4] have reflected in the most concise and clear form all the basic imperfections of GRT and of some alternative approaches. Therefore we shall often refer to this work occasionally quoting its specific statements. However, we do not think it expedient to dwell on every fallacious statement and shall bring only those relating to the criticism of GRT and to the construction of the field theory of gravitation.

Here $g \equiv \det|g_{\mu\nu}|$, $g_{\mu\nu}$ is the metric tensor of the Riemannian space, $T_M^{\mu\nu}$ is the energy-momentum tensor of the matter with the gravitational field excluded, $h^{\mu\nu\lambda}$ is the spin pseudotensor and dS_k is an element of the surface tending to "infinity". Putting into this expression $h^{\mu\nu\lambda}$ in the form introduced, for example in [5] we get

$$P^0 = -\frac{1}{16\pi G} \lim_{r\to\infty} \int r^2 \frac{x^k}{r} \frac{\partial}{\partial x^n}(g g^{00} g^{kn}) \sin\theta \, d\theta \, d\varphi \,, \qquad (I.2)$$

where $r^2 \equiv x^k x^n \delta_{kn}$.

A general solution to the Hilbert-Einstein equation

$$R^{\mu\nu} - \frac{1}{2} g^{\mu\nu} R = 8\pi G T_M^{\mu\nu} \,, \qquad (I.3)$$

where $R^{\mu\nu}$ is the Ricci tensor, $R = R^{\mu\nu} g_{\mu\nu}$, is (in the exterior of a body) the one obtained in [6]:

$$g_{kn} = -C\delta_{kn} + (C-A)\frac{x^k x^n}{r^2} \,, \quad g_{00} = B \,, \qquad (I.4)$$
$$g = -BAC^2 \,,$$
$$g^{kn} = -\frac{\delta_{kn}}{C} + \left(\frac{1}{C} - \frac{1}{A}\right)\frac{x^k x^n}{r^2} \,, \quad g^{00} = \frac{1}{B} \,,$$

in which

$$B = 1 - \frac{2GM}{r\sqrt{C}} \,, \quad A = C\left(1 + \frac{rC'}{2C}\right)^2 \cdot \left(1 - \frac{2GM}{r\sqrt{C}}\right)^{-1} \,, \quad C' \equiv \frac{\partial C}{\partial r} \,. \qquad (I.5)$$

As to the function $C(r)$, the only requirement is that it should be smooth and $\lim_{r\to\infty} C(r) \to I$. Putting (I.4) into (I.2) we obtain

$$P^0 = -\frac{1}{2G} \lim_{r\to\infty} r^2 C\left(\frac{\partial C}{\partial r} + \frac{C-A}{r}\right) . \qquad (I.6)$$

Taking, for example,

$$C = \left(1 + \frac{GM}{r}\right)^2 \,, \quad A = \frac{r+GM}{r-GM} \,, \quad B = \frac{r-GM}{r+GM} \qquad (I.7)$$

we obtain that the total energy P^0 is exactly equal to the active gravitational mass M. Should one take*)

$$C = \left(1 + \alpha^2 \sqrt{\frac{8GM}{r}}\right)^2 \,, \qquad (I.8)$$

---

*) In this case $A = (1 + \alpha^2\sqrt{2GM/r})^2 \cdot (1 - 2GM/r\sqrt{C})^{-1}$ and $\lim_{r\to\infty} C(r) = \lim_{r\to\infty} B(r) = \lim_{r\to\infty} A(r) \neq 1$, i.e. the metric remains asymptotically free. Within the frames of GRT, the choice of metrics $g_{\mu\nu}$ defined by fixing $C(r)$ in the form of (I.7) and (I.8) is equally admissible.

where α is a free parameter, then

$$P^{0} = (1+\alpha^{4})M. \qquad (I.9)$$

This points plainly to the nonuniqueness of the GRT predictions for the value of the total energy for the closed system */ under consideration and does not agree with the experimentally confirmed fact of "gravitational" and "inertial" masses being equal, which, by the way, was considered by Einstein as the one underlying the theory. Thus, the statements, even with the above reserves accepted, on the unique definition of the total energy in GRT and on an arbitrariness being absent in the GRT observational predictions must be considered as fallacious. As seen from the above example, neither local nor integral Laws hold true in GRT. Still, there has been no single experimental fact both in micro- and macrocosm questioning directly or indirectly the validity of these laws. Besides, Einstein himself said that "... one should undoubtedly require that the matter and gravitational field together satisfy the momentum and energy conservation laws" (ref. [1], p. 299). Moreover, just these laws have always been and remain to be the basic criteria of the validity of the theory under development, with all the known physical theories satisfying them. The universality of conservation laws should unquestionably reflect some global property of the matter independent of its nature [8]. This property of the matter has manifested itself as the one of space-time homogeneity and isotropy. One may even say that the universality of the conservation laws (for energy, momentum and angular momentum separately) following from generalization of the whole set of the experimental data available reflects the fundamental property of our material world, which manifests itself in the physical structure of space-time. It favours incontrovertibly the global nature of the Minkowski space and of the pseudo-Euclidean physical geometry of space-time which possesses the ten-parameter group of motion. The real space in GRT identifying the role of GF with that of the metric tensor $g_{\mu\nu}(r, GM)$ is the one with the Riemannian geometry not admitting the above group of motion corresponding to shifts and rotations. That is the reason why the conservation laws cannot hold in GRT, or vice versa, the absence of these laws is but an evidence of the Riemannian structure of its space-time. The Riemannian structure of space-time is considered to reflect the experimentally observed nature of the motion of test bodies and light signals in the gravitational field. However, it would be rash to make a conclusion on the geometric structure of space-time proceeding only from the study of the motion of test bodies and light signals because the structure of physi-

*/ For more details about the criticism of EMPTGF of GTR refer to [7].

cal geometry is determined not by particular physical manifestations of the moving matter but its global dynamic properties which are the conservation laws [8].

Really, assuming that the observational data on the nature of this motion reflect directly the properties of space-time we would naturally arrive at the notion of the Riemannian structure of space. But this would entail giving up the universality of the conservation laws, which would be wrong judging from the same observations. Consequently, it would be more reasonable to explain the observed behaviour of test bodies and light by the effect of the dynamic action of the gravitational field on these objects considering them in some space-time whose structure is determined by the global properties of the whole matter independently of its nature, with the gravitational field included, i.e. by the universality of the conservation laws. Since there are no experimental grounds for giving it up, it is necessary to recognize the universality of both the Minkowski space and of the pseudo-Euclidean physical geometry of space-time. Therefore, the list of the fundamental statements of any theory including the gravity theory should begin with the statement on the Minkowski space as a fundamental global space [8], because it is a consequence of the universality of the conservations laws.

2. ON NONUNIQUENESS OF GRT PREDICTIONS FOR GRAVITATIONAL EFFECTS

The lack of the conservation laws in GRT is not its only basic imperfection. As to be seen below and can be seen from the above Section, the nonuniqueness of the GRT predictions is a feature inherent in it and applies to all the gravitational effects. Let us prove it using as an example the effect of gravitational radiosignal retardation in the field of a static spherically symmetric body of mass M.

Proceeding from Hilbert-Einstein equation (I.3) we should take into account that the space-time coordinates x^μ used in it range their values within a manifold fixed by the chosen arithmetization of space-time. Therefore without losing the generality we may, in terms of the variables $x^\mu = (t, r, \theta, \varphi)$ and with the agreed arithmetization of space, put into correspondence the point $r = r_S = 0$ to the centre S of the body M, attribute the value $r = r_f$ to any surface point of this body (considering it to be spherical), assign the point $e(r_e, \varphi_e, \theta_e = \pi/2)$ to the position of the radiosignal source and the point $p\ (r_p, \varphi_p, \theta_p = \pi/2)$ to the position of the receiver or reflector sending the signals back to the point e. A general exterior, with respect to the body M, solution to equation (I.3) is (I.4) with an arbitrary function C(r). It is the arbitrariness resulting in the nonuniqueness of the

GRT predictions for gravitational effects and to the indefiniteness of the GF energy-momentum in the field of a central symmetric source. Really, taking, for example,

$$C(r)=I, \text{ or } C(r)=(I+GM/r)^2 , \qquad (2.I)$$

we obtain the following two different partial solutions *) for $g_{\mu\nu}(r, GM)$ defining the element ds^2:

$$ds_a^2=\left(1-\frac{2GM}{r}\right)dt^2-\left(1-\frac{2GM}{r}\right)^{-1}dr^2-r^2\left(d\theta^2+\sin^2\theta\, d\varphi^2\right) \qquad (2.2)$$

in the first case and

$$ds_b^2=\frac{r-GM}{r+GM}dt^2-\frac{r+GM}{r-GM}dr^2-r^2\left(1+\frac{GM}{r}\right)^2\left(d\theta^2+\sin^2\theta\, d\varphi^2\right) \qquad (2.3)$$

in the second one. In the two solutions the coordinate r is the same (as well as t, θ, φ), i.e. the positions of the centre S of the body M, of its surface and of the source and receiver (or reflector) of radiosignals correspond to the points $r=r_S=0$, $r=r_f$, e $(r_e, \varphi_e, \theta_e=\pi/2)$ and p$(r_p, \varphi_p, \theta_p=\pi/2)$.

Change-over from the coordinate r to variable $\rho = r + GM$ in (2.3) transforms it to the form adequate to (2.2):

$$ds_b^2=\left(1-\frac{2GM}{\rho}\right)dt^2-\left(1-\frac{2GM}{\rho}\right)^{-1}d\rho^2-\rho^2\left(d\theta^2+\sin^2\theta\, d\varphi^2\right). \qquad (2.2a)$$

However, these two solutions differ essentially since the value $\rho_S =$ GM corresponds to the centre S of the body M in (2.2a) and $r_S = 0$ in (2.2). A similar is the case with the change-over from r to $\rho = r-GM$ in (2.2) reducing it to the form similar to (2.3) rather than equivalent to it because this change-over**/ replaces $r_S = 0$ for $\rho_S = -GM$.

---

*) The authors of /4/ conclude on the absence of non-uniqueness in the GRT predictions for the effect of gravitational radiosignal retardation proceeding just from these solutions.

**) Note that such a transformation, $\rho = r \pm GM$, is actually based on the explicit use of the solution to the field equations because it involves the contribution GM due to the presence of the source taken into account only in the process of constructing the equation but not earlier. Therefore it cannot be interpreted as a simple coordinate transformation specifying another initial manifold of the values of x^μ independently of the problem being solved.

Using standard methods and confining in our calculations to the first order in G we obtain the following expressions/9, 10/ for the time for a signal to propagate to one end (with $\varphi_p - \varphi_e > \pi/2$ and the pericentre coinciding with r_f)

$$t_a = \sqrt{r_p^2 - r_f^2} + \sqrt{r_e^2 - r_f^2} + \tag{2.4}$$
$$+ GM\left\{2\ln\frac{r_p + \sqrt{r_p^2 - r_f^2}}{r_e - \sqrt{r_e^2 - r_f^2}} + \left(\frac{r_p - r_f}{r_p + r_f}\right)^{1/2} + \left(\frac{r_e - r_f}{r_e + r_f}\right)^{1/2}\right\}$$

in the case of solution (2.2) and

$$t_b = \sqrt{r_p^2 - r_f^2} + \sqrt{r_e^2 - r_f^2} +$$
$$+ 2GM\left\{\ln\frac{r_p + \sqrt{r_p^2 - r_f^2}}{r_e - \sqrt{r_e^2 - r_f^2}} + \left(\frac{r_p - r_f}{r_p + r_f}\right)^{1/2} + \left(\frac{r_e - r_f}{r_e + r_f}\right)^{1/2}\right\} \tag{2.5}$$

in the case of solution (2.3). Let us go over in (2.4) and (2.5) from the quantities $r_{e,p,f}$ to physically observed ones. Using for this purpose metrics (2.2) and (2.3), let us calculate, in the same first order in G, the radial physical distances (measured experimentally) from r_f to r_e and r_p which will be the same for these two metrics:

$$l_{e,p} = \int_{r_f}^{r_{e,p}} \sqrt{-g_{rr}}\, dr = r_{e,p} - r_f + GM\ln\left(r_{e,p}/r_f\right). \tag{2.6}$$

In addition, let us calculate, again in the first order in G, a relative frequency shift (measured experimentally) in the gravitational field of the source M which will also be the same for the two metrics:

$$\left.\frac{\Delta\omega}{\omega}\right|_{e,p}^{f} \equiv \delta_{e,p} = GM\left(\frac{1}{r_f} - \frac{1}{r_{e,p}}\right). \tag{2.7}$$

Using (2.6) and (2.7) we can now express the quantities $r_{e,p,f}$ in (2.4), (2.5) through the measurable quantities $l_{e,p}$ and $\delta_{e,p}$. The comparison of the times t for a radio signal to propagate from e to p, which follow from (2.4) and (2.5), demonstrates clearly their discrepancy pointing to the nonuniqueness of the GRT predictions for the time t exhibiting in this effect in the first-order quantities in G */.

For $r_f \ll r_{e,p}$ and with account (for the chosen accuracy) of the effect of signal deflection by the gravitational field of the source, from (2.4), (2.5) there follow the expressions

$$t_a = R + 2GM \ln \frac{r_e + r_p + R}{r_e + r_p - R} - 2GM, \qquad (2.4a)$$

$$t_b = R + 2GM \ln \frac{r_e + r_p + R}{r_e + r_p - R}, \qquad (2.5a)$$

in which

$$R = \sqrt{r_e^2 - r_\perp^2} + \sqrt{r_p^2 - r_\perp^2} \qquad (2.8)$$

is a relative slant distance between the points e and p, and $r_\perp$ is the coordinate of the crossing point of two straight lines connecting e and p, on the one hand, and S and the signal trajectory pericentre, on the other hand.

The analysis of other known gravitational effects performed in /10-13/ shows that in the class of solutions $C(r) = [1+(\lambda+1)(GM/r)]^2$, where λ is a free parameter, the nonuniqueness of the GRT predictions exhibits in all the effects without exception , manifesting itself in some of them in the first order in G and in others in the second one. For example, in the effects of ray deflection by the Sun field and in those of the Mercury perihelion shifts this nonuniqueness exhibits in the second-order terms in G. This became the basis (with the measurement accuracy available) for the triumphal "confirmation" of the relevant GRT predictions which consolidated soundly the faith in its being firm.

Yet, it should be noted that if the full time t is expressed through the Earth and Mercury orbiting periods $T_{e,p}$ calculated in the relevant metric, it will be the same for either case. This is explained, as shown, for instance in /13/, by the fact that the orbiting periods $T_{e,p}$ of bodies calculated with the help of (2.2) and (2.3) (see /11/)appear to be different and after putting into t instead of $r_{e,p}$ their expressions through $T_{e,p}$ the nonuniqueness present in t and $T_{e,p}$ are mutually compensated for with the final result becoming unique. However, such data as orbiting periods

---

*) Therefore the statement of the authors of /4/ (p.706) that "... there are no nonuniquenesses in the GRT predictions and contradictions to the theory in this effect" should be considered erroneous.

are not always available. If, for example, we have to measure the propagation time of a radiosignal reflected from a satellite or natural object orbiting "on the other side" of the Sun and leaving the solar system afterwards it would be impossible to apply this technique. But even if we manage to establish the conclusive connection between some measurable quantities this will not necessarily mean the uniqueness of the GRT predictions because every separate quantity, as seen from the calculation of t and $T_{e,p}$, is not unique.

Let us carry out a gedanken experiment clearly demonstrating the nonuniqueness of the GRT predictions for gravitational retardation time. Suppose, two test bodies spaced at some distance and containing a source, reflector and radiosignal detector are fixed at points A and B. In addition, a "needle" capable of reflecting radiosignals and, if necessary, of accomodating a massive small-sized body M is fixed at point S very close to the line AB but approximately equidistant from A and B. Let us perform two experiments using this "setup design". First, swinging the body M away from the "setup " to a long distance (to "infinity") find the distance between the points of different pairs from the time for radiosignals to propagate between them and, introducing the coordinates of the points this way put $r_S=0$. Then, moving the body back and "fixing" its centre on the "needle" determine the time t for a radiosignal to propagate from A to B (or there and back). Then the difference $\Delta t = t-t_o$, where t_o is the propagation time with M absent, just yields the experimental value of the gravitational retardation time. As to the theoretical values of Δt, in GRT they are given depending upon whether we use in the calculations metric (2.2) or(2.3), formulas (2.4a) or (2.5a) subtracting from their right-hand sides the term R, where the values for the quantities under the logarithm sign should be borrowed from the first experiment. Therefore we will have two different predictions for Δt, with at least one obviously being incorrect.

If we make a deeper insight into the nonuniqueness of the GRT predictions and try to find its initial cause we shall face the space-time problem again. Indeed, as well known, the Riemannian geometry in some coordinate system (with the agreed arithmetization of space) should be determined by specifying ten components of the metric tensor at each point of space-time. The ten Hilbert-Einstein equations contain fourteen unknown independent quantities, their solution will, generally speaking, be satisfied with four metric coefficients arbitrary. Putting it in other words, in the specified coordinate system some Hilbert-Einstein equations cannot determine the Riemannian geometry. This has already been confirmed above in Section I and in the present Section (see also /12,13/). To determine the unknown metric coefficients, with the agreed arithmetization of space, in GRT they introduce the so-called coordinate conditions and make the metric coefficients obey four additional, basically noncovariant, equations, with no constraints imposed on the choice of these conditions. The Hilbert-Einstein equations alongside with the chosen coordinate conditions may definitely make it possible to determine the metric tensor of the Riemannian space-time in the specified coordinate system and to calculate the relevant gravitational effects. But, what is the most important point, at the same time it is alleged in GRT that the results of theoretical predictions for gravitational effects will be independent of the choice of coordinate conditions. However, with the agreed arithmetization of space, the functional structure of the metric coefficients and, in particular, their functional connection with GM is heavily dependent on such a choice, i.e. different metric tensors of the Riemannian space-time will correspond to different coordinate conditions (see the above example and /12,13/). Yet, it is easy to see that this statement of GRT as well as those of the authors of /4/ (p.706) are at direct variance with the Weyl-Lorentz-Petrov theorem. According to this theorem, with the known equations for all time-like and all isotropic geodesic lines in any coordinate system, the metric tensor of space-time is determined in this system to an accuracy of a constant factor */ and, consequently, different metric tensors of the Riemannian geometry should lead to different predictions for the motion of test bodies. Since the choice of coordinate conditions is not confined by anything in GRT and different metric tensors correspond to different conditions, in virtue of what has just been said, this theory definitely cannot claim to give unique predictions for gravitational effects. All of it reflects the fundamental fact that GRT due to its intrinsic conceptual nature does not draw a distinction between the manifestations of noninertiality and of the gravitational field. Therefore when constructing a gravitation theory one should rest not only upon the Minkowski space (as required by the consequences of SectionI) but also irremovably take in it into account the metric properties of this space **/. Putting it in different words,

---

*) From the viewpoint of physics this means that the Riemannian geometry of space-time can basically be established by studying the motion of test bodies having zero (light) and nonzero rest mass.

**) The physical equations describing the properties of the matter always contain intrinsically also the space-time structure determined by the metric tensor. The essence of the statement that this or that phenomenon takes place, for example, in the Minkowski space is reflected exactly just by the Minkowski metric contained intrinsically in the relevant equations.

a gravitational theory should possess such a physically motivated (mathematically covariant) equation which would make it possible to distinguish uniquely between the manifestations of noninertiality and those of gravitational field /8/.

3. ON ALTERNATIVE FORMULATIONS OF GRT

Some aspects of gravitational theory using the metric tensor of the Minkowski space were treated earlier (see, for example, /15/. In some papers such approaches are termed "field" while in others they are called "bimetric". However, their essence does not go beyond the frames of GRT. This is clearly seen from the "field" approach discussed by the authors of /4/. By the way, these authors themselves maintain that their "... theory is completely equivalent to the customary "geometric" GRT" (p.703), while "... the versions of the field gravitational theory treated here for a flat world are the field formulation of GRT" (p.707), i.e. their theory should be interpreted as "... alternative to GRT in the form rather than in the content..." (p.699).

Analysing these approaches we could have (after thanking the authors of /4/) confined ourselves to the above statement in order to make the same remarks concerning them as in Sections 1,2. But since these approaches have been disseminated rather widely and might seem at first sight to have no fundamental problems we shall dwell on them in detail.

Such approaches are actually reduced to introducing formally the metric $\gamma^{\mu\nu}$ of a flat space and the gravitational field $h^{\mu\nu}$ whose sum determines the metric tensor $g^{\mu\nu}$ of the Riemannian space:

$$\sqrt{-\gamma}\left(\gamma^{\mu\nu}+h^{\mu\nu}\right)\equiv\sqrt{-g}\,g^{\mu\nu}. \tag{3.1}$$

In this case no physical hypotheses allowing to distinguish between the field $h^{\mu\nu}$ and the metric $\gamma^{\mu\nu}$ are introduced into the theory. Then the field equations *)

$$h_{\mu\nu;\alpha}{}^{;\alpha}+\gamma_{\mu\nu}h^{\alpha\beta}{}_{;\alpha;\beta}-h^{\alpha}{}_{\nu;\mu;\alpha}-h^{\alpha}{}_{\mu;\nu;\alpha}=16\pi G\left(t_{\mu\nu}+\tau_{\mu\nu}\right), \tag{3.2}$$

*) More than strange is the statement of the authors of /4/ on p.704 that "In the relativistric theory of gravitation developed in ..."/6,9,10,12/, "... there figure exactly the same Lagrangians L^g and L^m as those discussed here and in..."/16/ because the gravitational field Lagrangian L^g used in /6/ in addition to the quadratic terms from the first - order derivatives of GF also contains a term with second - order ones missing in the relativistic theory of gravitation (cf.(4.12a) with (2.3) from /16/).

where the covariant differentiation and shift of indices are carried out with the help of $\gamma_{\mu\nu}$, and $t_{\mu\nu}$ and $\tau_{\mu\nu}$ are the tensors of GF energy-momentum and of the remaining matter, respectively, are found from the variation principle proceeding from the Lagrangians L^g of the gravitational field *) and of the remaining matter, L^m, constructed specifically (see, for example[74,167]). Whereof these authors conclude that equations (3.2) "... contain the differential conservation laws

$$(t_{\mu\nu} + \tau_{\mu\nu})^{;\nu} = 0 \tag{3.3}$$

with following from them integral conservation laws reflecting the fact that a flat world allows the ten-parameter group of motion (Poincaré group)" (ref.[74], p.703).

An illusive nature of these conclusions and of just a possibility itself to introduce irremovably into the theory a global flat space-time with the help of the above technique (without additional physical hypotheses involved) is seen clearly. Indeed, taking into account (3.I) and the definition of $t_{\mu\nu}$ and $\tau_{\mu\nu}$ it makes no difficulty to get convinced that the metric $\gamma_{\mu\nu}$ of a flat space in (3.2) is cancelled out identically and the equations themselves are reduced identically to Hilbert-Einstein equations (I.3) not containing the metric $\gamma_{\mu\nu}$ any longer. Consequently, equations (3.2) can be nothing but another form of writing equations (I.3). Similarly, in virtue of the fact that the covariant derivatives in (3.2) taken with respect to the Minkowski metric from the left and from the right are transformed identically into ordinary partial derivatives and the energy- momentum conservation law (3.3) will take the form identical with GRT:

$$\partial^{\nu}\left(T_{\mu\nu} + \tau_{\mu\nu}\right) = 0, \tag{3.3a}$$

where $T_{\mu\nu}$ is not a tensor any longer because it contains ordinary derivatives only. So, the flat space-time in (3.2) and (3.3) does turn out to be "unobservable, ghostly, and one may say, fictitious..." (ref.[74], p.695) with no "smile" mentioned in [74] (p.695) left any longer. Consequently, the approaches suggested cannot contain either local or integral conservation laws and the theoretical predictions for gravitational effects obtained with the help of them cannot be unique.

All of it should have been expected from the very beginning because a formal, on the level of naive mathematical redefinition of (3.I), introduction of the metric $\gamma_{\mu\nu}$ of a

---

*) Equations (3.2) written in this or that form were discussed a long time ago (see, for instance[15]).

flat space without qualitatively new physical ideas involved cannot modify the physical content of a theory.

Therefore such an approach cannot alter essentially the physical conclusions of GRT regarding either the conservation laws or its predictions*). The identity of (3.2) and (I.3) just testifies to that. Such approaches cannot be called other than an artificial and far-fetched alternative formulation of GRT. Since this theory is definitely incompatible (see Sections I,2) with the notion of a global flat space-time /8/, every attempt to construct it on the basis of the Minkowski space is doomed to failure. If the metric of a flat space-time enters the equations of the theory irremovably, then such a theory and its conclusions will unavoidably differ from GRT and its conclusions.

4. THE RELATIVISTIC THEORY OF GRAVITATION (RTG)

The field approach developed in the relativistic theory of gravitation/6,17/ differs essentially from those considered in Section 3. Its basic initial statements can be formulated concisely in the following way.

I. The Minkowski space x^μ (with the metric $\gamma^{\mu\nu}$) is accepted to be the fundamental base space of RTG. This statement reflects the property inherent in the whole matter, irrespective of its nature, of the universality of the conservation laws separately for energy-momentum and angular momentum.

II. The gravitational field is considered in RTG to be a real physical field in this space with all attributes inherent in other physical fields: it is identified with the second-rank symmetric tensor field $\Phi^{\mu\nu}$ with representations corresponding to spin states two and zero.

III. In virtue of the universality of gravitational interactions and tensor nature of GF, the Lagrangian density of other forms of the matter (gravitational field excluded) is constructed in RTG on the basis of contractions with the effective tensor $g^{\mu\nu}$ determined by joining the gravitational field $\Phi^{\mu\nu}$ to the metric tensor $\gamma^{\mu\nu}$ according to the rule

$$\tilde{g}^{\mu\nu} \equiv \sqrt{-g}\, g^{\mu\nu} \equiv \sqrt{-\gamma}\, \gamma^{\mu\nu} + \sqrt{-\gamma}\, \Phi^{\mu\nu} \equiv \tilde{\gamma}^{\mu\nu} + \tilde{\Phi}^{\mu\nu}. \qquad (4.I)$$

Here the derivatives of nongravitational physical fields contained in the Lagrangian are set to be covariant derivatives ∇_μ taken with respect to the effective metric $g^{\mu\nu}$.

---

*) Using direct calculations it is easy to get convinced that the same nonuniquenesses are peculiar to the solutions of equations (3.2) and (I.3).

Statement III which should be called the "geometrization principle" introduces into the theory, due to the universality of gravitational interactions and tensor nature of GF, a secondary notion of the effective Riemannian space (with the metric $g^{\mu\nu}$ specified in one chart) having, as seen, a purely field origin. The Minkowski space (with the metric $\gamma^{\mu\nu}$) and the gravitational field $\Phi^{\mu\nu}$ in this space remain to be the initial notions of the theory. The geometrization principle of RTG is not identical to the equivalence principle of GRT because in the former theory as well as in other physical field theories, by virtue of the tensor rather than pseudotensor nature of all the physical quantities obtained on its basis, the GF energy density at a point cannot be reduced to zero by any coordinate transformations though the force action of GF can be compensated for at this point.

IV. The Lagrangian density for the gravitational field is considered in RTG to be a quadratic function of covariant first-order derivatives $D_\lambda g_{\mu\nu}$ taken with respect to the metric $\gamma^{\mu\nu}$ of the Minkowski space.

Based on these concepts, the relativistic theory of gravitation is constructed uniquely.

The most direct way to construct the scalar density $\mathcal{L}_g(\tilde{\gamma}^{\mu\nu}, \tilde{g}^{\mu\nu}, D_\lambda g_{\mu\nu})$ of the Lagrangian for a free GF in the Minkowski space, which should satisfy statement IV (with account of I-III), would be to represent it as the general superposition of all sorts of contractions of quadratic in first-order derivatives $D_\lambda \tilde{g}^{\mu\nu}$ with the tensors $\tilde{g}^{\alpha\beta}$ and $\tilde{\gamma}^{\alpha\beta}$ (see *), for example /6, 17/). But we shall act in a somewhat different way, i.e., when looking for the structure $\mathcal{L}_g$ we shall primarily proceed from the gauge principle /17/ requiring that the Lagrangian density of a free GF should vary by not more than a divergence**),

$$\mathcal{L}_g \to \mathcal{L}_g + D_\nu Q^\nu(x),$$

under transformations

$$\delta_\varepsilon \tilde{\Phi}^{\mu\nu} = \delta_\varepsilon \tilde{g}^{\mu\nu} = \tilde{g}^{\mu\lambda} D_\lambda \varepsilon^\nu(x) + \tilde{g}^{\nu\lambda} D_\lambda \varepsilon^\mu(x) - D_\lambda(\varepsilon^\lambda \tilde{g}^{\mu\nu}). \tag{4.2}$$

---

*) By using this method and the gauge principle the Lagrangian $\mathcal{L}_g$ density is determined uniquely.

**)) The gauge transformation (4.2) of the field $\tilde{\Phi}^{\mu\nu}$ differs essentially from its coordinate ($x \to x + \xi$) one:

$$\delta_\xi \tilde{\Phi}^{\mu\nu} = \tilde{\Phi}^{\mu\lambda} D_\lambda \xi^\nu(x) + \tilde{\Phi}^{\nu\lambda} D_\lambda \xi^\mu(x) - D_\lambda(\xi^\lambda \tilde{\Phi}^{\mu\nu}).$$

Here $\varepsilon^{\nu}(x)$ are the infinitesimal parameters of the gauge transformation while the operators δ_{ε} satisfying the equalities

$$(\delta_{\varepsilon_1}\delta_{\varepsilon_2}-\delta_{\varepsilon_2}\delta_{\varepsilon_1})\tilde{g}^{\mu\nu}(x)=\delta_{\varepsilon_3}\tilde{g}^{\mu\nu}(x)\ ,$$

where

$$\varepsilon_3^{\nu}=\varepsilon_1^{\mu}D_{\mu}\varepsilon_2^{\nu}-\varepsilon_2^{\mu}D_{\mu}\varepsilon_1^{\nu}\ ,$$

form the Lie algebra. One should also take into account the requirement of statement II to eliminate from the states of the field $\Phi^{\mu\nu}$ its representations corresponding to spins I and O'. This can be done by making $\Phi^{\mu\nu}$ obey the field equation

$$D_{\mu}\Phi^{\mu\nu}\equiv\frac{1}{\sqrt{-\gamma}}D_{\mu}\tilde{g}^{\mu\nu}=0\ . \qquad (4.3)$$

It should be stressed first of all that equation (4.3) in addition to excluding the nonphysical spin states of the gravitational field $\Phi^{\mu\nu}$ from consideration also makes the metric $\gamma^{\mu\nu}$ of the Minkowski space irremovable from the theory *), thus allowing to distinguish between the manifestations of noninertiality and GF. Besides, these equations reduce the class of possible gauge transformations (4.2) to the manifold $\varepsilon^{\nu}(x)$ satisfying the equation

$$g^{\alpha\beta}D_{\alpha}D_{\beta}\varepsilon^{\nu}(x)=0\ . \qquad (4.4)$$

When going over to the construction of equations which together with (4.3) form the system of basic equations for GF we should take into account that

$$\sqrt{-g}\rightarrow\sqrt{-g}-D_{\nu}(\varepsilon^{\nu}\sqrt{-g})$$

and

$$\sqrt{-g}R\rightarrow\sqrt{-g}R-D_{\nu}(\varepsilon^{\nu}\sqrt{-g}R)$$

are the simplest densities varying by a divergence quantity

---

*) By virtue of what has been said, equation (4.3) cannot have anything to do with the coordinate conditions.

under transformations (4.2). Here R is the scalar curvature of the effective Riemannian space defined by the equality

$$\sqrt{-g}R = \sqrt{-g}\, g^{\mu\nu}\cdot\left[\Gamma^{\lambda}_{\mu\nu}\Gamma^{\sigma}_{\lambda\sigma} - \Gamma^{\lambda}_{\mu\sigma}\Gamma^{\sigma}_{\nu\lambda} + \partial_{\sigma}\Gamma^{\sigma}_{\mu\nu} - \partial_{\nu}\Gamma^{\sigma}_{\mu\sigma}\right] =$$

$$= -\tilde{g}^{\mu\nu}\left(G^{\lambda}_{\mu\nu}G^{\sigma}_{\lambda\sigma} - G^{\lambda}_{\mu\sigma}G^{\sigma}_{\nu\lambda}\right) - D_{\nu}\left(\tilde{g}^{\mu\nu}G^{\sigma}_{\mu\sigma} - \tilde{g}^{\mu\sigma}G^{\nu}_{\mu\sigma}\right),$$

in which

$$\Gamma^{\lambda}_{\mu\nu} = \frac{1}{2}g^{\lambda\sigma}\left(\partial_{\mu}g_{\sigma\nu} + \partial_{\nu}g_{\sigma\mu} - \partial_{\sigma}g_{\mu\nu}\right),$$

$$G^{\lambda}_{\mu\nu} = \frac{1}{2}g^{\lambda\sigma}\left(D_{\mu}g_{\sigma\nu} + D_{\nu}g_{\sigma\mu} - D_{\sigma}g_{\mu\nu}\right).$$

Here we can add the term $\sim \gamma_{\mu\nu}\tilde{g}^{\mu\nu}$, also satisfying the gauge principle by virtue of (4.3). Then in the general case $\mathcal{L}_g$ can be presented as

$$\mathcal{L}_g = \lambda_1\left[\sqrt{-g}R + D_{\nu}Q^{\nu}(x)\right] + \lambda_2\sqrt{-g} + \lambda_3\gamma_{\mu\nu}\tilde{g}^{\mu\nu} + \lambda_4\sqrt{-\gamma}\,. \tag{4.5}$$

The divergence term

$$D_{\nu}Q^{\nu}(x) = D_{\nu}\left(\tilde{g}^{\mu\nu}G^{\sigma}_{\mu\sigma} - \tilde{g}^{\mu\sigma}G^{\nu}_{\mu\sigma}\right)$$

is added here (using the gauge principle) with a view to eliminate in $\mathcal{L}_g$ terms with second-order derivatives which enter $\sqrt{-g}R$. The meaning of the remaining quantities will become clearer later on. Lagrangian (4.5) yields the following expression for the GF energy-momentum tensor in the Minkowski space:

$$t^{\mu\nu}_{g} \equiv -2\frac{\delta\mathcal{L}_g}{\delta\gamma_{\mu\nu}} = 2\sqrt{-\gamma}\left(\gamma^{\mu\alpha}\gamma^{\nu\beta} - \frac{1}{2}\gamma^{\mu\nu}\gamma^{\alpha\beta}\right)\frac{\delta\mathcal{L}_g}{\delta\tilde{g}^{\alpha\beta}} +$$
$$+\lambda_1 J^{\mu\nu} - 2\lambda_3\tilde{g}^{\mu\nu} - \lambda_4\tilde{\gamma}^{\mu\nu}, \tag{4.6}$$

where

$$J^{\mu\nu} \equiv D_\alpha D_\beta \left(\gamma^{\alpha\mu} \tilde{g}^{\beta\nu} + \gamma^{\alpha\nu} \tilde{g}^{\beta\mu} - \gamma^{\alpha\beta} \tilde{g}^{\mu\nu} - \gamma^{\mu\nu} \tilde{g}^{\alpha\beta} \right). \quad (4.7)$$

Considering the least action principle we obtain from (4.6) two different in the form but identical in the content forms of the dynamic equation for GF:

$$\frac{\delta \mathcal{L}_g}{\delta \tilde{g}^{\mu\nu}} = \lambda_1 R_{\mu\nu} + \frac{1}{2} \lambda_2 g_{\mu\nu} + \lambda_3 \gamma_{\mu\nu} = 0 , \quad (4.8)$$

where

$$R_{\mu\nu} \equiv D_\lambda G^\lambda_{\mu\nu} - D_\mu G^\lambda_{\nu\lambda} + G^\sigma_{\mu\nu} G^\lambda_{\sigma\lambda} - G^\sigma_{\mu\lambda} G^\lambda_{\nu\sigma} , \quad (4.9)$$

and

$$\lambda_1 J^{\mu\nu} - 2\lambda_3 \tilde{g}^{\mu\nu} - \lambda_4 \tilde{\gamma}^{\mu\nu} = t^{\mu\nu}_g . \quad (4.10)$$

For equation (4.8) to be satisfied identically, with the gravitational field absent, and for $t^{\mu\nu}(g) = 0$ it is necessary to put

$$\lambda_4 = -2\lambda_3 , \quad \lambda_2 = -2\lambda_3 .$$

The values of λ_1 and λ_3 are easy to identify writing (4.10) with (4.3), (4.1) taken into account:

$$\gamma^{\alpha\beta} D_\alpha D_\beta \tilde{\Phi}^{\mu\nu} + 2\frac{\lambda_3}{\lambda_1} \tilde{\Phi}^{\mu\nu} = -\frac{1}{\lambda_1} t^{\mu\nu}_g . \quad (4.11)$$

It takes an easy-to-interpret form in Galilean coordinates:

$$\Box \Phi^{\mu\nu} + 2\frac{\lambda_3}{\lambda_1} \Phi^{\mu\nu} = -\frac{1}{\lambda_1} t^{\mu\nu}_g . \quad (4.11a)$$

Clearly, the factor $(2\lambda_3/\lambda_1) \equiv m^2$ should be interpreted as the rest mass squared of GF and the value of $(-1/\lambda_1)$, in virtue of the correspondence principle, should be set equal to 16π, i.e.:

$$\lambda_1 = -\frac{1}{16\pi} , \quad \lambda_2 = \lambda_4 = \frac{m^2}{16\pi} , \quad \lambda_3 = -\frac{m^2}{32\pi} .$$

Thus, the Lagrangian for a free GF in the Minkowski space constructed on the basis of the gauge principle will have the following general form:

$$\mathcal{L}_g = \frac{1}{16\pi}\tilde{g}^{\mu\nu}\left(G^{\lambda}_{\mu\nu}G^{\sigma}_{\lambda\sigma} - G^{\lambda}_{\mu\sigma}G^{\sigma}_{\nu\lambda}\right) - \frac{m^2}{16\pi}\left(\frac{1}{2}\gamma_{\mu\nu}\tilde{g}^{\mu\nu} - \sqrt{-g} - \sqrt{-\gamma}\right). \quad (4.12)$$

The relevant dynamic equations of GF, additional to equations (4.3), may be presented in two completely equivalent forms:

$$R^{\mu\nu} - \frac{m^2}{2}\left(g^{\mu\nu} - g^{\mu\alpha}g^{\nu\beta}\gamma_{\alpha\beta}\right) = 0 \ , \quad (4.13)$$

or

$$D_\alpha D_\beta\left(\gamma^{\alpha\beta}\tilde{g}^{\mu\nu} + \gamma^{\mu\nu}\tilde{g}^{\alpha\beta} - \gamma^{\alpha\mu}\tilde{g}^{\beta\nu} - \gamma^{\alpha\nu}\tilde{g}^{\beta\mu}\right) +$$
$$+ m^2\left(\tilde{g}^{\mu\nu} - \tilde{\gamma}^{\mu\nu}\right) = 16\pi t^{\mu\nu}_g . \quad (4.14)$$

This suggests that for the gravitational field possessing a rest mass, the energy-momentum conservation laws

$$D_\mu t^{\mu\nu}_g = 0$$

will be satisfied only with equations (4.3) held true. It should be stressed specifically that equations (4.13) or (4.14) are not gauge invariant even with $\varepsilon^\nu(x)$ satisfying (4.4). This means that a mass term introduced into the Lagrangian removes degeneration *) and determines uniquely the space-time geometry as well as the energy-momentum tensor density of GF.

The complete system of equations for a free GF is

$$R^{\mu\nu} - \frac{m^2}{2}\left(g^{\mu\nu} - g^{\mu\alpha}g^{\nu\beta}\gamma_{\alpha\beta}\right) = 0 , \quad (4.15)$$
$$D_\mu \tilde{g}^{\mu\nu} = 0 ,$$

or

$$\gamma^{\alpha\beta}D_\alpha D_\beta\tilde{\Phi}^{\mu\nu} + m^2\tilde{\Phi}^{\mu\nu} = 16\pi t^{\mu\nu}_g ,$$
$$D_\mu\tilde{\Phi}^{\mu\nu} = 0 \quad (4.16)$$

*) Without this mass term equation (4.14) is gauge invariant.

if written in another, equivalent, form.

With other forms of the matter available, the total Lagrangian density, by virtue of statements III, IV is presented as

$$\mathcal{L} = \mathcal{L}_M(\tilde{g}^{\mu\nu}, \Phi_a) + \mathcal{L}_g(\tilde{\gamma}^{\mu\nu}, \tilde{g}^{\mu\nu}, D_\lambda g_{\mu\nu}), \tag{4.17}$$

where Φ_a are the fields of the matter (GF excluding) and $\mathcal{L}_g$ is given by (4.12). This leads to the following dynamic equations:

$$D_\alpha D_\beta(\gamma^{\alpha\beta}\tilde{g}^{\mu\nu} + \gamma^{\mu\nu}\tilde{g}^{\alpha\beta} - \gamma^{\alpha\mu}\tilde{g}^{\beta\nu} - \gamma^{\alpha\nu}\tilde{g}^{\beta\mu}) + \\ + m^2\tilde{\Phi}^{\mu\nu} = 16\pi t^{\mu\nu}. \tag{4.18}$$

Here $t^{\mu\nu}$ is the symmetric energy-momentum tensor density of the whole matter, $t^{\mu\nu} = t^{\mu\nu}_{(g)} + t^{\mu\nu}_{(M)}$, in the Minkowski space. Equations (4.18) are reduced identically to the equations

$$R^{\mu\nu} - \frac{m^2}{2}(g^{\mu\nu} - g^{\mu\alpha}g^{\nu\beta}\gamma_{\alpha\beta}) = \frac{8\pi}{\sqrt{-g}}\left(T^{\mu\nu} - \frac{1}{2}g^{\mu\nu}T\right), \tag{4.19}$$

where $T^{\mu\nu} = -2(\delta\mathcal{L}_M/\delta g_{\mu\nu})$ is the energy-momentum tensor density of nongravitational forms of the matter in the effective Riemannian space. Taking into account field equations (4.3) we, finally, arrive at the following, fundamentally different from GRT, system of equally important basic dynamic equations of RTG:

$$\gamma^{\alpha\beta}D_\alpha D_\beta\tilde{g}^{\mu\nu} + m^2\tilde{\Phi}^{\mu\nu} \equiv \gamma^{\alpha\beta}D_\alpha D_\beta\tilde{\Phi}^{\mu\nu} + m^2\tilde{\Phi}^{\mu\nu} = 16\pi t^{\mu\nu}, \tag{4.20}$$

$$D_\mu\tilde{g}^{\mu\nu} \equiv D_\mu\tilde{\Phi}^{\mu\nu} = 0, \tag{4.21}$$

or, if written in the equivalent form,

$$R^{\mu\nu} - \frac{m^2}{2}(g^{\mu\nu} - g^{\mu\alpha}g^{\nu\beta}\gamma_{\alpha\beta}) = \frac{8\pi}{\sqrt{-g}}\left(T^{\mu\nu} - \frac{1}{2}g^{\mu\nu}T\right), \tag{4.20a}$$

$$D_\mu\tilde{g}^{\mu\nu} = 0. \tag{4.21a}$$

All the field variables of equations (4.20), (4.21) or (4.20a), (4.21a) are functions of the Minkowski space coordinates with its metric $\gamma^{\mu\nu}$ entering the equations irremovably. With boundary and initial conditions imposed, the solution to the basic system of RTG will have the property of uniqueness *). Owing to this, the physical quantities and predictions obtained with its help will also be unique. In virtue of field equations (4.2I), from (4.20) there directly follows the energy-momentum conservation law

$$D_\mu t^{\mu\nu} = 0$$

not having, in contrast to (3.3), any nonuniquenesses because the Minkowski space metric $\gamma^{\mu\nu}$ is intrinsic in the RTG equations. Note once again that this conservation law holds only with field equations (4.2I) fulfilled. Should we proceed from (4.20a) then, with account of the equalities

$$\nabla_\lambda \gamma_{\mu\nu} = -G^\sigma_{\lambda\mu}\gamma_{\sigma\nu} - G^\sigma_{\lambda\nu}\gamma_{\mu\sigma},$$

$$D_\mu \tilde{g}^{\mu\nu} = \sqrt{-g}\left(D_\mu g^{\mu\nu} + G^\lambda_{\mu\lambda} g^{\mu\nu}\right) = 0,$$

where ∇_λ is a covariant derivative taken with respect to the effective Riemannian metric $g_{\mu\nu}$, we will get another form of putting down the conservation law:

$$\nabla_\mu T^{\mu\nu} = 0.$$

Now set formally the GF rest mass equal to zero. Then

$$\mathcal{L}_g = \frac{1}{16\pi}\tilde{g}^{\mu\nu}\left(G^\lambda_{\mu\nu}G^\sigma_{\lambda\sigma} - G^\lambda_{\mu\sigma}G^\sigma_{\nu\lambda}\right), \qquad (4.12a)$$

while the dynamic equations of a free GF take the form:

$$D_\alpha D_\beta\left(\gamma^{\alpha\beta}\tilde{g}^{\mu\nu} + \gamma^{\mu\nu}\tilde{g}^{\alpha\beta} - \gamma^{\alpha\mu}\tilde{g}^{\beta\nu} - \gamma^{\alpha\nu}\tilde{g}^{\beta\mu}\right) = \qquad (4.14a)$$
$$= 16\pi t^{\mu\nu}_g .$$

This equation is invariant under acceptable gauge transformations while the field tensor $t^{\mu\nu}_{(g)}$ is not gauge invariant.

---

*) With allowance for the equation for the matter state, system (4.20), (4.2I) or (4.20a), (4.2Ia) becomes a closed system of equations determining the dynamics of both the field and matter.

Yet, since its variation $\delta_\varepsilon t^{\mu\nu}_{(g)}$ is, as easily seen with (4.14a) used, reduced under transformation (4.2) to the divergence of an antisymmetric third-rank tensor,

$$\delta_\varepsilon t^{\mu\nu}_g = -\frac{1}{4\pi} D_\lambda D_\sigma \left\{ \delta_\varepsilon \Pi^{[\mu\sigma][\nu\lambda]} \right\},$$

$$\Pi^{[\mu\sigma][\nu\lambda]} = \frac{1}{4}\left(\gamma^{\lambda\mu}\tilde{g}^{\nu\sigma} + \gamma^{\sigma\nu}\tilde{g}^{\mu\lambda} - \gamma^{\lambda\sigma}\tilde{g}^{\mu\nu} - \gamma^{\mu\nu}\tilde{g}^{\lambda\sigma}\right),$$

the gauge arbitrariness $t^{\mu\nu}_{(g)}$ will not affect the integral physical characteristics to be determined /6/. Neither will be gauge invariant such quantities as the effective Riemannian space-time interval and its geometric characteristics /17/.

With other forms of the matter ("substance") present, the total Lagrangian density will be determined by expression (4.17) with $\mathcal{L}_g$ taken from (4.12a), and its relevant dynamic equations will take the form

$$D_\alpha D_\beta\left(\gamma^{\alpha\beta}\tilde{g}^{\mu\nu} + \gamma^{\mu\nu}\tilde{g}^{\alpha\beta} - \gamma^{\alpha\mu}\tilde{g}^{\beta\nu} - \gamma^{\alpha\nu}\tilde{g}^{\beta\mu}\right) = 16\pi t^{\mu\nu}.$$

Due to the presence of"substance" these equations are not gauge invariant any longer and, as a result, the theory will be deprived of the gauge arbitrariness /17/.

Together with (4.3) the system of basic RTG equations will take this form

$$\gamma^{\alpha\beta} D_\alpha D_\beta \tilde{g}^{\mu\nu} \equiv \gamma^{\alpha\beta} D_\alpha D_\beta \tilde{\Phi}^{\mu\nu} = 16\pi t^{\mu\nu}, \qquad (4.22)$$

$$D_\mu \tilde{g}^{\mu\nu} \equiv D_\mu \tilde{\Phi}^{\mu\nu} = 0, \qquad (4.23)$$

or the equivalent one:

$$\sqrt{-g}\, R^{\mu\nu} = 8\pi\left(T^{\mu\nu} - \frac{1}{2} g^{\mu\nu} T\right), \qquad (4.22a)$$

$$D_\mu \tilde{g}^{\mu\nu} = 0. \qquad (4.23a)$$

Though equations (4.22a) nominally coincide in the form with those of Hilbert-Einstein, they should be interpreted differently because their field variables depend on the Minkowski space coordinates while the joint system contains the Minkowski metric irremovably. The complete system of RTG equations (including both the equations for matter and those for GF) in addition to the field variables of the matter and the metric effective Riemannian space tensor also contains intrinsically the metric Minkowski space tensor. It is for this paramount reason why RTG can treat all the physical fields, the gravitational one included, in the unique Minkowski space. This cannot be done remaining within the frames of GRT because its equations do not contain the metric tensor of the Minkowski space *). Though this idea is quite trivial still many experts in gravitational theories do not comprehend it for some reason.

RTG attaching a great importance to field equation (4.23), we shall dwell at length on it. As mentioned above, this equation has nothing to do with GRT coordinate conditions, and in RTG the freedom to choose coordinates remains. This equation not only selects, as required by statement II, the physical field states corresponding to the representations with spins 2 and 0, but due to fixation of the Minkowski space, also separates everything manifesting itself in the presence of GF from all having nothing in common with the action of field. For example, if in the case of a static spherically symmetric massive source M we agree in advance on a arithmetization of space with which the point $r_S = 0$ is put into correspondence with its centre S (this is natural and generally adopted), then the only joint solution to equations (4.22), (4.23) or (4.22a), (4.23a) will be the solution (I.4) with a function $C(r)=(I+GM/r)^2$, i.e. metric (2.3). The coefficients $g_{\mu\nu}$ determining metric (2.2) will not satisfy(4.23) and cannot therefore be regarded as the solutions of RTG. This suggests that all the RTG predictions for gravitational effects will be unique. In particular, for t, the time for a radiosignal to propagate from point e to point p in the field of M (see Sections I,2) RTG yields /9/ the result (2.5) rather than (2.4).

Field equation (4.23) introduced into the theory and making the Minkowski space metric irremovable from it, is inherent in the description of all physical phenomena and, as a result, leads to the consequences qualitatively different from GRT regarding not only the conservation laws or the uniqueness of the predictions for gravitational effects. For example, it is well known that according to GRT, there may exist three models of Friedmann's homogeneous and isotropic Universe. One of them is a closed Universe having a finite volume. In GRT, the choice of a model is heavily dependent on the outstanding problem of the mean matter density. According

*) The system of equations characterizing the matter will always contain, by virtue of the structure of the Lagrangian $\mathcal{L}_M(\tilde{g}^{\mu\nu}, \Phi_a)$ density, only the metric tensor Riemannian space-time density. Consequently, proceeding from only this system of equations it would be utterly impossible to introduce the notion of the Minkowski space.

to RTG, Friedmann's homogeneous and isotropic Universe can only be infinite and "flat" because its three-dimensional geometry is Euclidean. This means that the mean density of the matter energy in this Universe should be equal to the critical density determined on the basis of the Hubble constant measurements. RTG predicts /6,8,17,18/ the existence in the Universe (if Friedmann's model is valid) of a "hidden mass" whose energy density exceeds noticeably the presently observable density. It also suggests that the total density of the matter and gravitational field energy should be equal to zero in this Universe.

Telling is the role of field equation (4.23) in considering the development of gravitational collapse of a dust-like matter (star) having a mass of $M_0 > 3M_\odot$. According to GRT, such a star evolves within a finite proper time into a "black hole" with a singular density. Such objects have no material surface, and from the viewpoint of an external observer, the region inside the Schwarzschild sphere (having a radius of $\tilde{r}_g = 2GM$) turns out to be completely "blocked" for him, i.e. gravitational "self-confinement" of the body occurs with no physical signals being able to escape from the interior region into the exterior one. This conclusion of GRT is actually equivalent to admitting the existence of such forms of the matter in which it is unknowable (on the processes taking place inside a "black hole"). Such a conclusion contradicts the statement on the possibility of knowing the world and therefore cannot be agreed with. RTG alters the picture of gravitational collapse crucially /6,19/ and this is done with the help of equation (4.23). Let us show how and when this equation exhibits its effect. Introduce the space-time coordinates $x^\mu = (t, r, \theta, \varphi)$ of the Minkowski space bringing the point r=0 into correspondence with the centre of the collapsing star. The unknown metric coefficients, on the one hand, related to the unknown gravitational field $\Phi^{\mu\nu}$ and, on the other hand, determining the interval ds^2 of the effective Riemannian space, should be, according to RTG, functions of the coordinates x^μ of the Minkowski space as of the initial variables. With account of the problem being symmetric, ds^2 can be presented as

$$ds^2 = g_{00}(r,t)dt^2 + 2g_{01}(r,t)dt\,dr + \qquad (4.24)$$

$$+ g_{11}(r,t)dr^2 - W^2(r,t)(d\theta^2 + \sin^2\theta\, d\varphi^2).$$

Going over then from the initial variables x^μ to the variables $\xi^\nu = (\tau, R, \theta, \varphi)$ of the co-moving reference frame, i.e. putting

$$\tau = \tau(r, t), \quad R = R(r,t) \qquad (4.25)$$

and

$$g_{00} = \left(\frac{\partial \tau}{\partial t}\right)^2 - \left(\frac{\partial R}{\partial t}\right)^2 e^{\omega(\tau, R)},$$

$$g_{01} = \left(\frac{\partial \tau}{\partial t}\right)\left(\frac{\partial \tau}{\partial r}\right) - \left(\frac{\partial R}{\partial t}\right)\left(\frac{\partial R}{\partial r}\right) e^{\omega(\tau, R)},$$

$$g_{11} = \left(\frac{\partial \tau}{\partial r}\right)^2 - \left(\frac{\partial R}{\partial r}\right)^2 e^{\omega(\tau, R)},$$

we obtain

$$ds^2 = d\tau^2 - e^{\omega(\tau,R)} dR^2 - W^2 \cdot (d\theta^2 + \sin^2\theta d\varphi^2). \quad (4.26)$$

In this case the domain of the admissible values of τ and R should be consistent with the range of the primary variables t and r. In the case of a collapsing dustlike matter having the general active mass M_0 (under the assumption on the zero pressure and spatial homogeneity of the matter energy density, $\rho = \rho(\tau)$) , the simplest nonstatic solution to equation (4.22a) for the metric coefficients entering (4.26) is /20/

$$W_{in} = R\left(1 - \frac{\tau}{\tau_0}\right)^{2/3} \quad (4.27)$$

in the interior region, $0 \leqslant R \leqslant R_0$, and

$$W_{ext} = \left(R^{3/2} - R_0^{3/2} \cdot \frac{\tau}{\tau_0}\right)^{2/3} \quad (4.28)$$

in the exterior one, $R \geqslant R_0$,

$$\exp\{\omega(\tau, R)\} = (\partial W/\partial R)^2, \quad (4.29)$$

where $R_0 = [(9/2)GM_0\tau_0^2]^{1/3}$ determines the position of the material surface for $\tau = 0$.

In GRT, W is interpreted as a radialcoordinate ranging any values from 0 to ∞. Applying the equality

$$W^3(\tau)\rho(\tau) = W^3(0)\rho(0) \quad (4.30)$$

and (4.27) they obtain in GRT the following law for the energy density variation with τ :

$$\rho(\tau) = \rho(0) \cdot \left(1 - \frac{\tau}{\tau_0}\right)^{-2}. \quad (4.31)$$

Whereof it is concluded in GRT that within the finite proper time τ_0 the whole matter collapses into the point W=0 with its energy density $\rho(\tau_0)$ tending to infinity.

In RTG, the ranges for τ and W as secondary quantities cannot be determined independently of those for t and r. Really, RTG in addition to equation (4.22a) also has equation (4.23a) and among the solutions (4.27) - (4.29), acceptable ones are those simultaneously satisfying (4.23a). This equation, when transformed to the variables in the co-moving reference frame, $\xi^\nu = (\tau, R, \theta, \varphi)$, takes the form

$$\frac{1}{\sqrt{-g(\xi)}}\frac{\partial}{\partial\xi^\mu}\left[\sqrt{-g(\xi)}\, g^{\mu\nu}(\xi)\frac{\partial x^\lambda}{\partial\xi^\nu}\right] + \tag{4.32}$$
$$+ \gamma^\lambda_{\mu\nu}(x)\frac{\partial x^\mu}{\partial\xi^\alpha}\frac{\partial x^\nu}{\partial\xi^\sigma} g^{\alpha\sigma}(\xi) = 0,$$

where the Christoffel symbols for the Minkowski space are

$$\gamma^\lambda_{\mu\nu} = \frac{1}{2}\gamma^{\lambda\sigma}\left(\partial_\mu\gamma_{\sigma\nu} + \partial_\nu\gamma_{\sigma\mu} - \partial_\sigma\gamma_{\mu\nu}\right).$$

Putting (4.27)-(4.29) into (4.32) we obtain

$$\frac{\partial}{\partial\tau}\left(W^2 W'\dot{t}\right) = \frac{\partial}{\partial R}\left(\frac{W^2}{W'}t'\right),$$
$$\frac{\partial}{\partial\tau}\left(W^2 W'\dot{r}\right) = \frac{\partial}{\partial R}\left(\frac{W^2}{W'}r'\right) - 2rW', \tag{4.33}$$

where $f' \equiv \partial f/\partial R$, $\dot{f} \equiv \partial f/\partial\tau$.

Let us follow the dynamics of the material surface of a collapsing object. For this purpose we put in (4.28) $R = R_0$, i.e. $W_{ext} = W_f$. The solution to equations (4.33) leads in this case to the constraints

$$r_f = W_f - GM_0, \tag{4.34}$$

$$t = \tau + t_0 - 2\sqrt{2GM_0 W_f} + 2GM_0 \ln\left|\frac{\sqrt{W_f} + \sqrt{2GM_0}}{\sqrt{W_f} - \sqrt{2GM_0}}\right|, \tag{4.35}$$

where the constant t_0 has been introduced to synchronize the count of times τ and t from zero. As seen from here, the field equations (4.23a) of RTG lead to the lowest bound on the range of W_f,

$$W_f^{min} = 2GM_c\ , \quad r_f^{min} = GM_o \equiv r_g\ , \tag{4.36}$$

reached only with $t \to \infty$. This, in its turn, suggests that, according to RTG, the surface of a collapsing object reaches its limiting position *), $r_f = GM_o$, not within the time τ_o, as it was the case with GRT, but within

$$\tau_c = \tau_o - (4GM_o/3), \tag{4.37}$$

less than τ_o.

Applying the conservation law for the total energy in RTG,

$$r_f^3(\tau)\varepsilon(\tau) = r_f^3(0)\varepsilon(0) \tag{4.38}$$

and taking into account that, by virtue of (4.28),

$$r_f(\tau) = R_o(1 - \frac{\tau}{\tau_o})^{2/3} - r_g\ , \tag{4.39}$$

we obtain the following expression for the mean energy density $\varepsilon(\tau)$

$$\varepsilon(\tau) = \frac{3M_o}{4\pi}\left\{R_o\left(1 - \frac{\tau}{\tau_o}\right)^{2/3} - GM_o\right\}^{-3}. \tag{4.40}$$

In RTG, the limiting mean density of the collapsed object is finite and equal to

$$\varepsilon(\tau_c) \equiv \varepsilon_{max} = 3M_o / 4\pi r_g^3\ . \tag{4.41}$$

What was said above demonstrates that, according to the RTG concept, in the Nature there can be neither static nor non-static spherically symmetric bodies of radii $W_f \leq 2GM_o$, i.e. the collapsing matter tends asymptotically to the state of a finite density with the collapsing object brightness de - creasing exponentially. The collapsed body will always bare a material surface and possess an inner structure, i.e. no gravitational self-confinement takes place and the matter does not escape our space. All of it refutes completely the allegation of the authors of /4/ presented on p. 704 that any solution to the Hilbert-Einstein equations satisfies the RTG field equations (4.3) and that field equations do not alter anything.

In the general case, the physical space-time range for the variables ξ^ν in which the solution to the Hilbert-Einstein equations has been found is determined in RTG with the help of establishing their connection with the Minkowski space coordinates x^μ by solving field equation (4.23). In doing this, only the solution establishing the one-to-one

---

*) Each inner layer will occupy its limiting position when $t \to \infty$.

correspondence between ξ^ν and x^μ (with the Jacobian of transformation nonzero *) everywhere) should be sought for. It is just this solution allowing to interpret the variables as a possible coordinate system of the Minkowski space. The range Ω^* for ξ^ν determined proceeding only from the solutions to the Hilbert-Einstein equations does not generally coincide with Ω. Thus, if $g_{\mu\nu}(x)$ is the solution to (4.22a) in terms of the variables x^μ with the metric tensor $\gamma_{\mu\nu}(x)$, then the solution $g'_{\mu\nu}(\xi)$ in terms of the variables ξ^ν with the metric tensor $\gamma_{\mu\nu}(\xi)$,

$$\gamma_{\mu\nu}(\xi) = \frac{\partial x^\alpha}{\partial \xi^\mu}\frac{\partial x^\beta}{\partial \xi^\nu}\gamma_{\alpha\beta}(x), \tag{4.42}$$

will hold true in the domain $\Omega \subset \Omega^*$ only.

Let us, at last, dwell on the possibility of observing the manifestation of the Minkowski space. The legitimacy of formulating it is explained by the fact that when speaking about, for instance, the effects of ray deflection and gravitational radiosignal retardation we unvoluntary compare different situations. Speaking about the deflection of rays we compare their actual propagation with slant propagation when the gravitational field is absent while when speaking about retardation we compare the time of a signal propagation in the gravitational field with the one required by this signal to cover the distance between two selected points, again with the field "switched off" **). This means that in order to interpret such gravitational effects reasonably it is necessary to introduce both the Minkowski space and physical gravitational field. This is what has been done in RTG. In this case the Minkowski space will manifest itself in gravitational experiments indirectly.

Then how should we explain, proceeding from the Minkowski space, the effect of frequency variation in the gravitational field? The fact is that a photon of an energy $\omega_1 = E_{n'n}$ emitted in the gravitational field (at point I) by an atom will not vary its frequency wherever it arrives. A device put at point 2 and detected it will record the discrepancy between the frequency of the detected photon and the ω_2 corres -

---

*) Multi-valued solutions of these equations do not establish the one-to-one correspondence under coordinate transformations and therefore cannot be accepted.

**) Should we confine ourselves to the notion of the Riemannian space only, then it would be senseless from the viewpoint of physics to speak about ray deflection and gravitational retardation because a signal would have nothing to deflect from and to be retarded with.

ponding to the given atomic transition n→n' at this point. This can be explained as follows. Putting down in the Minkowski space the Dirac equation (for simplicity for a hydrogen atom in the approximation of a fixed nucleus) in the gravitational field of a static spherically symmetric source with $r_S=0$ (see /21/) it makes no difficulty to notice that the quantized quantity is now not the electron energy E but $P_0=E(f_+/f_-)^{1/2}$, where $f_\pm \equiv 1\pm GM_0/r$, and r can be interpreted as the coordinate of the atom nucleus in the gravitational field (with the GF inhomogeneities not exhibiting at distances of the order of atom sizes). Consequently, the energy spectrum of a hydrogen atom (in the Shrödinger approximation) will take the form

$$E_n = -\frac{R}{n^2}\left[\frac{f_-(r)}{f_+(r)}\right]^{1/2} . \tag{4.43}$$

This result could have been expected earlier taking into account the fact that in the gravitational field the role of the rest mass of an electron, which is present in the Rydberg constant, should be played by its rest energy $m(f_-/f_+)^{1/2}$ in GF. Consequently, the energy (or frequency) of a photon emitted by an atom (under specific transition) proves to be the following at this point:

$$\omega_1 = \omega_0\left[f_-(r_1)/f_+(r_1)\right]^{1/2} , \tag{4.44}$$

where ω_0 is the corresponding frequency in the absence of the field. This electron transition will give at point r_2 the frequency

$$\omega_2 = \omega_0\left[f_-(r_2)/f_+(r_2)\right]^{1/2} .$$

Since the experiment compares the frequencies corresponding to the same transitions, then an observer placed at point r_2 will record the frequency ω_1 of the photon arrived from point r_1 which will not coinside with ω_2. Besides,

$$\frac{\omega_2-\omega_1}{\omega_1} \equiv \frac{\Delta\omega}{\omega} \simeq GM\left(\frac{1}{r_1}-\frac{1}{r_2}\right) . \tag{4.45}$$

This is usually interpreted as a frequency variation in the gravitational field, and no contradiction with the experiment arises.

The effective variation of the lengths of intervals in the gravitational field can be explained in a similar way. Really, the length of an interval is measured experimentally by the number of emitted at some frequency atomic wavelengths which could be inserted into this interval.

At a given point, the atomic emission wavelength is determined by the quantity $\lambda = 2\pi v/\omega$, where v is the photon velocity at this point of the Minkowski space. Using the dynamic equations for test bodies in the gravitational field (see /22/) it is easy to obtain the equation for photon propagation

$$\dot{r}^2 + r^2\dot{\varphi}^2 f_+ f_- = f_-^2 / f_+^2 , \qquad (4.46)$$

where $\dot{r}$ and $r\dot{\varphi}$ are the radial and azimuthal velocities in the Minkowski space. If a photon is emitted along r (for simplicity we consider here, as before, the case of a static spherically symmetric source of GF) then $|\dot{r}| = f_-/f_+$ and $\lambda_{\parallel} = \lambda_0 (f_+/f_-)^{1/2}$, where λ_0 is the corresponding wavelength in vacuum, i.e. $dl_{\parallel} = (f_+/f_-)^{1/2} dl_0$. If it is emitted perpendicular to r, then $r\dot{\varphi} = (1/f_+)(f_-/f_+)^{1/2}$ and $\lambda_{\perp} = \lambda_0 / f_+$, i.e. $dl_{\perp} = f_+ dl_0$. Either result is consistent with the RTG solution (2.3).

When analysing the observability of the Minkowski space from principal positions /8/ it should be noted that the unique and profound connection between the conservation laws verified experimentally and the space-time structure is just an evidence of the observability of the properties inherent in this space. This means that the experimentally confirmed fact of the universality of the conservation laws points to the reality and observability of the Minkowski space. GRT has no conservation laws and, consequently, has no concept of the Minkowski space. Therefore to discuss it remaining within GRT is senseless.

The reality of the Minkowski space-time can also be supported by the following gedanken experiment. Let there be an observer N completely isolated from the external world in a lift (satellite) moving along the geodesic of the effective Riemannian space produced by the field of a static spherically symmetric massive body M. Will he, having any devices at his disposal, be able to registrate his motion or all his measurements testify to the state equivalent to rest (or uniform translational motion)? Putting it in other words, can N treat his system as inertial? GRT favours unambiguously the latter idea. According to RTG, this is not so.

Let in the lift (satellite) moving together with N there be two bodies having the only distinction that one of them is neutral and another possesses charge Q. Then an external observer K (motionless with respect to the body M) will registrate electromagnetic emission of the charge moving with acceleration with respect to him and a gradual retardation of the charged body as compared with the neutral one. The inner observer, having no means to communicate with the external world, should nevertheless also registrate the emission (because the density of electromagnetic energy cannot be made zero under any reference frame transformation, including transition from K to N) and a gradually accelerated motion of the charged body with respect to the neutral one.All of it could not be observed if the reference frame N were really inertial. Thus, the inner observer can never identify his state with that of being in inertial frame while studying the differences between them he can draw certain conclusions on the external conditions.

The above gedanken experiment is explained naturally in RTG. Indeed, according to RTG, the body motion along the geodesic of the Riemannian space is nothing but its motion in the Minkowski space (inertial reference frame) under the action of the gravitational field, i.e. the motion along the geodesic is due to the action of gravitational forces in inertial reference frame. In virtue of the last point, a charge moving under the action of gravitational forces in inertial reference frame will always emit electromagnetic energy nonvanishing with transition to the noninertial system N. This explains all the phenomena observed by N. As easily seen, the fundamentality of the Minkowski space played the decisive role in these reasonings based on RTG. Therefore the reality of the Minkowski space follows from here and with account of the observations made by N.

5. THE DYNAMICS OF TEST BODIES IN RTG

Using the Lagrangian /22/

$$\mathcal{L} = -m\left(g_{\mu\nu}\frac{dx^\mu}{d\sigma}\frac{dx^\nu}{d\sigma}\right)^{1/2}, \tag{5.1}$$

where x^μ are the space-time coordinates of a body m in the Minkowski space and σ is a canonical parameter, as the basis to construct the dynamics of test bodies m in the gravitational field $\Phi^{\mu\nu}(x)$, we obtain the following dynamic equations:

$$\frac{dp_\mu}{d\sigma} = -\frac{m}{2}\left(g_{\alpha\beta}\frac{dx^\alpha}{d\sigma}\frac{dx^\beta}{d\sigma}\right)^{-1/2}\cdot\frac{dx^\varepsilon}{d\sigma}\frac{dx^\lambda}{d\sigma}\frac{\partial g_{\varepsilon\lambda}}{\partial x^\mu} \equiv F_\mu . \tag{5.2}$$

Here the canonical momenta are

$$p_\mu = -m\left(g_{\alpha\beta}\frac{dx^\alpha}{d\sigma}\frac{dx^\beta}{d\sigma}\right)^{-1/2}\cdot\frac{dx^\nu}{d\sigma}g_{\mu\nu} \quad , \tag{5.3}$$

and the canonical parameter σ can be equated to the time t. As seen, the momenta p_μ satisfy the equality

$$g^{\mu\nu}p_\mu p_\nu = m^2 . \tag{5.4}$$

Using (5.I), (5.3) it is easy to construct the Hamiltonian function for a test body:

$$H(p,x) = \frac{1}{g^{00}}\left\{g^{0k}p_k + \left[g^{0k}g^{0n}p_k p_n + g^{00}\left(m^2 - g^{kn}p_k p_n\right)\right]^{1/2}\right\}. \tag{5.5}$$

For the case of the static spherically symmetric source M of the gravitational field, when according to RTG, in its exterior

$$g_{00} \equiv B = \frac{f_-}{f_+}, \quad g_{11} \equiv -A = -\frac{f_+}{f_-}, \quad g_{22} \equiv -Cr^2 = -r^2 f_+^2 ,$$

$$g_{33} = -r^2 f_+^2 \sin^2\theta \quad , \quad f_\pm \equiv 1 \pm M/r , \tag{5.6}$$

the Lagrange equations (5.2) will take the form[22]

$$\dot{\vec{p}} \equiv \frac{d\vec{p}}{dt} = -\frac{2EM}{r^3 f_+ f_-}\left\{\vec{r}\left(1 - \frac{m^2 f_-}{2E^2 f_+} - \frac{ML^2 f_-}{r^3 E^2 f_+^3}\right) + \frac{M[\vec{v}\vec{L}]}{2Er}\right\}. \tag{5.7}$$

Here the following first integrals of motion (in the equa - torial plane $\theta = \pi/2$) have been taken into consideration: the first one,

$$E = \sqrt{\frac{f_-}{f_+}}\left\{m^2 + \frac{1}{f_+^2}\left[\vec{p}^2 - (1 - f_+ f_-)\frac{(\vec{p}\vec{r})^2}{r^2}\right]\right\}^{1/2} =$$

$$= \sqrt{\frac{f_-}{f_+}}\left\{m^2 + \frac{f_-}{f_+}p_r^2 + \frac{p_\varphi^2}{r^2 f_+^2}\right\}^{1/2} \tag{5.8}$$

plays the role of the energy integral while another one,

$$L = \frac{f_+^3}{f_-} E r^2 \dot{\varphi} \tag{5.9}$$

plays the role of the integral of angular momentum ($\vec{L} = [\vec{r}\vec{p}]$). Equations (5.7)-(5.9) are also valid for test bodies with m=0. The relation between the velocity $\vec{v}=d\vec{r}/dt$ of the body and its momentum $\vec{p}$ is given in this case by

$$\vec{v} = \frac{f_-}{E f_+^3}\left\{\vec{p} - (1-f_+f_-)\frac{(\vec{p}\vec{r})\vec{r}}{r^2}\right\},$$

$$\vec{p} = \frac{f_+^2}{f_-^2}\left\{\vec{v} + (1-f_+f_-)\frac{[\vec{r}[\vec{r}\vec{v}]]}{r^2}\right\}, \tag{5.10}$$

or

$$p_r = \frac{f_+^2}{f_-^2} E \dot{r}, \quad p_\varphi \equiv L = \frac{f_+^3}{f_-} E r^2 \dot{\varphi}. \tag{5.10a}$$

As seen from (5.7), the force $\vec{F}$ of the central-symmetric source, having a nonzero angular momentum and acting on the test body, in addition to a radial component also contains a hyroscopic component noncollinear with $\vec{r}$. If the source M is ball-shaped with a radius of the order of M, then, according to (5.8), (5.9), the test body will approach it with a velocity decreasing (reaching zero) according to the following law:

$$\dot{r}\Big|_{r\to M} \to -\frac{f_-}{f_+}\Big|_{r\to M} \to 0, \quad \dot{\varphi}\Big|_{r\to M} = \frac{L f_-}{E r^2 f_+^3}\Big|_{r\to M} \to 0,$$

with the value of the radial momentum p_r tending to infinity,

$$\sqrt{p_r^2}\Big|_{r\to M} \to E\frac{f_+}{f_-}\Big|_{r\to M} \to \infty.$$

This suggests that r=M is the limiting surface, which any test body (the one having zero mass including) can reach. With radial falling of a body having an energy of $E<m\sqrt{3/2}$ its velocity

$$\dot{r} = -\frac{f_-}{f_+}\sqrt{1-\frac{m^2}{E^2}\frac{f_-}{f_+}} \qquad (5.11)$$

will be maximum at the point

$$r_0 = M\left(1+\frac{2E^2}{3m^2}\right)\left(1-\frac{2E^2}{3m^2}\right)^{-1}. \qquad (5.12)$$

For E=m r_0=5M (cf. /23/) and the body will approach the source further with a decreasing (reaching zero) velocity. In the case of $E>m\sqrt{3/2}$ (acceptable for photons) the value of $\dot{r}$ will decrease everywhere.

Considering also the co-moving tetrads h^a_μ

$$g_{\mu\nu} = h^a_\mu h^b_\nu \eta_{ab}, \quad h^{\mu a} = g^{\mu\nu} h^a_\nu, \quad h_{\mu a} = h^b_\mu \eta_{ab}, \qquad (5.13)$$

$$\mathrm{diag}\,\eta_{ab} = (1,-1,-1,-1), \quad a,b = 0,1,2,3,$$

we may go over from the canonical momenta p_μ to the generalized ones,

$$P_a = -h^\mu_a p_\mu, \quad P^a = P_b \eta^{ab}, \qquad (5.14)$$

obeying, in virtue of (5.4), (5.13), the constraint

$$P_0^2 - \vec{P}^2 = m^2. \qquad (5.15)$$

It is convenient to apply the generalized momenta P_a because the quantity

$$u_k \equiv \frac{P_k}{P_0} = -\frac{h^k_\mu \dot{x}^\mu}{h^0_\nu \dot{x}^\nu} \qquad (5.16)$$

determined with the help of them will yield the "true" velocity of a test body, i.e. the velocity with respect to the point the body crosses at the given instant of time (recorded by the clock put here). For the solution (5.6)

$$\vec{u} \equiv \frac{\vec{P}}{P_0} = f_+\sqrt{\frac{f_+}{f_-}}\left\{\vec{v} + \frac{1-\sqrt{f_+f_-}}{\sqrt{f_+f_-}}\frac{(\vec{v}\vec{r})\vec{r}}{r^2}\right\} \qquad (5.17)$$

yielding with account of (5.I5)

$$\vec{u}^2 = 1 - \frac{m^2}{E^2}\frac{f_-}{f_+} \quad \text{or} \quad E = \frac{m}{\sqrt{1-\vec{u}^2}}\sqrt{\frac{f_-}{f_+}} \,. \qquad (5.I8)$$

As seen, the "true" velocity of a test body when it approaches the region $r \to M$ will always tend to unity (for photons $|\vec{u}| = I$ always). It is, on the one hand, another evidence for a physical impossibility to reach this region and, on the other hand, it points to an obligatory account of quantum effects in it.

In terms of the generalized momenta, dynamic equations (5.2) are reduced to the form /22/

$$\frac{dP_a}{d\sigma} = -m\left(g_{\varepsilon\lambda}\frac{dx^\varepsilon}{d\sigma}\frac{dx^\lambda}{d\sigma}\right)^{-1/2} \cdot \frac{dx^\alpha}{d\sigma}\frac{dx^\beta}{d\sigma} \cdot$$
$$\cdot h^\mu_a h_{\beta b}\left(\frac{\partial h^b_\mu}{\partial x^\alpha} - \frac{\partial h^b_\alpha}{\partial x^\mu}\right) \equiv \mathcal{F}_a \,. \qquad (5.I9)$$

For (5.6)

$$\frac{d\vec{P}}{dt} = -\frac{EM\vec{r}}{r^3 f_+^2}\sqrt{\frac{f_+}{f_-}} - \frac{[\vec{P}\vec{L}]f_-}{Er^2 f_+^3}\left(1 - \sqrt{\frac{f_-}{f_+}}\right) \equiv \vec{\mathcal{F}},$$
$$\frac{dP_0}{dt} = \frac{(\vec{P}\vec{\mathcal{F}})}{P_0}, \qquad (5.20)$$

where

$$P_0 = E\sqrt{\frac{f_+}{f_-}}\,, \quad \vec{P} = \frac{1}{f_+}\left\{\vec{p} - (1 - \sqrt{f_+ f_-})\frac{(\vec{p}\vec{r})\vec{r}}{r^2}\right\}. \qquad (5.2I)$$

Taking into account the theory invariance under transformations of the Lorentz group it makes no difficulty to obtain the four-force of the moving (along z at a constant velocity β_0) gravitational source (5.6) acting upon a test body:

$$F_\lambda = -\frac{m}{2\sqrt{Q}}\frac{\partial Q}{\partial x^\lambda}, \qquad (5.22)$$

$$Q \equiv B(r')\gamma_o^2(1-\beta_o\dot{z})^2 - C(r')\dot{\vec{r}}'^2 - A(r')\frac{M^2(\dot{\vec{r}}'\vec{r}')^2}{r'^4}, \tag{5.22}$$

$$\vec{r}' = x\vec{n}_x + y\vec{n}_y + \gamma_o(z-\beta_o t)\vec{n}_z, \quad \gamma_o \equiv (1-\beta_o^2)^{-1/2}.$$

If the test body were moving (during some time at least) parallel to z at a distance constant from the source, $\vec{a}_o = \vec{r} - \vec{\beta}_o t$, then

$$\vec{F}(\vec{a}_o) = -\frac{mM}{a_o^3 f_+'^2}\sqrt{\frac{f_+'}{f_-'}}\frac{(1-\beta_o^2)^2}{(1-\beta_o^2\sin^2\theta)^{3/2}}(x_o\vec{n}_x + y_o\vec{n}_y + \gamma_o^2 z_o\vec{n}_z), \tag{5.23}$$

where θ is the angle between $\vec{a}_o$ and z-axis in the "fixed" reference frame. For example, with $\vec{a}_o \perp \vec{n}_z$, i.e. $\theta = \pi/2$, this yields

$$\vec{F}(\theta=\frac{\pi}{2}) = -\frac{mM}{\rho_o^2 f_+'^2}\sqrt{\frac{f_+'}{f_-'}}\sqrt{1-\beta_o^2}\cdot\vec{n}_\rho, \quad \rho_o^2 \equiv x_o^2 + y_o^2, \tag{5.24}$$

pointing to a similar to electrodynamics law of relativistic cancellation of transverse interaction force. Let us also bring the distribution of the strength field of the moving gravitational source (5.6) determing the strength of the action upon motionless test bodies placed at the relevant space points. In the first order in G and under the condition $(M/r') \ll \beta_o^2$ we obtain from (5.22)

$$\vec{F}(\vec{v}=0) \simeq -\frac{mM}{R^2}\frac{(1+\beta_o^2)\sqrt{1-\beta_o^2}}{(1-\beta_o^2\sin^2\theta)^{3/2}}(\vec{n}_\rho\sin\theta + \vec{n}_z\gamma_o^2\cos\theta), \tag{5.25}$$

where $\vec{R} = \vec{r} - \beta_o t\vec{n}_z$. As seen, the natures of relativistic variation of gravitational and electromagnetic force fields differ noticeably. In particular, from (5.25) we obtain

$$\vec{F}\left(\theta=\frac{\pi}{2}\right)=-\frac{mM}{R^2}\frac{1+\beta_0^2}{1-\beta_0^2}\vec{n}_\rho\ ,\ \vec{F}(\theta=0,\pi)=\mp\frac{mM}{R^2}\frac{1+\beta_0^2}{\sqrt{1-\beta_0^2}}\vec{n}_z\ ,$$

while the relevant electromagnetic forces have the form

$$\vec{F}_e\left(\theta=\frac{\pi}{2}\right)=\frac{ee_1}{R^2}\frac{1}{\sqrt{1-\beta_0^2}}\vec{n}_\rho\ ,\ \vec{F}_e(\theta=0,\pi)=\pm\frac{ee_1}{R^2}(1-\beta_0^2)\vec{n}_z\ .$$

Verification of the obtained results with the help of an experiment would help answer the question if the gravitational field has a tensor nature.

6. RELATIVISTIC QUANTUM EQUATION FOR SPINOR PARTICLES IN GRAVITATIONAL FIELD

The Dirac equation generalized for a curvilinear Riemannian space has been treated in many papers, the earliest ones being /24,25/. However, for various reasons they have not been developed further. Below we shall construct the relativistic quantum equation for spinor test particles in the gravitational field on the basis of the Minkowski space. In doing this, we proceed from the known expression for the density of the Lagrange function $\mathcal{L}$ in the effective Riemannian space,

$$\mathcal{L}=\frac{1}{2}\sqrt{-g}\{\bar{\psi}\Gamma^\mu(\nabla_\mu\psi)-(\nabla_\mu\bar{\psi})\Gamma^\mu\psi\}-\sqrt{-g}\,m\bar{\psi}\psi, \tag{6.1}$$

in which

$$\bar{\psi}=\psi^+\gamma^0\ ,\quad \Gamma^\mu=h^\mu_a\gamma^a\ ,$$

$$(\nabla_\mu\psi)=\partial_\mu\psi+\frac{1}{4}(\nabla_\mu h_{bv})h^v_c\gamma^c\gamma^b\psi\ , \tag{6.2}$$

$$(\nabla_\mu h_{bv})=\partial_\mu h_{bv}+\Gamma^\sigma_{\mu v}h_{b\sigma}\ ,$$

and the matrices with Latin indices a=0,1,2,3 satisfy

the condition

$$\gamma^a \gamma^b + \gamma^b \gamma^a = 2\eta^{ab}, \tag{6.3}$$

while those Γ^μ with the Greek indices $\mu = 0, 1, 2, 3$ obey the relation

$$\Gamma^\mu \Gamma^\nu + \Gamma^\nu \Gamma^\mu = 2g^{\mu\nu}. \tag{6.4}$$

Varying $\mathcal{L}$ over $\bar{\psi}$ we obtain, after a number of transformations (see the details in /2I/), the equation

$$\left\{\frac{i}{2}(\Gamma^\mu \partial_\mu + \partial_\mu \Gamma^\mu) - m + \frac{i}{8}(\partial_\mu h_{b\nu})(\Gamma^\mu \Gamma^\nu \gamma^b - \gamma^b \Gamma^\nu \Gamma^\mu)\right\}\psi = 0 \tag{6.5}$$

In the class of diagonal metrics (diag($g^{\mu\nu}$) $\neq 0$) this equation takes the following, fairly simple, form:

$$(\gamma^a \hat{P}_a - m)\psi = 0, \tag{6.6}$$

where the operator

$$\hat{P}_a = \frac{1}{2}(h^\mu_a \hat{p}_\mu + \hat{p}_\mu h^\mu_a) \tag{6.7}$$

plays the role of the operator of the generalized momentum for a particle in the gravitational field. In this case

$$\gamma^a \hat{P}_a = \gamma^0 \hat{P}_0 - \vec{\gamma}\hat{\vec{P}}, \tag{6.8}$$

and, for example, the combination $\bar{\psi}\gamma^a\psi$ behaves, under the Lorentz group transformations, as a four-vector.

Note also that equation (6.6) can be obtained by taking the square root from (5.4) and by correlating the classical momenta p_μ with the operators $\hat{p}_\mu = i\partial_\mu$ in the Galilean coordinates, simultaneously fulfilling the hermiticity requirement.

The continuity equation

$$\partial_\mu(\bar{\psi}\gamma^a h^\mu_a \psi) = 0$$

following from (6.6) to the following normalization condition for the ψ-function:

$$\int \bar{\psi}\gamma^a h^0_a \psi \, d^3x = 1.$$

$$\langle\hat{Q}\rangle = \frac{1}{2}\int d^3x\, \psi^+\left(\sqrt{\frac{f_+}{f_-}}\,\hat{Q} + \hat{Q}\sqrt{\frac{f_+}{f_-}}\right)\psi ,$$

$$\frac{d\langle\hat{Q}\rangle}{dt} = \left\langle\frac{\partial\hat{Q}}{\partial t}\right\rangle + \frac{i}{2}\int d^3x\, \psi^+\Big\{2[\hat{H}_0, \hat{Q}] +$$

$$+ \hat{H}_0\sqrt{\frac{f_-}{f_+}}\left[\hat{Q}, \sqrt{\frac{f_+}{f_-}}\right] + \left[\hat{Q}, \sqrt{\frac{f_+}{f_-}}\right]\sqrt{\frac{f_-}{f_+}}\,\hat{H}_0\Big\}\psi , \qquad (6.14)$$

where

$$\hat{H}_0 = (\vec{\alpha}\hat{\vec{P}}) + m\rho_3 . \qquad (6.11a)$$

Using (6.12)-(6.14) it can be shown (see /21/) that in the quasiclassical limit the quantum motion law for $\langle\hat{\vec{P}}\rangle$ is transformed exactly into the corresponding classical law (5.20).

It is helpful to apply in a number of cases the squared spinor equation

$$\left\{\frac{f_+}{f_-}E^2 - \hat{\vec{P}}^2 - m^2 - \frac{1}{r^2 f_+^2}\left(1-\sqrt{\frac{f_-}{f_+}}\right)(\vec{\sigma}\hat{\vec{L}}) - i\sqrt{\frac{f_+}{f_-}}(\vec{\alpha}\vec{F}_0)\right\}\psi = 0 \qquad (6.15)$$

easily obtained from (6.12) with account of the equality

$$[\hat{\vec{P}}\hat{\vec{P}}] = -\frac{i}{r^2 f_+^2}\left(1-\sqrt{\frac{f_-}{f_+}}\right)\hat{\vec{L}} \quad , \quad \hat{\vec{L}} \equiv [\vec{r}\hat{\vec{p}}].$$

In (6.15)

$$\vec{F}_0 = -\frac{ME\vec{r}}{r^3 f_+^2}\sqrt{\frac{f_+}{f_-}} ,$$

$$\hat{\vec{P}}^2 = \frac{1}{f_+^2}\Big\{-\nabla^2 + \Phi^2\partial_r^2 - \Phi(1-\Phi)\frac{2}{r}\partial_r -$$

$$-\frac{\Phi^2}{r^2}\Big[\sqrt{\frac{f_-}{f_+}}\left(1+\sqrt{f_+f_-}\right)^{-1} + \frac{3-4\Phi}{4f_+f_-}\Big]\Big\} ,$$

In accordance with that, the mean value of an observable $\hat{Q}$ in the state ψ will be determined, with the gravitational field present, by the expression

$$\langle\hat{Q}\rangle = \frac{1}{2}\int d^3x\, \psi^+(\tilde{\Gamma}_0\hat{Q} + \hat{Q}\tilde{\Gamma}_0)\psi , \qquad (6.9)$$

where $\tilde{\Gamma}_0 \equiv \gamma^0\gamma^a h^0_a$, while the dynamic law for physical quantities will take the form

$$\frac{d\langle\hat{Q}\rangle}{dt} = \langle\frac{\partial\hat{Q}}{\partial t}\rangle + \frac{i}{2}\int d^3x\, \psi^+\{2[\hat{H}_0, \hat{Q}] + \\ +(\hat{H}_0 + \frac{i}{2}\frac{\partial\tilde{\Gamma}_0}{\partial t})\tilde{\Gamma}_0^{-1}[\hat{Q}, \tilde{\Gamma}_0] + [\hat{Q}, \tilde{\Gamma}_0]\tilde{\Gamma}_0^{-1}(\hat{H}_0 - \frac{i}{2}\frac{\partial\tilde{\Gamma}_0}{\partial t})\}\psi , \qquad (6.10)$$

where

$$[\hat{H}_0, \hat{Q}] \equiv \hat{H}_0\hat{Q} - \hat{Q}\hat{H}_0 , \\ \hat{H}_0 \equiv \frac{1}{2}\gamma^0\gamma^a(h^k_a\hat{p}_k + \hat{p}_k h^k_a) + m\gamma^0 . \qquad (6.11)$$

In the case of metric (5.6) equation (6.6) after replacement $\psi(t)=\psi\exp(-iEt)$ is reduced to the form /21/

$$\{E\sqrt{\frac{f_+}{f_-}} - (\vec{\alpha}\hat{\vec{P}}) - m\rho_3\}\psi = 0, \qquad (6.12)$$

$$\hat{\vec{P}} \equiv \frac{1}{2}\{\frac{1}{f_+}\hat{\vec{p}} + \hat{\vec{p}}\frac{1}{f_+} - \frac{a\vec{r}}{r^2 f_+}(\vec{r}\hat{\vec{p}}) - (\hat{\vec{p}}\vec{r})\frac{a\vec{r}}{r^2 f_+}\} = \\ = \frac{1}{f_+}\{\hat{\vec{p}} - \frac{a\vec{r}}{r^2}(\vec{r}\hat{\vec{p}}) + \frac{ia\vec{r}}{r^2} - \frac{i\vec{r}\Phi}{2r^2\sqrt{f_+f_-}}\} , \qquad (6.13)$$

$$a \equiv 1 - \sqrt{f_+f_-} , \quad f_\pm \equiv 1 \pm \Phi , \quad \Phi \equiv GM/r .$$

The motion law for $\langle\hat{Q}\rangle$ will be written in this case as the term with $(\vec{\sigma}\hat{\vec{L}})$ corresponds to spin-orbital interaction, while the last quantum (dissipative) one appears due to vacuum effect pointing to its instability during particle production in the gravitational field.

By going over to a quasiclassical approach it makes no difficulty to find the law for spin procession in the gravitational field /21/ :

$$\dot{\vec{\zeta}} = -[\vec{\zeta}\,\vec{\Omega}],$$

$$\vec{\Omega} = \frac{\Phi\vec{L}}{Er^2 f_+^2}\sqrt{\frac{f_-}{f_+}}\left\{2\left(1+\sqrt{\frac{f_+}{f_-}}\right)^{-1} + \left(1+\frac{m}{E}\sqrt{\frac{f_-}{f_+}}\right)^{-1}\right\}. \tag{6.16}$$

For a nonrelativistic top, it yields in a weak field the known result, $\vec{\Omega}_{non} \simeq (3/2)(\Phi\vec{L}/mr^2)$, while for ultrarelativistic or massless particles the result is

$$\vec{\Omega}_{rel} \simeq 2\Phi\vec{L}/Er^2.$$

Comparison of the calculated angular precession velocity of the generalized momentum $(\vec{\omega}=[\dot{\vec{P}}\cdot\vec{P}]/P^2)$, with that of $\vec{\Omega}$ shows that

$$\vec{\omega} - \vec{\Omega} = \frac{\Phi\vec{L}f_-}{Er^2 f_+^4}\left(\frac{m}{E}\right)\left(1-\frac{m^2}{E^2}\frac{f_-}{f_+}\right)^{-1}. \tag{6.17}$$

Applying it to particle scattering on a weak potential, $\Phi \ll 1$, and taking into account (5.9) we obtain the following expression:

$$\frac{d\chi}{dt} = \left(\frac{m}{E}\right)\frac{\Phi}{1-(m/E)^2(1-2\Phi)}\frac{d\varphi}{dt} \tag{6.18}$$

for variation of angle χ between the spin and generalized particle momentum, where φ is a polar particle angle. This suggests that massless particles conserve their helicity in GF. The helicity of massive particles will not be the integral of motion. With the condition $\beta_\infty^2 \equiv 1-(m/E)^2 \gg 2\Phi$ fulfilled over the whole trajectory, the value of angle $\Delta\chi$ of particle spin rotation with respect to its generalized momentum will be after scattering

$$\Delta\chi \simeq (m/E)(2M/r_0\beta_\infty^2),$$

where r_0 is the closest distance between the particle and the centre of the scattering body. For $\beta_\infty^2 \ll 2M/\rho$, where ρ is an impact distance, $\Delta\chi \simeq \pi$. Assuming the neutrino rest mass m_ν to be about 20 eV we shall obtain that for relict neutrinos scattered on the Sun (with $r_0 \sim R_\odot$) the angle with respect to their spin rotation will be about 2°, and during their flight nearby a neutron star about π .

References

I. Einstein A. Complete Works. - M.: Nauka, I965, v.I.
2. Shrödinger E. - Phys. Zs., I9I8, N.I9, s.4.
3. Hilbert D. - Göttingen Nachrichten, I9I7, B4, s. 2I.
4. Zel'dovich Ya.B., Grishchuk L.P. - Usp. Fiz. Nauk, I986, v.I49, p. 695.
5. Landau L.D., Lifshitz E.M. Field Theory. - M. : Nauka, I973.
6. Logunov A.A., Mestvirishvili M.A. Foundations of Relativistic Theory of Gravitation. - M.: MSU Publishers, I985.
7. Denisov V.I., Logunov A.A. 'Modern Problems of Mathematics. Review of Science and Technology'.- M.: VINITI, I982, v.2I, p.49; Denisov V.I., Soloviev V.O. Teor. i Mat. Fiz., I983, v.56, p.30I.
8. Logunov A.A. Relativistic Theory of Gravitation and New Notions of Space Time. - M.: MSU Publishers, I986.
9. Logunov A.A., Loskutov Yu.M. - Dokl. Acad. Nauk SSSR, I985, v.785, p. 6I5; Teor. i Mat. Fiz., I986, v.66, p. I50.
I0. Logunov A.A., Loskutov Yu.M. - Teor. i Mat. Fiz., I986, v.67, p. I63.
II. Logunov A.A., Loskutov Yu.M. - Teor. i Mat. Fiz., I986, v. 67, p. 3.
I2. Logunov A.A., Loskutov Yu.M., Chugreyev Yu.V. - Teor. i Mat. Fiz., I986, v. 69, p. 328.
I3. Logunov A.A., Loskutov Yu.M. Nonuniqueness of Predictions of General Relativity Theory and Relativistic Theory of Gravitation. - M.: MSU Publishers, I986.
I4. Petrov A.Z. New Approaches in GRT. - M.: Nauka, I966.
I5. Rosen N. - Phys. Rev., I940, v.57, p. I47; Ann. of Phys., I963, v. 22, p.I; Kohler M.-Z. fur Phys., I952, v. I3I, s. 57I; I953, v. I34, s. 286; I953, v. I34, s. 306; Papapetrou A. - Pros. Roy. Irish Acad., I948,

A52, p. II; Cupta S. - Pros.Phys.Soc., I952, A65, p. 608; Thirring W. - Ann. Phys., I96I, v.I6, p. 69.

I6. Grishchuk L.P., Petrov A.N., Popova A.D. - Comm. Math. Phys.,I984, v. 94, p.379.

I7. Logunov A.A., Mestvirishvili M.A. - Fiz. Elem. Chas. i At. Yadra. I986, v. I7, p. 5; Teor. i Mat. Fiz., I986, v. 67, p. 323.

I8. Logunov A.A., Mestvirishvili M.A. - Teor. i Mat. Fiz., I984, v. 6I, p. 327; Progr. of Theor. Phys., I985, v. 74, p. 3I; Vlasov A.A., Logunov A.A., Mestvirishvili M.A. - Teor. i Mat. Fiz., I984, v. 6I, p. 223.

I9. Vlasov A.A., Logunov A.A. - Teor. i Mat. Fiz., I985, v. 63, p. 3; I986, v. 66, p. I63; 'Difference between Gravitational Collapse in Relativistic Theory of Gravitation and in that of GRT'. M.: MSU Publishers, I986.

20. Oppenheimer J.R., Snyder H. - Phys. Rev., I939, v. 56, p. 455.

2I. Logunov A.A., Loskutov Yu.M. -Teor. i Mat. Fiz., I986, v. 67, p. 9.

22. Logunov A.A., Loskutov Yu. M. - Teor. i Mat. Fiz., I987, v. 70, p.I63.

23. Kaplan S.A. - Zh. Exp. i Teor. Fiz., I949, v. I9,p.95I.

24. Fock V., Ivanenko D. - Phys. Zs., I929, v. 30, p.648.

25. Fock V. - Zs. f. Phys., I929, v. 57, p. 26I; ZRFXO, p. Fiz., I930, v. 62, p. I33; Application of some Ideas of Lobachevsky in Mechanics and Physics. M.-L. GITTL, I950.

GRAVITATIONAL-RELATIVISTIC METROLOGY

V.N. Melnikov
USSR STATE COMMITTEE FOR STANDARDS
Leninsky Prospect 9
117049 Moscow
USSR

ABSTRACT. Relativistic-gravitational effects in space, time and angle measurements, stability of standards and fundamental physical constants and other practical applications of GR are considered.

1. INTRODUCTION

Modern system of standards is based mainly on one of the most investigated interactions - the electromagnetic one. Besides, there are other fundamental physical interactions weak, strong and gravitational ones.

Gravitational interaction plays now an important role in solving the main task of metrology - ensuring the unity of measurements and achievment of their high precision. We mean not only gravimetry, which methods are widely used in geophysics, navigation etc., not only celestial mechanics, but precision measurements in general. Relativistic theory of gravitation (RTG) describes fundamental properties of space-time and this fact causes its particular role in physical phenomena and precision measurements.

Though gravitational interaction was the first known interaction it is still less investigated up till now. There are difficulties with quantization of the gravitational field, its unification with other physical interactions etc.

Here we dwell upon the applications of general relativity (GR) mainly though there are several alternative theories of gravitation. First, because experimental data confirm GR at the level of 10^{-3} and, second, they do not differ from GR in practical applications too much where weak field approximation is valid.

During many years GR was the least checked by experiment theory, being far from practical applications. This was caused by small precisions of measuring devices and

V. De Sabbata and V. N. Melnikov (eds.),
Gravitational Measurements, Fundamental Metrology and Constants, 283–297.

rather small effects of GR in gravitational fields of the Earth and the Solar System planets. Only three classical tests of GR were known: light deflection, perigelia precession, red shift and their modifications.

From 1960th the situation changed drastically. Nowadays RTG (we mean GR and viable alternative theories) become a really widely used in practice theory. And first of all in metrology, astrometry, navigation, geo and cosmic dynamics, geodesy etc.

An increasing role of the gravitational interaction in metrology is caused by/1,2/ :

1. Rising of precision of measuring devices, whereby the relativistic gravitational effects become detectable. Really, the accuracy of the time-frequency standard now is $\sim 10^{14}$, there are secondary standards with accuracies $\sim 10^{16}$, some with accuracies $\sim 10^{17} - 10^{18}$ are under project. In order to synchronize these standards at different points on the Earth and in the space one should take into account gravitational relativistic effects.

2. The fact that precise measurements are now increasingly carried out over large distances and time intervals as a result of the progress in a space technology, VLBI, laser and radiointerferometry, gravimetry, gradiometry etc./3/ For example in trajectory calculations of spacecraft flights to Venus, Mars and other planets, in exact monitoring of Vega flight to the Halley Comet gravitational relativistic effects were taken into account.

3. Development of modern physics towards the inification of fundamental interactions including gravitation that leads to a better understanding of the nature and relations between fundamental constants and solving the problem of their stability in space and time. It is known that in each new theory there appear new constants and it is necessary to relate them with old ones /4/. The knowledge of exact FPC values and their stability is necessary for the functioning of modern standards systems based on FPC.

After quantum metrology a new branch of fundamental metrology is beginning to form /1/ -gravitational-relativistic, or space-time metrology. Its basic aim is the ensuring of rising precision and unity of measurements over large space-time scales, in stationary, variable and strong gravitational fields. Its main branches are:

1) Use of GR models for improving the frequency-time-length service on the Earth, in the near and far space, setting up of a pulsar time scale /5/.

2) Elaboration of GR theory of astrometric measurements for ensuring the necessary precision of radar, laser interferometric and VLBI measurements on the Earth and in the near space (relativistic astrometry).

3) Study of atomic spectra and search for stability of

standards in real gravitational fields - local and cosmological /6/ at different epochs.
4) Study of possible variations and relations between FPC /7/, particularly of the gravitational constant G in the frames of unified, multidimensional, string and other alternative theories and in the experiment, search for the fifth force and deviations from the inverse square law.
5) GR models for precise geodetic, geodynamical, gravimetric and gradiometric measurements on the Earth and in the near space using VLBI, laser interferometers and space - craft ranging as well /8/.
6) Using highest precision measuring devices for testing fundamental theories including RTG particularly on a quantum level /9/.
7) Defining fundamental cosmological parameters by comparing theoretical cosmological models with observational data /10/.
8) Creation of matched hyerarchial GR models for different astronomical scales from planetary to cosmological ones.

With further growth of precision the role of GR methods in practical fields such as metrology, navigation, geodesy, geodynamics and geophysics will be increasing and become an everyday instrument for not only scientists but for engineers as well.

Now we dwell upon these topics in more detail.

2. CLOCKS SYNCHRONIZATION

It is known that in Newtonian physics rates of all clocks are the same because the time is absolute. In SR the rate is the same in each inertial reference frame, but in GR rates of clocks are different in different points even within one reference frame. In order to synchronize atomic clocks with necessary precision on the Earth or in the outer space one should have a definite procedure based on GR. A firm basis for that was established by classical tests of GR and particularly by series of experiments with clocks in 1972-77.

Among them are experiments of Hafele and Keating with clocks on board of commercial planes /11/: theoretical calculations: (-40 ± 23)ns, experimental results : (-59 ± 23) ns for west-to-east flights and (275 ± 21)ns - theoretical against (253 ± 7) - experimental - for east-to-west flight around the Earth.

One should mention also experiments with clocks placed on different heights (3500 and 250m above the sea level) observed time difference was $(33,8 \pm 6,8)$ns per day /12/, that was in a good agreement with theoretical estimations of 30,6ns per day/12/; experiments with three Loran C stations /13/ where clocks placed at different latitudes on the Earth surface were compared and the effect of dozens of ns per day was observed in an excelent agreement with the SR and GR predictions.

But the most precise was the experiment with the clocks put on board of a rocket /14/in which predictions of SR and GR were confirmed with the accuracy of 1 part in 10000.

The necessary procedures for clocks synchronizations using planes and satellites around the Earth are available /15/ and function (GPS and similar types). So the problem of time transfer with the accuracy of modern time-frequency standards of 10^{-14} and even higher on the Earth and in the space up to 50000km becomes only computational and technical one. We must model the real time transfer scheme using GR methods and develope more sophisticated GR models in future if necessary.

The following table is an illustration of the importance of GR corrections even for the USSR national time-frequency service /16/(plane flights with the speed v = 1000km/h, height is 12,8km in the west-east direction)

N	Time of flight in hours	$\Delta\lambda$ in radians	Δ T in ns
1	2	$\pi/9$	-9
2	4.8	$2\pi/9$	-23
3	6.7	$\pi/3$	-25
4	11	$\pi/2$	-48
5	15.5	$2\pi/3$	-68

Δ T is the sum of corrections due to gravitational potential differences at different heights, speed of flights and rotation of the Earth.

Another illustration is the GR correction to the time-frequency standard placed on board of a sputnik in a circle equatorial plane /17/ :

r/R_{earth}	Δ T in 10^{-10}s
2	-1.74
4	-4.32
7	-5.43

So total GR corrections due to a height and speed of a satellite are of the order of 1 ns for the time interval of$\sim$1 s.

3. RELATIVISTIC ASTROMETRY

Astrometry studies positions and motions of cosmic sources on the celestial sphere using angle coordinates of natural and artificial celestial bodies. The main aim of astrometry is a fundamental system of space-time coordinates at different cosmic scales and self-consistent complex of astronomical constants which realize its relations with the terrestrial ones. Nowadays radar, laser and VLBI methods besides standard optical ones are widely used.

The change in methods and rising of their accuracy from 0.1-0.01" to 0.001" and higher caused the imployment of GR methods. The problem lies now not only in calculations of GR corrections to measured quantities. It is necessary to revise all the models and algorithms within the frames of GR.

For a solution of present astrometric problems it is necessary to model with a high precision :

1. Space-time in the region including a source and a reciever.
2. Relativistic dynamics of the source and the reciever.
3. The spread of the signal from the source to the reciever

GR methods of astrometric data reduction are not self-consistent and comprise corrections for the following effects: light deflection in the gravitational field of the Sun, quadratic light abberation, terrestrial dynamic time retardation due to the orbital motion of the Earth, relativistic effects in the light retardation and planets motions, relativistic geodetic precession. These effects are calculated partially within the frames of SR, others - within GR and then are added to corresponding Newtonian quantities /18/.

A simple GR self-consistent model of astrometric optical and VLBI observations is worked out in /19/. All possible GR contributions to measured quantities are calculated within the accuracy of $5 \cdot 10^{-9}$ which is enough for the most of modern observations. Its shortcomings are: treatment of GR contributions without full Newtonian part, lack of dynamical interpretation of the relativistic orbital precession and separation of astrometric measurements from measurements on smaller space scales, e.g. time synchronization on the Earth and in the near space.

They may be overcome through the construction of a general theory of astrometric measurements which explanes all sets of terrestrial and Solar System observations. It is enough to elaborate this theory within the PPN formalism ensuring the accuracy of calculations at $5 \cdot 10^{-16}$ level. This theory must comprise:

1. Construction of the global inertial reference frame.
2. Reference frame of the body of reference from which measurements are done.
3. Relations between coordinates of these reference systems
4. Mathematical description of real physical procedures which realizes the construction of a space-time network on

the surface of the body of reference.
5. Theoretical formulae for reductions of all sets of observational data and
6. Elaboration of data processing algorithms.

A thorough GR study of astrometric (especially VLBI) measurements was done in /20/ and reported at this school by R.Hellings. With the discovery of millisecond pulsars using VLBI and time standards technique the problem of a pulsar time scale suggested by Gosstandart group (Ilyin et. al.) became a very important one.

There is one more wide field of VLBI and range measurements - the applications to the geodesy and geodynamics /18/ and some reports at the International School of Geodesy "A.Marussi", Erice: 15-25 june, 1987. But it is a separate question which deserves a thorough treatment elsewhere.

4. ATOMIC SPECTRA IN GRAVITATIONAL FIELDS

During last decades the treatment of quantum systems in strong electromagnetic and gravitational fields developed intensively in connection with different problems of cosmology, astrophysics and high energy physics. The study of atomic spectra in gravitational fields became also actual. It was caused by several reasons. First, modern time-frequency-length standards are based on atomic spectra properties, on their stability in all observed up till now processes and phenomena. From another point, external fields in principle must influence atomic spectra as they contribute to the total energy of a system in these fields.

Contrary to the case of electromagnetic field the gravitational one can not be shielded and this is related to the single "charge" (mass) in the gravitational interactions. Second, these effects may serve as tests of the gravitational interaction on a quantum level. Up till now GR tests were realised mainly with macro bodies at macro distances.

Of course here we mean the influence of external gravitational fields on proper energies of atomic transitions (in comoving system of reference). It is nothing to do with the doppler and red gravitational shifts.

The problem of atomic spectra in local and cosmological gravitational fields was treated by many authors /21, 22/who used both perturbation technique and exact solutions. At small (atomic) distances of the order of 10^{-8} cm one may expect that gravitational effects are negligible because of the smallness of gravitational to electromagnetic strengths ratio. Really, calculations within the perturbation theory /21/ for the relative shift of lower states

give $\Delta E/E = r_B^2(\text{curvature terms})/\alpha^2$. For a Schwarzchild metric $R_{oo} \sim r_g/R$ and the shift will be $\Delta E/E \sim (\alpha^{-2}) r_B^2 \cdot r_g/R^3$, where α is the fine structure constant, r_B is the Bohr radius, r_g is the gravitational radius and R is a radius of the body. This result coincides with a simple estimation even within the Newtonian approach. Numerical values of the relative shift are:
10^{-39}(Earth surface), 10^{-40}(Sun), 10^{-26}(neutron star), 10^{-68}(cosmological field at present epoch), 10^{-39}(proton gravitational field).

These results were confirmed by exact solutions /22/ obtained for atom-like systems in cosmological metrics. In a closed model the external field of the Universe decreases the binding energy as compared with that for a similar system in a flat space-time but in an open model it it increases this energy. Besides, new qualitative effects arise - ionization for large values of the main quantum number in the case of the closed cosmological model and decay in the case of the open model at the quantum number $N_{max} = \alpha^{1/2}((\mu^2 + 1)^{1/2} - \alpha)^{1/2}$, $\mu = mca/h$, where a is a scale factor. For the present value of $a = 10^{28}$cm $N_{max} \sim 10^{18}$, for smaller a N_{max} decreases. Of course, these effects are not observable because real atoms will be destroyed by thermal ionization if we follow the evolution of the Universe back in time.

But, there is one case when gravitational effects in local fields may be detectable /23/. Previous calculations are valid only for small $v \ll c$ of the atom in a reference frame related to the source of the gravitational field. For large velocities $\Delta E/E = (36/\alpha^2)(r_B/R)^2(1-v^2/c^2) \cdot (r_g/R)(1/R)^2$, where 1 is an impact parameter. This shift will be equal to the Lamb shift and so detectable if $1 \sim R$, $E \sim 10^9$GeV (neutron star). In cosmic rays there are particles with even higher energies. In a more strong gravitational field (e.g. black hole) this energy level may be lowered up to 10^3 GeV if an atom is moving along a conical trajectory in the Kerr black hole.

The conclusion is that in the Solar System and in the present cosmological background the influence of gravitational fields on atomic spectra is negligible and so time-frequency-length standards are stable to a high degree of accuracy.

Closely related to this problem is the problem of physical constants stability in space and time.

5. FUNDAMENTAL PHYSICAL CONSTANTS AND THEIR POSSIBLE VARIATIONS

In any physical theory we have constants which characterise the stability properties of different types of matter - of objects, processes, classes of processes etc. These constants are important because they arise in different situations and have the same values, at any rate within the modern precision of measurements. They are called the fundamental physical constants (FPC).

Constants arise in each physical theory. With the progress of science some theoties substitute others, so new FPC appear, some relations between constants are being established. So now we may speak only about present choice of FPC.

Really, quite recently before the creation of the electroweek interaction theory and grand unified models such constants were considered as FPC: $c, \hbar, \alpha, m_p$ or $m_e, G_F, G_S, G, H, \rho, \Lambda, k, J$ /4/, where G_F, G_S, G are constants of weak, strong and gravitational interactions, H, ρ, Λ are cosmological parameters (Hubble constant, mean density of the Universe and cosmological constant), k and J are the Boltzmann constant and mechanical equivalent of heat which serve as conversion factors. After the new definition of a meter was established in 1983 the speed of light c became a convesion factor also.

Nowadays a new set of FPC seems more preferable: $\hbar$, e, m_e, the Weinberg angle θ_W, Cabbibo angle θ_c, Λ_{QCD}, G, H, ρ, Λ etc./7/. Cosmological constants remain the same, though there are many attempts to connect them with atomic constants. If the unified theory of all interactions will be created then we probably shall know about relations between macro and microphysics constants, their number and nature.

One of the first hypothesis on the nature of FPC belongs to D.A.Wheeler: the set of FPC appears in each cycle of the Universe evolution together with physical laws that govern it. Less ambitious view on the nature of FPC is that they serve as measures of asymptotic states (e.g. $\hbar$ is a minimal quantum of action, e is a minimal observed charge etc.) or to make physical laws and relations dimensionless (e.g. v/c) or both. Finally, FPC and their combinations may serve as natural scales generating basic units.

Exact values of FPC are necessary for tests of fundamental theories, enlargement of our knowledge about nature and, in the end, for practical applications of fundamental theories.

As to the number of FPC there are two opposite tendencies: the number of "old" FPC is usually diminishing when a new more general theory is created, but new areas of science are generated, new processes appear where new constants may arise. If one believes in a simple nature of our world then one may arrive to a minimal choice of FPC related, possibly, to the Planck parameters: $m_L = (c\hbar/2G)^{1/2} \sim 10^{19}$GeV, $L = (\hbar G/c^3)^{1/2} \sim 10^{-33}$cm, as m_L charactarises the unification energy of all interactions and L is a scale where the classical notion of space-time loses its sense.

As to classification of FPC then one may devide them according to a generality as follows: 1) universal, $\hbar$, which devide the world on quantum and nonquantum (macro and micro) parts; and partially c (relativistic and nonrelativistic), 2) constants of interactions - α, θ_W, Λ_{QCD}, G, 3) constants of elementary constituences of matter - m_e, m_W, m_X,... and 4) conversion factors - k, J, partially c.

The verification of FPC stability has not only a fundamental but also a metrological aspect. The fact of FPC "constancy" was got empirically when physical laws were set up and checked afterwards. These main laws were obtained during last 2-3 centuries in experiments on the Earth and in the near space, i.e. in a limited space region and time interval (if we compare them with the Universe age and its radius). So, the extention of the fact of FPC constancy on all space and time scales is an extrapolation which must be tested every time and everywhere.

The question of FPC possible variations arose with the attempts to explane and find relations between macro and micro fenomena in papers by Miln, Eddington and Dirac /24/. According to Dirac's Large Number Hypothesis (LNH) G may vary with the age of the Universe (or cosmological time).

In the development of this trend one may single out two stages: 1. Study of hypotheses and models, their consequences and comparisons with experimental and observational data using the change $G(t) \rightarrow G$ in GR predictions (1937 - 1977). 2. Elaboration of consistent theories with variations of G and data analysis within these theories (from 1977 though some ideas were put forward by Dirac and

Stanukovich earlier).

At the first stage the conclusion was made that astronomical, astrophysical and geophysical data exclude time variations of atomic constants with a high degree of accuracy /4,25/,(see also R.Cohen's lecture).

At the second stage theories were suggested /26-28/, particularly with a conformal scalar field, and also the conception that FPC variations are not absolute but depend on a way of measurements, i.e. choice of standards, units and devices based on a definite tipe of interactions.

Each interaction via dynamics described by a corresponding theory defines a system of units and its basic standards. Yet there is no theory of all four interactions. So one may construct a system of measurements based on any interaction. But, practically it is done on a basis of the electromagnetic one (namely quantum electrodynamics - QED). Surely, one may define units, introduce standards based on the gravitational interaction as it was earlier with the old second. Then different standards may coexist, i.e. atomic second as represented by TAI and defined via Cs frequency, and the gravitational second, defined through the mean period of the Earth motion around the Sun (ephemeris time). These time units defined via different interactions and so FPC may not be in principle synchronized in space and time. Possibly they could change with respect to each other, particularly with the rate of the evolution of the Universe. That is why variations of the gravitational constant are possible in the atomic system of measurements (e, $\hbar$, m are constants by definition) and variations of masses are possible in the gravitational one (e, $\hbar$, G are constants). In practice one may check up only the first possibility as the modern system of units is based on atomic phenomena mainly.

Mathematically mentioned above two systems of measurements may be realised as two conformally related metrics (connected by a nonnegative scalar function). Arbitrary conformal transformations correspond to a choice of an arbitrary system of measurements /29/.

One way of the description of a variable effective G is the introduction of a scalar field in RTG . It may be considered as a fundamental field, as a material field or as a manifestation of grand unified models (Higgs field). This introduction may be done in many ways. One of them is the introduction of the conformal scalar field /4,27/. The advantages of this variant are that the corresponding cosmologies are nonsingular and variations of effective G are of the order of $10^{-12} - 10^{-13}$ year^{-1}/29/depending on the type of a cosmological model. This result do agree with present experimental data /30/.

Finally, the development of GUT and multidimensional

theories also leads to the FPC variations. Renormalization group methods /31/ give changes of electromagnetic and weak interaction constants with energies. The very idea of GUT is based on the fact that due to the growth of the electroweak interaction and the decrease of the strong interaction strengths with energies at 10^{15} GeV they become equal.

As to cosmological evolution the dependance of constants on energies is equivalent to their dependance on the temperature and, in the end, also on the cosmological time.

Multidimensional theories of the Kalutza-Klein type also lead to variations of certain FPC, namely of effective G, as far as it depends on the cosmological scale factor and a scalar field which in general are functions of time. In particular, cosmological solutions were analysed in the 10-dimensional supergravity with Gauss-Bonet and Lorentz-Chern-Simons terms of the superstring origin /32/. It was proved that for any equation of state there is no solution with a constant radius of the internal compact space. This leads to unevitable temporal variations of the effective G in this model. It depends on the Hubble constant and the number of extra dimensions (more than 4) as in estimates of other authors (see also Hellings's lecture).

As to experimental tests of G variations with time there are some results supporting it at a level 10^{-11} y^{-1} (van Flandern) and contradicting it at a level 10^{-12} (Hellings)per year. Future spacecraft ranging and possible laboratory tests will answer the question of G variations.

6. GRAVIMETRIC, GRADOIMETRIC AND GRAVITATIONAL WAVES MEASUREMENTS

The unity and high precision of gravitational measurements on the Earth, mechanical and electromagnetic ones depend also on exact values of the acceleration of gravity g. It is measured wth much better accuracy then G, though it depends on many factors such as the lattitude , time etc. and their number increase wth the growth of accuracy:

$$g = g_0 + g_1 \text{Sin}\,\varphi + g_2 \text{Sin}2\varphi + g_3(t) + g_4 \quad /3/$$

As the standard of g here serves the Earth itself but this standard is not stable. Besides rather well known changes with φ and a tidal action of the Moon and the Sun g_3 there are pooly known or unknown random variations g_4. So the study of the Earth gravitational field is an important task. Now different methods are used: ballistic gravimeters based on a laser interferometric technique, radar and laser ranging of satellites etc.

One could in principle use as an acceleration of gravity standard the acceleration caused by a rather massive body. But for all possible laboratory masses this acceleration is very small that leads to a great loss in accuracy.

Quite different situation exists with gravitational field gradients, i.e. with second derivatives of a gravitational potential. The value of this gradient is defined not by the mass of the body, but by its density. The Earth with the density of $5.5 gr/cm^3$ creates the gradient of 3000E. If very dense materials are used then one may obtain a wide range of gradients up to 6000E (Pb) and 12000E (Pt-Ir alloy).

Gravitational wave antennas being developed now are, according to the principle of their functioning, gradiometers of the dynamical field. Laser interferometric detectors possess the best potential sensitivity. They may register gradients of the order of 10^{-7}E. The stability and big values of artificial gradients probably may be used also for the ajustment of mechanical and electromagnetic units, in the ampere realization istead of the balance technique untill the new quantum standard will be set up. This realization may rise the accuracy by a factor 10.

High precision laser interferometric detectors serve not only as gravitational wave antennas but lead to a class of precise geophysical devices: they may act as a seismograph or deformograph also /3/. The development and new ideas in laser interferometric gravitational wave antennas were discussed in detail by V.Rudenko and R.Drever at the school.

7. PRECISION MEASUREMENTS AND ITS ROLE IN TESTING RTG

Due to very small values of gravitational relativistic effects GR and other alternative theories became stimulators in the development and employment of the highest precision technique. Examples are time-frequency standards, radar and laser ranging devices, VLBI, gravitational waves detectors, gravimeters, SQUIDs etc. Due also to the progress of space technique all the Solar System became a laboratory for testing fundamental physical theories.

Here we dwell upon some most important experiments suggested for testing RTG as a very wide literature exist on this subject /9/. Among them one should mention experiments for testing 2nd order effects and the so called "gravimagnetic" ones connected with the rotation of gravitational field sources and of test bodies as well. First is the "POINTS" project reported at the school by R.Reasenberg. The author believes that optical interferometers in space will allow to measure angles with the accuracy of 5μarcsec i.e. the 2nd order light deflection effect in the field of the Sun.

Nowadays, VLBI leads in the technique of angle measurements on the Earth. VLBI in space may increase the accuracy up to μ arcsec level, so they may compete in space with the optical interferometry and may be used to test the 2nd order effect in a light deflection also (/33/ and Soviet "ASTRON" project).

Another important 2nd order effect to be measured remains the experiment with clocks on board of a spacecraft around the Sun which could verify the red shift predictions of the 2nd order (Solar Probe type).

Among relativistic effects with rotation we must single out such as:

geodetic precession of a gyroscope axis on board of a sputnic in a close polar orbit /34/ with the rate: $(1 + 2\gamma) \cdot 8''/3$ per year.

Lense-Thirring precession connected with the drag of inertial referance frames in the gravitational field of a rotating body /35/ with the rate $(4\gamma + 4 + \alpha_1)\ 0.05''/8$ per year, where γ and α_1 are the PPN-parameters.

spin-orbital effect of the center of gyro mass periodic shift if the spin of a gyro lies in the radial direction /33,36/.

Other important experiments to test RTG are gravitational wave experiments, astrometric range measurements of spacecraft and planets of the "FOBOS" type to test variations of FPC(especialy G), defining astronomical constants etc.

8. FUNDAMENTAL COSMOLOGICAL PARAMETERS

Cosmological constants $\Omega = \rho / \rho_{crit.}$ (parameter of the mean density of the Universe), H and Λ are the worst defined FPC : H is known up to the factor 2 and Λ may be even zero. Though many modern theoretical schemes (especially inflationary ones) lead to $\Omega = 1$ (quasiflat cosmological model) observational data favour the open model. Really, from the barion density calculations based on light elements nucleosynthesis it follows that $\Omega \leq 0.12$ /37/ for $H \geqslant 40$km/s Mpc, limits from dynamics of clusters and superclusters are: $0.3 \leq \Omega \leq 0.7$ /38/. The same results ($\Omega < 1$) are given also in /39/. Any reliable data on the hidden mass which may bring to $\Omega = 1$ are absent, though there are many speculations of this sort.

As to H values there are several groups data ranging from 50 to 100km/s Mpc, each with about 10% accuracy.

The problem of Λ has a long history. It was first introduced by Einstein in his static cosmological model, then it was abondoned after the creation of friedmann models but during last decades was reintroduced with the ideas of quantum vacuum /27/. One of the last estimates of was obtained in /40/. Within the Friedmann cosmology with $\Lambda \neq 0$ the statistical analysis for the sample of 96 first-ranked cluster galaxes with redshifts $0 < z < 0.4$ was done. For H = 50km/s Mpc the following limits were got :

$-0.5 \cdot 10^{-56} \leqslant \Lambda \leqslant 1.4 \cdot 10^{-56} cm^{-2}$. It was obtained also that the value H = 100km/s Mpc is not compatible with all restrictions on the density parameter, age of the Universe and so on.

9. CONCLUSIONS

As we already stressed applications of RTG methods become necessary for precise coordinate and time measurements on the Earth, in the near and far space, in solving cosmological problems and also some problems concerning the unification of physical interactions. While achieving more presision we shall be in need of more complex relativistic models even for the use in the Solar System.

The matched hyerarchial models of different time and space scales (local, Solar, galactic, galactic cluster and cosmological) should be elaborated in order to have a unified self-consistent reduction scheme.

For the realisation of all these important tasks a very wide cooperation of specialists in the gravitational theory and experiment must be welcomed. One of them may be the Project on Relativistic-Gravitational Metrology within the World Laboratory which was elaborated and discussed during the work of the school. It comprises all the topics treated in this paper.

10. REFERENCES

1. Melnikov V.N. In: Quantum Metrology and FPC. Leningrad, VNIIM, 1985, p.210, in Russian.
2. Melnikov V.N. In: Abstr. GR-11. Stockholm, 1986, p.139.
3. Kolosnitsyn N.I., Melnikov V.N. Izmeritelnaya Technika. 1985, n3, p.6.
4. Stanukovich K.P., Melnikov V.N. Hydrodynamics, Fields and Constants in the Theory of Gravitation. Moscow, Energoatomizdat, 1983, in Russian; also in: Grav. Radiation and Relat., v.3, Singapore, W.S., 1986, p.418.
5. Ilyin V.G. et.al. Izmer. Techn. 1983, N12, p.34; Metrologia, 1986, v.22, N1, p.65.

6. Melnikov V.N., Shikin G.N. Izvestiya Vusov, Fizika. 1985, N1, p.55; in: Abstr. GR-10, Padova, 1983, p.93.
7. Melnikov V.N. In: Problems of Gravitation. Moscow, MGU, 1986, p.46, in Russian.
8. Borisova L.B., Melnikov V.N. Izmer. Techn. 1985, N4, p. 16; In: Gravitational Waves. P2-85-667. Dubna, 1985, p. 111; in Russian.
9. Bronnikov K.A., Melnikov V.N. Izmer. Techn. 1988, N3.
10. Yakimov V.E. In:Grav. Radiation and Relat., v.3, Singapore, W.S. 1986, p. 283.
11. Hafel I.C., Keating R.E. Science, 1972, v.177, p.166.
12. Briatore L., Leschutta. Nuovo Cim. 1977, v.37B, N2.
13. See in /3/.
14. Vessot R. et. al. Phys.Rev. Lett. 1980, v.45, p.2081.
15. Allan D., Ashby N. In: Relat. in Celest. Mech. and Astrometry. Reidel. 1986, p. 299.
16. Borisova L.B., Melnikov V.N. Metrologia, USSR, 1988, N3.
17. Borisova L.B., Bronnikov K.A., Melnikov V.N. Ibid. 1988.
18. Relat. in Celest. Mech. and Astrometry. Reidel, 1986.
19. Pavlov N.V. Astron Zh. 1984, v.61, p.385; p.600; 1985, v.62, p.172.
20. Hellings R. Report GR-10, Padova, 1983, in this book.
21. Parker L. GRG. 1981, v.13, N1.
22. Melnikov V.N., Shikin G.N. In: Problems of Gravitation and Elementary Particle Theory (PGEPT), Moscow, Energoatomizdat, v.15, 1984, in Russian.
23. Gorbatsevich A.K., Priebe A. Ibid. 1985, v.16, p.98.
24. Dirac P. Nature, 1937, v.139,p.323.
25. Melnikov V.N., Stanukovich K.P. In PGEPT, 1978, v.9,p.3.
26. Canuto V. et.al. Phys.Rev.D, 1977, v.16, p.1643.
27. Bronnikov K.A. , Melnikov V.N., Stanukovich K.P. Preprint ITP-68-69, Kiev, 1968.
28. Zaitsev N.A., Melnikov V.N. In: PTGEPT, 1979, v.10,p.131
29. Melnilov V.N., Radynov A.G. In: On relat. theory, v.2 Singapore, W.S., 1985, p.196.
30. Hellings R. Report at this school.
31. Bogolyubov N.N., Shirkov D.V. Introd. to the Theory of Quant. Fields, Moscow, Nauka, 1976.
32. Ivashuk V.D., Melnikov V.N. Abstr. Semin. Exact Solut. Tartu, TGU, 1988.
33. Vessot R. Contemp. Phys.1984, v.25, p.355.
34. Schiff L. Phys.Rev. Lett. 1960, v.4, p.215.
35. Will K. Theory and Exper. in Grav. Phys. Cambr. Un.Press 1981.
36. Karpov O.B. Abstr.GR-10, Padova, 1983, p.965.
37. Schramm D., Steigman G. Astroph.J.1982, v.243, p.1.
38. Davis M., Hucra J. Ibid.1983, v.254, p.437.
39. Davis M., Peebles P. Ibid. v.267, p.465.
40. Yakimov V.E. Abstr.GR-11, Stockholm, 1986, p.159.

[illegible]

GRAVITATIONAL WAVES

S.N.Pandey
Department of Applied Sciences
M.M.M.Engineering College
Gorakhpur
India

1. The Possibility of Gravitational Waves

There are great many similarities between gravitation and electromagnetism. Also, Einstein's equations, like Maxwell's equations, have radiative solutions.

No one has yet certainly detected gravitational radiation. However, in the very early days of relativity, Einstein had pointed out the difficulty of generating or detecting this radiation, due tothe weakness of gravitational coupling constant. He found out for a rod that spins about one of its axes, the radiated power

$$P = \frac{32 G I^2 \omega^6}{5c^5} \qquad (1.1)$$
$$= 1.73 \times 10^{-59} I^2 \omega^6 \text{ ergs/sec.}$$

where I is the moment of inertia of the rod about theaxis of spin, ω is the angular velocity, G is the gravitational constant and c is the velocity of light. The frequency of the radiated wave is 2ω because of the symmetry of the rod.

Apart from this, there are other considerations as well that point out this difficulty. For instance, the ratio

$$\frac{Gm^2}{e^2} \sim 10^{-43} \qquad (1.2)$$

implies that a graviton emission is forty-three order smaller than a photon emission. Moreover, if we anticipateto make this ratio close to unity then it requires energy of the order of 10^{30} eV. However, this high value of energy is not possible to achieve by the presently available technology.

But, theoretical considerations lead to believe that gravitational waves always exists in empty space. For instance consider an empty space and assume that $g_{\mu\nu}$ are approxima-

V. De Sabbata and V. N. Melnikov (eds.),
Gravitational Measurements, Fundamental Metrology and Constants, 299–305.

tely constant. This means that $g_{\mu\nu,\sigma}$ and all christoffel symbols are small. Further, if we count only the first order terms, the vacuum Einstein equations lead to

$$\simeq \quad \Gamma^{\alpha}_{\mu\alpha,\nu} - \Gamma^{\alpha}_{\mu\nu,\alpha} = 0 \qquad (1.3)$$

Empty space means no matter, no physical fields except gravitation because gravitational field does not disturb the emptyness. Now we use in (1.3) a harmonic condition

$$g^{\mu\nu} \Gamma^{\lambda}_{\mu\nu} = 0 \qquad (1.4)$$

that leads to on simplification

$$g^{\mu\nu} g_{\rho\sigma,\mu\nu} = 0 \qquad (1.5)$$

Therefore, one can see that each $g_{\rho\sigma}$ satisfies the d'Alembert equation, and its solution will consist of waves travelling with velocity of light. These are gravitational waves.

2. Gravitational Waves on a Background Metric

Now we consider them on a background which may not necessarily be a flat background. So, we suppose

$$g_{\mu\nu} = \overset{B}{g}_{\mu\nu} + h_{\mu\nu} \qquad (2.1)$$

such that $|h_{\mu\nu}| \ll 1$. The background metric or unperturbed metric $\overset{B}{g}_{\mu\nu}$ dominates the perturbation $h_{\mu\nu}$. Then we can treat $h_{\mu\nu}$ as a tensor field on the manifold whose metric is $\overset{B}{g}_{\mu\nu}$. If $\overset{B}{g}_{\mu\nu}$ is a vacuum metric ($\overset{B}{R}_{\mu\nu} = 0$) then the wave equation, in Lorentz gauge, is

$$\nabla^{\alpha}\nabla_{\alpha} \Psi_{\mu\nu} - 2 \overset{B}{R}_{\alpha\mu\nu\beta} \Psi^{\alpha\beta} = 0 \qquad (2.2)$$

where

$$\Psi_{\mu\nu} = h_{\mu\nu} - \tfrac{1}{2} \overset{B}{g}_{\mu\nu} h \ , \quad \Psi^{\mu\nu}{}_{;\nu} = 0 \qquad (2.3)$$

If $g^{B}_{\mu\nu} = \eta_{\mu\nu}$, the Minkowski metric, then

$$\psi = 0, \quad h_{\mu\nu} u^{\nu} = 0, \quad \psi^{\mu\nu}{}_{;\nu} = 0 \qquad (2.4)$$

where u^{ν} is the four velocity vector.

Even if the Universe is filled with matter having energy momentum tensor $T_{\mu\nu}$, one can define a sourceless gravitational waves by demanding the perturbed value of $T_{\mu}{}^{\nu(1)} = 0$. This means that $R_{\mu}{}^{\nu(1)} = 0$ that yields

$$\nabla^{\alpha}\nabla_{\alpha} h_{\mu\nu} - 2 R^{B}_{\alpha\mu\nu\beta} h^{\alpha\beta} = 0 \qquad (2.5)$$

with

$$T_{\mu}{}^{\nu(1)} = 0, \quad \psi^{\mu\nu}{}_{;\nu} = 0, \quad \psi = 0, \quad h_{\mu\nu} u^{\nu} = 0$$

as the gravitational wave equations.

Further, if the waves are with wavelength much shorter than the background radius of curvature, that is, if

$$|\nabla_{\alpha}\nabla_{\beta} h_{\mu\nu}| \gg |R^{B}_{\alpha\mu\nu\beta} h_{\lambda\sigma}| \qquad (2.6)$$

then they propagate on null geodesics and admits geometrical optics. It is in this context we can now consider them on an isotropic background.

3. Conformally Invariant Gravitational Wave Equations

Even if there are no chance of ever detecting gravitational waves, theoretically it is important as it provides a link between general relativity and the microscopic frontiers of physics. The smoothening of initial inhomogenieties and anisotropies in the early Universe are thought to be associated with particle creation[1]. Graviton creation can easily be understood by studying the gravitational wave equations because they are not conformally invariant [2,3]. This means that these equations describing the gravitational waves can can be amplified under certain adiabatic conditions obtainable in the early Universe. The amplification of waves in classical sense amounts to creation of particles in quantum sense.

However this is not possible in electromagnetism.

To this effect, consider the background metric in the form

$$ds^2 = g_{\mu\nu}\, dx^\mu dx^\nu \\ = a^2(\eta)(d\eta^2 - dx^2 - dy^2 - dz^2) \tag{3.1}$$

where $cdt = a(\eta)\,d\eta$. Equation (2.5) in (3.1) becomes

$$h_i^{k\,\prime\prime} + \frac{2a'}{a} h_i^{k\,\prime} + a^2 g^{lm} h^k_{i,l,m} = 0 \tag{3.2}$$

where a prime denotes derivative with respect to η . This on being conformally transformed to flat background assumes the form

$$\mu'' + \mu\left(n^2 - \frac{a''}{a}\right) = 0 \tag{3.3}$$

Evidently (3.3) distinguishes itself differently from other massless wave equations in Minkowski background. The manifestation of gravitation appears as an additional potential a''/a. This is not zero, in general, and this leads to amplification of gravitational waves for some adiabatic law of expansion of a .

The energy momentum tensor of an assembly of gravitons each having four momenta $p^\mu = \hbar k^\mu$ can be given by

$$T_{\mu\nu} = \frac{\hbar k_\mu k_\nu}{\omega} N \tag{3.4}$$

where N is the number of gravitons per unit volume. The energy momentum tensor $T^{\lambda\mu}(k, \omega)$ can be taken as a matrix element of an energy-momentum tensor operator between final and initial states. This has been used to show [4] the rate of decay of the 3d (m = 2) state of hydrogen atom into the 1s state with emission of one graviton as

$$\Gamma(3d \rightarrow 1s) = \frac{2^{23} G m_e^3 c}{3^7 5^{15} (137)^6 \hbar^2} \\ = 2.5 \times 10^{-44} \ \mathrm{sec}^{-1} \tag{3.5}$$

That is, there is no chance of observing such a transition.

If we view (3.3) in the context of early Universe when it was with energy $\sim 10^{19}$ Gev indicates the fact of a background of gravitons as the Universe has a background of photons (microwave background radiation). But, unfortunately, it is also not detected.

4. The Problem of Detecting Gravitational Radiation

It is natural to ask, what sort of experiments would we like to do in the field of gravitational radiation ? The classic example of radiation experiment is that of Hertz to generate and to detect electromagnetic waves in the same laboratory.

Another type of experiment involves the detection of natural gravitational radiation. Suppose these are sources generated outside the Earth, or even outside the solar system. Is there any way to detect the existence of gravitational waves incident on the Earth from some cosmic source ? Is it possible that such sources exist ?

If the Universe was created with an enormous bang, there might have been a lot of gravitational radiation generated initially. If we calculate the absorption cross-section for these waves by ordinary matter, we find it is incredibly small. Consider, for example, a microwave graviton is incident on a molecule with a large moment of inertia, say, ammonia molecule. If the graviton frequency is resonant with ammonia and has a line breadth of 10^4 cycles, the absorption cross-section is roughly 10^{-60} cm^2 . So, these gravitational waves are not easily absorbed by atomic and molecular particles. If such waves were generated at the time of creation, and if the Universe is really finite and unbounded there are some reasons to believe that these waves would still be with us. Wheeler has estimated that the interstellar gravitational radiation could be as high as 10^{-29} to 10^{-28} gram/cm^3 and still be consistent with present data on the expansion of the Universe. If Fourier transform is considered, it would be peaked at 1 cycle per million years. Hopefully there is a high frequency tail that might be detectable.

5. Linear Theory of Gravitation

The major defect of the linearized theory is the absence of self gravitational field at lowest order. This may be remedied either by beginning with Newtonian approximation at first order or by perturbing not flat space but an existing curved metric. The Newtonian alternative I shall not discuss but a few words on linear perturbation theory are appropriate here. Thereafter it would be appropriate to investigate briefly the question of conditions under which the wave equation $\nabla^\alpha \nabla_\alpha h_{\mu\nu} = \lambda \eta_{\mu\nu}$ is unique. It is because it is expected to throw some light for experiments concerning ultrarelativistic motions like gravitational waves, Shapirio motion and so on.

When source is considered we have

$$\nabla^{\alpha}\nabla_{\alpha} h_{\mu\nu} - 2R_{\alpha\mu\nu\beta}^{\ \ \ \ \ \ B} h^{\alpha\beta} = \varkappa\, \delta T_{\mu\nu} \tag{5.1}$$

where $\delta T_{\mu\nu}$ is the first order change in $T_{\mu\nu}$ at a given coordinate position. Thus, the metric perturbation generated in equation (5.1) acts back on its source $\delta T^{\mu\nu}$ via the perturbed christoffel symbols because at first order the equation of motion of the matter are

$$\nabla_{\rho}\, \delta T^{\rho\sigma} = - T^{\mu\sigma} \delta \Gamma^{\rho}_{\mu\rho} - T^{\mu\rho} \delta \Gamma^{\sigma}_{\mu\rho} \tag{5.2}$$

Self-gravitational effects of $\delta T^{\rho\sigma}$ can be coupled to $\delta T^{\rho\sigma}$ at first order because $T^{\rho\sigma}$ is not zero at zero order.

With this, we now turn to examine the conditions under which the wave equation

$$\nabla^{\alpha}\nabla_{\alpha} h_{\mu\nu} = \lambda\, \eta_{\mu\nu} \tag{5.3}$$

is unique. For this, we consider only a second order tensor field. This yields the field equation in linear approximation as

$$\begin{aligned} \Lambda \eta_{\mu\nu} &+ \alpha_1 h\, \eta_{\mu\nu} + \alpha_2 h_{\mu\nu} + \alpha_3 h_{,\mu\nu} \\ &+ \alpha_4 \left(h^{\lambda}_{\nu,\mu\lambda} + h^{\lambda}_{\mu,\nu\lambda} \right) \\ &+ \alpha_5 h^{\lambda\sigma}_{\ \ ,\lambda\sigma\mu\nu} + \alpha_6 h^{\lambda\sigma}_{\ \ ,\lambda\sigma}\, \eta_{\mu\nu} = T_{\mu\nu} \end{aligned} \tag{5.4}$$

Here Λ is a constant and raising and lowering of indices are done by $\eta_{\mu\nu}$. When this satisfies the weak equivalence principle, we find

$$\begin{aligned} \alpha_1 h_{,\mu} &+ \alpha_2 h^{\nu}_{\mu,\nu} + \alpha_3 D h_{,\mu} \\ &+ \alpha_4 D h^{\lambda}_{\mu,\lambda} + \alpha_4 h^{\lambda\sigma}_{\ \ ,\mu\lambda\sigma} + \end{aligned} \tag{5.5}$$

$$+ \alpha_5 \mathcal{D} h^{\lambda\rho}{}_{,\lambda\rho\mu} + \alpha_6 h^{\lambda\rho}{}_{,\lambda\rho\mu} = 0$$

where $\mathcal{D}$ stands for D'Alembertian operator.

As shown by us [5], the equation (5.5) is equivalent to (5.3) under suitable coordinate conditions. This means that any linear Lorentz invariant theory of gravitation based on a tensor field of second order that satisfies the weak equivalence principle is equivalent to the linear approximation of general relativity. That is, the experimental verifications of the predictions ofgeneral relativity in such circumstances, say, for example, the gravitational waves connected with ultrarelativistic motion of particles, desires its description by only a single tensor field of second order. It, therefore, allows us to understand the general sense of such experiments.

Acknowledgement

The author wishes to thank Prof.Venzo de Sabbata for providing financial support to participate in the Erice program, and to Prof.Pande B.B.Lal for encouragement.

REFERENCES

1. L.P.Grishchuk and Ya.B.Zel'dovich (1982) in "Complete Cosmological Theories in Quantum Structure of Space Time" eds. M.J.Duff and C.J.Isham, Cambridge University Press, 1982, p.409
2. L.P.Grishchuk (1977) Sov.Phys. USP, 20 ,319
3. S.N.Pandey (1983) Intern.J. Theor. Phys. , 22 ,209
4. S.Weinberg (1972) Gravitation and Cosmology, Wiley and Sons
5. S.N.Pandey, A.M.Finkelstein and Ya.Kreinovich (1983) Nuovo Cimento Lett. 38 ,270

THE GRAVITATIONAL WAVE EXPERIMENT OF THE ROME GROUP. DATA RECORDED DURING SN1987A.

G. Pizzella
Department of Physics, La Sapienza University,
P.le A. Moro 2, 00185 Rome, Italy
and
Istituto Nazionale di Fisica Nucleare, Sezione di Roma

ABSTRACT. The cryogenic g.w. antenna of the Rome group located in Geneva at CERN is briefly described. The antenna has operated for nine months with a noise temperature of the order of 18 mk corresponding to an h value of the order of 10^{-18}. In order to improve the performance of the antenna the experiment was turned off for a few months; unfortunately it was off at the time of the SN1987A.

The data recorded with the room temperature gravitational wave antenna, located in Rome, during the supernova SN1987A are shown. The antenna gives a signal with an amplitude of 135 K, against a background noise of 29 K, about 1.4 seconds before the first neutrino, of the five neutrino event, observed by the LSD collaboration with a detector located under Mont Blanc. The probability that such a signal is due to chance is computed to be 3%.

The 135 K event is related to a total g.w. energy, emitted by the supernova, of the order of 2400 solar masses, as calculated with the standard formulae of General Relativity. This abnormal figure throws much doubt on the interpretation of the phenomenon in terms of gravitational waves with the classical cross-section.

1. THE CRYOGENIC ANTENNA

The Rome group has realized a resonant cryogenic antenna that consists of an alluminium cylinder with lenght L=3 m and mass M=2270 kg, cooled with liquid helium to a temperature T=4.2 K [1]. The mechanical vibrations are detected by means of a resonant capacitive transducer operating with constant electrical charge; the electrical signal is amplified with a very low noise amplifier based on the use of a dc SQUID.

The antenna has been installed in Geneva, at CERN, and it has operated from November 1985 until July 1986 [2]. During these nine months of operation the antenna has produced data for about 70% of the time with a noise temperature T_{eff} on the average between 15 and 20 mK. Short burst of gravitati-onal radiation were searched. Assuming the duration of these bursts to be of the order of τ_g=1 ms we calculate

V. De Sabbata and V. N. Melnikov (eds.),
Gravitational Measurements, Fundamental Metrology and Constants, 307–310.

the minimum (SNR=1) value of h(t) for the most favorable conditions of direction and polarization, from the formula $h(t)=(1/\tau g)(L/v^2)\sqrt{kT_{eff}/M}$ where v is the sound velocity in alluminium. For T_{eff}=15 mK we obtain $h(t)=1\times10^{-18}$. The behaviour of the hourly averages h(t) versus time is shown in Fig.1.

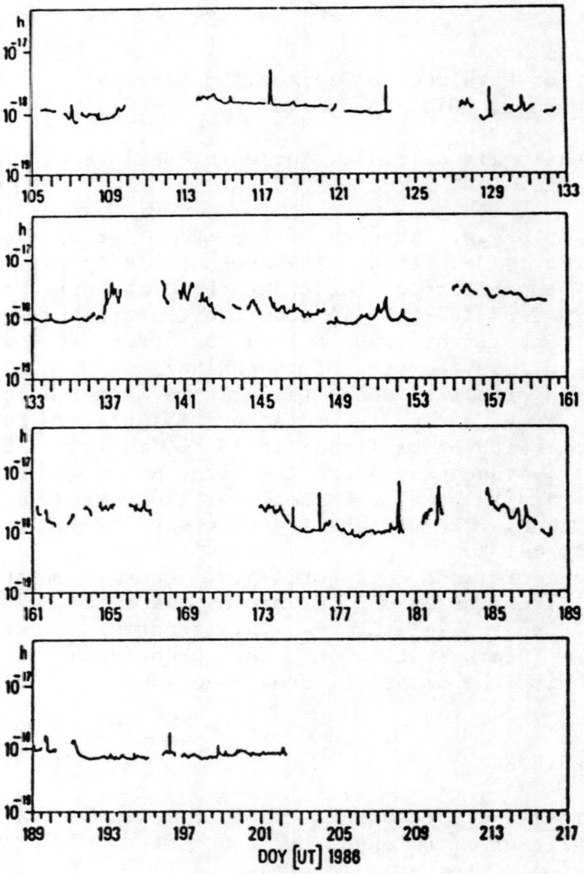

Work is in progress to correlate these data with those recorded by similar antennas of the Universities of Louisiana and Stanford during a period of about three months in 1986.

2. DATA RECORDED DURING SN1987A

At the time of the supernova 1987A in the Large Magellanic cloud the above cryogenic antenna at CERN was, unfortunately, not in operation because we wanted to introduce some changes concerning the mechanical filters. At the University of Rome we had in operation, however, a room temperature antenna with equal mass equipped with a piezoelectric ceramic and a FET amplifier [3]. This antenna has a noise temperature $T_{eff} \simeq 29$ K, more than three orders of magnitude worse than that of the cryogenic antenna. The use of such a low sensitivity detector clearly leaves, a priori, very little hopes of detecting gravitational waves. Since, however, this instrument was the only one we had in operation during the SN1987A we decided to analyze the recorded data.

The analysis is based on the information we received by the Mont-Blanc group [4], that a neutrino burst had been observed, beginning at $2^h52^m36^s.8$ U.T., by the LSU collaboration.

The signal from the FET amplifier is processed with a Phase Sensitive Detector which extracts the Fourier component at ν_R=858 Hz and then is filtered with a Wiener filter for obtaining the energy innovation. In looking for a possible gravitational signal we take into account that, according to present information provided by beta and double beta decay, the neutrino mass seems to be of the order or less than 10 eV. As a consequence we expect a delay with respect to the g.w. of the order of 2.7 s for neutrinos with 10 MeV of energy travelling from a distance of 52 kpc. Thus we looked for a signal, within a few seconds, say 3 seconds, before the first detected neutrino. The results of our analysis are shown in Fig.2

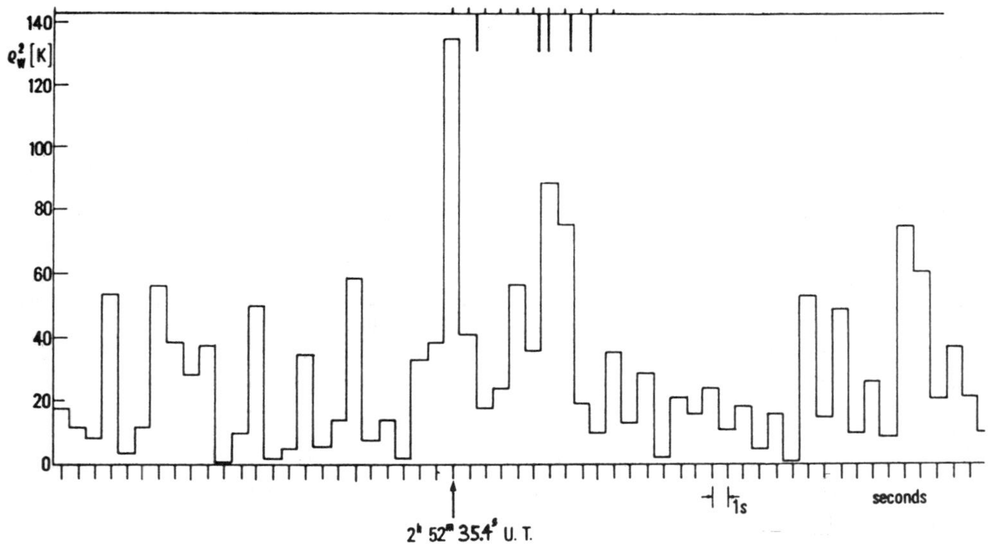

where the energy innovations ϱ^2_w, expressed in kelvin, is shown versus time [5]. For comparison the times of the five neutrinos detected by the LSD group are also shown. It is seen that an energy innovation of 135 K occurs 1.4 ± 0.5 s before the first neutrino.

The probability that an innovation of 135 K occurs by chance in a given second has been measured to be 10^{-2} .

If we tentatively assume that our signal is due to gravitational waves we can calculate the total energy emitted by the supernova as gravitational waves under the reasonable hypoghesis of an isotropic emission of unpolarized waves with bandwidth of 1000 Hz. We obtain the abnormal figure of 4×10^{50} joule corresponding to 2400 solar masses. This result throws serious doubts on the interpretation of our signal in terms of gravitational waves using the standard cross section.

REFERENCES

1. E.Amaldi, P.Bonifazi, P.Carelli, M.G.Castellano, G.Cavallari, E.Coccia, C.Cosmelli, V.Foglietti, R.Habel, I.Modena, G.V.Pallottino, G.Pizzella, P.Rapagnani, F.Ricci, 'Preliminary Results on the Operation of a 2270 kg Cryogenic Gravitational-Wave Antenna with a Resonant Capacitive Transducer and a dc SQUID Amplifier', Il Nuovo Cimento 9C, N.4, luglio-agosto 1986.
2. E.Amaldi, P.Bonifazi, P.Carelli, M.G.Castellano, G.Cavallari, E.Coccia, C.Cosmelli, V.Foglietti, S.Frasca, R.Habel, I.Modena, R.Onofrio, G.V.Pallottino, G.Pizzella, P.Rapagnani, F.Ricci, 'Operation of the 2270 kg Gravitational Wave Resonant Antenna of the Rome Group', Proc. 13th Texas Symposium on Relativistic Astrophysics, Chicago, December 1986.
3. F.Bronzini, S.Frasca, G.V.Pallottino, G.Pizzella, G.Vannaroni, 'An Experimental Apparatus for Studying the Backgroud of Gravitational Wave Antennas and its Correlation with Geophysical Phenomena', Il Nuovo Cimento 8C, N.8, maggio-giugno 1983.
4. M.Aglietta, G.Badino, G.F.Bologna, C.Castagnoli, A.Castellana, W.Fulgione, P.Galeotti, C.Saavedra, G.Trinchero, S.Vernetto, V.L.Dadykin, V.B.Kortchaguin, A.S.Malguin, V.C.Ryassny, O.G.Ryazhskaya, V.P.Talochkin, G.T.Zatsepin, V.F.Yakushev, 'Characteristics of a Neutrino Burst Detected during the Occurence of Supernova SN1987A'. Europhysics Letters, June 1987.
5. E.Amaldi, P.Bonifazi, M.G.Castellano, E.Coccia, C.Cosmelli, S.Frasca, M.Gabellieri, I.Modena, G.V.Pallottino, G.Pizzella, P.Rapagnani, F.Ricci, G.Vannaroni, 'Data Recorded by the Rome room temperature gravitational wave antenna during the supernova SN1987A in the large Magellanic cloud' Europhysics Letters, June 1987.

SOLAR-SYSTEM TESTS OF GENERAL RELATIVITY, THE TRANSITION TO SECOND ORDER

Robert D. Reasenberg
Radio and Geoastronomy Division
Smithsonian Astrophysical Observatory
Harvard-Smithsonian Center for Astrophysics
Cambridge, Massachusetts 02138, USA

ABSTRACT. The solar system is the traditional laboratory for gravity research. In this laboratory, experiments of heretofore undreamed of accuracy are made possible by modern technologies. Among these are atomic clocks, planetary radar, interplanetary spacecraft, and laser ranging. The results of these experiments are consistent with the predictions of general relativity. The differences between the predictions of general relativity and of classical physics are confirmed to accuracies as great as one part in a thousand. However, the solar system tests examine these differences only to first order in post-Newtonian effects.

One of the classical tests of general relativity is based on the deflection of light by the Sun. At the solar limb, this deflection is 1.75 arcseconds. The second-order contribution to this deflection is 11 microarcseconds (μas) at the solar limb; it is proportional to d^{-2}, where d is the impact parameter of the light ray. To measure the second-order contribution to the deflection, we are developing an astrometric optical interferometer, POINTS, which could operate from the bay of the Space Shuttle, mounted on the proposed Space Station, or supported by an independent spacecraft. POINTS would measure the separation of stars about 90^0 apart with an uncertainty of a few μas. It could be used for many types of astrophysical research in addition to the measurement of the deflection of starlight. Among these uses are the direct determination of the Cepheid distance scale and a deep search for extra-solar planetary systems.

I. INTRODUCTION

General relativity, the standard theory of gravitation, is now well supported by several experimental tests that make use of modern technology. As a result, the role of the experimentalist must shift. Initially, the task was to test the putative superiority of the new theory over the Newtonian approach. Now the task should be to find the limits of applicability of the standard theory. As competitive

V. De Sabbata and V. N. Melnikov (eds.),
Gravitational Measurements, Fundamental Metrology and Constants, 311–344.

theories are developed, there is also the need to compare their predictions to experiment. In the past, these competing theories created the need for a framework within which the tests and theories could be compared. Known as the Parameterized Post-Newtonian (PPN) framework (see Will [1981] and references therein), its ad hoc parameters have developed a life of their own. Some of these will be mentioned in later sections.

In Section II, we consider some of the technologies that make possible the present and pending solar-system tests. The status of the CfA program is discussed in Section III and the status of several tests, both past and prospective, is the subject of Sections IV and V. The results of all credible tests are consistent with general relativity and the post-Newtonian corrections have been tested to 10^{-3}. In Section VI we ask the question: What direction should be taken for future tests? It is suggested that small improvements in existing tests are of limited value and that major efforts in the future should address previously unaccessed aspects of the theory. A deflection experiment sufficiently sensitive to measure the post-post-Newtonian terms is proposed as meeting that criterion and is the subject of Section IX. The experiment would be conducted with a space-born dual astrometric optical interferometer, POINTS. A discussion is presented in Sections VII, VIII, and X through XIII of the present concepts for the instrument and its mission.

II. TECHNOLOGIES

Clocks. There are now several types of clocks that derive their stability from an atomic resonance. Species used include cesium 133 (≈9193 MHz), rubidium 87 (≈6835 MHz), and hydrogen (≈1420 MHz). The cesium resonance forms the basis for the current definition of the second of atomic time. For further discussion, see Audoin and Vanier [1976], and Vessot [1976]. Other clocks are being or have been developed that obtain their stability from a microwave cavity at cryogenic temperatures; Turneaure and Stein have developed a superconducting cavity stabilized oscillator and Braginsky is developing an oscillator that uses a crystaline Al_2O_3 cavity.

For relativity experiments, hydrogen masers have long been preferred because of their greater stability over time scales of hundreds to tens of thousands of seconds. Substantial further improvements in these clocks can be expected in the next one to two years. Vessot, Silvera, et al. [Vessot et al. 1987; Walsworth et al. 1986] have operated a maser at 0.4K using superfluid ^{4}He as a wall coating. Since there is only one such clock, it is not yet possible to determine experimentally the stability of this kind of clock. However, a noise model predicts it has a stability of 10^{-18} for a 1000 sec time. Further increases in stability may be possible with lower temperature operation and a ^{3}He wall coating (Vessot, private communication, 1987).

At room temperature, stability to below 10^{-16} on time scales of

1000 to 5000 sec are expected by a technique known as adiabatic fast passage which delivers H_2 in a single state (cf. the usual dual state H_2) entering the oscillator cavity. Vessot estimates a further threefold improvement (at longer time scales) will come from new room-temperature wall materials that he and his colleagues are developing.

Delay Observable. The delay observable is the prime source of information for most of the solar-system tests. Planetary radar, the first of the modern observables, was the basis for the initial time delay tests and is still critical for the study of perihelion advance. Spacecraft tracking, which has superseded radar for Mars (for which both are available), is the basis for the most stringent of the solar-system tests, the Viking time delay test. Lunar laser ranging provides the best available measure of the Moon's motion and the parameters on which it depends. It is the basis for the test of Nordtvedt's hypothesized violation of the principle of equivalence. These observables are discussed by Reasenberg [1983, 1986c].

Radio Interferometry. Interferometry is a technique by which it is possible to achieve the resolution of a large aperture by sampling the signal at discrete points within the aperture. In all cases, a signal from a distant source is intercepted by two or more receiving antennas and, either by recording it on magnetic tape or by sending it directly over cable, that signal is brought to a cross-correlator which produces fringes. From those fringes, and the delay introduced by the correlator to produce them, it is possible to determine the difference in the path length for each pair of receiving antennas and therefore the apparent direction to the source.

In the radio domain, there are two principal techniques: (1) Very-Long Baseline Interferometry (VLBI) makes use of an atomic clock at each of two or more stations to preserve on magnetic tape the temporal record of the received phase from a distant source. Cross correlation is performed later. (2) Connected element interferometry is a close analogue of its classical optical counterpart. The master clock and cross correlator are directly connected to the receivers and the cross correlation is performed in real time.

Astrometric interferometry [Shapiro, 1976] has applications in astrophysics, geophysics, and geodesy. In differential interferometry, observations are of sources which are close together in the sky. Many errors are correlated in the observations of the two sources, thus a high accuracy differential measurement is possible. This differential approach is important in the application of radio interferometry to the deflection tests as discussed in Section V.

III. STATUS OF THE CFA PROGRAM

The objective of our analysis is to use the solar system as a "test bed" for theories of gravitation. Most of this work was transplanted from MIT to CfA at the beginning of 1983; however, much of the lunar

Table I. Solar-System Measurements

Source[1]	Number of Measurements	Approximate Time Span of Observations: From	To
VIKING SPACECRAFT (Mars)			
Lander delays (plasma corrected)	1,041	July 1976	Aug. 1980
Lander delays (not plasma corrected)[2]	239	Aug. 1980	Nov. 1982
Lander Doppler shifts	14,003	July 1976	Nov. 1982
Orbiter delays[3]	4,060	June 1976	Aug. 1977
MARINER 9 SPACECRAFT (Mars)			
Delays (Normal Points)[3,4]	185	Nov. 1971	Oct. 1972
PIONEER SPACECRAFT (Jupiter)			
Delays (Normal Points)[3]	2	Dec. 1973	Dec. 1974
VOYAGER SPACECRAFT (Jupiter)			
Delays (Normal Points)[3]	2	Mar. 1979	July 1979
VOYAGER SPACECRAFT (Saturn)			
Delays (Normal Points)[3]	2	Nov. 1980	Aug. 1981
RADAR RANGING			
Mercury	5,239	1969	present
Venus	5,464	1969	present
Mars[5]	30,000	1966	present
LUNAR LASER RANGING			
Retroreflector delays (Normal Points)[6]	4,401	Oct. 1970	present
MILLISECOND PULSAR TIMING[7]			
Single frequency	196	Nov. 1982	Aug. 1984
Dual frequency	114	Sep. 1984	present
MERIDIAN-CIRCLE OPTICAL OBSERVATIONS[8]			
Sun	1,023	1970	1978
Moon	212	1970	1978
Inner planets (Me,V,Ma)	1,518	1970	1978
Outer planets (J,S,U,N)	1,643	1970	1978
Sun and planets	~320,000	1750	1970

Notes to Table I

---

[1] All observables are time delays except for (i) Viking Lander Doppler shifts, (ii) the meridian-circle and photographic optical observations, and (iii) the arrival times of pulses from pulsars.

[2] Plasma corrections could only be made with reasonable accuracy using the dual-band time-delay "downlink" measurements involving the Viking orbiters (see, for example, Reasenberg et al., 1979), the last of which ran out of attitude control gas in August 1980.

[3] The spacecraft Normal Point (NP) is a compressed datum: the equivalent Earth-planet time delay that would have been measured between the centers of mass of the Earth and the planet. All spacecraft NP's were obtained from the Jet Propulsion Laboratory where they were derived directly from the tracking data.

[4] The Mariner 9 NP's have received an additional level of compression: All data from each tracking "pass" were combined by us to form one NP.

[5] In recent analyses, these data have been omitted because of the superiority of the Viking data. As time passes, new radar data become increasingly valuable so these data (and a topography model) will again be used.

[6] Each lunar laser ranging (LLR) NP is a single estimate, from the observers, of the round-trip propagation time between a ground station (McDonald Observatory in Texas or CERGA in France) and one of the lunar retroreflectors. Each NP is an average based on all photons received from many laser pulses.

[7] Each pulsar NP, as in the case of LLR data, is an average over many timings. These estimates are provided by J.H. Taylor from Arecibo observations, usually lasting 2 hours on each of 2 days each month. Until September 1984 there were 4 NP's per day at a single frequency. After that date, 1 NP is obtained at each of 2 frequencies.

[8] The data are right ascension and declination measurements, approximately half each. All data since 1970 were provided by the U.S. Naval Observatory.

---

analysis is still done at MIT by R. King et al. with whom we collaborate.

Table I lists the current working data set -- 39,344 observations, excluding both the 30,000 Mars radar data and the 320,000 optical observations from before 1970, neither of which is included in our recent studies. It is a varigated ensemble of observations of different types and, within each type, of different accuracies. How does one use such a data set to study theories of gravitation? Our approach has been to build a large parametric model of the solar system and the observables. In order for this model to be useful, we must numerically integrate the equations of motion for the planets, the Moon, and the Moon's rotation. These ephemerides can be generated by means of an analytic series expansion, but that approach has proved to be too unwieldy and we now estimate that to evaluate such series would be more expensive than to perform our numerical integrations.

The solar-system model contains a few ad hoc parameters whose estimated values are a measure of the correctness of the underlying gravitational theory. However, in addition, there are over a thousand other parameters representing such things as planetary masses, orbital elements, the Moon's mass and orbital elements, its low-order gravitational harmonic coefficients, its elastic Q factor, and its rotational orbital elements. The model also includes the masses of asteroids, the locations of tracking stations, and the locations of the lunar retroreflectors and Viking Landers. A major fraction of all the parameters in this model is associated with the topography of the inner planets. These topographic models are necessary for proper reduction of the radar data. Finally, there is an ad hoc parametric description of irregularities in the rotation of the Earth and a series of biases that we assume to exist in all the observables. At the present time, the postfit RMS weighted residual from each of our subsets of data is approximately one. The modeling and weighting scheme seems to be working reasonably well.

IV. SCIENCE OF THE CFA PROGRAM

This section contains results from four of our tests in the solar system: the perihelion advance, Nordtvedt's principle of equivalence, the secular variation of G, and the jewel of the solar-system tests -- the Shapiro time delay. The discussion of these studies in the proceedings (Reasenberg [1983, 1986c]) of the 1982 and 1985 Schools is still applicable and is not reproduced here. Some of the effects are discussed within the Parameterized Post Newtonian (PPN) framework (Will [1981]).

The portion of the perihelion advance that is not caused by planetary perturbation depends on the PPN parameters β and γ as well as on the solar quadrupole moment. Determinations of the latter by means of observations of the Sun had caused considerable controversy, but more recently the results seem to have been accepted as sufficiently reliable to constrain J_2 in the solar-system solutions. When we assume $J_2 = 0$, we obtain $\beta = 0.99 \pm 0.02$. When we estimate J_2, we obtain $\beta = 0.975 \pm 0.05$ and $J_2 = (-3 \pm 3)10^{-6}$. Thus, from our perspective, a reliable determination of J_2 from solar observations is very useful.

Nordtvedt [1968a,b] has described the observable effect in the Earth-Sun-Moon system of a possible violation of the Principle of Equivalence. An ad hoc parameter η is introduced into the equations of motion such that $\eta = 0$ if general relativity is correct. Within a fully conservative PPN framework, $\eta = 4\beta - \gamma - 3$. The classical result from lunar laser ranging is $\eta = 0.0 \pm 1.5\%$ (or $\pm 3\%$) [Shapiro et al., 1976]. Our current analysis shows that with the presently available larger data set the uncertainty can be reduced to under 0.5%. We have also investigated the use of the Earth-Mars-Jupiter-Sun system for determining η. We find, regrettably, that the uncertainty from this determination would be about 3%.

Dirac's [1937, 1938] large numbers hypothesis gives rise to the notion that the gravitational constant may appear to change in time when measured in atomic units. Thus, for example, if we compare a gravitational clock -- the period of a planet's orbit -- to an atomic clock, we may detect a drift rate. Although a "back of the envelope" calculation suggests that our data should be exceedingly sensitive to this question, the high correlations between a parameter representing $\dot{G}$ and the other parameters decreases this sensitivity by a factor of about 250. Numerical studies in which we investigate the possible effects of model errors on the estimate of this parameter indicate that it is particularly sensitive; thus, the ratio of the realistic error to the statistical error is particularly high. Nonetheless, we find $\dot{G}/G = (0 \pm 2)10^{-11}$ per year and we expect a reduction in the uncertainty by at least a factor of 2 by the time we finish our analysis.

The Shapiro [1964] time-delay effect provides the basis for the fourth test, the most stringent solar-system test of relativity (cf. tests of the underlying principle of equivalence by means of the gravitational redshift). Consider signals that are emitted from Earth, pass close to the Sun, and are returned from a spacecraft or planet. The round-trip propagation time is that expected from Euclidean geometry plus an additional term,

$$\Delta\tau = \frac{2r_o}{c} S \,\ell n \left[\frac{r_e + r_p + R}{r_e + r_p - R}\right] \tag{1}$$

where $r_o = 3$ km, $2r_o/c = 20\ \mu s$, and $S = (1 + \gamma)/2$. In Eq. (1), r_e, r_p, and R are, respectively, the distances from the Sun to the Earth, from the Sun to the target, and from the Earth to the target.

Equation (1) has been the basis for several tests of general relativity. The Viking mission provided a dramatic improvement over previous time delay experiments. There were four Vikings, two orbiters and two landers. Each orbiter was equipped with both a high-gain antenna which contributes to a high signal-to-noise ratio, and a transponder which would receive an S-band signal from the ground and send back coherent S-band and X-band signals. Given the S-band and X-band signals coming down, one can determine the plasma columnar content of the path between Earth and Mars. Each of the landers also has a high-gain antenna. Their orbits with respect to the center-of-mass of Mars are relatively easy to model to high accuracy; the landers do not execute a substantial random walk. For both the orbiters and the landers, the time delay observable far from superior conjunction has a precision of about 10 ns (in round-trip time), that is, the distance measurements are uncertain by roughly the height of a person. (For comparison, some one-way light-times are: the astronomical unit, about the distance between Earth and the Sun, 499s; the Earth's radius, 21ms; the distance between Earth and the Earth-Moon barycenter, 15ms; the radius of Mars, 11ms.)

A preliminary analysis, based on observations made during 40 days surrounding the first superior conjunction on 25 November 1976, yielded $(1+\gamma)/2 = 1.0 \pm 0.005$ [Shapiro et al., 1977], a factor of 4 improvement over the results of the Mariner-9 (relativity) experiment. From a formal analysis based on 14 months of observation, we determined that the data are consistent with relativity, and that $(1+\gamma)/2$ is 1.000, with an uncertainty of 0.001 [Reasenberg et al., 1979]. In terms of the Brans-Dicke theory, this result implies $\omega > 500$. For this analysis, we had developed a systematic procedure for correcting the effect of the plasma. All aspects of the analysis were under computer control, with no potentially prejudicial hand intervention. Most of the residuals fit nicely within the band, $0. \pm 0.1\,\mu s$. The residuals get worse near superior conjunction as was to be expected.

We have done some numerical experiments to determine what sort of accuracies we ought to expect from a complete analysis of all of the data now available. These suggest that we can decrease the uncertainty in the estimate of $(1+\gamma)/2$ by a factor of two below our best published results.

V. DEFLECTION EXPERIMENTS

One of the classical tests of relativity is a measurement of the predicted deflection of light by the solar mass: $1.75\ R_o/d$ sec where R_o is the solar radius and d is the impact parameter. In the PPN framework, the expected amplitude is multiplied by a factor of $(1+\gamma)/2$, the same factor that multiplies the Shapiro delay effect. In 1967, Irwin Shapiro suggested that the then emerging technology of radio interferometry might be used to perform a bending test of relativity with a precision considerably greater than that which had been possible using classical optical techniques. Here we consider two examples of such a test.

The most successful of the interferometry experiments was conducted in two successive years by Fomalont and Sramek [1977] using a connected element interferometer with a 35 kilometer baseline. In this experiment, it was technologically feasible for the first time to use observations at two frequencies, S-band (2.7 GHz) and X-band (8.1 GHz). Since the contribution of plasma to time delay or phase delay varies with f^{-2}, where f is the frequency of the signal, given measurements made at two frequencies, in principle, one can remove most of the effect of the plasma. That technique was very important in their experiment.

There were three sources observed; these fell nearly on a great circle arc 10 degrees long. Observations were made in each of two years on 12 days during the same one month period surrounding the occultation of the central source by the Sun. Data were taken when the impact parameter was as small as 5 solar radii. When the two experiments were combined, several systematic effects could not be combined in the usual root-sum-square fashion. The authors concluded that the combined experiment had yielded $(1+\gamma)/2 = 1.007 \pm 0.009$.

There is now an experiment under development at CfA and the Haystack Observatory to again do the light bending test, but to substantially increased precision. Here four antennas will be employed: two in California, two in Massachusetts; two looking at the source, two looking at a reference star. The antennas looking at the source will be making observations at 23 GHz, a much higher frequency than previously possible. The antennas looking at the reference star will make observations at both S-band and at X-band, that is, at about 2 GHz and 8 GHz. In addition, the antennas looking at the reference star will be directed periodically toward the main star to make a measurement from which it will be possible to determine approximately the plasma contribution to the observations of the principal source.

Based on their detailed error analysis, the experimenters expect to measure the deflection with an uncertainty of 0.1% [B. E. Corey and I. I. Shapiro, private communication, 1983]. A major uncertainty in this estimate of the experiment's accuracy is the condition of the near solar region (i.e., the weather on the Sun). As of this writing, the data have been taken and a preliminary examination of some of the data shows that there are fringes [I. I. Shapiro, private communication, 1987].

VI. SOLAR-SYSTEM TESTS IN TRANSITION

The results of all accepted experiments are consistent with the predictions of general relativity, and, with the exception of the Binary Pulsar, all the experiments depend on first-order post-Newtonian effects. One must begin to wonder whether it is particularly fruitful at this time to invest in further improvement of the present set of tests. I believe that it is no longer reasonable to make substantial investments in small improvements to old first-order tests.

What kinds of tests seem promising? The Stanford Gyroscope Experiment (GP-B) will investigate effects governed by the off-diagonal terms of the metric. As such, it will explore new territory and thus be interesting. Tests that are related to singularities would provide a dramatic new approach. However, none has been found. The detection of gravity waves would also provide a significant advance. The work with the Binary Pulsar (PSR 1913+16) provides such a detection, albeit an indirect one. "The ratio of the observed rate of decay to that predicted by the quadrupole formula is presently 1.00 $\pm$0.04" [Taylor, 1987]. Although one can still raise questions about the nature of the companion object, the good fit of the data to the present model makes the "cosmic conspiracy" argument uncompelling: gravity waves have been detected.

Finally, there is the possibility of doing one of the familiar solar-system tests to post-post-Newtonian order. The problem is difficult because the gravitational potential at the limb of the Sun is 2×10^{-6}. A second-order experiment must be at least three orders of magnitude more sensitive than the present solar-system tests, and such an advance is generally made only with great difficulty.

A second-order test, if it is to be done within the solar system, will likely utilize the potential close to the limb of the Sun. Putting an instrument package close to the Sun causes substantial engineering problems. Several years ago, an ambitious mission, called STARPROBE [Underwood and Randolph 1982], was being considered by NASA. But that mission, which would have placed a spacecraft within four solar radii of the Sun without providing an opportunity for a second-order test, is now not actively being considered. Thus, we must find a way of doing an experiment in which we use photons that pass close to the solar limb. Two possibilities come to mind, the light deflection experiment and the time-delay experiment. Of course, these are really two ways of looking at the same phenomenon. One would expect the time delay experiment to involve spacecraft tracking and the NASA Deep Space Network which, for a combination of technical and institutional reasons, seem unlikely in the foreseeable future to be able to yield data with the accuracy necessary to measure the second-order time delay. Thus, we are limited to considering a light deflection experiment.

At the solar limb, the second-order light deflection is 10.9 microarcsec (μas) [Epstein and Shapiro, 1980; Fishbach and Freeman, 1980; Richter and Matzner, 1981], cf. 1.75 sec for the first-order deflection. The plasma-related problems, which are a limiting factor in first-order experiments using microwave signals, demand the use of substantially higher frequency signals for a second-order test. A reasonable solution is to use signals in the optical region. Since the resolution of any angle-determining instrument is closely related to the ratio of the wavelength to the size of the instrument, the optical region offers an additional advantage. Any device that would perform such an experiment would have to be above Earth's atmosphere because of the large corrupting effect of fluctuations of the atmospheric refraction.

An astrometric instrument capable of making tens of measurements per day, each with an uncertainty of a few μas, would have numerous applications to astrophysical work of current interest, in addition to the second-order light deflection test. Over a ten-year period, a search for stars that wobble around the star-planet center of mass would either discover other planetary systems or set stringent bounds on their existence. (For a discussion of the search for planetary systems, see Black [1980].) Applications to solar-system research, stellar structure, the cepheid distance scale, and the structure of the galaxy have been described by Reasenberg [1984] and are discussed in Section VIII. New lines of research developed in the wake of the availability of the data would be expected.

What criteria should be applied in selecting among candidate instruments? The "scientific throughput" $T = \rho/\sigma^2$ is one important measure, where ρ, the observation rate, is the number of observations per day and σ is the corresponding measurement uncertainty. For an instrument operating in a regime where it is limited by noise (including photon statistics), T is an invariant if we neglect housekeeping, repointing, and target-acquisition times.

Recent work in ground-based astrometry has yielded exciting gains in scientific throughput on individual targets. Telescopes using Ronchi rulings have demonstrated a differential precision of 4 milliarcseconds (mas) in a 20-minute observation over fields of angular size 0.1 to 0.5 degrees [Gatewood 1987 and references therein; Buffington 1986]. In addition, Monet and Dahn [1983] report close to a milliarcsecond precision over a very narrow field, using a CCD array. Atmospheric turbulence limits such ground-based astrometric measurements, but the seeing-induced image motion is correlated for stars in close angular proximity, thus permitting this improved differential accuracy [Lindegren 1980]. Future large-aperture astrometric telescopes at good observing sites may achieve differential measurement precision of about 1 mas with long integration times.

An astrometric interferometer on Mt. Wilson [Mozurkewich et al., 1988; Shao et al. 1988] has recently achieved day-to-day reproducibility of 50 mas in right ascension for stars separated by up to 60 deg. Declination uncertainties are much larger because only the North-South baseline is operating. Here again, the limiting factor is the effect of the atmosphere. The interferometer has been operated in a "two color mode" in which the small dispersion of the atmosphere is used to determine and correct for the refraction of the atmosphere. Preliminary results with stars no fainter than m = 3 suggest that a fivefold improvement in precision should be possible [Colavita et al. 1987].

The European astrometric instrument HIPPARCOS, which is scheduled for launch in 1989, will usher in the age of space astrometry. This survey instrument will observe about 10^5 stars over a 2.5 year nominal mission life. The resulting catalog will have a nominal accuracy of 2 mas at epoch and a proper motion uncertainty of a few mas/y [Kovalevsky, 1982].

For any astrometric instrument, the so-called diffraction limit is of order λ/D where λ is the observing wavelength and D is the linear scale of the instrument perpendicular to the line of sight.

For $\lambda/D = 5$ μas and $\lambda = 0.5$ μm, D = 20 km. Thus, for any practical instrument, one must plan to use statistics to split the fringe or Airy disk. Below we consider the statistics of this splitting.

In general, we assume an astrometric instrument collects light and causes it to pass through two exit ports with intensities I_1 and I_2 such that $\Delta = (I_1 - I_2)/2$ changes as the instrument is rotated by δ with respect to the source direction. (For some kinds of instruments, the rotation must be in the "instrument plane.") Then

$$\sigma(\delta) = I_0/\left[2\sqrt{N_d}(\partial\Delta/\partial\delta)\right] \tag{2}$$

where $I_0 = I_1 + I_2$ and N_d is the number of detected photons [Reasenberg, 1986a]. We can apply this to an astrometric interferometer having two telescopes:

$$\sigma_I = \lambda/\left[2\pi L\sqrt{N_d}\right] \tag{3}$$

where L is the baseline length.

Similarly, Eq. (2) can be applied to a single-telescope instrument. We assume a knife-edge detector. Although more efficient detection is possible using a filter matched to the shape of the Airy spot, practical instruments cannot make use of this advantage. The best normally done is for the knife edge to divide the light so as to fall on two detectors. We find that

$$\sigma_T = 3\pi\varsigma\lambda/16D\sqrt{N_d} \tag{4}$$

$$\varsigma \geq 1$$

where D is the telescope diameter and the equality holds if the knife edge passes through the center of the Airy disk. We can use Eqs. (3) and (4) to compare a telescope to an interferometer under the assumption that the knife edge is centered and that the collecting area, optical efficiency, and detector efficiency result in an equal number of detected photons in the two cases.

$$\sigma_T/\sigma_I = 3\pi^2 L/8D \tag{5}$$

In a Ronchi ruling type of single-field astrometric telescope, the uncertainty is in practice larger than that of Eq. 4 (with $\varsigma = 1$). The ruling line widths must be chosen larger than the diffraction-limited stellar image size to control systematic errors coming from aberrations of off-axis images and thus the knife edge spends little time near the center of the image where the maximum positional sensitivity resides. (The off-axis aberration is substantially reduced in a two-mirror telescope, but this approach introduces other systematic errors that are comparably as serious as the ones it removes.) Thus, the telescope suffers an inefficiency or effective loss of integration time in calculating N_d. A further reduction of N_d occurs because the Ronchi ruling is typically 50% opaque. When these effects are included, the uncertainty, σ_T, is increased by a factor not likely to be less than two. If the instruments are to be of comparable size, then we might take the focal length of the telescope to be equal to the baseline of the interferometer. Since a typical astrometric telescope has a focal ratio of about 15, the resulting ratio of σ's (including the above efficiency factor taken as 2) is about 110 in favor of the interferometer.

For both types of instruments, the uncertainties discussed above are for the observation of a single object with no consideration of the reference object or objects. For POINTS, two target stars are observed simultaneously yielding an estimate of the angle ($\approx$90 deg) between them. In a Ronchi telescope observation, the reference stars must be found in the small (say ϕ = 10 arc min) field around the target. Since such stars are statistically faint ($m \geq 15$), the time required to make an observation of a m = 10 target will be dominated by the integration time required for the reference. This time can be shortened by simultaneously observing a large number of the brightest stars in the field. By observing 30 stars in a field of ϕ = 10 arc min, one collects the equivalent photon flux of an m = 13 star.

It is intended that most of the POINTS targets will be at least as bright as magnitude 10. To obtain the angle between a pair of such stars to a given accuracy would require a little longer than twice the time implied by Eq. 3 (i.e., the time required to collect the required number of photons). The small extra time would compensate for the additional error introduced by the metrology system.

For a Ronchi telescope, the time required to observe an m = 10 star would be dominated by the photon rate of the collective reference objects, and would be about 17 times that implied by Eq. 4 with ς = 2. The factor would be considerably larger in a Ronchi telescope not designed to look simultaneously at a large number of reference stars. Of course, the speed advantage that an interferometer has from using a bright reference star is lost when the target is fainter than about m = 13, but the intrinsic ratio of errors (or the equivalent ratio of required observing times) is independent of target or reference star brightness.

VII. AN ASTROMETRIC INTERFEROMETER

Figure 1 shows an artist's rendition of the present design for POINTS mounted on the Multimission Modular Spacecraft (MMS). There are two baselines, each 2 m long, and four telescopes, each with primary mirrors 25 cm in diameter. The baselines are separated by an angle $\varphi = \varphi_0 + \Delta_\varphi$ where φ_0 is approximately 90° and Δ_φ is the articulation angle: $|\Delta_\varphi| < \Delta_m$, where Δ_m, the articulation range, will probably be between 0.5° and 5°. The articulation of the two interferometers allows the observed stars to lie within a few arcsec of the optical axes of their respective interferometers, thus essentially eliminating off-axis distortions and the attendant biases. POINTS measures θ, the angular separation between a pair of target stars, by measuring φ and

Fig 1. An artist's rendition of POINTS with 2 m separations between pairs of telescopes of 25 cm diameter. The instrument comprises two U-shaped interferometers joined by a bearing which permits φ, the angle between the principal axes of the interferometers, to vary from its nominal of 90 deg. The Multimission Modular Spacecraft is shown mounted under the instrument.

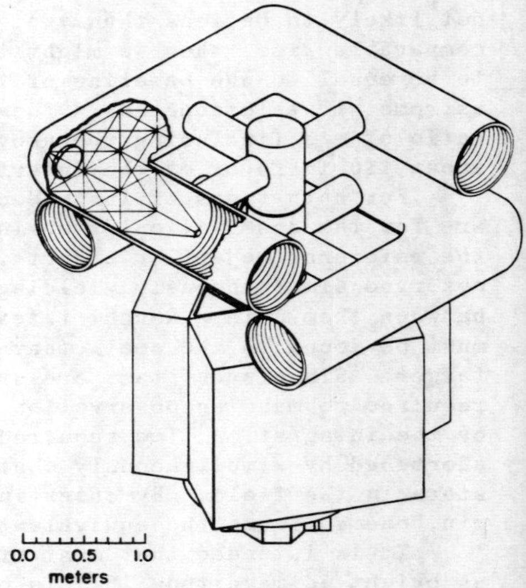

adding with proper sign δ_1 and δ_2, where δ_i is the angular separation between the apparent direction to the star and the principal axis of the interferometer. See Fig. 2.

The approximately right angle between the interferometers provides three advantages: (1) For a given target star, the reference star is chosen from an equatorial band. If Δ_m, the articulation range, is as little as 1°, then the band has an area of 720 sq. deg and can be expected to contain several stars as bright as magnitude 5.

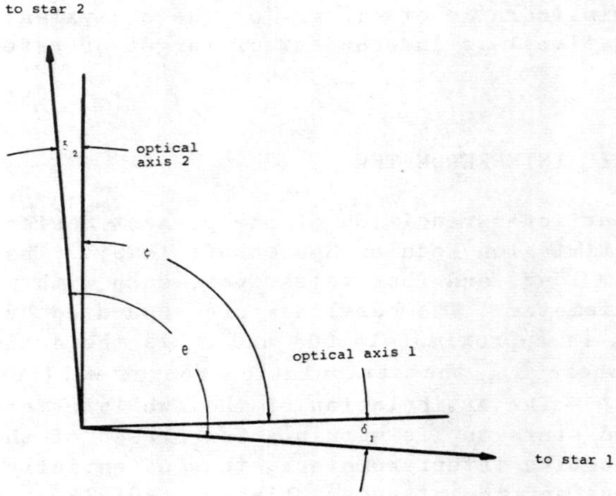

Fig. 2. POINTS observing geometry.

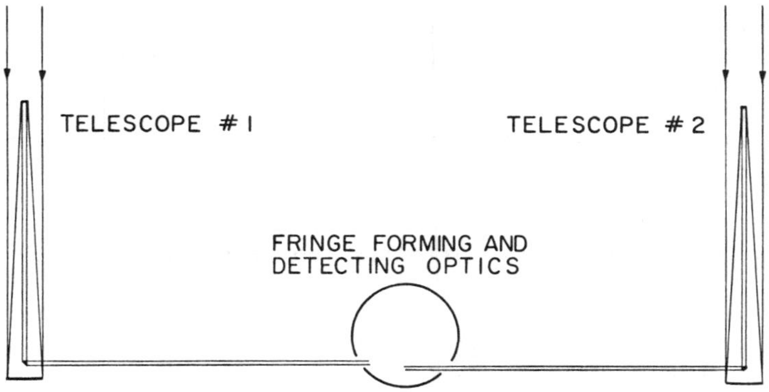

Fig. 3. An optical interferometer using telescopes.

Thus, the observation time need not be made artificially long to compensate for a dearth of bright reference stars. (2) Parallax measurements are absolute, not relative. (3) Instrument bias can be determined by 360° closure. Such determinations can be made often and require no special observing sequence. (See Section XII.)

In each of the two interferometers, a pair of afocal telescopes collects samples of the starlight and directs them toward the fringe-forming and detecting assembly, Fig. 3. How should the fringes be formed and analyzed? After considering several possibilities, we find that we are most comfortable with a beamsplitter and a combination of a lens and a prism which yields a spectrum of the interference, i.e., a channelled spectrum, Fig. 4. When the instrument's principal axis

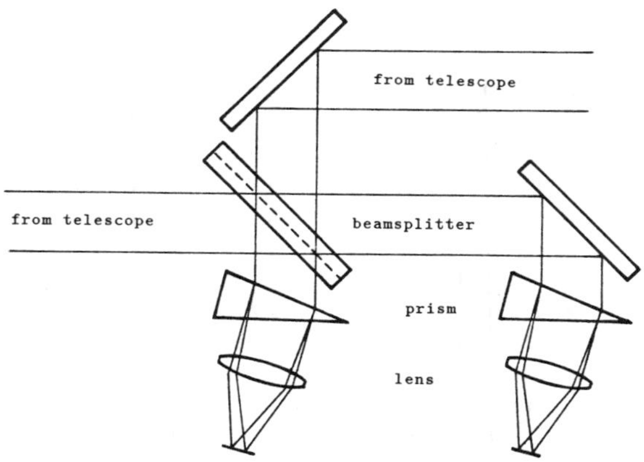

Fig. 4. Fringe forming optics.

is aligned with the apparent direction to the star ($\delta = 0$), the signal at a given wavelength has equal intensity at the two beamsplitter exit ports. With misalignment ($\delta \neq 0$), constructive interference for a given wavelength at one port is complemented by destructive interference at the other port. At each port, an alternating pattern of constructive and destructive interference is found. The resulting complementary channelled spectra, which are detected by linear arrays, form the basis for determining δ.

For an optical interferometer as described above, the measurement uncertainty based on photon statistics can be shown to be [Reasenberg, 1986a]

$$\frac{\sigma(\delta)}{5\mu\mathrm{as}} = \frac{10^{(m_b-10)/5}}{\left[\frac{\tau/W}{240\,\mathrm{sec}}\ \frac{T}{7000\mathrm{K}}\ \frac{\eta}{0.02}\right]^{1/2} \frac{D}{25\mathrm{cm}}\ \frac{L}{2\mathrm{m}}} \tag{6}$$

where the target is assumed to be a black body. In Eq. (6), m_b is the bolometric magnitude of the target star, τ is the observation time, T is the star's color temperature, and η is the optical efficiency which is assumed to be wavelength independent; W is the optical-bandpass factor, where W = 1 for an unbounded optical passband. If we limit the bandpass to the range from 0.25 to 0.8 μm, then W = 1.2 at T = 7000K. For stars at 4000K or 11500K, W = 2.0. (N.B. At 11500K, T/W(T) has about the same value as at 7000K.)

Equation (6) applies to a single interferometer. For an instrument observing a pair of stars, the required observation time is the sum of the times calculated from Eq. (6) for the individual stars to be observed with the desired precision plus an additional contribution due to the internal metrology. The metrology contribution is made small by design.

For a 2 m baseline, the nominal 5 μas accuracy corresponds to a displacement of one end of the interferometer toward the source by 0.5Å. Displacements of internal optical elements can be similarly important. Thus, the instrument requires real-time metrology at the sub-angstrom level. That metrology is discussed in Section XI.

If POINTS is to achieve a measurement accuracy of a few μas, does it need comparably fine pointing? Fortunately, the answer is "no." Because of the dispersed-fringe detection system, the fringe is observable for a large instrument pointing offset without the need for mechanical path-length compensation. In the nominal instrument design, each detector array would have 128 elements and the optical bandpass would be from 0.25 to 0.8 μm; a rotation of the instrument by 1 arcsec would change the number of fringes on the detector array by 27. Thus, the fringes on the arrays would reach the Nyquist limit at $|\delta| = \delta_N = 2.4$ sec. At $\delta = 0.5\,\delta_N$, the fringe visibility is 0.97; at

$\delta = \delta_N$, 0.90; at $\delta = 1.5\,\delta_N$, 0.78; and at $\delta = 2\,\delta_N$, zero. In principle, the instrument could continue to gather information at a useful rate even for $|\delta|$ greater than δ_N. In practice, the pointing system could allow a (one-σ) error of $\approx 0.5\delta_N$ or a deadband of $\approx 2\delta_N$. Such a modest pointing requirement presents two advantages. First, an "off the shelf" pointing system can be used for the instrument. Second, there should not be excessive vibration introduced by the pointing system as might be the case if the pointing uncertainty needed to be much smaller.

VIII. MICROARCSECOND ASTROMETRY

There is a regrettable erroneous belief among some astrophysicists that astrometry is a dead subject. They are apparently unaware of the dramatic progress that has been made in this field during the last few years, as discussed in Section VI. The HIPPARCOS Mission, which will survey 10^5 stars at an estimated astrometric accuracy of 2 mas, should make the astrophysical community aware of the importance of modern astrometry. POINTS is a natural interferometric extension of HIPPARCOS.

Table II lists seven areas of astronomy that will receive important contributions from POINTS astrometry. Many of the present list of scientific uses for POINTS were suggested and elaborated on by the members of a team (see Table III) that has applied to NASA/Astrophysics to perform a Phase-A study of POINTS under the Explorer Program. In the next section, we consider the potential for a test of general relativity using POINTS. Here we briefly consider some of the other applications in the table. A more detailed discussion is given by Reasenberg et al. (in preparation by members of the Phase-A Team for AJ, 1988). Of course, targets of opportunity and serendipitous discoveries will likely add to the scientific richness of a POINTS mission.

Table II. POINTS Science

General Relativity
Extra-Solar Planetary Systems
Global Reference Frames
Extragalactic Astronomy
Galactic Astronomy
Solar-System Astronomy
Cosmic Distance Scale

Table III. Phase A Investigators

Name	Affiliation
Dr. Robert W. Babcock	Smithsonian Astrophysical Observatory
Dr. Peter L. Bender	University of Colorado Joint Inst. for Lab. Astrophys.
Dr. Andrew Buffington	Center for Astrophysics & Space Science University of California at San Diego
Dr. Bruce Carney	Department of Physics and Astronomy University of North Carolina
Dr. John F. Chandler	Smithsonian Astrophysical Observatory
Dr. Marc V. Gorenstein	Smithsonian Astrophysical Observatory
Dr. Paul Horowitz	Department of Physics Harvard University
Dr. John P. Huchra	Smithsonian Astrophysical Observatory
Dr. James A. Hughes	US Naval Observatory
Dr. Kenneth J. Johnston	Naval Research Laboratory
Dr. Burton F. Jones	Lick Observatory University of California
Dr. Eugene H. Levy	Department of Planetary Science U. of Arizona Lunar & Planetary Lab.
Dr. Michael R. Pearlman	Smithsonian Astrophysical Observatory
Dr. Robert D. Reasenberg (PI)	Smithsonian Astrophysical Observatory
Dr. Irwin I. Shapiro	Smithsonian Astrophysical Observatory
Dr Myles Standish	Jet Propulsion Laboratory
Dr. Stuart Weidenschilling	Planetary Science Institute
Dr. Clifford M. Will	Department of Physics Washington University

"One of the major unsolved questions in science is 'How was the solar system formed?' It has become clear that the answer to this fundamental question will not be forthcoming until a comprehensive search for other planetary systems has been undertaken" [Black, 1984]. Until then, the underlying generalities of the origin of such systems remain inaccessible. An astrometric instrument can detect the motion of a star around the star-planet center of mass. As seen from 100 pc, the Sun is tenth magnitude. The wobbles induced by the planets and their orbital periods are listed in Table IV.

There are more suitable target stars then could be observed using even all of the observing time during a ten-year POINTS mission. However, the study of as few as 100 targets would yield an important result. Our present understanding of star formation requires planetary systems to be common. After ten years, either a POINTS survey of 100 promising targets would have found planetary systems or it would have been shown reliably that such systems are rare and that our present understanding of star formation is not correct.

POINTS is better suited to work within the Galaxy than beyond. However, as shown in Table V, there are areas of extragalactic astronomy that could be addressed with the instrument. In addition, POINTS will provide an improved distance scale based on the direct determination of parallax for Cepheid and RR Lyrae variable stars. This would be of considerable value to extragalactic astronomy.

Table IV. Sun at 100 pc

Planet	Astrometric Wobble (μas)	Period (years)
Earth	0.03	1
Jupiter	50	11.9
Saturn	27	29.5
Uranus	8.4	84.
Neptune	15	164.8
J.C.*	2.5	11.9

*Jovian Core. From models of the formation of the solar system, it is believed that the nonvolatile core of Jupiter is about 5% of the planet's mass, i.e., 15 Earth masses. Other estimates have the core several times more massive. It is possible that planetary systems may form without giant gas planets but with their corresponding cores.

Table V. Extragalactic Astronomy

Bound or measurement of quasar relative motion
Proper motions of nearby galaxies Angular momentum of local group
Proper motions within nearby galaxies Distance estimates via velocity statistics

Galactic astronomy would be revolutionized by POINTS measurements. Distances could be determined to all parts of the Galaxy. When combined with interstellar absorption lines, distance data would yield a 3-D map of the interstellar medium. Proper motions of stars in the spiral arms would permit us to "see" the density waves that are believed to be responsible for the arms. The galactic mass distribution could be determined in several ways. For example, the proper motion of the LMC and SMC should be about 15 mas/y if there is no "dark matter" in the Galaxy. This motion could be measured in the POINTS reference frame. The use of POINTS to study solar-system bodies has been discussed by Reasenberg [1984]. There are 16 natural satellites and numerous asteroids that could be observed. The former would yield satellite orbits, improved planet orbits, planet masses and perhaps low-order terms in the zonal harmonic expansion of the planets' mass distribution, and some satellite masses. The latter would yield asteroid orbits and, via mutual perturbation, asteroid masses.

IX. THE POINTS RELATIVITY EXPERIMENT

The microarcsecond accuracy achievable with POINTS raises the possibility of significant new tests of general relativity. The first test would come simply from improvements in accuracy in the PPN parameter γ. The highest accuracy to date comes from measurements of the Shapiro time delay using radio ranging to the Viking landers, which show that $(1 + \gamma)/2$ must be unity to within one part in a thousand [Reasenberg et al., 1979]. POINTS could achieve a two to three order of magnitude improvement in accuracy. More importantly, POINTS may provide the opportunity to detect "higher order" effects in the deflection of light that have not been detectable to date. The most important higher order deflection comes from the next approximation to gravitational theory beyond the post-Newtonian, called the post-post-Newtonian approximation. Whereas the "first-order" deflection depends linearly on the potential GM/rc^2 (where G is the constant of gravity, M is the mass of the Sun, r is the distance from the Sun, and c is the speed of light), the "second-order" deflection depends quadratically on it.

The second-order contribution depends on the PPN parameters β and γ, as well as on a parameter Λ that is explicitly post-post-Newtonian. In the usual interpretation of the PPN formalism, γ measures how much spatial curvature a given theory of gravity predicts at first post-Newtonian order; Λ measures spatial curvature predicted at the second post-Newtonian order. Current experiments provide no information about this important parameter, which is the coefficient of the second-order terms in the metric that contribute to the deflection. (The remaining part of the 11 μas comes from geometric contributions and not from the higher order part of the metric. Although many existing gravitational theories have values for γ equal to or close to unity, in agreement with experiment, they are likely to differ from general relativity in their values for Λ. In general relativity, $\Lambda = 1$. A determination of the value of Λ would be the first measurement of a gravitational effect in the solar system beyond the post-Newtonian level, and would represent an important new test of general relativity.

Our numerical studies have shown that the second-order deflection experiment is marginal, yielding an uncertainty that could be as much as ten times, or as little as one tenth of the effect. The two-order range comes from uncertainties about the degradation of instrument accuracy for near-Sun observations and the possibility that the instrument accuracy would be substantially better when observing stars brighter than $m = 10$. The latter requires that the control of systematic error be correspondingly better than normally would be required. At this time, judgements on these matters would be premature.

X. POINTS, THE INSTRUMENT

There are three basic questions that need to be addressed for an astrometric instrument operating at the μas level: (1) measurement precision and scientific throughput; (2) fringe stability and limiting magnitude; and (3) systematic error. Similar questions would need to be asked about an astrometric telescope.

The first (and easiest) question was addressed in Section VII in terms of photon statistics, which should dominate observing time. Fringe stability is closely related to the behavior of the pointing system. The question is discussed below. The question of systematic error has been central to the design of the instrument. Hardware aspects of this question are discussed in the next section. The correction of instrument biases is discussed in the following section, Mission Considerations.

We assume that in order to "see a fringe," it is necessary to collect 25 photons before the uncertainty in the fringe rate accumulates a phase error of π. However, the instrument will experience unmodelled angular accelerations due, for example, to Earth gravity gradient and solar radiation-pressure imbalance in the case of a free flier. For a Shuttle-mounted instrument, the disturbance is

Table VI. Process noise covariance density and its relation to both the nominal limiting magnitude and the pointing error.

P_o $\frac{(\mu\ \mathrm{rad})^2}{\mathrm{sec}^5}$	Notes*	m_o stellar magnitude	$m_I(\max)^+$ stellar magnitude	rms pointing error$^{+\S}$ mas	fringes$^\#$
5×10^{-3}	Corresponds to a thin rod in low Earth orbit with angular acceleration due to gravity gradient and time scale due to orbital period.	15.3	17.4	8	.3
10	Disturbance from the Shuttle through a suspension with a 10 sec period.	13.6	15.2	29	1.1
10^4	Disturbance from the Shuttle, direct connection.	12.1	13.4	90	3.5

* In all three cases one would know the acceleration _a priori_ and this information could be used to reduce S_o. However, for the purpose of calculating quantities for this table, such information was excluded. In that respect, the results in the table are pessimistic. The results do not include provision for actuator error. In principle, this error would be considered part of the process noise. In this respect, the results in the table are optimistic.

\+ Assumes that $m_T = 7$.

§ Pointing error scales as $10^{m_T/6}\ D_T^{-5/3}$. Calculation assumes that $P = P_o$ as given in column 1.

\# Assumes $\lambda = 0.5\ \mu$m.

orders of magnitude larger. For our analysis, we assume that an independent star-tracking telescope of diameter D_T observes a star of magnitude m_T. In the presence of disturbance, how well does this star tracker determine the angular velocity of the interferometer? Equation (7) defines the fringe tracking threshold under the assumption that a Kalman filter (e.g., Jazwinski [1970]) is used to estimate the angular position, velocity, and acceleration of the instrument [Reasenberg, 1984].

$$\frac{D_T}{10\ \mathrm{cm}}\left[\frac{D_I}{25\ \mathrm{cm}}\right]^2 = \frac{L}{2\ \mathrm{m}}\left[\frac{P}{P_o}\right]^{1/4}\ 10^{\frac{1}{2}\left[\frac{4m_I + m_T}{5} - m_o\right]} \qquad (7)$$

In this expression, L, D_I, and m_I are the baseline length, mirror diameter, and stellar magnitude for the interferometer; P is the process noise assumed to be a white angular acceleration variance density. Given P_o, we can calculate m_o as shown in Table VI. The table also shows the rms pointing error based on the same estimator.

Although the guide telescope substantially simplifies the initial acquisition of fringes by the first interferometer, it may play a relatively less important role in fringe tracking. In the case of a bright target star, the working fringe tracker would provide the Kalman filter with better angle data than would the telescope. In turn, these data would assure that the post-acquisition fringe rate error would be small on a time scale required to collect 25 photons from the target star.

Just as the star tracking telescope provides an estimate of the fringe rate for the first interferometer, allowing it to detect fringes, the first interferometer in turn can provide a fringe rate estimate for the second interferometer, allowing it to detect fringes with a fainter star. For the latter case, the corresponding values of m_0 are larger by about 5. For the smallest process noise considered in Table VI, this brings the instrument to about the threshold of detector noise, a factor not considered in the above analysis. For the largest process noise, quasar observations are still possible.

XI. CONTROL OF SYSTEMATIC ERROR

The control of systematic error is central to achieving the nominal accuracy of 5 μas. We address this problem at three levels: (1) stable materials, structural design, and thermal control; (2) real-time metrology; and (3) the detection and correction of systematic error in conjunction with the global data analysis. The best materials fail by orders of magnitude to provide the long-term dimensional stability required by POINTS in the absence of other means of controlling systematic error. We therefore limit our reliance on stable materials as much as possible. Stable materials for the structural elements of the instrument serve principally to limit the dimensional changes (and change rates) that the metrology system must determine. In a few places, we are forced to rely on material stability (over short times, i.e., up to a few hours). The instrument is designed so that such metering elements are small (i.e., of order 1 cm). This facilitates thermal control and reduces the contribution to the critical dimensions of the instrument of the unavoidable changes in size of these metering elements.

The instrument relies on two kinds of laser-driven optical interferometers to determine changes in critical dimensions. "Full-Aperture Metrology" (FAM) surveys the optical components that transfer the starlight from the primary mirrors to the beamsplitter in each stellar interferometer. Specialized, point-to-point laser gauges are used in support of FAM and to measure φ, the angle between the two baselines.

The high-precision star position measurement is made with respect to the optical axis of the interferometer, whose position is determined by the positions of the optical elements used to transfer the starlight. In order to achieve the required pathlength measurement accuracy, variations in the optical pathlength through the system must be monitored to about 10 pm. The metrology system must determine the average change in the starlight delay induced by all motions and distortions of all optical elements. FAM, a novel technique to measure such average changes, is discussed below. It provides three significant advantages over conventional approaches.

(a) FAM removes complexity. The usual metrology systems use a large number of laser gauges to determine the locations of the elements individually. From these measurements, the optical path through the system is computed. FAM directly measures the optical path through the system.

(b) FAM measures the correct quantity. Because the metrology signal fully illuminates the surface of each optical element that determines the starlight phase at the beamsplitter, the phase of the metrology signal is representative of the average starlight path through the system.

(c) FAM provides the basis for an operational definition of the direction of the interferometer baseline. It results in a a pair of "fiducial points" located in front of each interferometer. These fiducial points, which lie on lines parallel to (or held at fixed small angles to) the interferometer baselines, are used to determine φ, the angle between the two interferometers' optical axes.

Figure 5 illustrates the FAM technique. The key element is the set of primary mirrors which have shallow (phase contrast) zone plates on their surfaces: alternate zones are depressed about 50 nm. The zones are approximately in the form of Newton's rings: each zone has about the same total area. We have shown that the required "Zone-Plate Mirror" (ZPM) can be made holographically. Our colleagues at the Perkin-Elmer Corporation (P-E) have developed and tested a technology for making a phase-contrast zone plate of high uniformity on a curved mirror surface. They are extending this work to a ZPM that could be used for a long-duration space mission. The ZPM will be the subject of a paper to be published in 1988.

The FAM technique requires two servos. The error signal from the auxiliary null interferometer (Fig. 5a) is used to shift the metrology beamsplitter (and associated hardware) so as to maintain a constant difference between the optical paths from the metrology beamsplitter to the two retroreflectors inside the fiducial blocks. The error signal from the FAM interferometer (Fig. 5b) is used to shift the starlight beamsplitter (and associated hardware) so as to maintain a constant difference between the two paths from one beamsplitter to the other (i.e., via the two telescopes and their fiducial blocks). With both servos working, there is a small (say under 2 mm) and constant difference between the distances from the starlight beamsplitter to the fiducial points which are the apices of the retroreflectors inside the fiducial blocks. Under the assumption that these servos do not need to track path changes due to vibration, they can have small bandwidths. Since it is desirable to limit the contamination (e.g., via scattering from the surfaces of the optical elements) of the starlight by the laser signal, a minimum of laser light is used in the

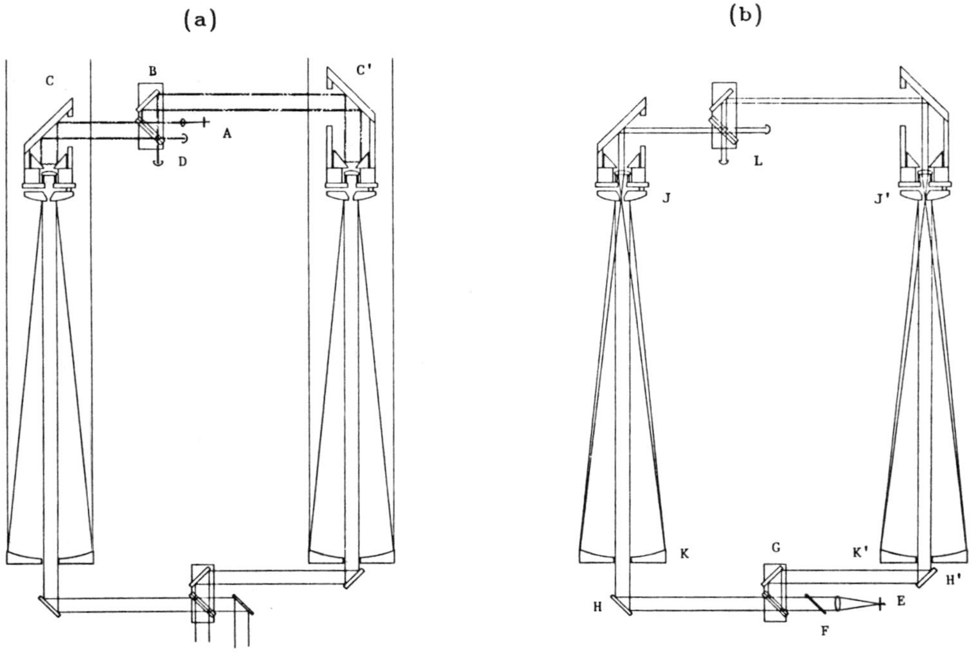

Figure 5. Interferometer optical paths: drawing not to scale.
(a) Optical paths for starlight (solid) and auxiliary null (dot-dash) interferometers. For the latter, a laser signal passes through the spatial filter A and is divided by the metrology beamsplitter B to form beams that enter the fiducial blocks C and C'. Within each fiducial block, the beam is deflected by the 45° mirror, retroreflected by the hollow corner cube, and returned toward the metrology beamsplitter by the 45° mirror. The returning signals are combined at the metrology beamsplitter and fall on detectors D.

(b) Optical paths for FAM (see text) interferometer. A laser signal passes through the spatial filter E and injection beamsplitter F and is divided by the starlight beamsplitter G. The separated beams are reflected from the tertiary mirrors H and H' and secondary mirrors J and J' to fully illuminate the primary mirrors K and K'. The zone plates on the primaries diffract the signal to focal points in the holes in the secondaries and the athermal lenses collimate the diffracted light. Finally, the signals are reflected from the 45° mirror, are recombined at the metrology beamsplitter, and fall on detectors L.

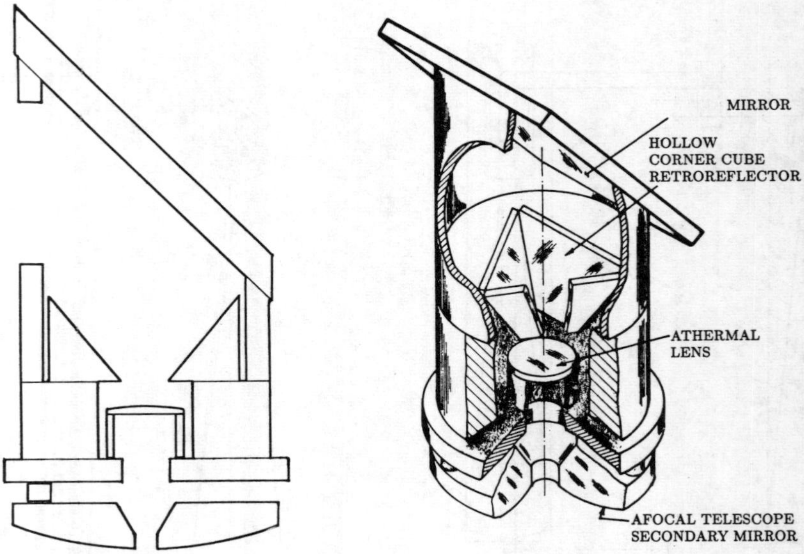

Fig. 6. A fiducial block is shown schematically (left) and pictorially (right, courtesy of Perkin-Elmer) without the thermal shields.

FAM servo and it is therefore the slower of the two. The fiducial blocks (Fig. 6) and the metrology beamsplitter assemblies are made of optically contacted high-stability material such as ULE (TM Corning Glass Works) and kept at a stabilized temperature. These devices pose the only critical materials problem identified. However, as shown at the end of this section, that problem does not result in a violation of the instrument error budget.

The angle φ between the baselines of the two interferometers (i.e., the instrument articulation) is determined according to Eq. (8) by the measurements of the six distances among four fiducial blocks in the system (see Figure 7).

$$\cos\varphi = (d_2^2 + d_3^2 - d_1^2 - d_4^2)/2b_m b_f \tag{8}$$

Here d_i are the four distances between the fiducial points of one interferometer and the fiducial points of the other and b_m, b_f are the baseline distances for each interferometer, i.e., each is the separation of the fiducial points of one interferometer. Each of the six distances is measured by a laser gauge which we assume produces a biased length estimate. We require that the bias in each of these measurements not change rapidly, although slow drifts are acceptable as discussed in the next section.

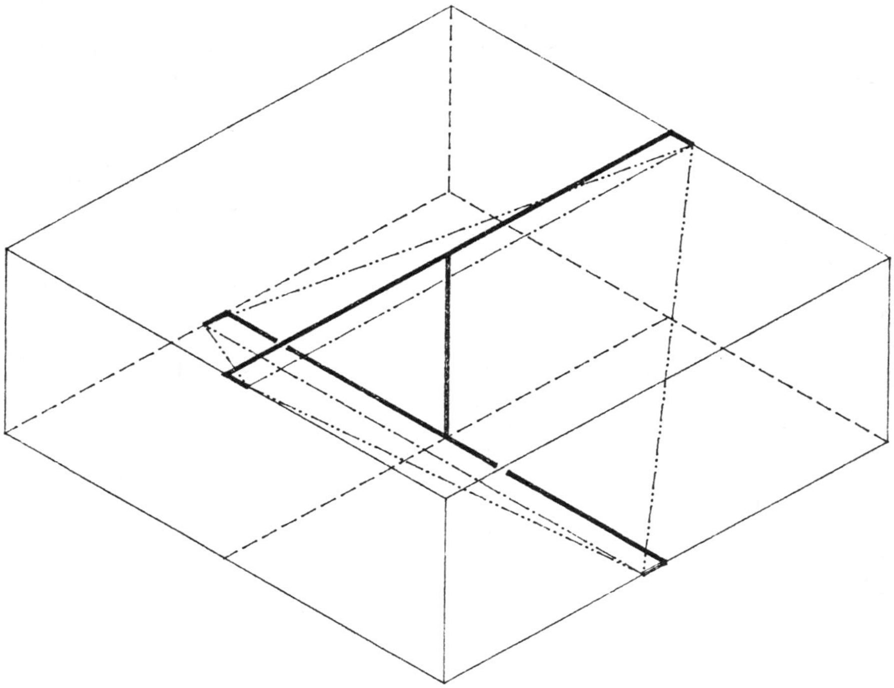

Fig. 7. The angle φ is determined by measuring the six distances among the four fiducial blocks at the ends of the short heavy lines (b_i shown "dot-dash"; d_i shown "dot-dot-dash"). The instrument is shown in the same orientation in Fig. 1.

On a time scale of ten minutes to a few hours, the instrument depends on the stability of the laser for accuracy. Of the nominal measurement uncertainty, 2.5×10^{-11} radians, we assign 0.5×10^{-11} radians to the laser. Fortunately, this does not require correspondingly high laser stability. In the determination of φ, we depend on the ratio of laser-determined distances and therefore laser frequency shifts cancel. (There are secondary effects that cause this cancellation to be imperfect at about the 10^{-3} level. The observing geometry adds another factor of nearly ten to the requirement.) In the FAM application, the quantity being measured (with 10 pm accuracy) is the difference between two nominally equal paths and laser frequency shifts would affect only this difference, which should be under 2 mm. Thus, the stability in either system need not be much better than 10^{-9}.

Presently available, inexpensive (≈$3000) stabilized HeNe lasers for laboratory work meet or exceed these stability requirements. However, HeNe lasers are not likely to have a sufficiently long life for the proposed ten-year mission. Recent success with Neodymium-doped YAG lasers suggests that a rugged, stabilized version will be available soon. These devices would not suffer the gas-diffusion problem that limits the HeNe laser life. They appear as a reasonable candidate for the metrology signal source. (For example, see Harrison and Mooradian [1986].)

To avoid change of the size or shape of either the fiducial-blocks or the metrology beamsplitter assembly, which would cause a corresponding change in the bias of the metrology system, these devices will be made of a stable material and kept in a thermally stable environment. The effective separation of metrology links within the fiducial blocks would be under 1 cm, and the temperature would be held stable to 10^{-2} K on a time scale of a few hours. As discussed below, the effects of temperature fluctuations on a longer time scale and of a long-term linear drift will be routinely removed in analysis of the data. The required thermal stability could be achieved with an active control of an outer region of the fiducial block enclosure to 10^{-1} K and a layer of insulation which, combined with the heat capacity of the fiducial block, would yield a thermal time constant of about one day. Using ULE (TM Corning Glass Works), which has a linear coefficient of thermal expansion $|\alpha| < 0.03 \times 10^{-6}$/deg C between 10 and 50 deg C, the resulting distance error would be ≤ 3 pm, well within the required limit.

XII. MISSION CONSIDERATIONS

Kovalevsky [1984] describes five classes of astrometry from "very narrow field" to "global." Since (by these definitions) POINTS is a global astrometric instrument ($\theta_o \sim 90°$), it is natural to consider a global reference frame. This is in sharp contrast to the local reference frames defined by the small fields of view of classical astrometric instruments. Even those classical instruments that Kovalevsky calls "wide field" are, by the present standard, limited to differential measurements. Once the reference frame is defined by a grid of reference stars, the 90° baseline separation provides a maximum choice of reference stars for each target star, thus allowing scientific objectives to drive mission design.

Consider the following scenario: A set of N grid stars is selected so as nominally to be distributed uniformly over the celestial sphere. The interferometer is used to observe each pair of stars within its articulation range Δ_m. The number of observations K will be proportional to $N^2\Delta_m$ and we define $M = K/N \propto N\Delta_m$, the number of observations per star. We have performed series of Monte Carlo covariance studies [Chandler and Reasenberg, unpublished, 1981, 1984] based on this scenario. We considered the _a posteriori_ uncertainty in the angular separation of each pair of stars whether directly observed or not. We found that the distribution of these uncertainties becomes essentially unimodal for large M: M greater than a critical value that seems to be 3.5 ±0.3, depending on the particular random set of stars.

The Monte Carlo covariance studies have shown that when the average number of measurements per star reaches a second critical value which is about 4.2 [Chandler and Reasenberg, unpublished notes, 1986], the a posteriori estimated separation between about half of the pairs of stars (including those not measured directly) has an uncertainty $\overline{\sigma}(\theta)$ less than the measurement uncertainty $\sigma(\theta)$; for less than 5% of the pairs, $\overline{\sigma}(\theta) > 3\sigma(\theta)$. The stars of such a grid would be used as the reference stars for the majority of the scientific applications of POINTS.

The sensitivity studies have further shown that when the observations are combined in a least-squares estimate of the individual stellar coordinates, it is possible to estimate simultaneously several instrument bias parameters per day without significantly degrading the stellar coordinate estimates [Chandler and Reasenberg, unpublished notes, 1986]. Thus, metrology biases and related errors can be allowed to change on a time scale of hours without significantly degrading the performance of the instrument. Such drifts could, in principle, come from slow changes in the shape of the optical components used in the metrology system. Since these components would be in the most precisely controlled thermal environments in the instrument, thermal effects due to the reorientation of the instrument with respect to the Sun can be made negligible. The residual drift would come from the internal relaxation or radiation damage of the optical components or similar changes to the thermal control system. Such changes are expected to be slow, even at the level of precision required for optical interferometry. However, should it prove necessary, the observation sequence could be designed to provide an even more frequent determination of instrumental bias. We give no further consideration to this question since there is no basis at present for anticipating such a problem.

It would be natural to include the observation of a small number of bright quasars ($12 < m \leq 15$) in the standard measurement series. The observations of quasars could be performed a large number of times per series to compensate for their larger magnitude. These quasar observations would serve two purposes. First, they would measure or bound the quasars' relative motions which are ordinarily assumed to be negligible. Second, they would stabilize the reference frame against uniform rotation and help identify in successive observation series the proper motions of the grid stars.

Aberration is the astronomer's term for change of apparent direction of a light source as a result of the motion of the observer. The largest component is proportional to v/c where v is the observer's velocity and c is the speed of light. For the Earth's orbital motion, $v/c \approx 10^{-4} \approx 21$ sec. The effect of aberration would be included (to third order in v/c) in the model of the observable used for data analysis. In this way, the star-grid frame would automatically be tied to the Earth-ephemeris frame with an uncertainty of ≈ 0.1 $\mathrm{sec}/\sqrt{N^*}$, where N^* is the total number of observations. At 60 observations per day for ten years at a duty factor of 75%, the coordinate tie would have an uncertainty of ≈ 250 μas. With the addition of an ad hoc parameter, the relativistic theory of aberration could be tested to third order at essentially no extra cost.

If we use 300 grid stars plus a few quasars and take M = 4.2, the observation series requires about one month at the nominal rate of 60 observations per day. Such a series could be repeated four times a year for ten years to provide not only coordinates but proper motions and parallaxes for the stars. Our Monte Carlo covariance studies show that after ten years the coordinate uncertainties are ~ 0.6 μas, the proper motion uncertainties are ~ 0.4 μas per year, and the parallax uncertainties are ~ 0.4 μas [Reasenberg, 1986b]. Note that this parallax determination is a factor of 3 better than one would naively calculate from the coordinate uncertainties in a single series. The reason for this enhancement is that the 90^0 nominal angle of POINTS results in a direct observation of absolute parallax.

In another series of Monte Carlo covariance studies, we investigated the ten-year observing sequence with fewer observations. Based on this study, it appears that if at least one of the observing series is complete (i.e., M = 5), then observations can be deleted from the other series by a variety of random or systematic procedures yielding an increase in the mean parameter-estimate uncertainty which depends only on the square root of the total number of observations. Further, additional stars can be added to the observation sequence with a minimal number of observations per star. Thus, if the instrument were run full-time on a star survey, the survey could comfortably include 5000 stars and would result in the knowledge of their positions, proper motions, and parallaxes below the 1 μas (1 μas per year) level.

After several years of observation, the postfit residuals from the data reduction could be used to investigate irregularities in the motions of individual stars. Their apparent motions would be analyzed both for the intrinsic scientific interest and to improve the stability of the resulting reference frame. Initially, the position of each star would be determined with respect to the mean of the positions of all the stars. After the initial modeling of the irregularities in stellar proper motions, the position of each star would be determined with respect to the nominal modeled reference positions of all the stars. The implicitly iterative procedure is expected to converge quickly as long as the number of stars in the reference grid is large (i.e., greater than 100). The iterative nature of the procedure may be obscured in practice as the work would be paced by the rate at which new data become available.

XIII. POINTS DEVELOPMENT PROGRAM

We are now developing many aspects of POINTS in the hope that it will eventually become a NASA flight project. The work is divided into a series of tasks which are listed in Table VII. These tasks are

Table VII. POINTS Development Tasks

A. <u>Technology Development</u>

1. Evaluation of optical performance of a 15 cm spherical zone-plate mirror in support of full-aperture metrology.

2. Development of a point-to-point laser gauge able to operate with a precision of a few pm (in vacuum, between POINTS fiducial blocks).

B. <u>Design of Critical Elements</u>

1. Design of the optical bench (without articulation) and determination of instrument mass and moments of inertia. Find stiffness, eigen frequencies, and optical path sensitivity for each eigenmode.

2. Design the fiducial blocks. Investigate materials (i.e., ULE and alternatives), their optical and thermal properties, and the methods of fabrication they permit.

3. Design the detector head. Investigate limitations to the optical bandwidth of the instrument. For a representative design, determine $\eta(\lambda)$ where η is the photon detection probability and λ is the optical wavelength.

C. <u>Systems Analysis for the Instrument and its Mission</u>

1. Investigate planetary-system search strategy appropriate to POINTS. Consider observing time per star as a function of type (magnitude) and distance as well as the density of stars by type.

2. Develop specifications for the principal optical components.

3. Formulate an error budget, including the effects of vibration, component aberrations, and alignment error.

4. Formulate and evaluate algorithms for the acquisition and tracking of fringes.

5. Investigate limiting magnitude via a full Kalman filter simulation of the tracking of targets in the presence of rotational process noise. (Include the results of task C4.)

6. Determine a workable set of procedures for the analysis of the mission data. Include the detection and certification of planetary systems.

7. Investigate the scheduling of observations. The algorithms should depend on: (1) the spacecraft orbit; (2) pointing, acquisition, and observing time; and (3) the sky restrictions (i.e., Earth limb limit, solar and lunar exclusions).

---

organized into three groups: (A) technology development; (B) design of critical elements; and (C) systems analysis of the instrument and its mission. Although the tasks within each group are not strictly ordered by priority, at least for group C, the tasks at the bottom of the list appear to us to be less critical at the present stage of effort than are those at the top of the list. Our objective is to start with the present design and both to extend the existing analyses in scope and detail and to develop the technologies critical to the instrument. Most significant among the latter are the metrology techniques comprising tasks A1 and A2.

As of the preparation of this paper, four (spherical) ZPM have been made by Perkin-Elmer Corporation. Two of them are on long-term loan to SAO where we have been characterizing them. The test interferometer for studying the phase tracking characteristics of the ZPM is working and we are beginning to make measurements.

The next step is to replace the more critical of the manual adjustments on the test interferometer with remotely (computer) controllable devices in preparation for placing the equipment in vacuum. The former will be done in 1988 an the latter either that year or the next.

The schedule for the remaining tasks depends on the degree of support that the project receives. However, even at a minimal level of support, we plan to do some work on most of the tasks.

ACKNOWLEDGEMENTS

I thank N.L. Murphy for her careful preparation of the text. Work on solar-system tests of general relativity and related matters is supported in part by NASA through grant NAGW-722, by NSF grant PHY-84-09671, and by the Smithsonian Institution. Work on POINTS is supported in part by NASA through its Innovative Research Program, by NSF grant PHY-84-09671, and by the Smithsonian Institution, both directly and through its Scholarly Studies Program. The collaboration of the Perkin-Elmer Corporation in the development of POINTS has been important in reaching our present state and we look forward to a continuation of this relationship.

REFERENCES

Audoin, C., and Vanier, J., 1976, Atomic frequency standards and clocks, J. Phys. E.: Sci. Instrum., 9:697.

Black, D. C., 1980, In search of other planetary systems, Sp. Sci. Rev., 25:35.

Black, D. C., 1984, Search for other planetary systems using space astrometry, in: "Proceedings of the Workshop on High Angular Resolution Optical Interferometry From Space," P. B. Boyce and R. D. Reasenberg, eds., BAAS, 16:767.

Buffington, A., 1986, BAAS, 18:941.

Colavita, M. M., Shao, M., Staelin, D. H., 1987, Two-color method for optical astrometry: theory and preliminary measurements with the Mark III stellar interferometer, Applied Optics, 26:4113.
Dirac, P. A. M., 1937, The cosmological constants, Nature, 139:323.
Dirac, P. A. M., 1938, Proc. R. Soc. Lond., A165:199.
Epstein, R., and Shapiro, I. I., 1980, Post-post-Newtonian deflection of light by the sun, Phys. Rev. D, 22:2947.
Fishbach E. and Freeman, B. S., 1980, Second-order contribution to the gravitational deflection of light, Phys. Rev. D, 22:2950.
Fomalont, E. B., and Sramek, R. A., 1977, The deflection of radio waves by the sun, Comments on Astrophys. 7:19.
Gatewood, G. D., 1987, The multichannel astrometric photometer and atmospheric limitations in the measurement of relative positions, Astron. J., 94:213.
Harrison, J. and Mooradian, A., 1986, Spectral Linewidth of Semiconductor Lasers, in Methods of Laser Spectroscopy, Y. Prior, A. Ben-Reuven, and M. Rosenbluh, Eds. Plenum Publishing.
Jazwinski, A. H., 1970, "Stochastic Processes and Filtering Theory," Academic Press, New York.
Kovalevsky, J., 1982, The Satellite Hipparcos, in Proceedings: The Scientific Aspects of the Hipparcos Space Astrometry Mission, ESA SP-177, Strasbourg, February 1982.
Kovalevsky, J., 1984, Prospects for space stellar astrometry, Sp. Sci. Rev., 39:3.
Lindegren, L., 1980, Astron. Astrophys., 89:41.
Monet, D. G. and Dahn, C. C., 1983, Astron. J., 88:1489.
Mozurkewich, D. et al., 1988. submitted to Astron. J.
Nordtvedt, K., 1968a, Equivalence principle for massive bodies I. Phenomenology, Phys. Rev., 169:1014.
Nordtvedt, K., 1968b, Equivalence principle for massive bodies II. Theory, Phys. Rev., 169:1017.
Reasenberg, R. D., Shapiro, I. I., MacNeil, P. E., Goldstein, R. B., Breidenthal, J. C., Brenkle, J. P., Cain, D. L., Kaufman, T. M., Komarek, T. A., and A. I. Zygielbaum, 1979, Viking relativity experiment: verification of signal retardation by solar gravity, Astrophys. J. Lett., 234:L219.
Reasenberg, R.D., 1983, Present state of the experimental verification of the four-dimensional general theory of relativity, in Unified Field Theories of More than 4 Dimensions Including Exact Solutions, V. deSabbata and E. Schmutzer, Eds., World Scientific Publishing, Singapore.
Reasenberg, R. D., 1984, Microarcsecond astrometric interferometry, in: "Proceedings of the Workshop on High Angular Resolution Optical Interferometry From Space," P. B. Boyce and R. D. Reasenberg, eds., BAAS, 16:758.
Reasenberg, R. D., 1986a, POINTS: A small astrometric interferometer, in: Large Optics Technology, Proc. SPIE 571, G. M. Sanger, Ed.
Reasenberg, R. D., 1986b, Microarcsecond astrometric interferometry, in: Astrometric Techniques, IAU Symposium 109, H. Eichhorn and R. J. Leacock, eds., Reidel, Dordrecht.

Reasenberg, R. D., 1986c, Solar system tests in transition, in Proceedings of the Ninth Course of the International School of Cosmology and Gravitation: Topological Properties and Global Structure of Space-Time, P. G. Bergmann and V. de Sabbata, Eds. Plenum Press, New York.

Richter, G. W., and Matzner, R. A., 1981, Gravitational deflection of light at 1 1/2 PPN order, Astrophys. and Space Sci., 79:119.

Shao, M. et al., 1988. The Mark III stellar interferometer, Astron. and Astrophys., in press.

Shapiro, I. I., 1964, Fourth test of general relativity, Phys. Rev. Lett., 13:789.

Shapiro, I. I., 1967, New method for the detection of light deflection by solar gravity, Science, 157:806.

Shapiro, I. I., 1976, Estimation of astrometric and geodetic parameters from VLBI observations, in: "Methods of Experimental Physics," M. L. Meeks, ed., Academic Press, New York.

Shapiro, I. I., Counselman, III, C. C., and King, R. W., 1976. Phys. Rev. Lett., 36:555.

Shapiro, I. I., Reasenberg, R. D., MacNeil, P. E., Goldstein, R. B., Brenkle, J. P., Cain, D. L., Komarek, T. A., Zygielbaum, A. I., Cuddihy, W. F., and Michael, W. H., Jr., 1977, The Viking relativity experiment, J. Geophys. Res., 82:4329.

Taylor, J. H., 1987. Astronomical and space experiments to test relativity, in General Relativity and Gravitation, Proceedings of the 11th International Conference on General Relativity and Gravitation, M. A. H. MacCallum, Ed. Cambridge University Press, Cambridge.

Underwood, J. H. and Randolph, J. E., 1982, STARPROBE Scientific Rationale, A Report of the Ad Hoc Working Groups, JPL Publication 82-49, Pasadena.

Vessot, R. F. C., Mattison, E. M., Walsworth, R. L. Jr., Silvera, I. F., Godfried, H. P., and Agosta, C. C. 1987, A hydrogen maser at temperatures below 1K, in: IEEE Tran. on Instrumentation and Measurement, Vol. 117-36, June 1987, #2, 588.

Vessot, R. F. C., 1976, Frequency and time standards, in: "Methods of Experimental Physics," M. L. Meeks, ed., Academic Press, New York.

Walsworth, R. L. Jr., Silvera, I. F., Godfried, A. P., Agosta, C. C., Vessot, R. F. C., and Mattison, E. M. 1986, Hydrogen maser at temperatures below 1K, Phys. Rev. A, 34:2550.

Will, C. M., "Theory and Experiment in Gravitational Physics," Cambridge University Press, Cambridge, 1981.

THE WEYL-DIRAC THEORY AND THE VARIATION OF THE GRAVITATIONAL CONSTANT

Nathan Rosen
Department of Physics
Technion-Israel Institute of Technology
32000 Haifa
Israel

ABSTRACT. The Weyl theory of gravitation and electromagnetism, as improved by Dirac, leads to the possibility of a gravitational constant varying with time, the variation depending on the choice of gauge. By introducing a background metric corresponding to a space-time of constant curvature one is led to a natural gauge which gives $\dot{G}_e/G_e \sim 5\times10^{-12}y^{-1}$, where G_e is the effective gravitational constant.

1. WEYL GEOMETRY

Let us begin with a brief description of Weyl's geometry.[1] It will be recalled that in Riemannian geometry, for a given coordinate system x^μ, one has the metric tensor $g_{\mu\nu}$ which determines the interval ds between two neighboring points separated by coordinate differentials dx^μ, where

$$ds^2 = g_{\mu\nu}dx^\mu dx^\nu \tag{1}$$

The length ξ of a vector ξ^μ is correspondingly given by

$$(\xi)^2 = g_{\mu\nu}\xi^\mu\xi^\nu \tag{2}$$

Under a coordinate transformation $x^\mu \to x'^\mu$ one has

$$g_{\mu\nu} \to g'_{\mu\nu} = g_{\alpha\beta}\,(\partial x^\alpha/\partial x'^\mu)(\partial x^\beta/\partial x'^\nu) \tag{3}$$

and

$$\xi^\mu \to \xi'^\mu = \xi^\alpha(\partial x'^\mu/\partial x^\alpha) \tag{4}$$

If the vector ξ^μ undergoes parallel displacement from x^μ to $x^\mu+dx^\mu$, one gets

$$d\xi^\mu = -\xi^\alpha\{\alpha{}^\mu\nu\}\,dx^\nu \tag{5}$$

V. De Sabbata and V. N. Melnikov (eds.),
Gravitational Measurements, Fundamental Metrology and Constants, 345–355.

where $\{\alpha^{\lambda}\nu\}$ is the Christoffel symbol,

$$\{\mu^{\lambda}\nu\} = \frac{1}{2} g^{\lambda\sigma}(g_{\mu\sigma,\nu} + g_{\nu\sigma,\mu} - g_{\mu\nu,\sigma}) \tag{6}$$

with a comma denoting an ordinary partial derivative. It is readily found that under such a displacement the length does not change,

$$d\xi = 0 \tag{7}$$

Eqs. (1) and (2) can be regarded as being associated with the existence of a standard length in the whole space. This is borne out by (6), according to which this length can be carried to any point by parallel displacement.

In Weyl's geometry the situation is different in that one can take an arbitrary standard of length at each point. One can change the standard of length by means of a gauge transformation (GT) under which

$$g_{\mu\nu} \rightarrow g'_{\mu\nu} = e^{2\lambda} g_{\mu\nu} \tag{8}$$

where λ is an arbitrary (smooth) function. From (1) and (2) it follows that

$$ds \rightarrow ds' = e^{\lambda} ds \tag{9}$$

$$\xi \rightarrow \xi' = e^{\lambda}\xi \tag{10}$$

It is assumed that under a parallel displacement, which is now given by

$$d\xi^{\mu} = -\xi^{\alpha}\Gamma^{\mu}{}_{\alpha\nu} dx^{\nu} \tag{11}$$

where $\Gamma^{\mu}{}_{\alpha\nu}$ is a connection different from (6), the length ξ is changed by an amount

$$d\xi = \xi\phi_{\mu}\, dx^{\mu} \tag{12}$$

where ϕ_{μ} is a vector that characterizes the geometry together with $g_{\mu\nu}$. After the GT (10) one gets in place of (12)

$$d\xi' = \xi'\phi'_{\mu}\, dx^{\mu} \tag{13}$$

provided one takes

$$\phi_{\mu} \rightarrow \phi'_{\mu} = \phi_{\mu} + \lambda_{,\mu} \tag{14}$$

One can now ask about the form of $\Gamma^{\lambda}{}_{\mu\nu}$. If it is assumed that

$$\Gamma^{\lambda}{}_{\mu\nu} = \Gamma^{\lambda}{}_{\nu\mu} \tag{15}$$

one can show that

$$\Gamma^{\lambda}_{\mu\nu} = \{\mu^{\lambda}\nu\} + g_{\mu\nu}\phi^{\lambda} - \delta^{\lambda}_{\mu}\phi_{\nu} - \delta^{\lambda}_{\nu}\phi_{\mu} \tag{16}$$

It follows that under a GT $\Gamma^{\lambda}_{\mu\nu}$ is invariant,

$$\Gamma'^{\lambda}_{\mu\nu} = \Gamma^{\lambda}_{\mu\nu} \tag{17}$$

A quantity Q (which can be a scalar or a tensor of any order) is said to be gauge covariant if, under the GT, it transforms according to the relation

$$Q \to Q' = e^{n\lambda}Q \tag{18}$$

One then says that Q has the power n, and one writes

$$\Pi[Q] = n \tag{19}$$

For example,

$$\Pi[g_{\mu\nu}] = 2, \quad \Pi[g^{\mu\nu}] = -2 \tag{20}$$

In view of the gauge invariance of $\Gamma^{\lambda}_{\mu\nu}$ the curvature tensor, given by

$$D^{\lambda}_{\mu\nu\sigma} = -\Gamma^{\lambda}_{\mu\nu,\sigma} + \Gamma^{\lambda}_{\mu\sigma,\nu} - \Gamma^{\alpha}_{\mu\nu}\Gamma^{\lambda}_{\alpha\sigma} + \Gamma^{\alpha}_{\mu\sigma}\Gamma^{\lambda}_{\alpha\nu} \tag{21}$$

is also gauge invariant. One finds that

$$D_{\mu\nu} \equiv D^{\lambda}_{\mu\nu\lambda} = R_{\mu\nu} - F_{\mu\nu} - 2\phi_{\mu;\nu} - g_{\mu\nu}\phi^{\alpha}{}_{;\alpha} - -2\phi_{\mu}\phi_{\nu} + 2g_{\mu\nu}\phi^{\alpha}\phi_{\alpha} \tag{22}$$

where a semicolon denotes the usual covariant derivative of Riemannian geometry. Here

$$F_{\mu\nu} = \phi_{\mu,\nu} - \phi_{\nu,\mu} \tag{23}$$

It should be remarked that Eqs. (14) and (23) make plausible Weyl's interpretation of ϕ_{μ} as the electromagnetic potential and $F_{\mu\nu}$ as the electromagnetic field. One also finds that

$$D_{\mu\nu} - D_{\nu\mu} = -4F_{\mu\nu} \tag{24}$$

$$D \equiv g^{\mu\nu}D_{\mu\nu} = R - 6\phi^{\alpha}{}_{;\alpha} + 6\phi^{\alpha}\phi_{\alpha} \tag{25}$$

Obviously

$$\Pi[D] = -2 \tag{26}$$

One would like to derive the field equations from a variational principle

$$\delta\int W(-g)^{1/2}\, d^4x = 0 \tag{27}$$

where W is a scalar and the integral is gauge invariant. For this one needs to have $\Pi[W] = -4$. If one wants W to contain D, the latter must appear as D^2 in view of (26). However this leads to fourth order field equations. Weyl tried to get around this difficulty by the use of a special gauge, but the situation was not considered to be satisfactory.

2. DIRAC MODIFICATION

Dirac[2] became interested in Weyl's geometry because it gave the possibility of a varying value of Newton's gravitational constant. On the basis of his Large Number Hypothesis Dirac concluded that the gravitational constant G should vary as 1/T, with T the age of the universe, so that

$$\dot{G}/G = -1/T \tag{28}$$

where $\dot{G}=dG/dT$. Since according to the general relativity theory G is a constant, Dirac turned to the Weyl theory. In order to avoid the difficulty with the variational principle he introduced a gauge-covariant scalar β of power -1, so that under the GT

$$\beta \to \beta' = \beta e^{-\lambda} \tag{29}$$

In order to get a simple expression for W in light of (25), he took (without the term L_m that has been added here to describe the matter)

$$W = F_{\mu\nu}F^{\mu\nu} - \beta^2 D + 6(\beta,_{\underline{\mu}}+\beta\phi^{\mu})(\beta,_{\mu}+\beta\phi_{\mu}) + 2\Lambda\beta^4 + L_m \tag{30}$$

Here an underlined index is to be raised with $g^{\mu\nu}$ and Λ is the cosmological constant which will be set equal to zero. To within a divergence W can be written

$$W \doteq F_{\mu\nu}F^{\mu\nu} - \beta^2 R + 6\beta,_{\underline{\mu}}\beta,_{\mu} + L_m \tag{31}$$

Putting (31) into the variational principle (27) and varying $g_{\mu\nu}$, one gets what corresponds to the Einstein equations,

$$G^{\mu\nu} = -(8\pi/\beta^2)(M^{\mu\nu}+T^{\mu\nu}) + (2/\beta)(g^{\mu\nu}\beta_{;\underline{\alpha}\alpha}-\beta_{;\underline{\mu\nu}}) +$$

$$+(1/\beta^2)(4\beta,_{\underline{\mu}}\beta,_{\underline{\nu}}-g^{\mu\nu}\beta,_{\underline{\alpha}}\beta,_{\alpha}) \tag{32}$$

where $T^{\mu\nu}$ is the energy-momentum density tensor of the matter (derived from L_m) and $M^{\mu\nu}$ that of the electromagnetic field,

$$4\pi M^{\mu\nu} = \frac{1}{4} g^{\mu\nu} F_{\alpha\beta} F^{\alpha\beta} - F^{\mu}{}_{\alpha} F^{\nu\alpha} \tag{33}$$

Varying ϕ_μ gives the Maxwell equations

$$F^{\mu\nu}{}_{;\nu} = 4\pi J^{\mu} \tag{34}$$

where J^{μ} is the charge-current density vector (obtained from the variation of L_m). Varying β leads to an equation which is essentially the condition for the matter sources of the field to be gauge invariant. This will not be considered here.

If one takes the divergence of (32), making use of the Bianchi identities and (34), one gets

$$T^{\mu\nu}{}_{;\nu} - T\beta_{,\underline{\mu}}/\beta = J_{\alpha} F^{\alpha\mu} \tag{35}$$

In the absence of an electromagnetic field $(F_{\mu\nu}=0)$, if one takes $T^{\mu\nu}$ for the case of a pressureless fluid (made up of dust particles),

$$T^{\mu\nu} = \rho u^{\mu} u^{\nu} \tag{36}$$

with $u^{\mu}=dx^{\mu}/ds$, where x^{μ} are the particle coordinates and ρ the rest-mass density, (35) yields the conservation equation

$$(\frac{\rho}{\beta} u^{\nu})_{;\nu} = 0 \tag{37}$$

and the equations of motion

$$du^{\mu}/ds + \{\alpha{}^{\mu}\beta\} u^{\alpha} u^{\beta} = (g^{\mu\alpha} - u^{\mu} u^{\alpha}) \beta_{,\alpha}/\beta \tag{38}$$

Let us write

$$\rho = m\rho_n \tag{39}$$

where m is the particle rest mass and ρ_n is the particle density. The conservation of the number of particles is given by

$$(\rho_n u^{\nu})_{;\nu} = 0 \tag{40}$$

Putting (39) into (37) and comparing this with (40) we see that

$$m = m_o \beta \quad (m_o = \text{const.}) \tag{41}$$

Now let us go back to the field equations (32). From the form of the first term on the right one concludes that the gravitational constant G is given by

$$G = G_o/\beta^2 \quad (G_o = \text{const.}) \tag{42}$$

In view of (41) we then have

$$Gm = G_o m_o/\beta \tag{43}$$

If we assume that β is a slowly varying function of time, then the motion of a planet around a star of mass m is determined in the Newtonian approximation by Gm. Let us define the effective gravitational constant G_e by the relation

$$Gm = G_e m_o \tag{44}$$

Then (43) gives

$$G_e = G_o/\beta \tag{45}$$

If G_e varies with time, this variation can be determined by observations on planetary motion. However, since β can be transformed arbitrarily by a GT according to (29), we see that in order to have G_e a definite function of time, we need some way of fixing the gauge. One possibility is based on making use of the bimetric general relativity theory.

3. BIMETRIC GENERAL RELATIVITY

Let us assume that, in addition to the "physical" metric $g_{\mu\nu}$, which is the fundamental tensor of general relativity theory, there is a second, "background", metric $\gamma_{\mu\nu}$ at each point of space-time. We can now write two kinds of covariant derivatives, the usual one formed with $g_{\mu\nu}$, denoted by a semicolon as above, and another one with $g_{\mu\nu}$ replaced by $\gamma_{\mu\nu}$, denoted by a bar ($|$) and to be referred to as a γ-derivative. Correspondingly we have two kinds of Christoffel symbols, $\{\mu^\lambda\nu\}$ formed with $g_{\mu\nu}$ and $\Gamma^\lambda_{\mu\nu}$ formed with $\gamma_{\mu\nu}$, and two curvature tensors $R^\lambda{}_{\mu\nu\sigma}$ and $P^\lambda{}_{\mu\nu\sigma}$ respectively.

One finds that

$$\{\mu^\lambda\nu\} = \Gamma^\lambda_{\mu\nu} + \Delta^\lambda_{\mu\nu} \tag{46}$$

where $\Delta^\lambda_{\mu\nu}$ is a tensor having the same form as $\{\mu^\lambda\nu\}$ but with ordinary derivatives of $g_{\mu\nu}$ replaced by γ-derivatives. Similarly it is found that

$$R^\lambda{}_{\mu\nu\sigma} = P^\lambda{}_{\mu\nu\sigma} + K^\lambda{}_{\mu\nu\sigma} \tag{47}$$

where again $K^\lambda{}_{\mu\nu\sigma}$ is obtained from $R^\lambda{}_{\mu\nu\sigma}$ by replacing ordinary derivatives of $g_{\mu\nu}$ by γ-derivatives. Contracting (47) gives

$$R_{\mu\nu} \equiv R^{\lambda}{}_{\mu\nu\lambda} = P_{\mu\nu} + K_{\mu\nu} \tag{48}$$

Some time ago,[3,4] when it was proposed to introduce the background metric, it was assumed that $\gamma_{\mu\nu}$ describes a flat space-time so that $P^{\lambda}{}_{\mu\nu\sigma}=0$. It follows from (48) that one can write the Einstein equations either in the usual form

$$R_{\mu\nu} - \frac{1}{2} g_{\mu\nu} R = -8\pi\, T_{\mu\nu} \tag{49}$$

or in the form

$$K_{\mu\nu} - \frac{1}{2} g_{\mu\nu} K = -8\pi\, T_{\mu\nu} \tag{50}$$

with $R_{\mu\nu}$ replaced by $K_{\mu\nu}$, since these are equal. Consequently the physical contents of the Einstein equations are not altered by the introduction of the background metric. However, there is an improvement in the formalism: one can describe the energy and momentum of the gravitational field by means of a tensor instead of the pseudo-tensor that one has in general relativity without a background metric.

For a given $\gamma_{\mu\nu}$ the solution of the field equations for $g_{\mu\nu}$ contains four arbitrary constants because of the Bianchi identities. One can therefore impose four conditions on $g_{\mu\nu}$ so that in the absence of a gravitational field one can have $g_{\mu\nu}=\gamma_{\mu\nu}$. One possibility is to take

$$Y^{\mu} \equiv (\kappa g^{\mu\nu})_{|\nu} = 0 \tag{51}$$

the covariant De Donder condition, with the scalar κ given by

$$\kappa = (g/\gamma)^{1/2} \tag{52}$$

where g and γ are the determinants of the two metrics. Another possibility is to take

$$\kappa = 1 \tag{53}$$

and to replace (51) by a weaker condition such as requiring Y_{μ} to be the gradient of a scalar function.

In order to modify the physical contents of the general relativity theory (for the purpose of removing some of the singularities occurring in it), it was recently proposed[5] to take $\gamma_{\mu\nu}$ as describing a space-time of constant curvature (hence of maximum symmetry) with positive spatial curvature, a space-time associated with the structure of the universe. The corresponding line-element can be written

$$d\sigma^2 = dt^2 - a^2\cosh^2(t/a)(d\chi^2 + \sin^2\chi\, d\Omega^2) \tag{54}$$

where

$$d\Omega^2 = d\theta^2 + \sin^2\theta\, d\phi^2 \tag{55}$$

and a is a constant of the order of the size of the universe. One has (for constant curvature)

$$P_{\mu\nu} = (3/a^2)\gamma_{\mu\nu} \tag{56}$$

so that, in view of (48), $R_{\mu\nu}$ and $K_{\mu\nu}$ are not equal. To modify the theory one now replaces (49) by (50) as the field equations. With the help of (48) and (56) one can write these in the Einstein form

$$G_{\mu\nu} = -8\pi\, T_{\mu\nu} + S_{\mu\nu} \tag{57}$$

where

$$S_{\mu\nu} = (3/a^2)(\gamma_{\mu\nu} - \frac{1}{2}\, g_{\mu\nu}\gamma_{\alpha\beta}g^{\alpha\beta}) \tag{58}$$

If we assume that matter interacts only with $g_{\mu\nu}$ and not with $\gamma_{\mu\nu}$, then we have

$$T^{\mu\nu}{}_{;\nu} = 0 \tag{59}$$

It follows from (57) and the Bianchi identities that

$$S^{\mu\nu}{}_{;\nu} = 0 \tag{60}$$

where indices are raised with $g^{\mu\nu}$. One finds that (60) is equivalent to the De Donder condition (51), so that the latter is forced on us by the field equations.

If we take the curvature parameter a of the order of 10^{28} cm, then the bimetric general relativity theory with the field equations (57) and (51) gives predictions that are indistinguishable from those of the Einstein theory for gravitational phenomena in the solar system. However, near a singularity of the physical metric the two theories may give different results. According to the bimetric theory there are no black holes; when a massive star undergoes gravitational collapse, it reaches a state in which it fills its Schwarzschild sphere and is in hydrostatic equilibrium.[6]

In cosmology the bimetric theory gives isotropic models without the "big bang". One can have a closed universe which first contracts to an extremely high, but finite, density and then expands. Corresponding to the background line-element (54) one takes for the physical line-element

$$ds^2 = e^{2\phi}dt^2 - a^2e^{2\psi}\cosh^2(t/a)(d\chi^2 + \sin^2\chi d\Omega^2) \tag{61}$$

where ϕ and ψ are functions of t.

Numerical calculations were carried out to obtain solutions of the field equations for ϕ and ψ in the case of a zero-pressure model, which corresponds fairly well to the present state of the universe. To get agreement with present observations of the Hubble effect ($T=1/H=2\text{x}10^{10}$y) and the mean density of matter ($\rho=4\text{x}10^{-31}\text{gm/cm}^3$), one obtains approximately

$$x \equiv t/a = 2.6, \quad a = 2.3\text{x}10^{28} \text{ cm} \tag{62}$$

with t=0 at maximum density. One also finds

$$\phi=0.017, \quad d\phi/dx=-0.008; \quad \psi=-0.073, \quad d\psi/dx=0.192$$

4. BIMETRIC WEYL-DIRAC THEORY

Let us now modify the Weyl-Dirac theory discussed perviously by introducing the constant-curvature background metric. This is done by replacing $R_{\mu\nu}$ by $K_{\mu\nu}$ in (31) and (32). The field equations (32) become

$$G^{\mu\nu} = -(8\pi/\beta^2)(M^{\mu\nu}+T^{\mu\nu}) + (2/\beta)(g^{\mu\nu}\beta_{;\underline{\alpha}\alpha}-\beta_{;\underline{\mu}\underline{\nu}}) +$$

$$+(1/\beta^2)(4\beta_{,\underline{\mu}}\beta_{,\underline{\nu}}-g^{\mu\nu}\beta_{,\underline{\alpha}}\beta_{,\alpha}) + S^{\mu\nu} \tag{64}$$

Taking the divergence of (64), one gets, corresponding to (35),

$$T^{\mu\nu}{}_{;\nu} - T\beta_{,\underline{\mu}}/\beta = J_\alpha F^{\alpha\mu} \tag{65}$$

with

$$T^{\mu\nu} = T^{\mu\nu}-(\beta^2/8\pi)S^{\mu\nu} \tag{66}$$

If we take $T^{\mu\nu}=J^\mu=0$, then (65) gives

$$(\beta^2 S^{\mu\nu})_{;\nu} - S\beta\beta_{,\underline{\mu}} = 0 \tag{67}$$

and this can be put into the form

$$(\beta^2\kappa g^{\mu\nu})_{|\nu} = 0 \tag{68}$$

so that we now have the gauge-invariant form of the De Donder condition (51).

Let us now consider the field equations (64) and (68) in the absence of an electromagnetic field, and let us apply them to an isotropic model of the universe as described by (61). For the Einstein gauge with $\beta=1$, the field equations are those of bimetric general relativity, (57) and (51). If one now carries out a GT so as to get an arbitrary gauge function of β, then

$$g_{\mu\nu} \rightarrow \bar{g}_{\mu\nu} = \frac{1}{\beta^2}\, g_{\mu\nu} \tag{69}$$

This metric satisfies (64) and (68).

In the bimetric general relativity theory the conditions (51) and (53) on the physical metric could not be imposed together. However, in the present theory with (51) replaced by (68) and the function β at our disposal it is possible to satisfy both (68) and (53). Taking $\kappa=1$, or $g=\gamma$, we have both metrics giving the same volume for a four-dimensional region, so that the scale of the universe provides a natural gauge for the physical metric.

With the line-elements (54) and (61), corresponding to $\beta=1$, we have

$$\kappa=e^{\phi+3\psi} \tag{70}$$

If we now go over to

$$\beta=e^{\frac{1}{4}(\phi+3\psi)} \tag{71}$$

and transform the metric according to (69), we get for the new line-element

$$d\bar{s}^2 = e^{\frac{3}{2}(\phi-\psi)}dt^2-a^2e^{\frac{1}{2}(\psi-\phi)}\cosh^2(t/a)(d\chi^2+\sin^2\chi d\Omega^2) \tag{72}$$

One can introduce a clock time T by the relation

$$dT = e^{\frac{3}{4}(\phi-\psi)}dt \tag{73}$$

We can now determine how the gravitational constant changes with time. From (42) we have

$$\dot{G}/G = -2\dot{\beta}/\beta \tag{74}$$

with $\dot{G} \equiv dG/dT$. Using (71) and (73) and writing $x=t/a$, we get

$$\dot{G}/G = -(1/2a)e^{\frac{3}{4}(\psi-\phi)}(d\phi/dx + 3d\psi/dx) \tag{75}$$

With the numerical values given by (62) and (63) for the zero-pressure universe one gets

$$\dot{G}/G = -1.1\text{x}10^{-11}y^{-1} \tag{76}$$

For the effective gravitational constant given by (45) one gets[7]

$$\dot{G}_e/G_e = -5.5\text{x}10^{-12}y^{-1} \tag{77}$$

This can be compared with the value of $\dot{G}/G$ given by (28) as proposed by Dirac. If we take $T = 1/H = 2\text{x}10^{10}y$, (28) gives

$$\dot{G}/G = -5\text{x}10^{-11}y^{-1} \tag{78}$$

so that the present theory gives a value smaller than that of Dirac by an order of magnitude.

REFERENCES

1. H. Weyl, Ann. Phys. (Lpz.) **59**, 101 (1919).
2. P.A.M. Dirac, Proc. R. Soc. Lond. **333**, 403 (1973).
3. N. Rosen, Phys. Rev. **57**, 147 (1940).
4. N. Rosen, Ann. Phys. (N.Y.) **22**, 1 (1963).
5. N. Rosen, Found. Phys. **10**, 673 (1980).
6. A. Harpaz and N. Rosen, Astrophys. J. **291**, 417 (1985).
7. N. Rosen, Found. Phys. **13**, 363 (1983).

SELECTED PROBLEMS OF GRAVITATIONAL WAVE EXPERIMENT

V.N.Rudenko
Dept. of Physics
Moscow State University
117234 Moscow
U.S.S.R.

INTRODUCTION.

The gravitation-wave experiment, as the field of interest and activity in modern physical science, exists for about twenty years. In spite of the fact that gravitational waves have not yet been registered, the considerable success has been achieved in construction and exploration of gravitational wave antennae. Today these devices have come up to the sensitivity level which is very reasonable from the point of view of the realistic astrophysical forecast. The event with Supernova 87 demonstrated clearly the needs in permanently operating gravity antennae. An episodic turning on antenna is now insufficient to detect some gravity-wave burst with unknown arrival time. The putting into operation of long baseline laser-gravitational antennae (which do not need liquid helium support) will probably make this task easy. Actually, this mean organization of the first gravity-wave observatory, and positive results may be awaited which look now quite realistic. Kip Thorne, the principal ideologist of GW-astronomy, has published recently a large review article on the subject of interest (Thorne,1987). So we don't need to repeat this work. Instead, in three lectures that follow, we concentrate our attention on possible new programs and ways of developing gravitational wave experiment in nearest future.

I. A NEW PROGRAM OF SEISMIC DETECTION OF GRAVITY WAVE-BURSTS

ABSTRACT. A promising astrophysical prediction of relatively intensive gravitational wave bursts at the frequency range 0.1-0.01 Hz, and an existing transparant window in the seismical noise spectrum in that region, lead to a new way to detect gravitational waves by means of a spe-

V. De Sabbata and V. N. Melnikov (eds.),
Gravitational Measurements, Fundamental Metrology and Constants, 357–385.

cial seismograph array which could be placed on a block of the earth's crust with the size 50-70 km. The gap between the signal and the noise levels is three orders of value less. The possibility to overcome the gap is discussed.

Generally, the problem of seismic detection of gravity-wave is not a new one. Earth has already been used as a gravity-wave detector. Weber and collaborators (Forward et al., 1961, Weber, (1967)) established the upper limit for stochastic gravity-wave background, having observed the lowest earth-quadruple-modes vibrations supposedly being excited by gravitation radiation. It was found that for frequency range 10^{-3}-10^{-4} Hz the spectral energy density of gravitational background is $t_w \lesssim 10^8$-10^9 erg/ /cm^2.Hz. Recently, that method of evaluating any stochastic gravity-wave background has been utilized again (Baughn and Kuhn (1984)) with the help of a more elaborated earth structure model.

Just after the discovery of pulsars Dyson (1969) speculated about the possibility to detect gravitational radiation by any proper seismograph placed on the earth's surface. He found that the earth's response depended on irregularities in the shear-wave modulus, so a large value of response had to be on the outer surface of earth. Those conclusions are true under long-wave limit when the gravitational wave length is much longer than the earth's radius, $\lambda_g \gg \ell \sim R_\oplus$. Assuming that the local dynamics is described by the flat-earth equations, Dyson has estimated the displacement of earth's crust induced by gravitational wave:

$$\Delta X_g \simeq \frac{1}{2} h \frac{v_s}{\omega} f(\theta) \sim h \lambda_s \tag{1}$$

where h, ω, the dimensionless amplitude and the frequency of a gravitational wave; v_s and λ_s, the velocity and the wavelength of a sound; $f(\theta)$, some angular factor of the order of unity. For the optimistic prediction of the pulsar gravitational radiation on the earth, $h \simeq 10^{-21}$, the formula (1) gives

$$\Delta X_g \simeq 2.10^{-17}\text{cm} \quad (\; v_s = 4.5\text{x}10^5\text{cm/s}, \; \omega/2\pi = 1 \text{ Hz} \tag{2}$$

An estimate of the background noise in a seismometer used by Dyson for the rms linear displacement was

$$\langle \Delta X_{sn} \rangle = 10^{-7} \text{ cm Hz}^{-1/2} \tag{3}$$

After integrating the measuring time, $\tau_m = 10^8$s, and averaging the signals coherently on the array of 100 seismometers one could obtain only

$$<\Delta X_{sn}> = 10^{-12} \text{ cm} \qquad (4)$$

Thus, there was the residual gap of five orders of magnitude between the signal (2) and noise (4) levels. So, Dyson was forced to make a pessimistic conclusion about the opportunity to register the pulsar gravitational radiation by such method. However, he remarked that the situation could be better in case of some acoustical resonance: "It is possible to imagine arrangements of the local topography, in which a block of the earth's crust may be seismically isolated by faults from its surroundings, and in which a gravitational wave may excite resonant standing vibrations". So that "the response of a seismometer may exceed our estimate by a large factor". Several attempts to realize these ideas have been made. For example, Wigging and Press (1969) analyzed 20.5 hours of seismic data recorded in the frequency region, and gave the upper limit for stimulated amplitude of the ground motion $\lesssim 10^{-19}$cm. In the early seventies the research group from Tel-Aviv University (D. Saden, A. Menahem, M. Meidav 1972) announced about some positive results obtained on seismic observatory in Eilat, but they were not confirmed afterwards.

In this lecture a new program of seismic detection of gravity wave-bursts proposed by Moscow University gravitational - wave group (Braginskii V.B., Gusev A.V., Mitrofanov V.P., Rudenko V.N. (1985)) is presented. What are principal motives for a new program?

a) It is a more propitious astrophysical prediction at frequencies one or two orders below pulsar's region. Figure 1 presents some fraction of more complete forecast from the recent review by Kip Thorne (1986). The right side of the Figure gives probable events (gravity-bursts) at frequency band ($\Delta\nu \sim 10^2-10^4$ Hz) typical of the bar-antenna. The more optimistic estimation here, in terms of dimensionless metric correction, is $h \sim 10^{-19}-10^{-20}$. It is proposed that the time duration of bursts, τ , should satisfy $\nu_0\tau \simeq 1$. The line from right corner down to the left corner up illustrates black hole coalescence events at Hubble distance. In the region of interest, $\nu \sim 10^{-2}$, the forecast is $h \sim 10^{-17}-10^{-18}$. It is a realistic estimation, but there is a nonzero probability for more optimistic level marked by cherished beliefs. Thus, we have, at least, more than two orders beneficial. Below we use value $h \sim 3.10^{-17}$ for numerical calculations.

b) Second. Just in the frequency band of interest there is a transparent window in the seismic noise spectrum. Figure 2 demonstrates the drop which corresponds to the period $\tau_0 \simeq (30-40)$s. The displacement of spectral density at the bottom of the drop is estimated as:

$$S_x \simeq 7.10^{-13} \text{ cm}^2/\text{Hz} \qquad (5)$$

These data are taken from the paper by Savino J. et al. (1972).

c) Third. During the last few years the idea about block structure of the earth crust was introduced. The soviet geophysicist Sadovskii M.A. is one of the main authors of this idea (Sadovskii et al. 1985). According to this theory, the earth crust looks like a net with cells included one into another, Figure 3. The boundaries of cells are, sometimes, the junctions of different densities; sometimes (Sadovskii 1986) the faults are; their depths vary and can have the size of lithosphere. These faults can be determined by helium-jets detectors sited at the aircraft. It was established that the size of cells, or blocks, had some preferrable values. Those preferrable values make a geometrical progression with approximately constant index, k. On the average, $k = 3.5 \pm 0.9$. Figure 4 (Sadovskii 1986) presents block's size distribution for different areas. For the european part of S.U. the average value, $k = 3.2$ and probable size of large blocks is $\ell = 50\text{-}70$ km. It corresponds to acoustical resonance just in the region of interest $\sim (20 \div 40)$s, ($v_s \simeq 4.5\text{x}10^5$ cm/s).

Thus, it is possible to formulate new program of seismic detection of gravitational waves as follows:

- a large block of earth crust ($\ell = 50\text{-}70$ km) is used as a gravitation detector, and
- some special seismometer array placed on the surface of the block is explored as a read-out system,
- the goal is the detection of gravity wave bursts ($\nu_g \tau \sim \sim 1$) in the frequency range, 0.1-0.01 Hz,
- the awaited dimensionless bursts amplitude is $h \sim 2.10^{-17}$, or more.

To understand reality of this program it is necessary to discuss several questions.

1. Requirements for seismometer.

For the GW-bursts the condition $\omega_g \tau \simeq 2\pi$ is valid. So there is no resonance gain in block's response, which by the order of magnitude, is:

$$\Delta \ell_g \simeq \frac{1}{2} h \ell \simeq 7.10^{-11} \text{cm} \; (h = 2.10^{17}. \; \ell = 70 \text{ km}) \qquad (6)$$

A seismometer has to register this displacement at the background of a) intrinsic thermal noise, b) external seismic noise.

The thermal standard deviation, $\Delta \ell_T$, in the time interval τ, shorter than the relaxation time, $\tau_r \sim \sim Q \omega_g^{-1}$ is:

$$\Delta \ell_T \simeq \left(\frac{kT}{m\omega_0^2} \frac{\omega_0 \tau}{Q} \right)^{1/2} \simeq \frac{10^{-6}}{\sqrt{mQ}} \; ; \qquad T = 300 \text{ K}, \; \omega_0 = 0.2$$

And to get $\Delta l_T \lesssim \Delta l_g$ one must have $mQ > 10^8$, or $Q \geq 10^4 \div 10^5$. The seismometer's reaction to seismic excitation in the same time interval, $\hat{\tau}$, is:

$$<\Delta l_s> \simeq \left(S_x \frac{\omega_0 Q}{2\pi} \frac{\hat{\tau}}{\tau_r} \right)^{1/2} \simeq 10^{-7} \text{ cm} \quad (7)$$

The difference between (6) and (7) is three orders. It is true only for seismometer's noise variations during the short time interval, $\hat{\tau} \ll \tau_r$, The stationary noise amplitude of the seismometer is, actually, by $\sqrt{Q}$ times larger, i.e.

$$<\Delta l_{st}> \simeq <\Delta l_s> \sqrt{Q} \sim 3.10^{-5} \text{cm} \quad (8)$$

$$(Q \simeq 10^5)$$

Let's summarize the main characteristics the new seismometer has satisfied.

The parameters: $\tau_0 = \frac{2\pi}{\omega_0} \sim 30$ s, $m \simeq (10^3-10^4)$g (9)

$Q \gtrsim 10^4-10^5$, i.e. $\tau_r \sim (10^7-10^8)$s

The thermal sensitivity: $\Delta X_{min} \simeq 7.10^{-11}$cm,

The dynamic region: $\mathcal{D} X \simeq 100-120$ dB.

It is not easy to construct such devices, but there are no any principal bans. And even such a device will have the signal-noise ratio equal to SNR $\simeq 10^{-3}$ on the seismic background. We shall explain our hope to overcome this lack of point III, but before that elucidate the role of blocks.

2. The role of blocks.

Let us consider the simplest one-dimensional model: a rod $[-l/2, l/2]$ on x-axis and the gravitational wave fall normally. The equations for the component of deformation θ are:

$$\begin{cases} \theta_{tt} - v_s^2 \theta_{xx} - 2\delta \theta_{xxt} = \frac{1}{2} h_{tt} \\ \theta(-l/2, t) = \theta(l/2, t) = 0 \end{cases} \quad (10)$$

i.e. gravity wave excites the rod (block) at its every point as distributed equivalent force $\sim h(x,t)$. But it is well known (Hirakawa 1973) that in long-wave limit, $l \ll \lambda_g$, the change of variable

$$\frac{\partial u}{\partial x} = \theta(x,t) - \frac{1}{2} h(x,t)$$

reduces Eq. (10) to the boundary problem:

$$\begin{cases} u_{tt} - v_s^2 u_{xx} - 2\delta u_{xxt} = 0 \\ u_x(-L/2,t) = 1/2\; h(-L/2,t) \\ u_x(L/2,t) = 1/2\; h(L/2,t) \end{cases} \qquad (11)$$

i.e. GW-excitation is presented by equivalent boundary forces. It reflects Dyson's remark that the main part of interaction of GW with elastic media occurs due to irregularities. That means also that it is very useful to know the position of the block boundaries to extract the signal effectively from the noise background. But the value of signal is practically independent of the presence of blocks, if one hopes to detect a short GW burst. The Fourier transformation (11) permits us to introduce an apparatus function of block:

$$\begin{aligned} u(x,\omega) &= H(x,\omega)\, h(\omega) \\ H(x,\omega) &= (-1)\left[2\mu \cos\frac{\mu L}{2}\right]^{-1} \sin\mu x \\ \mu &= \frac{\omega}{v_s}\left(1+j\frac{\omega}{\omega_c}\right)^{1/2}, \quad \omega_c = v_s^2/2\delta \end{aligned} \qquad (12)$$

We are interested in low frequency limit, $\omega << \omega_c$. There are two possibilities here:

a) The acoustic dissipation is small, $\mu_o L << 1$, $\mu_o = (\omega/\omega_o)^2 (v_s/2\delta)$; the block displays itself as a resonance-bar detector,

$$\omega_m = \frac{\pi(1+2m)}{L} v_s \;, \quad H(x, \omega_m + \Omega) \simeq \frac{2 v_s^2}{\omega_m L} \cdot \cos\frac{\omega_m}{v} x \cdot \left[\Omega - j\frac{\omega_m}{2Q_m}\right]^{-1} \qquad (13)$$

so that $|u(L/2)|_{max} \simeq \frac{1}{4} h_o L Q_o$ (or $Q_o \to \omega_o \tau^1/2$ for short bursts)

b) The acoustic dissipation is large, $\mu_o L >> 1$. That means there is no any block as the resonance structure,

$$H_b(x,\omega) \simeq \frac{v_s}{2\omega} \begin{cases} e^{-\mu(x+L/2)} & -L/2 \le x \le 0 \\ e^{\mu(x-L/2)} & 0 \le x \le L/2 \end{cases}$$

$$|H_b| \simeq \frac{v_s}{2\omega} \rightarrow |u| \simeq \frac{v_s}{2\omega} h \simeq \frac{1}{4\pi} \lambda_s h \qquad (14)$$

It is clear(this is the Dyson result (1)) that for the short bursts, $\omega_o \tau^1 \sim 2\pi$, both cases (13) and (14) yield almost the same result (so $L \simeq \lambda_s/2$),and the signal response does not depend on the block-structure hypothesis.

3. The lack of SNR.

Come back to the principal question: is it possible to overcome three order lack of SNR value? If the background is white normal noise nothing can be done. But for other cases the situation is not hopeless. Let's consider a simple example of boundary "pulse noise" $F(t)$:

$$U_x(-L/2,t) = n_1(t) = 1/2 \cdot h(-L/2,t) + \alpha F_1(t)$$
$$U_x(L/2,t) = n_2(t) = 1/2 \cdot h(L/2,t) + \alpha F_2(t)$$

With the accuracy of v_S/c, $h(-L/2,t)$ and $h(L/2,t)$ are the simultaneous excitations, while F_1 and F_2 are not. So the problem of noise suppressing is the problem of coincidences. One might measure $U(x,t)$ by some seismometer array, and then (due to the apparatus function $H(x,\omega)$ reconstruct the boundary conditions $U_x(-L/2,t) = n_1(t)$ and $U_x(L/2,t) = n_2(t)$. If $n_1(t) = n_2(t)$, one concludes "yes" for GW-excitation, and if $n_1(t) \neq n_2(t)$ the conclusion is "no". The total probability of error is $P = (1-P_0)\gamma + P_0\beta$, where P_0, the probability of signal arrival taken a priori; γ -, the probability of "false alarm", and β -, the probability of "signal omission". So, P_0 = 1/2, the suppression noise factor is $q = P/P_0 = (\gamma+\beta)^{-1}$. Let $F(t)$ be the casual pulse sequence with the average number N/sec, the pulse duration $\hat{\tau}$, the amplitude α_m. For the coincidences with time and amplitude resolutions Δt and $\Delta\alpha$ it is easy to find

$$\gamma = 4\Delta t\,(N_1N_2/(N_1+N_2))\,(\Delta\alpha/\alpha_m)\ ,\quad \beta = 2N_1N_2\hat{\tau}/(N_1+N_2),$$

i.e.

$$q \simeq (N_1+N_2)/2N_1N_2\hat{\tau},$$

if $N_1 \sim N_2 \sim N_0$, $q \simeq (N_0\hat{\tau})^{-1}$;
for the time duration of pulse, - $\hat{\tau} \sim \tilde{\tau}_0 \sim 30$ s, and the average pause interval $N_0^{-1} \gtrsim 3000$ s, and one obtains $q = 10^2$.

In the two-dimensional block-model there are two practically independent directions to count coincidences xx and yy ; so the suppressive factor will be $\sim q^2$, and $q = 30$ would be enough to remove a three-order deficit in SNR. For the internal block noise sources the additional depression is possible due to the spatial filtering procedure. Two seismometers placed near the block's boundary with small separation, $\Delta X \ll L$, increase SNR by the factor $q_i \simeq 3L/\Delta X$ for internal noise, if one measures the value $\tilde{U}_x = [U(\Delta X,t) - U(0,t)]\,\Delta X^{-1}$.
We gave that trivila example to illustrate an idea that "special colour" of GW response and the hypothesis about

pulse nature of noise permits us to hope to succeed in our program of seismic detection of GW bursts. Of course, the exact conclusion will be possible after more careful investigations of the question.

II. A BEAM AND BAR GRAVITATIONAL ANTENNA LIVING TOGETHER

ABSTRACT. A comparative analysis of both types of gravitational antennae (bar and beam detectors) is presented. It is shown that present technological progress provides some advantages for a beam antenna. A gravitational antenna, containing acoustic and optical degrees of freedom, is also considered. Expressions are obtained for the corresponding parts of the antenna in response to gravitational bursts of various structure. Also a calculation is made of the sensitivity of a long-base interferometric antenna as a substitution of bar detectors for free test masses. The sensitivity level is determined for such a substitution. In particular, it should be possible to detect metric perturbations $h \sim 10^{-18}$ without deep cooling and $h \sim 10^{-19}$ to 10^{-20} on cooling to helium temperatures. The novel qualities of the combined antenna are discussed.

At present, two types of gravitational antennae are under operation: a resonance solid-body gravity-wave detector with radio frequency read-out system, so called bar antenna, and free mass separated at long distance with optical-interferometric read-out system, so called beam antenna. Although a bar-antenna has longer history and experience, the beam antenna looks more preferrable at the moment, for some technological reasons. To explain that, consider the level of sensitivity which is reasonable now for each device mentioned above.

1. The Resonance Bar-Antenna.

General structure of this antenna includes a mechanical bar detector coupled with some electro-mechanical transducer, then an amplifier, some recorder (a computer data processor, and so on), see Fig. 5. In literature the sensitivity limited by thermal noise of mechanical bar-detector is cited only as "potential sensitivity" of the bar-antenna. It may be estimated by a detectable amplitude of the bar's oscillations, ΔX_{τ}, during the measurement time, τ.

a) The potential sensitivity:

$$\Delta X_{\tau}\Big|_{min} \gtrsim \xi \left[(\kappa T / M\omega_{\mu}^{2})(\omega_{\mu}\hat{\tau}/Q) \right]^{1/2} \qquad (1)$$

For operating bar-detectors the product MQ (effective mass to quality factor) achieved $\sim 10^{13}$ g (3.10^{6} g $\times 3.10^{6}$ for aluminium bars; 10^{4} g $\times 10^{9}$ for supphire bars); it is difficult to increase this number significantly. Another product, $\omega_\mu \tau$ (resonance bar frequency to time of signal) for short GW bursts is approximately 2π ($\tau \sim 10^{-3}$s, $\omega_\mu \sim 2\pi \times 10^3$). So, the last free parameter in (1) is bar's temperature $T \simeq 4$K ; ξ –, the true statistical factor. After replacing the number in (1), one gets:

$$\Delta X_\tau \big|_{min} \simeq 10^{-18} \text{ cm} \tag{2}$$

and if the bar's length is of the order of $\sim 10^2$.

$$h_{min} \simeq (\Delta X_\tau / \ell) \sim 10^{-20} \text{ for } \tau \sim 10^{-3}\text{s} \tag{3}$$

To exclude some possible speculations (in particular, about cooling to 10^{-3} K°) and to force the parameters in Eq. (1), let's calculate the value of standard quantum limit of the oscillator's amplitude,

$$\Delta X_q = \left(\frac{\hbar}{2M\omega_\mu}\right)^{1/2} \simeq \left[\frac{10^{-27}}{2.10^6 . 10^4}\right]^{1/2} \sim 2.10^{-19}\text{cm} \tag{4}$$

i.e. there is no large reserve to increase the potential sensitivity (2) to the quantum level (4).

b) The real sensitivity:

In practice, the real sensitivity is defined by the noise temperature, T_a , of electronic included in antenna (transducer, amplifier). The rule of detection, by the order of magnitude, is:

$$\left[(F_0 \tau)^2 / M \tau\right] \alpha_p \gtrsim \kappa T_a \, \Delta f \tag{5}$$

where F_0 , the external force equivalent to GW burst; it is connected with the amplitude stimulated by $\Delta X_S = (F_0 \tau)/2M\omega_\mu^{-1}$. Under optimal coupling bar with a parametric transducer, the converting power coefficient, α_p , goes to maximum value, $(\alpha_p)_{max} = \omega_p/\omega_\mu$; ω_p , the pump frequency; so the real sensitivity level, for bandwidth $\Delta f = \tau^{-1}$ is:

$$\Delta X_{r.s} \gtrsim \xi \left[(\kappa T_a / M\omega_\mu^2) \cdot (\omega_\mu/\omega_p)\right]^{1/2} \tag{6}$$

The reasonable upper value for the pump frequency is $\omega_p \simeq 10^{10}$ (for d.c. squid it is possible that $\omega_p \simeq 10^{11}$); the modern state of art for high-frequency amplifiers permits one to have $T_a \simeq 10$ K . In practice, however, the coupling factor is normally less than the optimal value, and the noise temperature, T_a , is closer to 10^2K than to 10 K (Braginskii, Mitrofanov, Panov 1985). All these things give the real sensitivity level somewhere between:

$$\Delta X_{rs} \gtrsim 3.10^{-18} - 3.10^{-16} \text{ cm, i.e.} \quad (7)$$

the potential sensitivity bar-detector is not, in practice, realized by radio frequency read-out systems.
Let's make some remarks about displacement transducer. Now all large aluminium bar-detectors operate with such a device (Stanford, Maryland, Roma). It makes the task of bar-displacement registration easier, but does narrow the receiving bandwidth (one order of sensitivity costs two orders of bandwidth). As the supphire-bar antenna is used, any practical displacement transducer looks unsuitable, due to the necessity of keeping value of Q-factor very large ($\sim 10^9$-10^{10}).

The final conclusion from the above discussion is: the modern and near-future bar-antennae with r.f. read-out system could have sensitivity limited by:

$$h \gtrsim 10^{-19}, \quad \Delta f \lesssim 10^3 \text{ Hz.} \quad (8)$$

2. A free-Mass Beam-Antenna.

A beam antenna is composed of free mass (Fig. 6) (in practice, the mass is on very soft suspension); the mirrors are fastened on the mass; a laser beam goes between mirrors; at the output , the interferometric picture is registered by a photodetector, The minimum detectable displacement of the mirror on the optical noise background is defined as:

$$\Delta l_{min} \simeq \frac{\lambda}{2\pi} \cdot \frac{1}{n} \left[A \; \frac{2\hbar\omega_e}{\eta W_0} \Delta f \right]^{1/2} \quad (9)$$

where n is the number of beam's round trips between mirrors; for a Fabry-Perot cavity, $n = (1-R)^{-1}$; $\lambda = 5.10^{-5}$ cm; the quantum efficiency of the photodetector is $\eta \simeq 0.5$. Practically, the laser power is $W_0 \lesssim 1$ mW. The baseline, l , is limited by the condition $l < \lambda_g/2$ only. The sensitivity defined by (9) depends on the quality of technology presented by two numbers:

R - the reflection coefficient of mirrors,

A - the factor of noise exceeding over short noise level.

The first model of a beam-antenna was constructed by Forward with colleagues (Forward et al. 1973). It was one meter per branch Michelson interferometer with data $W_0 = 80\ \mu$W, $n = 1$, $\eta = 0.7$; the sensitivity on the frequency displacements $f \sim 1$ kHz was got, $\Delta l_{min} \simeq 1.10^{-12}$ cm/$\sqrt{HZ}$; it corresponds to the short noise level, or $A = 1$. If it is a Fabry-Perot cavity with $R = 0.99$ ($n = 10^2$) the sensitivity would occur at $\Delta l_{min} \simeq 10^{-14}$ cm / $\sqrt{HZ}$, i.e. in absolute values, the resolution of optical read-out had lost in comparison with r.f. transducers.

Recently, there occured the skip in technology of mirror manufacturing. It is possible now to have R = 0.9999 (see CalTech antenna in paper by Drever et al. (1985)). On the short noise background it gives $\Delta l_{min} \simeq 3.10^{-17}$ cm (W_0 = 1 mW). Really, in 1984 the forty-meter antenna in CalTech had the sensitivity $\Delta l_{min} \simeq 3.10^{-15}$ cm$/\sqrt{Hz}$ i.e. the factor $A = 10^4$ (in his lecture R.Driver reported the new sensitivity for CalTech of $\Delta l_{min} \sim 8.10^{-16}$ cm$/\sqrt{Hz}$ for 40-meter antenna). It is clear that such a resolution for a long baseline antenna, l = 3.10^5 cm (the project LIGO Caltech - MIT) provides the sensitivity to gravitational radiation in bandwidth Δf = 1000 Hz even now at the level of:

$$h = (\Delta l_{min}/l)\cdot\sqrt{\Delta f} = 3.10^{-19}, \quad (10)$$

the obvious reserves, $W \to 1$ wt and $A \to 1$, could yield

$$h \simeq 10^{-22}, \qquad \Delta f = 1000 \text{ Hz} \qquad (11)$$

These arguments show why we believe the beam-antenna is a more perspective device than a bar-antenna. It is also clear that an optical read-out system can now compete successfully with r.f. transducers. This circumstance stimulates us to consider the following example:

3. A Bar-Detector with Optical F.P.-Cavity.

The device we would like to discuss is given in Fig. 7. Mechanical vibrations of the bar-detector produce some phase shift is optical Fabry -Perot cavity (mirrors fixed at the ends of the bar), whose output signal is registered by photodetector. A new feature of optical read-out system is its space size equalized to size of the bar. It means that a gravity wave interacts with optical degree of freedom as well as with acoustical one. So, there is an interesting methodological question: what is the structure of response for such kind of antenna ? Are there two different responses, or only one ? Below we give schematic qualitative answers to these questions, but their analysis is given in our paper of 1986 (Kulagin,Polnarev, Rudenko).

Let's take a special synchronous reference frame where the points O'O (Fig.7) are fixed: O is the middle bar's point; O', the point where an optical source and a receiver are placed. One can find the phase shift of interest, $\Delta\psi$, integrating the photon equation in the gravity field ($k_4 \sim \omega_e$) :

$$\frac{dk_\alpha}{dp} = \frac{1}{2}\frac{\partial g_{\beta\delta}}{\partial x^\alpha} k^\beta k^\delta \; ; \quad g_{\beta\delta} = \eta_{\beta\delta} + h_{\beta\delta} \qquad (12)$$

The phase shift calculated in this way is obviously an "electromagnetic response", $\Delta\psi_{em}$. Value $\Delta\psi_{em}$ consists of two integrals, each of them corresponding to forward and backward paths of the photon between mirrors,

$$\Delta\psi_{em} \simeq A_0 \int_{X(1)}^{X(2)} h_{\beta\delta}(x^\alpha)dp + B_0 \int_{X(2)}^{X(1)} h_{\beta\delta}(x^\alpha)dp \tag{13}$$

where A_0 and B_0 are the normalization constants. It is clear, however, that points $X(1)$ and $X(2)$ are not motionless in the chosen reference frame. They perform nongeodesic motion stimulated by elasticity and gravity-wave forces together.

The corresponding corrections for value $\Delta\psi_{em}$ in Eq. (13) can be neglected. But it is necessary to add some shift, $\Delta\psi_{ac}$, arising due to the Doppler effect for reflection from the moving mirrors (an acoustical part of response)

$$\Delta\psi_{ac} \simeq \int \frac{v_s}{c}\omega dt = \frac{\omega}{c}\left[2\xi\left(\frac{\ell}{2}, t-\frac{\ell}{c}\right) - \xi\left(-\frac{\ell}{2}, t-\frac{2\ell}{c}\right) - \xi\left(-\frac{\ell}{2}, t\right)\right] \tag{14}$$

where $\xi(x,t)$ are the coordinate displacements of the bar's ends. Thus, the total response of such antenna contains electromagnetic (optical) and acoustical parts:

$$\Delta\psi = \Delta\psi_{em} + \Delta\psi_{ac} \tag{15}$$

if $v_s \to 0$ (the limit of "free-mass antenna") the acoustical part, $\Delta\psi_{ac} = 0$, and the response are only electromagnetic. So, $\Delta\psi_{em}$ exists always; it prevails for a very short gravity-wave burst, $\hat{\tau} \ll \ell/c$, and has the same order as $\Delta\psi_{ac}$ for a long-wave limit, $\hat{\tau} \gg \ell/c$. It is more easy to estimate relative values of two mentioned parts taking the frequency-shift response. At long-wave limit one has (for $\omega_g\hat{\tau} = 2\pi n$):

$$\begin{aligned}(\Delta\omega/\omega)_{em} &\simeq h(\ell/c\hat{\tau}) \simeq h(\ell/\lambda_g)\\ (\Delta\omega/\omega)_{ac} &\simeq (v_s^2\hat{\tau}/\ell c)h \simeq \omega_g\hat{\tau}(v_s/c)h \simeq n\, h(\ell/\lambda_g)\\ [\Delta\omega_{ac}/\Delta\omega_{em}] &\simeq n(\lambda_s/\ell) \simeq 2n\end{aligned} \tag{16}$$

One can see from (16) that both parts of response have the same order, because $n \simeq (1-2)$ for short-gravity bursts. However, there is the principal difference between them: the lifetime of $\Delta\psi_{em}$ is of the order of burst duration, while the lifetime of $\Delta\psi_{ac}$ is of $\sim Q_\mu\omega_\mu^{-1} = \tau_r \gg \hat{\tau}$ due to "acoustical memory" of a bar-detector.

4. At last, we come up to the main question of this lecture: is it possible to unite free mass and a bar gra-

vitational antenna? The simple method to arrange so is to change free test mass of a beam antenna ($L \simeq 3\cdot10^5$ cm) by a resonance bar-detector ($\ell = 3$ m), and to organize read out as shown in Fig. 8. There one laser and three independent outputs: one represents long baseline free-mass antenna, two others - resonance bars with optical F.P.-cavity. For measurement time, $\hat{\tau}_m \simeq \hat{\tau}$, the free-mass output is by $(L/\ell) \sim 10^3$ times more sensitive than bar's outputs.

But for large integrating time, $\hat{\tau}_2 \simeq Q/\omega_\mu$, the sensitivity of the bar's output may be increased by $(Q/\omega_\mu\hat{\tau})^{1/2} \sim 10^3$ times ($Q \simeq 10^6$, $\omega_\mu\hat{\tau} \sim 1$).

So, it is reasonable to study the sensitivity of such a combined antenna more carefully.

A gravitational wave pulse will cause optical phase shift with the total spectrum

$$\Delta\Psi_S(\omega) = \frac{N_0\omega_e}{c}\left[L + \omega^2\ell\left(\omega^2-\omega_\mu^2 - j\,\omega\omega_\mu Q_\mu^{-1}\right)^{-1}\right] h_s(\omega) \qquad (17)$$

where $h_s(\omega)$ is the spectral amplitude of the gravitational wave.

Three main sources of fluctuations with constant spectral power densities perturb the combined antenna:

the resonance-bar thermal noise, G_B;

the light beam noise, G_L;

and the noise of the light pressure, G_p, so that:

$$G_B = \frac{2}{\pi}\,\frac{kT\omega_\mu}{m\,Q_\mu}, \qquad G_p = \frac{2\hbar\omega_e W_0 n^2}{m^2c^2}, \quad G_L = A\,\frac{2\hbar\omega_e}{\eta W_0}\left(\frac{\lambda}{2\pi n}\right)^2 \qquad (18)$$

Taking into account the transfer function of the combined antenna from (17) one can convert these sourses into equivalent fluctuating metric background: $h^2_{fl}(\omega) = h_B^2(\omega) + h_p^2(\omega) + h_L^2(\omega)$

$$\text{a)}\quad h_B^2(\omega) = (G_B/L^2\omega_\mu^4)\left\{\left[(1+\ell/L)\nu^2-1\right]^2 + \nu^2/Q_\mu^2\right\}^{-1} \qquad (19)$$

$$\text{b)}\quad h_p^2(\omega) = (G_p/L^2\omega_\mu^4)\left\{\left[(1+\ell/L)\nu^2-1\right]^2 + \nu^2/Q_\mu^2\right\}^{-1}$$

c) $h_L^2(\omega) = (G_L/L^2)\,[(\nu^2-1)^2 + \nu^2/Q_\mu^2]\,\{[(1+l/L)\,\nu^2-1]^2 + \nu^2/Q_\mu^2\}^{-1}$

where $\nu = \omega/\omega_\mu$.

The maximum of a signal-to-noise ratio corresponds to the optimal filtration and is defined by the integral:

$$\mu = (2\pi)^{-1} \int_{-\infty}^{\infty} [\,|h_s(\omega)|^2 / h_{fl}^2(\omega)\,]\, d\omega \tag{20}$$

Let's take a gravitation wave pulse in the form $h_s(t) = h_0 \cos\omega_\mu t,\ -\tau \le t \le \tau,\ (\omega_\mu \tau/2\pi) \gtrsim 1$, so that $h_s(\omega) = (h_0\tau/\pi)\,[\sin(\omega_\mu-\omega)\tau/(\omega_\mu-\omega)\tau]$. If one does not consider large optical power, $W_0 \lesssim 1\,W$, then $G_B \gg G_P$ and there are three different situations for estimating the integral (20).

1. First. $(G_B + G_P)\,Q_\mu^2 \ll G_L\,\omega_\mu^4$, or $G_B\,Q_\mu^2/\omega_\mu^4 \ll G_L$, or

$$\frac{2\pi}{\lambda}\, n_0 \left(\frac{kT}{m\omega_\mu^2}\right)^{1/2} \ll \left(A\, \frac{2\hbar\omega_0}{\eta W_0} \cdot \frac{\omega_\mu}{Q_\mu}\right)^{1/2} \tag{21}$$

That means the spectral density of light noise exceeds the response peak density of thermal bar noise. The integral (20) results in:

$$\mu \simeq \frac{L^2 h_0^2 \tau}{G_L}\left[1 + \frac{l^2}{L^2}\, Q_\mu (\omega_\mu \tau/2\pi)\right] \tag{22}$$

The first term in Eq. (22) corresponds to a free-mass antenna response, the second, to the resonance-bar reaction. So, as $(\omega_\mu\tau/2\pi) \sim 1$, the condition of "equivalent sensitivity" $Q_\mu (l^2/L^2) = 1$ is fulfilled for typical parameters: $Q_\mu = 10^6$, $l = 3.10^2$cm, $L = 3.10^5$cm. However, the condition of the first case (21) might be satisfied at room temperature, if W_0 = 1mW, $A = 10^4$, $n = 10^2$. Weakness of the light noise, or increasing of the reflection number might be compensated only by the bar cooling; but here the reserve is not large: for $n = 10^3$ the temperature 3 K is needed.

The filtration process corresponding to Eq. (22) supposes a wide bandwidth channel, $\Delta f = 10^3$ Hz to observe a free-mass antenna response and a narrow bandwidth channel, $\Delta f = \delta_\mu/\pi$ (time of measurement, $\tau_m \sim Q_\mu/\omega_\mu$) for bars response. The band frequency reduction permits one to increase the bar antenna sensitivity up to the sensitivity of free-mass antenna. However, the total "equivalent sensitivity" level is not higher than $h_{min} = 10^{-17}$ for T = 300 K and $h_{min} = 10^{-18}$ for T = 3 K.

2. Second. $(G_B + G_P) \gg G_L\,\omega_\mu^4$, or $G_B/\omega_\mu^4 \gg G_L$,

or

$$\frac{2\pi}{\lambda} n_0 \left[\frac{KT}{m\omega_\mu^2} \cdot \frac{1}{Q_\mu} \right]^{1/2} >> \left(A \, \frac{2\hbar\omega_e}{\eta W_0} \cdot \omega_\mu \right)^{1/2} \tag{23}$$

This situation is contrary to the previous one: the spectral density of the light noise is less than the thermal spectral density at fringes of the resonance curve. We have from (20):

$$\mu \simeq \frac{L^2 h_0^2 \tau}{G_L} \cdot \frac{G_L \omega_\mu^4}{G_B} \left[1 + \left(\frac{\ell}{L}\right)^2 \omega_\mu \tau \right] \tag{24}$$

The filtration procedure (24) results in the "cutting out" of resonance thermal noise components ("noise whiting" operation) and after that gathering signal-noise in bandwidth $\Delta f = \tau^{-1}$. Thus, if $\omega_\mu \tau \sim 1$ the sensitivity is limited by thermal noise and it may achieve $h_{min} \simeq (10^{-19}-10^{-20})$ for T = 300 K - 3 K. In principle, for very short pulses ($\omega_\mu \tau \ll 1$) it is possible to realize the regime of "equivalent sensitivity" due to expanding bandwidth to $\Delta\omega = \tau^{-1} = (L/\ell)(G_B/G_L \omega_\mu^2)^{1/2}$. However, the second case has a small probability; so to fulfill Eq.(23) one has to suppose that $\eta = 10^2$, $W_0 \gtrsim 1_w$, and $A = 1$.

3. Third. $(G_B+G_P) Q_\mu^2 >> G_L \omega_\mu^4$, but $(G_B+G_P) \lesssim G_L \omega_\mu^4$
These inequalities are contrary to conditions (21) and (23). That means the spectral density of the light noise is less than the resonance peak density of the thermal noise, but larger than thermal density at fringes. It is a more practical, and thus, a more interesting case. The signal-to-noise ratio (20) is:

$$\mu = \frac{L^2 h_0^2 \tau}{\pi G_L} \left[1 + \frac{\omega_\mu \tau}{2\pi} (\ell/L)^2 \left(G_L \omega_\mu^4 / G_B\right)^{1/2} \right] \tag{25}$$

For "equivalent sensitivity" regime one has to compensate the small ratio, $(\ell/L)^2 \sim 10^{-6}$ admitting that $G_L \omega_\mu^4 / G_B \simeq 10^{12}$. However, this value has to be decrease by at least one order to satify the first condition of the case, $G_L \omega_\mu^4 / G_B \ll \ll Q_\mu^2$, if $Q_\mu \gtrsim 10^6$. That means the high level of "equivalent sensitivity" for the resonance-bars response might be achieved only approximately with accuracy of 0.3 inside the bandwidth $\Delta\omega \simeq 2\delta_\mu (G_B Q^2 / G_L \omega_\mu^4)^{1/2}$, meanwhile, moderate sensitivity levels might be obtained exactly ($Q_\mu < 10^6$). The total possible performance is presented in Table 1 ($\eta_0 = 10^2$, $\Delta f = 10^3$ Hz).

Now we can draw some conclusions. It is possible to construct a combined antenna at the sensitivity level, $h_{min} \sim 10^{-18}$ without deep cooling of bar components (T $\gtrsim$ 70 K°). At helium temperature one can achieve the sensitivity $h_{min} \sim 10^{-19}$. Some technical advantages might be associated with a "transformator of a displacement", a niobium diaphragm attached to the bar's end (see Kulagin, Polnarev, Rudenko 1986). Then the second case considered above would become a real one with sensitivity $h_{min} \sim 10^{-20}$. =

The reasons for the combined antenna proposal are the following:

i) the multiplicative response - three independent outputs - provides a better resistance to nongravitational perturbations,

ii) the possibility to discover a wave-character of a perturbation by measuring the delay time of bar-component reactions. Due to acoustical memory one needs to measure only a phase-shift between two relaxation tails of bars; a technique for that is well known. If $L = 3.10^5$ cm, the phase-shift is of the order of $\Delta\psi = \omega_\mu L/c = 10^{-1}$ rad. The upper estimation by Rao-Kramer inequality (see Van Tress 1968) defines the measurement accuracy, $(\Delta\psi^2)^{1/2} \gtrsim \mu^{-1/2}$ thus, $\mu \gtrsim 10^2$ is demanded. So one can definitely register the delay time effect one order of magnitude worse than the absolute sensitivity presented by (22-)-(23).

Finally, we should like to mention that the addition of bar-components to a free-mass laser gravitational antenna does not, practically, increase the project cost, if deep cooling is not used: but when it is used the cost will probably be increased, but not more than 20% of the total amount. It is interesting to emphasize also that a combined antenna is not equivalent to the two antennae of Weber type under operation at present, because only refined interferometric optical readout system is satisfactory for the sensitivity required.

III. QND-MEASURING TECHNIQUE FOR GRAVITATIONAL WAVE EXPERIMENT

ABSTRACT. The problem of quantum non-demolition measurement in application to gravitational antennae has been widely discussed during the last few year. It has been realized that there were no principal limitations for detection of an arbitrarily small force distributed on a probe-oscillator. Accordingly, the traditional measuring procedure has to be changed. The general principles of the new procedure (QND-measurement) were formulated, and a number of gedanken experimental devices were proposed. Almost all of them explored a non-classical state of oscillator, so called "squeezed state", with non-equal noise in its quadrature amplitudes. Recently, the first experiments were carried out where a squeezed state of an electromagnetic field was observed. It is reasonable to expect, in the near future, progress in the squeezed state technique. From this reasoning, a new complete QND scheme, including a bar detector, an optical Fabry-Perot cavity read out system and a meter, measuring a quadrative amplitude of output e.m. field, is proposed. A detectable value of force in this system is by "g" times smaller than the standard quantum limit, where "g" is the squeezing factor.

Let us come back to the resonance-bar antenna to discuss the problem of quantum limit of sensitivity. Our goal in this lecture is to propose and analyze some closed scheme, which permis precise measurement of the force (equivalent to GW-bursts) applied to the gravitational detector. But at first, we shall give a brief outline of QND-measurements.

The problem of quantum limits of sensitivity in macroscopical experiments with test bodies (in particular, GW experiment) is now well realized (see review articles: Caves, Thorne, Drever et al. (1980) and Braginskii, Vorontsov, Thorne (1980)). An experimental device consists, as a rule, of a probe-oscillator, a transducer to convert mechanical energy into electromagnetical form, an electronic amplifier with some recorder. It is supposed that a probe-oscillator is excited by some force which is to be measured. If a transducer picks up, continuously, the oscillator's coordinate, $x(t)$, the "SQL", standard quantum limit of sensitivity, arises (ξ -, statistical factor):

$$\Delta x_q = \xi \left(\hbar / 2m\omega_\mu \right)^{1/2} ; \quad F_q = \xi \, \tau^{-1} \left(m \hbar \omega_\mu \right)^{1/2} \tag{26}$$

Δx_q is the "absolute limit"; it reflects the absence of body's trajectory in quantum mechanic. But the limit for value of force is not obligatory; it only results from the failure of measuring procedure, when one tries to measure the observable force variations of the oscillator's coordinate.

Caves, Thorne, Drever et al. (1978) were the first who pointed at the necessity to measure the oscillator's quadrature amplitudes, X_1 and X_2, so that

$$x(t) = X_1 \sin \omega_\mu t + X_2 \cos \omega_\mu t \tag{27}$$

then any principal limit for measuring value of force is absent. Simultaneously, Braginskii, Vorontsov and Khalili (1978) proposed a stroboscopic measuring procedure: one had measured oscillator's coordinate only at definite moments of time: $\omega_\mu t_i = \pi n$, or $\omega_\mu t_j = (\pi/2)(2n+1)$. It is obvious that the proposal is the particular case of quadrature amplitudes measurement. It is interesting to emphasize now that both procedures mean operation with oscillator in special quantum state, so called "squeezed state". The last one has wave function which collapses periodically in coordinate space (or momentum space). Thus, the principal role of "squeezed states" was realized already at first steps of non-demolition measurement theory.

The second link of antenna setup, the transducer (or energy transformer) is also an oscillator, but of electromagnetical nature. So, one has to measure the field quadrature amplitudes, E_1 and E_2 . Besides, one has to provide a special interaction between mechanical and electromagnetical degrees of freedom, so that E_1 and E_2 would depend on only one of quadrature amplitudes, X_1 or X_2 , i.e. $E_i = E_i(X_j)$, $i,j = 1.2$.

The last step of experimental setup, the measuring device, or simply, the measurer, is the link where information is read out really, and where the process of dequantization takes place: all subjects are treated classically after that step. It is a more difficult point of description. One needs to define the inner structure of a measurer by taking into account measurer's back action of the foregoing links. That back action has to be good enough in order not to destroy QND properties of measurement.

Only the two first cascades, a probe-oscillator and a transducer, used to be considered in quantum mechanics framework (see reviews by Caves et al. (1980); Braginskii et al.(1980), and Milburn, Lane, Walls (1983)). The action of the last link, the measurer, was modelled by gedanken state reduction of the system. At present, however, some comprehension principles and possible organization of a QND measurer occurred in sphere of quantum optics. Papers by Shapiro (1980), Yurke, Denker (1981), and Yurke (1985) described general ideas, and gave some examples of a device for QND measurement of quadrature amplitude of e.m. field. The main feature of such device is a squeezed state of its back action radiation field. It measures a definite quadrature amplitude of field in some signal channel, and sends back the noise field squeezed at conjugate quadrature amplitude. Thus, for the resonance bar detector with optical F.P.-cavity we have possibility now to consider a complete closed measuring setup. This step must have all three structure elements: a probe-oscillator, an optical transducer and QND-measurer (see Fig. 9).

The QND-measurer consists of three circulators, homodyne detector and two degenerate parametric amplifiers (or four-wave-mixers) with different pump frequencies. The homodyne detector measures the quadrature component of the optical readout system output radiation, two degenerate parametric amplifiers (or four-wave-mixers) squeeze in a special way the vacuum fluctuations of the load and the circulators direct the fields into appropriate branches of the system. More concretely, the system output field is directed through circular C1, C2 and C3 into the load (absorber black body with zero temperature) and dissipates in it. Zero fluctuations of the load enter through circulator C3 the degenerate parametric amplifier DPA2 with pump frequency $2(\omega_0+2\Delta)$; $[\omega_0 \equiv \omega_e]$ and bandwidth $\Delta\omega \lesssim |\Delta|$, then get through circulators C3 and C3 into DPA1 with pump frequency $2\omega_0$ and bandwidth $\Delta\omega \lesssim |\Delta|$, and then get through circulators C2 and C1 into the system. Detailed analysis of this model is carried out in our paper, Kulagin, Rudenko (1987); here we briefly discuss only the main points.

The system of motion equations (Fig. 9) is:

$$\ddot{\hat{x}} + \omega_\mu^2 \hat{x} = -2\gamma\,[\hat{E}_1 \sin(\Delta t+\varphi) + \hat{E}_2 \cos(\Delta t+\varphi)] + f_g(t)$$

$$\dot{\hat{E}}_1 + \delta_e \hat{E}_1 + 2\beta \hat{X}\cos(\Delta t+\varphi) = 2\delta_e \hat{E}_{b1} \qquad (28)$$

$$\dot{\hat{E}}_2 + \delta_e \hat{E}_2 - 2\beta \hat{X}\sin(\Delta t+\varphi) = 2\delta_e \hat{E}_{b2}$$

$$\beta = \frac{\omega_0 E_0}{2\ell}; \quad \gamma = \frac{S E_0}{4\pi m t_0^2}; \quad \Delta = \omega_p - \omega_0$$

where E_0, the amplitude of the field inside optical resonator under the action of "laser force" with mirrors fixed; S, the cross-section of the beam; m, ω_μ - the mass and frequency of mechanical oscillator; ℓ, ω_0 - the length and eigenfrequency of optical resonator; δ_e - , the damping of the optical resonator due to the right mirror (with $R < 1$); $\hat{x}$ -, the coordinate operator of moving mirror; ω_p -, the frequency of the pump; $f_g(t) = F(t)/m$, the external (gravitational) force); $\hat{E}_1$, $\hat{E}_2$ and $\hat{E}_{b1}$, $\hat{E}_{b2}$ -, the quadrature component operators of the measuring field, $\hat{E}$

$$\hat{E} = \hat{E}_1 \cos\omega_0 t + \hat{E}_2 \sin\omega_0 t \qquad (29)$$

and back action field, $\hat{E}_{ba}$

$$\hat{E}_{ba} = \hat{E}_{b1}\cos\omega_0 t + \hat{E}_{b2}\sin\omega_0 t \qquad (30)$$

with continuum commutation relations (Caves 1982.)

Laser field inside the cavity has the form:

$$E_p = E_0 \sin(\omega_p t + \varphi), \qquad (31)$$

φ is now the arbitrary phase, and we take, for simplicity, $|t_{input}| \to 0$, but $E_L |t_{input}|$ = const; E_L -, the input laser field.

The system like (28) was considered earlier with $\hat{E}_{b1}$ and $\hat{E}_{b2}$ to be the white stationary noise (Gisev, Rudenko (1979)).

In our analysis E_{bq} is squeezed noise with two points of squeezing, ω_0 and $\omega_0+2\Delta$. Let the back action noise have the form:

$$\hat{E}_{bq} = \hat{E}_c \cos\omega_0 t + \hat{E}_s \sin\omega_0 t + \hat{E}_{2c}\cos(\omega_0+2\Delta)t + \hat{E}_{2s}\sin(\omega_0+2\Delta)t \tag{32}$$

where $\hat{E}_c, \hat{E}_s$; $\hat{E}_{2c}, \hat{E}_{2s}$ – , the quadrature components of fluctuations in narrow bandwidths ($\Delta\omega \lesssim |\Delta|$) near left and right sidebands which are detuned from the pump frequency ω_p: $\omega_0 = \omega_p - \Delta$, and $\omega_0 + 2\Delta = \omega_p + \Delta$ (the signal and idle frequencies, in terms of the theory of parameterical amplifiers in radio engineering).

By taking into account that $R = -R_0 \simeq -1$, we get for the measuring field:

$$\hat{E}_m = (\hat{E}_1 - \hat{E}_{b1})\cos\omega_0 t + (\hat{E}_2 - \hat{E}_{b2})\sin\omega_0 t \tag{33}$$

Using the spectral notion of multimode signals we get from Eq. (28-33), for the measured variable $[\hat{E}_2(\omega)]_m$ the near zero frequencies: ($\omega \sim 0$, $\psi = \beta\gamma$)

$$[\hat{E}_2(\omega)]_m = \hat{E}_2(\omega) - \hat{E}_{b2}(\omega) =$$

$$\begin{aligned}
&= \frac{E_s(\omega)}{\delta_e - i\omega}\left[\delta_e + i\omega + i\frac{2\delta_e\psi}{\delta_e - i\omega}\left(\frac{1}{d_+} - \frac{1}{d_-}\right)\right] - \\
&- \frac{2\delta_e\psi\,\hat{E}_c(\omega)}{(\delta_e - i\omega)^2}\left(\frac{1}{d_+} + \frac{1}{d_-}\right) + \\
&+ \frac{2\delta_e\psi\,\hat{E}_{2c}(\omega)}{\delta_e - i\omega}\left[\frac{e^{i2\varphi}}{d_+(\delta_e - i(\omega+2\Delta))} + \frac{e^{-i2\varphi}}{d_-(\delta_e - i(\omega-2\Delta))}\right] + \\
&+ i\frac{2\delta_e\psi\,E_{2s}(\omega)}{\delta_e - i\omega}\left[\frac{e^{i2\varphi}}{d_+(\delta_e - i(\omega+2\Delta))} - \frac{e^{-i2\varphi}}{d_-(\delta_e - i(\omega-2\Delta))}\right]
\end{aligned} \tag{34a}$$

$$d = d(\omega) = \omega_\mu^2 - \omega^2 - \frac{4\psi\Delta}{(\delta_e - i\omega)^2 + \Delta^2}\ ; \quad d_\pm = d(\omega \pm \Delta) \tag{34b}$$

where $\hat{E}_c(\omega), \hat{E}_s(\omega)$ and $\hat{E}_{2c}(\omega), \hat{E}_{2s}(\omega)$ belong to different not overlapping frequency pieces of back action noise. Their correlation matrix can be (see Caves 1982,1985) ($q, p = s, c$; for simplicity, we assume that $\langle \hat{E}_p(\omega) \rangle = 0$):

$$\begin{aligned}
&\langle \hat{E}_s^+(\omega)\hat{E}_s(\omega') + \hat{E}_s(\omega')\hat{E}^+(\omega) \rangle = 2\pi(N_0/g_0)\,\delta(\omega - \omega') \\
&\langle \hat{E}_c^+(\omega)\hat{E}_c(\omega') + \hat{E}_c(\omega')\hat{E}_c^+(\omega) \rangle = 2\pi(N_0/g_0)\,\delta(\omega - \omega')
\end{aligned} \tag{35}$$

$$< \hat{E}_p(\omega)\, \hat{E}_q(\omega) > \; = 0$$

$$< \hat{E}_c^+(\omega)\, \hat{E}_s(\omega') + \hat{E}_s(\omega')\, \hat{E}_c^+(\omega) > \; = \; < \hat{E}_s^+(\omega) \hat{E}_c(\omega) + \hat{E}_c(\omega) \hat{E}_s^+(\omega) > \; \simeq 0$$

where we use the fact that $2|\Delta| \simeq 2\omega_\mu \ll \omega_0$. The same formulas valid for $\hat{E}_{2c}$ and $\hat{E}_{2s}$ with obvious substitution, $g_0 \to g_2$, and correlations of $\hat{E}_c\ \hat{E}_s$ and $\hat{E}_{2c}, \hat{E}_{2s}$ are zero. Factors g_0 and g_2 depend on the structure of degenerate parametric amplifiers: g_0 is the gain of DPA 1 and g_2 is the gain of DPA 2.

Using Eqs. (34,35) one can calculate signal-to-noise ratio for the system. Let the external (gravitational) force have the form of a short radio-pulse, $f_g(t) = f_0 \sin\omega_\mu t$, $0 < t \le \hat{\tau}$, $\omega_\mu \hat{\tau} \gtrsim 2\pi$ then for the signal in $[\hat{E}_2(\omega)]_m$ we obtain:

$$[E_2(\omega)]_s \simeq i \frac{f_0 \hat{\tau} \beta}{4(\delta_e - i\omega)} \left(\frac{1}{d_-} - \frac{1}{d_+} \right) \tag{36}$$

The signal-to-noise ratio will be estimated according to optimal filtration formula:

$$\mu = \pi^{-1} \int_\Omega \frac{|E_2(\omega)|_s^2 \, d\omega}{[N_2(\omega)]_m} \tag{37}$$

where $N_2(\omega) \to 2\pi [N_2(\omega)]_m\, \delta(\omega - \omega') =$

$$= \frac{1}{2} < [\hat{E}_2(\omega)]_m [\hat{E}_2^+(\omega')]_m + [\hat{E}_2^+(\omega')]_m [\hat{E}_2(\omega)]_m >$$

Ω -, the bandwidth of interest.

We assume that a photodetector is ideal, i.e. quantum efficiency $\eta \simeq 1$ and the statistics of arriving field and photocurrent are identical, then the homodyne detector measures one of the quadrature phase of the falling field, and does not add any own noise to the existing one (Yuen, Shapiro 1980).

The measurement would be quantum non-demolition only, if special conditions are met. The first condition concerns the detuning Δ of the system: we must choose Δ according to the equation:

$$\frac{1}{d_+(0)} + \frac{1}{d_-(0)} = 0 \tag{38}$$

i.e. $$(\delta_e^2 + 4\Delta^2)(\omega_\mu^2 - \Delta^2) - 4\psi\Delta = 0$$

In this case contibution of $\hat{E}_c$ into $[\hat{E}_2(\omega)]_m$ vanishes at zero frequencies (see Eq. (34a)). For the contribution of $\hat{E}_{2c}$ (noise or idle frequency), to be also important at zero frequencies, we must also choose the phase φ according to the equation:

$$e^{i2\varphi} = \frac{\delta_e - i2\Delta}{(\delta_e^2 + 4\Delta^2)^{1/2}} \tag{39}$$

The physical meaning of Eq. (39) is the following: the phase φ of the laser pump is defined by the phase of DPA2 pump and the time dealy in the system.

Let us consider the sensitivity of the system in two cases: $\delta_e \gg \omega_\mu$ and $\delta_e \ll \omega_\mu$. In the first case, $(\delta_e \gg \omega_\mu)$, the optimal squeezings are:

$$(g_2)_{opt} \simeq (g_0)_{opt} = g \tag{40}$$

and the minimal detectable force with optimal value g is (clearly, Eq. (26)):

$$F_{min} = F_q\, g^{-1/2} \tag{41}$$

In the second case, $(\delta_e \ll \omega_\mu)$ Eq. (41) is also correct, but optimal squeezings are:

$$(g_0)_{opt} \simeq \frac{\omega_\mu^2}{\delta_e^2} (g_2)_{opt} = g \tag{42}$$

because the narrow resonance curve of the optical resonator decreases the fluctuations at idle frequency $\omega_0 + 2\Delta$, $|\Delta| \simeq \omega_\mu \gg \delta_e$ Thus, we show that the minimal detectable force is smaller than the standard quantum limit (26), if we can achieve considerable squeezing g. It must be stressed that δ_e can be greater than ω_μ, but Eq. (41) is valid, in contract to the case of modulating pump (see reviews Caves et al. (1980), and Braginskii et al. (1980)).

Losses in the mechanical system and the optical resonator will destroy the squeezed states. If the temperature of Weber resonator with quality factor, Q_μ, is zero, then the minimal detectable force becomes

$$F_{min} = F_q\, Q_\mu^{-1/4} \tag{43}$$

For expression (43) to be valid, we must meet the condition

$$\delta_d < \delta_e / g_m$$

where $g_m = max\,(g_0, g_2)$ and δ_d -, the damping of optical resonator when $R_0 = 1$ (losses in mirrors, diffraction and so on. For $Q_\mu \simeq 10^8$ (sapphire resonator) the gain over standard quantum limit can be about $\sim 10^{-2}$.

Let's makes some conclusions. Today we can construct a reasonable closed setup (a gravitational bar-antenna) with actually unlimited precision for measuring force (i.e.

for GW bursts). The question is only of practical possibility to generate squeezed state of e.m. field with the factor q large enough. The first experimental steps in that direction have already been made (Yamamoto, Imoto, Machida 1986; Levenson et al. 1986). Of course, our measuring device will be also good for a bar-antenna with modulating pump, which certainly will result in formula (41) (see papers Johnson, Bocko 1981, Bocko,Bordoni et al.,1986; Bordoni, Panfilis et al. 1986).

REFERENCES

Bocko M.F., Bordoni F.,Fuligni F., Johnson W.W. (1986) In:"Noise in Physical Systems and Noise 1985, p.47; ed. A.D'Amico and P.Hazzerri, Elsevier Science Publishers B.V.

Bordoni F., De Panfilis S., Fuligni F., Iafolla V., Nozzoli S. (1986). Proc. of Fourth M.G.H. on G.R. ed. R.Ruffini. Elsevier Science Publishers B.V.

Boughn S.P., Kuhn J.R. (1984) Ap.J. 286, 387.

Braginskii V.B., Vorontsov Yu.I., Thorne K.S., 1980, Science 209, 547.

Braginskii V.B., Vorontsov Yu.I., Khalili F.Ya. (1978) Soviet Physics - JETP Letters 27, 276

Braginskii V.B., Gusev A.V., Mitrofanov V.P., Rudenko V.N., Yakimov V.N. (1985), Uspekhi Fiz.Nauk, 147, 422 (in Russ.

Braginskii V.B., Mitrofanov V.P., Panov V.I. (1985). Systems with small Dissipation. University of Chicago Press Chicago.

Caves C.M., Thorne K.S., Drever R.N.P., Sandberg B.D., Zimermann M. (1980) Rev.Mod.Phys. 52, 341.

Drever R.W.P., Weiss R., Linsay P.S., Saulson P.R., Spero R., Schutz F. (1985). A detailed engineering design study and development and testing of components for a laser interferometer gravitational wave observatory. CalTech: Pasadena, California, and MIT: Cambridge,Mass.

Dyson F.J. (1969), Ap.J. 156, 529.

Forward Forward F.L., Zipoy D., Weber J., Smith S. and Benioff (1961) Nature 189, 473.

Gusev A.V., Rudenko V.N. (1976). Radio Eng. and Electron. Phys. 21, 40

Gusev A.V., Rudenko V.N. (1979). Sov.Phys.JETP 49, 755.

Hirakawa H. (1973). J.Phys.Soc.Japan 35, 295.

Johnson W.W., Bocko M. (1984). Phys.Rev.Lett.,47P,1184.

Kulagin V.V., Polnarev A.G., Rudenko V.N. (1986). Zh.Exp. Teor.Fiz.,91, 1553

Levenson M.D., Shelby R.M., Reid M., Walls D.F. (1986). Phys.Rev.Lett., 57, 2473.

Milburn G.J., Lane A.S., Walls D.F. (1983). Phys.Rev. A27, 2804.

Saden D., Menachen A.B., Meidav (1972), Preprint Tel-Aviv University.
Sadovskii M.A. (1985). Uspekhi Fiz.Nauk 147, 421 (and refs)
Sadovskii M.A. (1986). Bull.Acad.Scien.USSR,4, 3 (in Russ.).
Savino J., McCamy K., Hade G. (1972). Bull.Seismol.Amer. 62, 141.
Shapiro J.H. (1980). Optics Letters 5, 351.
Thorne K.S., Drever R.W.P., Caves C.M., Zimmermann M. and Sandberg V.D. (1978). Phys.Rev.Lett.,40, 667.
Thorne K.S. (1987). Gravitational radiation. CalTech Preprint, GRP-109.
Van Treos H. (1968). Detection, estimation and modulation theory. ed. J.Wiley and Sons, New-York,London-Sydney.
Weber J. (1967). Phys.Rev.Letts,18, 498.
Wiggins R.A., Press F. (1969). J.Geophys.Res.,14, 5351.
Yamamoto Y., Imoto N., Machida M. (1986). Phys.Rev. A33, 3243.
Yuen H.P., Shapiro J.H. (1980). IEEE Trans. IT-26, 78.
Yurke B., Denker J. (1984). Physica,108B, 1359.
Yurke B. (1985). J.Opt.Soc.Am.B. 2, 732.
Kulagin V.V., Rudenko V.N. (1987). "Nondemolishing measurement in sistems with spectral selected squeezing" to be published in Zhurnal Eksperimentalnoi i teoreticheskoi fiziki (November - December).
Caves C.M. (1982) Phys.Rev.D 26, 1817.

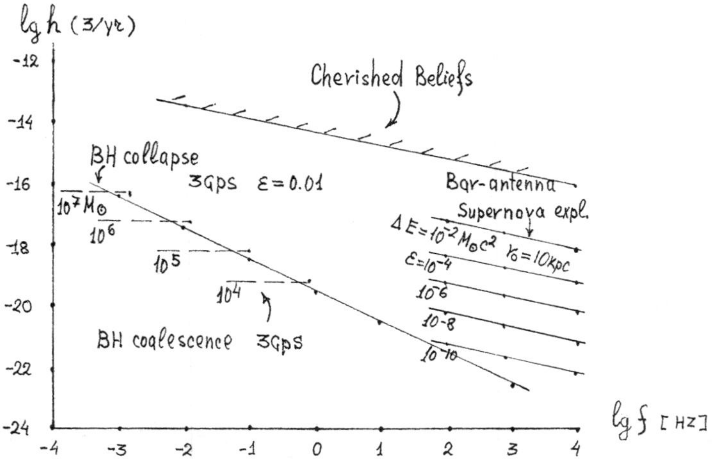

Fig. 1 Astrophysical forecast for GW-bursts(Thorne 1987). The estimation of "h" corresponds to the events, which occures three times per year. The right side represents the frequency region of a bar-antenna. The long inclined line below gives events at the Hubble distance 3 Gpc. The line "Cherished Beliefs" is an optimistic prediction level.

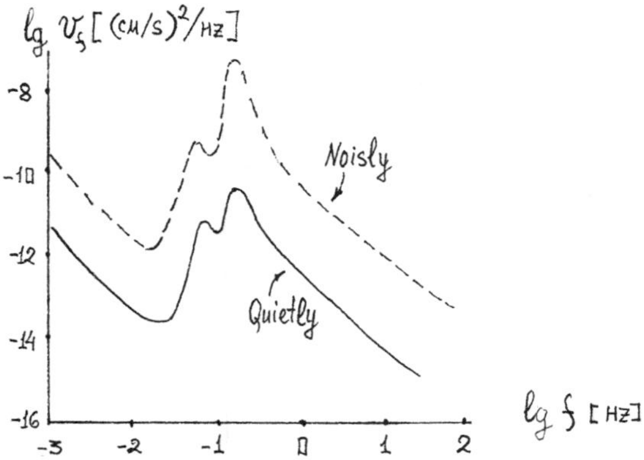

Fig. 2 The spectral density of seismical noise (velocity of points on the Earth surface squared per unit of frequency band). There is downfall, corresponding to period of 30 - 40 sec.

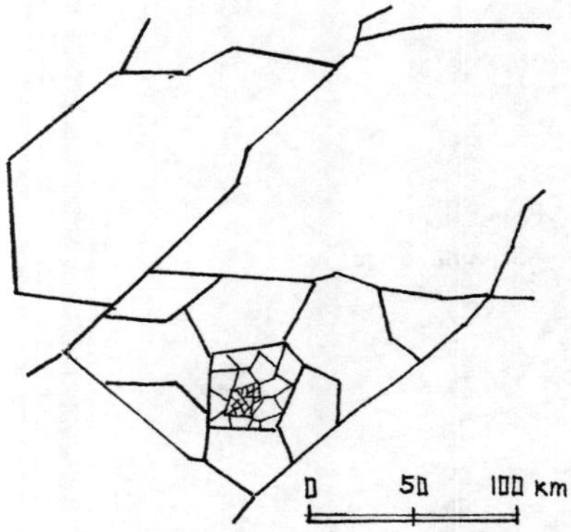

Fig.3 A scheme of Earth crust structure. Large blocks consist of similar cells by smoller scale, enclosed one into another.

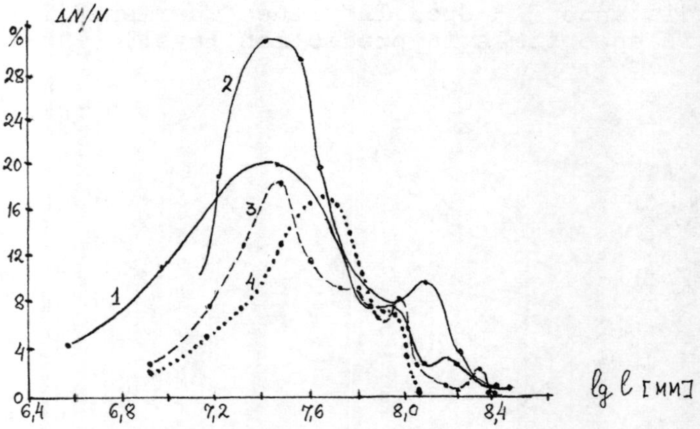

Fig.4 A block size distribution on different regions

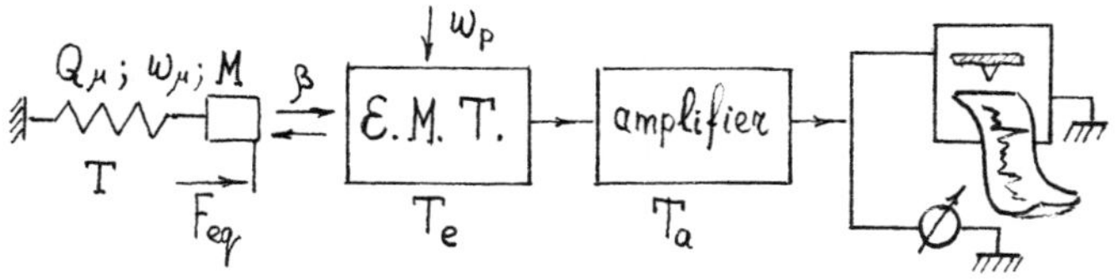

Fig.5 A block-scheme of a bar-antenna: a prob-oscillatore (a gravitational detector); E.M.T. - an electro-mechanical transducer; an amplifier and a recorder. The equivalent noise temperatures are marked as T, T_e , T_a .

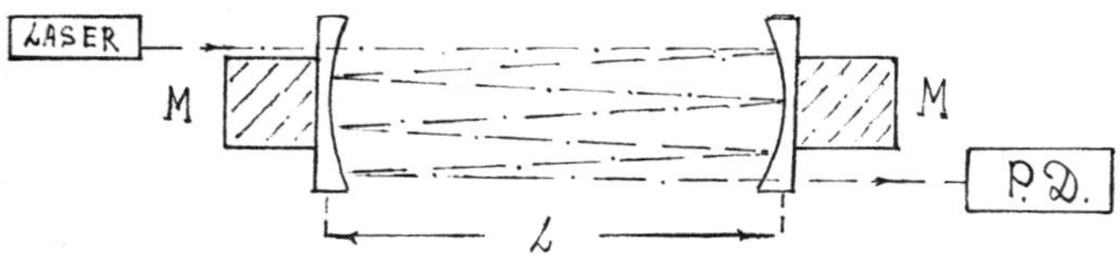

Fig.6 A schematic diagram of a laser interferometric free mass antenna (a beam-antenna): P.D. - photodetector; M - test masses with mirrors of F.P.-cavity; L - base distance.

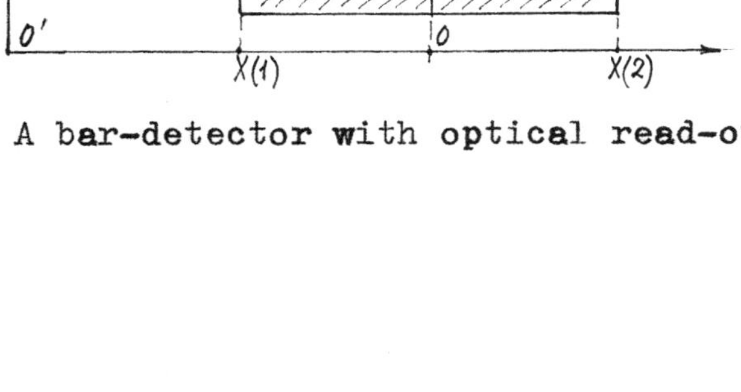

Fig.7 A bar-detector with optical read-out sistem.

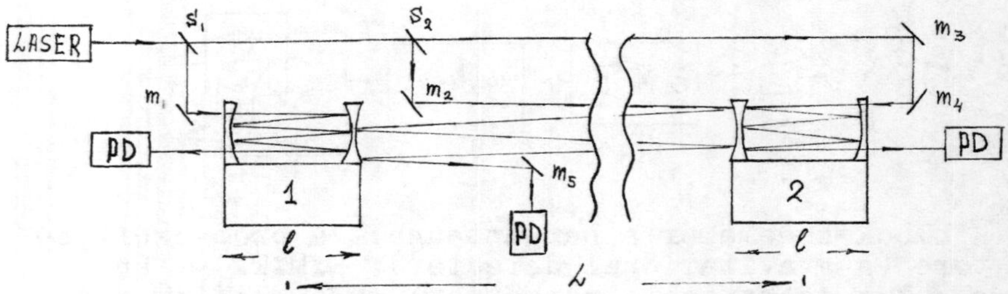

Fig.8 A schematic diagram of combined optoacoustical gravitational antenna: PD - photodetectors; $S_{1,2}$ - beam splitters; $m_{1,2}$ - mirrors to direct beams; 1, 2 - bar-detectors with FP-cavities.

TABLE 1

Parameters \ Data	G_L	$\frac{G_B}{\omega_\mu^4}$	$\xi=\sqrt{\frac{G_L \omega_\mu^4}{10^{12} G_B}}$	h_{min}
$A = 10^4$ $W_o = 10^{-3}W$ $T = 300K$ $Q = 10^5$	$2.5\cdot 10^{-25}$	$3\ 10^{-37}$	1.0	$5\ 10^{-17}$
$A = 10^4$ $W_o = 10^{-1}W$ $T = 70\ K$ $Q = 5\ 10^6$	$2.5\ 10^{-27}$	$2\ 10^{-38}$	0.4	$5\ 10^{-18}$
$A = 10^2$ $W_o = 10^{-1}W$ $T = 3\ K$ $Q = 10^7$	$2.5\ 10^{-29}$	$3\ 10^{-40}$	0.3	$5\ 10^{-19}$
$A = 1$ $W_o = 10^{-1}W$ $T = 3\ K$ $Q = 3\ 10^8$	$2.5\ 10^{-31}$	$3\ 10^{-42}$	0.3	$5\ 10^{-20}$

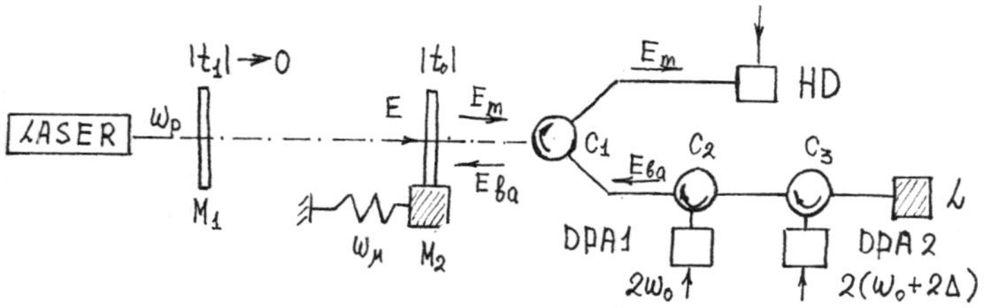

Fig.9 A possible closed QND scheme for bar-antenna with FP-cavity as read-out sistem: M_1 - input fixed mirror; M_2 - output mirror , attached to a probe-oscillator (model of a bar-antenna ;; a measuring device consists ofcirculators C_1,C_2,C_3, narrow bandwidth degenerate parametric amplifiers DPA1, DPA2, homodyne detector HD and black body load L.

INTRODUCTORY LECTURES ON THE PHYSICS OF HIGH ENERGY DENSITIES: THEORIES, MODELS, MEASUREMENTS

V.P. Shelest, M.Yu. Novikov, A.S. Sorin
USSR State Committee for Standards
Leninsky prospekt 9
Moscow 117 049
USSR

ABSTRACT. These lectures are devoted to a review of the contemporary high energy density physics. Pre-quark models and the quark conception are briefly treated, and a sketch of quantum chromodynamics is given. Semi-phenomenological models of hadron matter under extreme conditions are then described, i.e. lattice calculations, quark-gluon bag model, hydrodynamic theory of multiple production, and experiments to detect signals of quark-gluon plasma formation. The last lecture looks at confinement criteria, thermodynamics of the quantum field model with confinement (QED_3), and the question of whether there is deconfinement in this model.

0. PRELIMINARIES

The last three decades saw a number of considerable achievements in understanding the entities of which the Universe is constructed. In these outline lectures we would dwell on some of these, as well as on new, and at present only hypothetical conceptions created over the recent years due to the study of the structure of strongly interacting matter at high energy densities. The inventory of issues we raised and that stand at the forefront of contemporary physics of elementary particles is by no means complete and to some extent it mirrors the authors' own scientific interests. Not all of these questions have been equally thoroughly studied or more or less settled. Moreover, some of these are likely to be reviewed in future, and we attempt to give a critical analysis of them. In our exposition we did mention the names of some pioneering researchers without pretending to give a historical background of this branch of knowledge that at present is forcefully developing.

V. De Sabbata and V. N. Melnikov (eds.),
Gravitational Measurements, Fundamental Metrology and Constants, 387–442.

LECTURE 1. FROM PRE-QUARK MODELS TO QUANTUM CHROMODYNAMICS

1.1. Experiment in the subnuclear physics and the pre-quark models

1.1.1. Now we would describe in general the picture produced by experiment in the subnuclear physics and provide a step-by-step exposition of the theory describing the world of elementary particles.

One of the essential achievements by the twentieth-century physics has been the discovery that the atom is not indivisible or elementary but has a composite structure. In 1911 Rutherford demonstrated that the atom consists of a small positively charged nucleus surrounded by a cloud of electrons. Then it was discovered that the nucleus itself could be decomposed into its constituents, the nucleons (the proton and the neutron). Vigorous development in high-energy physics over the last three decades related to the construction of particle accelerators has led to the discovery of a great number of new elementary particles. Most of these are short-lived strongly interacting particles, the so-called resonances. In 1960 , it was found (by Hofstaedter) that the proton is an extensive, and not a point-like object with the root mean square charge radius of approximately 1 F. The experiments carried out at the Stanford Linear Accelerator Centre on high-energy proton scattering and later experiments on head-on collision of high-energy protons in intersecting storage rings at CERN and at a number of other laboratories showed that in protons there are discrete scattering centres which do not reveal their internal structure at the levels of energy used in experiment, and which are quasi-free objects (the so-called Feynman parton model). The discovery of an eventual internal structure in the hadron constituents calls for consideraly higher energy beams. Now this is very much a thing of the future. However, from the experimental material we now have it is clear that some new fundamental essences are buried in the hadrons.

As the number of elementary particles surpasses by far that of chemical elements the very notion of an elementary particle as an essential structural element of matter has lost its original meaning. The most important problem arose of how to order into a clear-cut system this considerable amount of particles which would remind in a sense the periodical table of elements.

1.1.2. These theoretical conceptions have been developed along two ways. On the one hand, the symmetry properties of strong interactions were studied that could lead to group ing the particles into families close to each other by their properties, or multiplets. Thus it was established that strong interaction is charge-invariant, that is, independent from charge. Charge invariance makes strongly interacting particles group into isotopic multiplets, that is, families of particles differing from each other by their electromagnetic characteristics. Isotopic multiplet members are different charge states of one and the same particle. Examples of isotopic multiplets are the proton-neutron doublet, the triplet of π^+ , π^- , and π^o mesons, as well as others. But the number of such isotopic multiplets is still too large.

Later it was made clear that isotopic multiplets are grouped into large families, or unitary supermultiplets . The difference between members of a single unitary supermultiplet comprising several isotopic multiplets is essentially greater than between particles within an isotopic multiplets. However, members of different unitary supermultiplets are clearly distinct from each other. As an example of a unitary supermultiplet a baryon octet may serve with spin parity $J^p = 1/2^+$ - proton, neutron, Λ - hyperon, Σ^+, Σ^-, and Σ^o hyperons, Ξ^- and Ξ^o - hyperons.

On the other hand, the so-called composite models of elementary particles were created using which attempts were made to reduce the whole variety to bound states with a minimal number of fundamental or true elementary particles. Composite models provided for a clear explanation of strong interaction symmetry as the manifestation of symmetry in interaction between the fundamental fields. In the first composite Fermi-Yang model proton, antiproton, neutron and antineutron were regarded as the fundamental particles. Later, after strange particles were discovered, a number of composite new models sprung up, the most successful of which was the Sakata model. In this latter the fundamental particles were the proton, the neutron and the Λ -hyperon (as well as their antiparticles). All mesons were baryon-antibaryon bound states, and baryons consisted of two fundamental baryons and one fundamental antibaryon. Using this model, the existence of a number of new particles was predicted , a good description of meson unitary supermultiplets given, and a lot of phenomena in physics of weak interactions explained.

But the Sakata model met with insurmountable difficulties in the analysis of baryon unitary supermultiplets as it showed a completely erroneous baryon family structure. These difficulties were of a fundamental character and let physicists to conclude that they should either give up the very attractive minimal composite model with three baryon fundamental fields or ascribe very unusual properties to these fundamental particles. It is this second way that Gell-Mann and Zweig, the creators of the quark model, chose to follow.

1.2. The quark conception

1.2.1. The experiments we mentioned earlier and the development of theory resulted in the realisation that hadrons were composite objects. The contemporary theoretical model describing hadron structure was named the quark model. According to this model, all hadrons consist of the more fundamental constituents or quarks called so by one of the model's creators, Murray Gell-Mann.

Quarks as many other microobjects have internal properties that are invariable in strong interactions. These properties are characterised by quantum numbers whose sets represent the given object in an unambiguous way. Among them are the electric charge, the baryon charge, the spin, and others. Quarks have special properties, as for instance in all particles observed without exception the charge is divisible by the electron charge; in quarks, however, the charge is a fraction of the latter.

All hadrons known at present, according to the quark model, seem to consist of six kinds of quarks, or as they are usually known, six flavours: u, d, s, c, b, t. To obtain a correct classification of hadrons a supplementary quantum number was introduced later called 'colour' (Bogolubov & Tavkhelidze; Han & Nambu, 1965). The quark of a given flavour may find itself in three quantum states which differ from each other in colour.

The quark model combinatorics is described by two simple rules used in the construction of hadrons:

(1) baryons consist of three quarks of different colours
(2) mesons are combined out of a quark and an antiquark, so that all the three colour-anticolour pairs are represented with an equal probability.

A consequence of these rules is that no hadron has a determined colour, but all of them are bleached (white). A baryon would inevitably contain quarks in all the three colour states, for instance red, yellow, blue (when superposed, the produce white). It could be imagined that in a meson the quark and the antiquark have a certain colour and an anticolour at a given moment so that at any measurable period of time they exist for equal time in three different colour modifications, in red, yellow, blue. Combinations according to these rules suffice to obtain all hadrons observed at present and to predict new ones.

Most of the observed properties of quarks could be deduced from these rules. It is established, for instance, that each quark should have the baryon charge +1/3 and each antiquark -1/3. Therefore any set of three quarks would have the baryon charge +1, and thus define a baryon. For mesons, quark baryon charges +1/3 and -1/3 are annihilated, and thus the meson, as required, has a zero baryon charge.

Hadron spin is described in the same way if it is assumed the quark spin is 1/2. A particle consisting of an odd number of quarks such as the baryon should therefore also have a half-integer spin which stands in agreement with the known property of baryons. A particle consisting of an even number of quarks such as the meson would have an integral spin.

1.2.2. Although we do not see colour particles with our own eyes, the colour quantum number which has been introduced is no formal construction with no predictable sequels. It could be found at least in two ways:

(1) The neutral pion π^o almost always decays into two photons. The rate at which it does so depends both on the number of quark colours and their charges. Experimentally observed π^o lifetime is obtained in case the quarks should be in the three colour states.

(2) As a result of the electron-positron annihilation, sometimes a group of hadrons makes its appearance and sometimes a muon and antimuon ($\mu \bar{\mu}$). At large enough energies the probabilities ratio of hadron creation and muon-antimuon pair creation reaches constant value equal to the sum of quark charges squared. The results of experiments at energies from 2 to 3 GeV stand in reasonable agreement with the theoretical predictions if there are three colour varieties of quarks, and if this is not so, they do not agree at all.

Finally, colour quantum number eliminates the problem of quantum statistics which demands that no two particles of half-integer spin should have the same set of quantum numbers (the Pauli principle).

1.2.3. Since the quark concept has been advanced there has been an intensive search for free-state quarks going on. In accelerator experiments attempts were made to detect the creation of quarks in proton-proton collisions. The results of all experiments were negative. The cross-section of light quark creation (with mass up to 1o GeV) was no greater than $10^{-39} \div 10^{-38} cm^2$, while for heavier quarks (with mass 25 GeV) the upper cross-section limit is close to $10^{-34} cm^2$. By way of comparison we could indicate that hadron creation cross-section is usually $10^{-25} \div 10^{-30} cm^2$.

No quarks have been detected in cosmic rays. With a ninety-per cent certainty, a quark flux in cosmic rays is no greater than ($1 \div 2$) x $10^{-11} cm^{-2} sec^{-1} str^{-1}$ while the flux of cosmic protons with the energy above 1 GeV is equal to $1 cm^{-2} sec^{-1} str^{-1}$ or 10^{11} times greater than the previous value.

The occurrence of quarks on earth has been studied in a great number of substances including water, air, limestone, lava, dust, plankton, oysters, meteorites, and others. The absence of quarks has been ascertained with the accuracy

$10^{-17} \div 10^{-23}$ as to the number of protons and neutrons.

The unsuccessful attempts to detect free quarks over the last two decades produced the following alternatives for the subnuclear physics:

(1) either the quark model and all theoretical conceptions based thereon are erroneous;

(2) or dynamic laws of nature exist prohibiting free-state existence of quarks.

At present the second point of view has gained support and is now being corroborated experimentally. Quarks do exist but they are kept under lock and key and so are not observable in a free state, just like phonons existing only in crystals.

1.2.4. The following simple model could be suggested to illustrate this possibility. Let two balls be connected by an elastic string. Then at small distances (string unstretched) they behave as if free, which is seen in experiment, but at large distances (string strongly stretched) they are strongly attracted to each other. We shall never be able to greatly increase the distance between them as we are prevented from it by the string they are attached to. At present it is supposed that such an elastic string (or simply, a string) is formed between the quarks as a result of their colour charges interaction.

In the hadron string model it is assumed that the quarks in hadrons are connected to each other by certain strings which resemble elastic strings. There is a quark at each of the string's ends. The string we are concerned with differs from the real string in that it is assumed its tension remains the same when stretched infinitely. In this sense the string reminds rather of a single cobweb with the spider at the end of it. The string might be broken, but at the breaking point the quark-antiquark pair is formed again.

What is then a string? In the view accepted at present the string is something rather like an isolated strength line of an electric field, but for the purposes of a heuristic explanation it is convenient using magnetic strength lines as analogy. The string is then like a constant magnet with the quarks as its poles. It is known that when a magnet bar is broken in two there are formed two new magnets each with the pair of poles. It is quite impossible to break off just one pole, southern or northern. And this is why there are no free quarks in nature.

To give more meaning to the analogy given here it should be assumed that the string is not a mathematical line, but has some thickness of its own. It is natural to suppose that in short strings , that is in low energy states (or non-excited states) length and thickness are approximately the same, and there is no need to separate the two. The respective model is called a bag. In this model the hadron is like a balloon inside which quarks are held. If such a balloon be quickly rotated, it would elongate and become melon-shaped. And this is just the string.

Recently a more fundamental theory of elementary particles strong interactions sprung up, a theory of colour gauge fields called quantum chromodynamics. We will look at it in more detail presently.

1.3. Quantum chromodynamics

1.3.1. The quark rules of hadron construction we cited above followed unambiguously from the demand that the hadron states observed should be bleached. But the logical question arises, why should all hadrons be white? The quark model as given above fails to answer this question. At present the point of view has been accepted according to which the clue to the mystery lies in the dynamics of colour interaction. It is considered that inside hadrons, besides quarks, there are quanta of eight vector massless fields or gluons, having, just like quarks, a quantum number, colour. Gluons and quarks are constituent parts of which the hadron is fashioned. The colour of quarks and gluons is not only their quantum number, but a characteristic of their interaction, or charge. Quarks have flavour and could be three colour states. This makes it possible for the quark (of any of the three colours) to be bound to other quarks (of any of the three colours) through gluon exchange. Mathematical theory of such interaction is constructed in full analogy to quantum electrodynamics (QED) , where charged particles interact by exchanging photons, and is called quantum chromodynamics (QCD). The concept of chromodynamics which is now predominant is as follows. There are several flavours of quarks $q^{(f)}$ $(f = 1,2, \ldots , \nu)$. Until very recently it was considered $\nu = 5$. But at present data have emerged (unfortunately, rather contradictory) which point to the need for the sixth flavour. Perhaps the number of flavours is to grow even further. Each of the quarks $q^{(f)}$ forms a triplet of the group SU(3), that is a quark of each colour may be in one of the three states $q_i^{(f)}$, $i = 1,2,3$. The index i is conventionally called 'colour' , for example i = 1, blue; i = 2, yellow; i = 3 , red. Combinatory rules of bleached baryons and mesons construction, described in 1,2.1, would look as follows in terms of quark states $q_i^{(f)}$:

$$B = (\sqrt{6})^{-1} \sum_{i,j,k=1}^{3} \varepsilon_{ijk}\, q_i q_j q_k ,$$

$$M = (\sqrt{3})^{-1} \sum_{i=1}^{3} \bar{q}_i q_i .$$

Here ε_{ijk} is the antisymmetrical tensor. The local (gauge) group SU(3) (it should be reminded in constructing electrodynamics the local (gauge) group U(1) is used) is usually called the colour $SU(3)_c$. Colour quarks $q_i^{(f)}$ interact with vector fields of gluons A^a_μ , a = 1, ..., 8 (the gauge field, in the electrodynamics case it is a photon), and form an octet (according to the number of parameters or generators of the group $SU(3)_c$). These are the Yang-Mills fields (first suggested by Yang and Mills in 1954). All the quarks have a charge determining their interaction with gluons. Quarks of different colour are attracted to each other, although attraction is weaker than that between a quark and antiquark of opposite colours. This has no analogy with QED where charge namesakes repulse. Thus we have three sets of colour objects: a quark colour triplet (and an antiquark colour triplet) as well as a gluon colour octet.

The basic field equations are analogous to the Maxwell (they are called the Yang-Mills equations) and the Dirac equations:

$$\partial^\mu F^a_{\mu\nu} + g f^{abc} A^{b\mu} F^c_{\mu\nu} = g \bar{q} \frac{\lambda^a}{2} \gamma_\gamma q ,$$

$$(-i\gamma^\mu \partial_\mu + m_q) q + g \frac{\lambda^a}{2} A^a_\mu \gamma^\mu q = 0,$$

where strengths $F^a_{\mu\nu}$ (a = 1, 2, ..., 8) are expressed through gluon fields (potentials) A^a_μ as following

$$F^a_{\mu\nu} = \partial_\mu A^a_\nu - \partial_\nu A^a_\mu + g f^{abc} A^b_\mu A^c_\nu .$$

In these formulas λ^a are the Gell-Mann matrices; f^{abc} - the structural constants of Gell-Mann matrices algebra, μ,ν - Lorentzians. Repeated indices imply summation over all values which they may accept; colour indices of the quark fields have been dropped, as for example

$$\lambda^a q \equiv \sum_{j=1}^{3} \lambda^a_{ij} q_j .$$

The constant g characterises the power of the quark -gluon and the gluon-gluon interactions. As it may be seen, the first equation differs from the Maxwell equations in that it has non-linear terms (which creates additional difficulties already in solving the classical problem for the gluon fields). We also see that the role of colour charges is performed by the matrices $g\lambda^a$. It must be remembered, of course, that they accept a certain numerical value only in acting upon certain quark colour states. Gell-Mann suggested for a theory of colour quarks and gluons system described by the equations given above the name of quantum chromodynamics. For a very long time , until quantum chromodynamics was created, many physicists considered quantum field theory inapplicable to the description of strong interactions. Previous difficulties are absent in QCD. Thus field theory may be considered rehabilitated.

Now there is every reason to construct quantum chromodynamics in analogy to quantum electrodynamics, but taking into account the peculiarities brought about by gluon self-interaction. For want of time, however, we would give here a purely descriptive exposition of these specific features.

1.3.2. The most remarkable trait of chromodynamics is the unusual dependence of interaction force on distance between quarks. The effective coupling constant is large enough at great distances and decreases (tending to zero) with decreasing distance (this property is called asymptotic freedom) . The situation here is the reverse of what is known from quantum electrodynamics where a charge in vacuum (or in a material medium) because of the vacuum (or the medium) polarisation effect is partly screened by the opposite charges so that with approach to the source its effective charge grows.

The weakening of charge at small distances (antiscreening effect) should be caused by the self-interaction terms in the Yang-Mills equation. Which of the factors, quark-antiquark screening or antiscreening brought about by gluon field self-interaction, could have the upper hand, depends on the number of colour degrees of freedom and the number of quark flavours. The following expression can be obtained for the effective charge g(r) as to the function of the distance r from charge,

$$g^2(r) = g^2(r_o)\left[1 - \frac{33-2\nu}{6\pi} g^2(r_o) \ln \frac{r}{r_o}\right]^{-1}, \quad r > r_o .$$

It should be reminded that ν is the number of quark flavours (for $\nu = 5,6 \quad 33 - 2\nu > 0$). The value of r_o could be changed arbitrarily. For instance, if the charge is point-like, then we may cut out a sphere around it with the radius r_o and include into $g(r_o)$ in addition to the initial bare charge, all bound charges brought about by medium polarisation inside the sphere with the radius r_o (prior to experiment we have no knowledge of what r_o and $g(r_o)$ should be fixed in this formula). Solving the expression obtained as regards g^2 (r_o) we obtain

$$g^2(r_o) = g(r)\left[1 + \frac{33 - 2\nu}{6\pi} g^2(r) \ln \frac{r}{r_o} \right]^{-1}, \quad r > r_o .$$

Let us now decrease r_o. At $r_o \ll r$ we obtain

$$g^2(r_o) \simeq \left[\frac{33 - 2\nu}{6\pi} \ln \frac{r}{r_o} \right]^{-1}$$

and therefore g^2 $(r_o) \to 0$ at $r_o \to 0$. This result is the direct opposite of what is known from the quantum electrodynamics: Landau and Pomeranchuk demonstrated that in QED the charge $g(r_o)$ is zero for a certain finite value of r at any value of the bare charge $g(r_o)$ ('nullification of theory'). Conversely, in QCD the practically zero bare charge $g(r_o)$ at $r_o \to 0$ corresponds to the finite charge g(r) at a finite distance r. At small distances interaction weakens and the particles become free. Vacuum polarisation leads not to charge decrease with growing distance, as in electrodynamics, but to its increase.

The existence of asymptotic freedom means chromodynamics can yield correct results (as distinguished from electrodynamics) in the range of distances as small as needed. Within this limit $(r_o \to 0)$ the effective charge $g(r_o) \to 0$ which means that quarks inside the hadron behave as quasi-free particles inside the small r region. This explains experiments on deep inelastic scattering of electrons and neutrinos which in a certain range of energy and transferred momenta are interpreted as scattering on independent point-like objects, the quark partons.

Asymptotic freedom is one of the most important qualities of quantum chromodynamics which attracted close attention towards it and stimulated its extensive study. Indeed, it is only the presence of these qualities that permits to speak about essential success in constructing strong interactions theory.

Thus at very high energies (or in other words deeply inside hadrons, at small distances) quarks turn out to be weakly bound or quasi-free, so that their behaviour could be easily described by mathematics, as perturbation theory works here. The problem of considering small distances is solved.

1.3.3. This is a great achievement. But now we should have a look at what is going on at elementary particle 'boundaries' at relatively large distances. We need first of all to ensure quark confinement, or in other words the impossibility for them to exist in a free state. The confinement problem in quantum chromodynamics is not, however, reducible only to the free quark problem. No states with non-zero colour charge have been observed in experiment. Therefore confinement in chromodynamics means, strictly speaking, the absence in a free state of any particle carrying a colour charge. It is this restriction of quark and gluon movement within a small spatial region that leads to the hadron spectrum observed in experiment.

Unfortunately, at present there is still no consistent quantum field solution for the confinement problem within the framework of quantum chromodynamics as there is no large -distance perturbation theory. But the belief exists that chromodynamics does include confinement. This belief is based on the fact the reverse side of asymptotic freedom is the growth of interaction at great distances. In fact, from the expression for the effective coupling constant g(r) it could be seen it increases with the increase of r. We do not know up to which values r this formula could be employed, but it is clear its use could not be limitless, as with decreased denominator we quit its applicability region $g(r) < 1$. The characteristic scale is the value r_{QCD} of order r_0 at which the constant becomes strong ($g(r) \sim 1$) and the simple formula for g(r) no longer applicable.

Assuming chromodynamics to be capable of describing quark retention in hadrons, we might expect r_{QCD} to coincide in order with the hadron radius, e.g. it is a value of order 1F. Experimental values of $g^2/4\pi = \alpha_s$ obtained in studying processes where short distances are essential approximately correspond to the value expected.

What could be said about quark interaction at large distances exceeding r_{QCD}? In the non-relativistic potential ansatz in order to provide for quark confinement in hadrons the potential V(r) should be taken which grows infinitely with increasing r (infinite ionisation energy). Usually linear growth of potential at large distances is assumed (which is confirmed as compared with the experiment of theoretical calculations for mesons consisting of heavy quark-antiquark pairs).

What are the reasons following which interaction energy may be obtained which grows linearly with distance? It is known that potential of the form $V(r) \sim r$ is characteristic for the one-dimensional Coulomb problem. Such a potential might spring up if the gluon field between two quark sources placed at sufficientlygreat distance from each other ($r > r_{QCD}$) is compressed by self-interaction into a string so that its strength lines become parallel thus imitating one-dimensional configuration. This picture emerges in chromodynamics within the framework of some model approximation. It could be compared with hadron string models which describe such important properties of these as their belonging to certain regge trajectories.

Qualitative reasoning on which confinement hopes are grounded could also be resumed as follows. A quark with a colour charge emits a gluon. But the gluon itself has a colour charge and could also emit gluons. As gluon is a massless particle the probability amplitude that it would be emitting other gluons in the forward direction tends to infinity (similarly to the way that tends to infinity the probability amplitude of emitting photons in the forward direction of a charged particle whose mass is vanishingly small. This is why there could be no massless charged particle in a free state). Successive gluons would also be emitting new gluons in the forward direction with the infinite probability amplitude, and so on. This could go on until gluons carrying a colour charge come up against another quark and a compensation of their colour charge with that quark's charge occurs. Such a flux of colour gluons directed towards one side behaves like a string with colour charges at its ends. Increasing distance between the quarks sitting at the string's ends leads to an increase of their potential energy.

In such a situation it should be expected that as the quark and antiquark are being brought farther apart, when string field energy becomes sufficient to produce new pairs, the string is broken and mesons are created. In reality this could take place, for example, in e^+e^- annihilation, $e^+e^- \to \gamma \to \bar{q}\, q \to$ hadrons when at an intermediate stage in the process a quark-antiquark pair is created which has high energy. If such a picture of long-distance interaction indeed takes place, quarks become bound in mesons, that is, only bleached particles are observed in a free state.

The formation of strings in chromodynamics could be also related to the properties of vacuum (the ground state of the quark-gluon system). Now, there are significant arguments for holding that the true vacuum does not coincide with the trivial vacuum of perturbation theory where the gluon field is absent.This certainty is based on, first, the study of the Yang-Mills equations in the semi-classical approximations where topologically non-trivial configurations of gauge field, or instantons, were used, first discovered by Polyakov et al. (1975). This last term originating with t' Hooft describes semi-classical trajectories of sub-barrier transitions between topologically different vacuum states of the gauge field (it is known from quantum mechanics that sub-barrier transitions lower the energy of the ground state (Fig.1.1.)). Second, important data have been

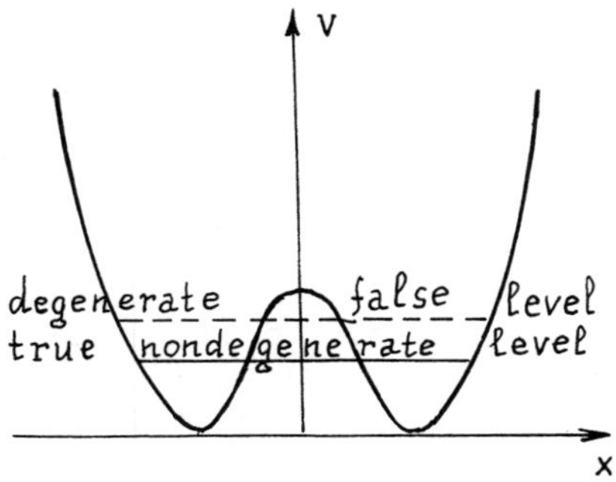

Fig.1.1.The potential minima are analogues of topologically non-equivalent vacuum states

obtained by Shifman et al. (1979) in the QCD sum rules method which established a connection between the theory of QCD vacuum and hadron phenomenology. Lying at its basis are standard relations between medium correlation functions (in this case, the QCD vacuum) and the parameters of its excitations (that is to say, hadrons).

We have no need of going into more detail, but one moment is important for the future. It has to do with the notion of non-perturbative vacuum energy density ε_{vac} . It is also known that quantum field theory is the theory of an infinite number of interacting oscillators. Even in electrodynamics where the oscillators are linear, the total vacuum energy is infinite, as each oscillator has zero point energy $\hbar\omega/2$ and their number is infinite. In QCD things are in a similar way but they are more complicated because of oscillator interaction (remember gluon self-interaction). This infinity is eliminated by renormalisation after which, however, the contribution remains of the soft modes with the wavelength $\lambda \sim r_{QCD} \sim 1$ F that are non-perturbative. Non-perturbative vacuum energy density was calculated on the basis of hadron phenomenology and has turned out to be

$$\varepsilon_{vac} = \frac{11-(2/3)\nu}{128\pi^2} \langle 0 | (g F^a_{\mu\nu})^2 | 0 \rangle \simeq -(0.5 \div 1)\, \text{Gev}/\text{F}^3.$$

It is negative as concerns the gluonless vacuum whose energy is conventionally taken to be zero.

Returning to strings, let us notice that gluon field sources (heavy quarks) could have broken the non-perturbative vacuum, creating in it lacunas with excessive energy density which is unprofitable energywise. Thus it might be expected that as the quark and antiquark are drawn apart, the gluon field they induce would be compressed to form a string not unlike a magnetic flux tube inside a second-kind superconductor. Besides, if the quarks are examined at short range, they should be manifested, as before, as quasi-free particles.

Quark unobservability is closely connected with colour unobservability. All hadrons are white states, that is, the colour group $SU(3)_c$ singlets.

At present, great efforts are being made for a stricter and more serious substantiation of the notion described and of quark confinement. In such situations illusions are possible. The difficulties of considering large distance remain. However, we would like to note that in a computer model of gluodynamics with static quark sources calculation it is shown that effective confinement does exist (Creutz, Jacobs, Rebbi, 1979). Naturally enough, we will not be talking about the details of this model. The general significance is that all calculations are done not in the Euclidean spacetime, but along a lattice in this continuum.

1.3.4. We have already mentioned confinement and the fact quarks and gluons are buried within the hadron whose radius is of order 1 F. But how would quarks and gluons behave in a system consisting of many hadrons, or hadronic strongly interacting matter? Could they spread in the volume taken up by the hadron matter, or would they still remain buried in the hadrons from which the system is constituted? Or putting it another way, the question arises: to what extent is confinement absolute? Could conditions be created where a colour current could exist, same as electrical current? Or still in other words, could hadrons be melted down and hadron matter transformed into quark-gluon matter?

These intriguing questions are by no means the product of theoreticians' idle curiosity. In fact, in the first instances of Big Bang and in the cores of neutron stars the energy density of substances could be considerable superior to the normal nuclear energy density. Moreover, at present extreme conditions could be produced in laboratory : we mean high-energy nucleus-nucleus (AA-) collisions brought about in accelerators. These problems are those of the frontline of contemporary elementary particle physics, theoretical as well as experimental, and we will attempt to discuss them in the following lectures.

LECTURE 2. COLLECTIVE PROPERTIES OF HADRON MATTER UNDER EXTREME CONDITIONS. SEMI-PHENOMENOLOGICAL MODELS AND EXPERIMENTS

2.1. Naive picture of deconfinement

2.1.1. Advancing along the scale of energy density (temperature) towards an increase of the latter is the macroscopic way of penetrating into matter. This has mainly to do with the natural idea of more complex composite formations determining the microstructure of a substance breaking down and simpler ones arising. As a result macroscopic properties of the system undergo dramatic changes, the study of which permits to draw conclusions about microstructure. As the simplest example we would like to mention the transition of electrically non-conductive atomic gas , at increased temperature, into a conductive plasma state as a result of atom ionisation with formation of positively charged ions and electrons. When energy density increases, the fine structure of physical theory is revealed characterised by an ever growing degree of fundamentality. In principle, macroscopic systems could be produced with energy densities greatly superior to the normal nuclear energy density $\mathcal{E}_A = Am_N / V_A \simeq 0.15\ Gev/F^3$. Under natural conditions they seem to exist as neutron stars as well as at certain stages of the evolution of the Universe. But under laboratory conditions the optimal way of obtaining them is related to the capacities of modern accelerator techniques in forming ultra-relativistic heavy ion beams. Estimates based on extrapolating existing experimental data onto pp- and pA - high energy collisions show that a strong interacting system created in the AA-collision region could exist for such a span of time that would be enough to reach local thermodynamical as well as chemical equilibrium, and its energy and compression would reach energy densities greatly superior to $\mathcal{E}_A$.

Suppose energy density is superior to the nucleon energy $\mathcal{E}_N \simeq m_N / \frac{4}{3}\pi R_N^3 \simeq 0.5\ Gev/F^3$ $(R_N \simeq 0.8\,F)$ and try to analyse system behaviour qualitatively. If in a very simplified way a free hadron could be pictured as a rubber balloon of a certain radius, this could mean that compartment walls between neighbours have been deformed, as the volume set aside for the hadrons is less than that necessary to pack them closely. As a result they lose their individuality, and medium description in hadronic terms loses any sense. Structural units they consist of, colour quarks and gluons (desribable in the balloon model as air molecules and elements of the rubber envelope) must enter into play. And finally, intuitively it might be expected that at some critical energy density hadrons would get dissociated into their constituents (the balloon would burst). The analogy with atom ionisation in the above example is evident.

At first glance it seems that such a naîve picture openly contradicts the current conception in strong interaction physics of elementary particles. As we have said already in the first lecture, numerous attempts to detect free-state hadron substructures have failed, although it is firmly established they are there. Colour quarks and gluons are buried inside hadrons just like the phonon dwelling only in solids. This property (confinement) is part and parcel of dynamic theory describing the real world of elementary particles, that is, of quantum chromodynamics.

2.1.2. But now, let us not jump to conclusions. It is well known that different communities are widely different by reason of the difference of profession, age, population, although they do have one thing in common: that man in community has little to do with a hermit. In the same way the hadron in a macroscopic system is quite unlike its free counterpart, quarks and gluons dwelling in the hadronic community have little similarity to those buried in an isolated hadron. There is nothing uncommon about this. We are not struck with wonder at the fact neutrons in not too heavy nuclei are stable, although a free neutron is known to have a lifetime of about 918 seconds, finally breaking down into $pe^-\bar{\nu}$.

It turns out that it is just the collective properties that provide the clue for settling the conflict between gluon and quark confinement in an isolated hadron and the possible dissociation of a hadronic system into quarks and gluons.

The hypothetic state of matter which could be formed as a result of hadrons dissociating into their constituents was given the name of the quark-gluon plasma. In this state colour quarks and gluons of all hadrons are brought together and could in principle quasi-freely travel within the macroscopic distances, restricted in this by the spatial region occupied by plasma matter. In normal hadronic matter these distances are characterised by the microscopic scale of order of characteristic hadron size, that is about 1 F. In connection with the considerable difference in the scales the supposed transition of normal hadronic matter into quark-gluon plasma was called deconfinement. The character of this transition has not been sufficiently studied as yet , although there are indications that it is abrupt, has great latent heat and is perhaps a first-kind phase transition.

There is a number of theoretical reasons to suggest the existence of one more transition having to do with the fact that the effective mass of quarks in the quark-gluon plasma could vanish at some energy density. The spontaneously broken chiral symmetry in the state of hadronic matter could be restored in this case (but not only in this; chiral symmetry may also exist in the massive fermions case, as there is a thermal bath that clearly breaks the Lorentzian invariance, and therefore the commonly used arguments of vacuum quantum field theory are not applicable in this con text). The fact that the effective mass of microstructures depends on the medium they dwell in is well known in solid-state physics. For instance, the mass of an electron propagating in an ion crystal (polaron) does not coincide with that of an isolated electron. The interrelation between phase transitions of deconfinement and restoration of spontaneously broken chiral symmetry is still not clear. It is not to be excluded that they coincide as in the opposite case the critical energy density needed to realise chiral phase transition must inevitably be superior to that corresponding to the deconfinement transition (Shuryak, Pisarski, 1981).

Roughly resuming the qualitative assessment of system behaviour at energy densities superior to the nucleonic, it could be stated that in this case the existence of hadronic matter seems impossible, and consequently the hypothesis emerges on a new state of strongly interacting matter, the quark-gluon plasma. Limited temperature and baryon charge density follow on the heels of limited energy density. Thus some critical (ultimate) temperature and baryon charge density could be expected, compatible, of course, with the critical energy density, above which hadronic matter cannot exist.

2.1.3. The idea of the ultimate temperature is not a new one, but has a long history behind it. It appeared much earlier than QCD (more than two decades ago) starting from a phenomenological description of elementary particle strong interaction (the so-called statistical bootstrap model). On the basis of this model, Hagedorn (1965) found that hadronic states density increases exponentially with the growth of mass, and concluded to the existence of an ultimate temperature above which hadronic matter cannot exist. At that time an extremist point of view was even held according to which this temperature was considered the insurmountable upper limit similar to the absolute zero. Later the suggestion was made (Shelest,Zinovjev,1974) that it characterises the temperature of hadronic matter transition to a new and as yet unknown state.

In 1975, Cabbibo and Parisi argued that within the framework of the hadron quark model (with the quarks enslaved in hadrons) exponential increase of state density spells the liberation of quarks as soon as a certain critical temperature is exceeded. We shall go back to this question later.

Research on macroscopic state of strongly-interacting matter becomes now one of the most urgent and interesting problems in high-energy physics. A significant role in this belongs to the relatively simple way in which the infinite medium problem is formulated in an equilibrium state as compared to, for example, the exclusive hadronic reactions. Its theoretical analysis calls for the use of the methods of statistical physics and quantum field theory of elementary particles. In order to solve this problem, the most advanced computer systems (of the Cray type) are being used. They perform calculations which allow to transcend perturbation theory and to describe in a unified way both phases of strongly interacting matter, thus getting an equation of state which is a fundamental characteristic of matter determining possibilities of applying formal thermodynamic and hydrodynamic machinery to concrete physical systems. The undoubted benefit of this approach is that it is hoped to solve the problem starting from the first principles, that is , from the QCD Hamiltonian. However, numerous fundamental and technical difficulties have to be overcome on this way.

2.2. Lattice calculations. Quark-gluon bag model

2.2.1. The logical question arises of how reliable are the predictions of deconfinement realisation? We would like to remind that in QCD there is no rigorous proof either confinement or deconfinement exists. The unshakeable basis on which the belief rests that these features do in fact take place is the asymptotic freedom of QCD, that is, colour effective charge decrease at decreasing distance and at increasing energy density (temperature and/or baryon charge density).

The dependence of the effective charge (running coupling constant) on the energy scale is a well-established fact in theory (renormalisation group technique). A number of essential peculiarities could be deduced knowing this dependence. Field theory at finite temperature T and chemical baryon potential μ_B has a complementary energy scale T and μ_B which regulate running coupling constant behaviour. It was shown by Collins and Perry as early as 1975 using renormalisation group arguments that in QCD at finite temperature the running coupling constant as distinct from QED at finite temperature tends to zero at high temperature and/or chemical potential

$$g^2(T,\mu_B)\Big|_{max(T,\mu_B)\to\infty} \sim \left[\ln\frac{max(T,\mu_B)}{\Lambda_{QCD}}\right]^{-1}, \quad \Lambda_{QCD}\sim r^{-1}_{QCD}\,.$$

It is the fact that led to the widespread current point of view that at high temperatures strongly interacting matter is a quasi-free quarks and gluons gas (quark-gluon deconfinement). But as opposed to QED , QCD at finite T and μ_B abounds in infrared divergencies which prevents standard perturbation theory from working. Part of these are eliminated by a respective analogue of the Debye effect, that is by screening static chromoelectric fields. As in QED fermionic (quark) and non-static chromoelectric fields are thermally suppressed at long distances. However, the problem remains of the static infrared chromomagnetic sector of theory - it is non-trivial and non-perturbative due to gluon self-interaction (Polyakov,1978; Linde,1980;Gross, Pisarski, Yaffe,1981; Kalashnikov,Klimov,1981; Kajantie,Kapusta,1982, and others).Nonetheless, the opinion exists that in the theory the non-perturbative infrared screening at distances of order $\sim(g^2T)^{-1}$ of static chromomagnetic fields may arise. If this is really so , perturbation theory expansion of thermodynamic potential is violated as first shown by Linde (1980) for although all higher terms of perturbation theory series are finite they have the same order g^6T^4 .Even if as a result the coefficient at g^6T^4 is finite it is impossible to calculate it in perturbation theory, and the thermodynamic potential could at best be calculated up to order g^6T^4. However, for high enough temperatures this is a good approximation, as asymptotically $g\to 0$ at $T\to\infty$. Thus the existence is predicted of a new high-temperature phase of strongly interating matter, a phase of quasi-free quarks and gluons (cf.3.1). As it is clear from the above discussion, this prediction is unfortunately true only as a result of strong model assumptions.

2.2.2. There are some achievements along the way to solving these complicated problems in the lattice version of the theory which is suited for computer calculation. Non-perturbative calculations are based here on numerical integration by the Monte-Carlo method of exact expressions given in theory. The most probable results are those of the gluon quantum dynamics (gluodynamics) where quarks are considered just like static sources. In this case testimonials of confinement and first-kind phase transition have been obtained at the temperature $T_c \simeq 200$ MeV. In other words, there is a static deconfinement: forces induced between static quark charges imbedded into gluonic medium undergo Debye screening (Engels, Karsch, Montvay, Satz 1981; Kajantie, Montonen, Pietarinen, 1981; Kuti, Potonyi, Szlachanyi, 1981; McLerran, Svetitsky, 1981). But calculations including a relativistic quark dynamic are not complete, and the question on the fate of phase transition in this case has not been solved as yet. Within the framework of a number of simplifying assumptions and at zero baryon charge it was found an abrupt change of energy density <u>versus</u> temperature in the vicinity of the temperature T_c (Fig.2.1.). At $T \ll T_c$ numerical data are close enough to the ideal hadron gas energy density with the spectrum of the hadronic masses known from experiment, and at $T \gg T_c$ by the Stephan-Boltzmann formula $\varepsilon(T) = \sigma T^4$ with the coefficient $\sigma = \frac{\pi^2}{30}(2 \times 8 \times 7/8 \times 2 \times 2 \times 3 \times 2) \simeq 12$ as in massless quark-gluon ideal gas (where account is taken of 2 spin and 8 colour degrees of freedom for gluons, 2 spin, 2 flavour, 3 colour and 2 particle-antiparticle degrees of freedom for quarks). Provisional Monte-Carlo calculation data also indicate that the phase transition of chiral symmetry restoration could take place at the temperature $T_{ch} \simeq T_c$.

Characterising in general the results of computer calculations it should be stressed that they are still far from perfect. Problems arise concerning improvement of calculation accuracy (increased lattice size, account taken of lattice artefacts and finite lattice size effects, fuller accounting for quark dynamics which calls for more perfect calculation equipment), and the more fundamental problems related to introducing baryon chemical potential and to clarifying and supplementing the confinement and deconfinement criteria which are at present only static (Wilson loop behaviour and Polyakov thermal line behaviour) and are the most significant elements of numerical results analysis now. This latter has to do, of course, not so much with computing problems as with the global problem of correct and complete understanding of what we call confinement and deconfinement.

2.2.3. One of the important steps in this direction are the attempts to formulate simple enough (in order to do the calculations to the end) and as vivid as possible semi-phenomenological models qualitatively reflecting properties of the real hadronic world. The most popular is the quark-gluon bag model. It is at least as vivid as the balloon model and is conceptually close to it, but even closer to reality than the latter.

In the simplest version of the model, the hadron is a bubble (or bag) with the volume V of perturbative QCD vacuum with the energy BV occupied by the ideal gas of massless quarks and gluons with the energy E(V), surrounded by the non-perturbative (physical) QCD vacuum. As a result the bag mass is M = E(V) + BV. As already noted in 1.3.3., in current view perturbative QCD vacuum is an excited state with energy density B as to the true physical vacuum whose energy density could be conventionally put at zero. That is why the bubble is subject to constant external pressure B from the physical vacuum which tends to do away with the energetically unprofitable configuration with surplus energy BV. The system is stabilised because the bubble is full of quark-gluon gas: the pressure of the gas inside the bubble and that of the outside vacuum become equal $p_{int} \equiv -\partial E(V)/\partial V = B$. Maintaining this equality is the same as the requirement for bag mass to be minimal $\partial M/\partial V = 0 \Leftrightarrow -\partial E(V)/\partial V = B$. As a result a strict relation is established between bag volume and mass. So for a spherically symmetrical bag, the equation holds V = M/4 B.

Recently a number of works have appeared devoted to the study of hadron system thermodynamics within the quark-gluon bag model with their non-zero spatial size taken into account (Baacke,1977; Hagedorn,Rafelski,1980; Kapusta,1981; Gorenstein,Petrov,Shelest,Zinovjev,1982). In calculating the partition function, the Van der Waals approximation is used. Thus the main deficiency of the formerly popular bootstrap model is made up which was neglecting hadron size. However, analysis of hadron gas models of quark-gluon bags with the Van der Waals exclusion of their own volumes where a strict relation is presupposed between bag volume and its mass, showed quite unrealistic behaviour of the thermodynamic function in the high-temperature region (Gorenstein,1983). This fact made necessary introducing physical circumstances extraneous to the model, that is manmade fixing of the desired high-temperature behaviour of the quark-gluon gas type. The decisive step towards obtaining realistic high-temperature behaviour was made by Gorenstein,Petrov,Shelest,Zinovjev (1982), and was to do away with the strict relation between bag volume and mass (in this approach volume is a fluctuating quantity).The thing is that a relation of the type

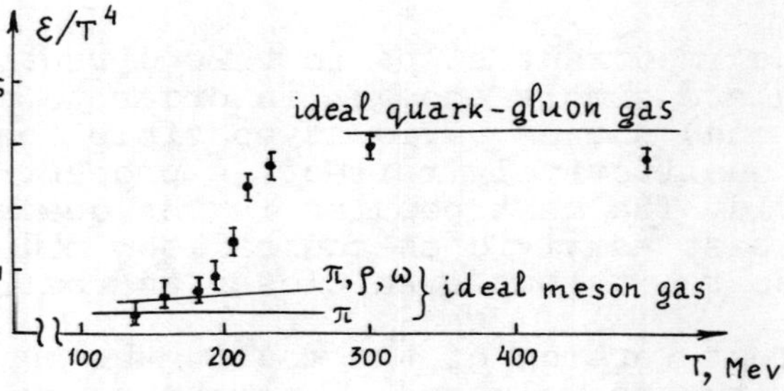

Fig.2.1. Energy density <u>versus</u> temperature

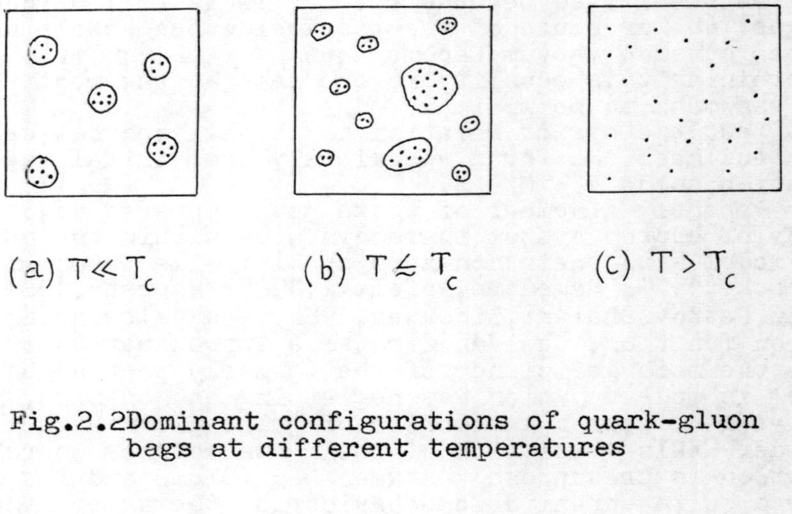

Fig.2.2Dominant configurations of quark-gluon bags at different temperatures

V = M/4B is completely broken with increased temperature as there is the pressure additional to that of the vacuum being sustained by the quark-gluon bags in the dense macroscopic system formed by the quark-gluon bags. Due to this equilibrium conditions of a single quark-gluon bag are changed as well. Partition function analysis taking account of this modification and the important demand that every bag in the system have a zero colour charge showed the temperature T_c exists at which the system is subject to the first-kind phase transition. At low temperatures ($T \ll T_c$) the main contribution to the partition comes from configurations of the lightest hadrons (or bags) with low number density of particles, so that the system behaves like π meson ideal gas (Fig.2.2a). With increasing temperature the contribution from more massive hadrons grows, and they predominate up to the temperature T_c and their average number is proportional to the system volume (Fig.2.2b). Phase transition at the point $T = T_c$ corresponds to the domination of a one-bag configuration at $T > T_c$ filling up the whole system, and thus the picture is created of the gas of bag constituents, the quark-gluon plasma (Fig.2.2c).In this way if we limit ourselves to the numerical accuracy of several dozen per cent, the quark-gluon bag model provides the basis for the following simple picture: the ideal π-meson gas at $T < T_c$ and the ideal quark-gluon gas at $T > T_c$:

$$P_h = \frac{\pi^2}{30} T^4, \qquad \varepsilon_h = \frac{\pi^2}{10} T^4 \qquad (T < T_c);$$

$$P_q = \frac{\pi^2}{90} 37\, T^4 - B, \quad \varepsilon_q = \frac{\pi^2}{30} 37\, T^4 + B \qquad (T > T_c),$$

where only the light u and d quarks are taken into account for the quark-gluon phase, and π-meson for the hadron phase, while at $T > m_\pi$ π-meson mass may be neglected. Extrapolating the expression for P_h and P_q into the intermediate temperature region and having found their intersection point (Fig.2.3), phase transition temperature could be estimated: $T_c \simeq (90B/34\pi^2)^{1/4}$. The equation of state obtained has been widely used recently for different quantitative estimates in the problems of looking for quark-gluon plasma formation signals.Its most vivid feature is the very high jump of energy density at $T = T_c$: $\Delta\varepsilon = \varepsilon_q(T_c) - \varepsilon_h(T_c) = 4B$. The current estimate of B is $B^{1/4} \simeq (200 \pm 50)$ MeV. This value of B is different from ε_{vac} (cf.1.3.3.). The explanation (Shuryak,1982) is that there are two vacuum fluctuation scales: the bag constant corresponds to the hadronic scale of order 1 F, while the energy ε_{vac} is caused by fluctuations with the characteristic scale of order of the constituent quark size $\sim 1/3$ F.

Up to now we have been describing the results obtained within the statistical quark-gluon bag model in the case of zero baryon chemical potential ($\mu_B = 0$). There are results as well for μ_B =0. In this case in the μ_B - T plane there is the curve $\mu_B = \mu_B(T)$ of the first-kind phase transition which separates the hadron and the quark-gluon phases of the strongly-interacting matter (Fig.2.4.)

Summing up the discussion of theoretical predictions on deconfinement, the conclusion may be drawn that inspite of their being incomplete, they should stimulate experimental research directed at detecting the quark-gluon plasma.

2.3. Hydrodynamic theory of multiple production

2.3.1. Under natural conditions quark-gluon plasma might have existed only in the first $10^{-6} - 10^{-5}$ seconds after the cosmological explosion when energy densities were attained by far superior to the nucleonic. It is not to be excluded that it might exist in the core of the more massive neutron stars. However, it is very tempting to create the conditions for it in laboratory. At present it is considered that collision of high-energy heavy ions are the short cut to it. In this case energy densities needed could be attained (estimated at from 1 GeV/F^3 to 2.5 GeV/F^3) using the existing accelerators refitted to accelerate heavy ions. The first experiment in this series is likely to be the one at CERN started in late 1986 in which O^{16} ions will be collided with Pb nuclei. By the end of 1987 it is scheduled to obtain ion beams of S^{32} and Ca^{40}. We would also like to note that the quark-gluon plasma could be produced in hadron-hadron collisions in a series of specially selected events (of great multiplicity per unit of secondary particle rapidity). In these experiments different characteristics of the particles observed at the final stage of the multiple production process will be measured such as the average transverse momentum, average multiplicity, relative yield, rapidity and transverse momentum distribution, and others. Using these data, the state should be reconstituted of the strongly-interacting matter produced due to collisions between heavy ions, and reliable signals of quark-gluon plasma formation selected. This latter is an important but still not finally resolved problems of theory. The difficulty is due mainly to the fact of quark-gluon plasma being produced against a substantial background brought about by the usual dynamic processes, and it exists only for a part of time of strongly-interacting matter evolution. Hence seeking for these signals must take place within the multiple production model correctly reflecting the space-time picture of system evolution. The macroscopic theory of multiparticle production processes, or thermo- and hydrodynamics, brought into being more than thirty years ago thanks

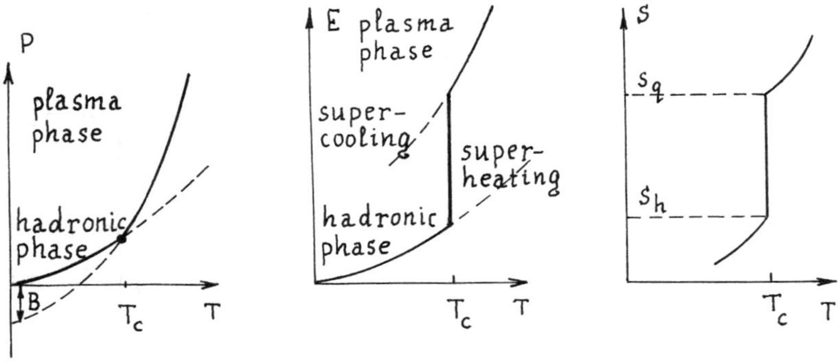

Fig. 2.3. Phase transition in quark-gluon bag model

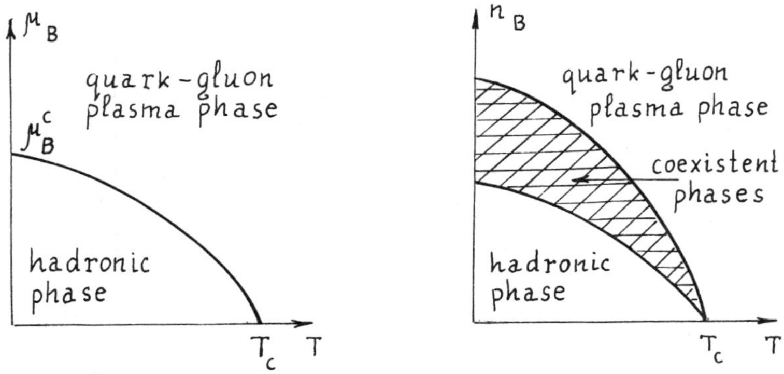

Fig. 2.4. Baryon chemical potential μ_B and baryon number density n_B versus T

to the efforts by Heisenberg, Fermi, Pomeranchuk and especially Landau, provides at present essentially the only basis for obtaining such a picture.

2.3.2. According to the hydrodynamic theory, multiparticle production process is divided into three stages in time (Fig.2.5.):
(1) hydrodynamic system formation stage;
(2) stage of the proper hydrodynamic expansion of the system
(3) stage of secondary particle formation due to decay of hydrodynamic system elements.

At the first stage following on the heels of a collision between two hadrons (ions) a composite system appears with spatial dimensions by far superior to the mean free path of a quark or a gluon, and as a result (in current estimation during $\tau_{form} \sim 1$ F) local thermodynamic equilibrium is brought about in the system and the initial and boundary conditions are formed for the second stage of the process, the hydrodynamic expansion proper (Fig.2.5). In Landau's fundamental work, the initial conditions implied a certain Lorentzian system of reference where at the initial moment of time the uniform relativistic fluid is at rest taking up the volume of a Lorentz-contracted disc. There are a number of reasons casting doubt on realisability of the initial Landau conditions. In present-day view, hydrodynamic expansion starts at a certain non-trivial velocity distribution of relativistic fluid elements which is created at the first non-hydrodynamic stage of system evolution and is subordinaate to the conditions of fluid movement being independent from the chosen Lorentzian system of reference (scaling) . The full theoretical solution of the problem of initial and boundary conditions which in fact dictate the character of the second stage in system evolution calls for quantum effects to be taken into account, which is of course beyond the scope of the hydrodynamic model and demands detailed quantitative consideration from QCD positions.

At the second stage of evolution the system expands according to the equations of relativistic hydrodynamics in the form of energy conservation laws, momentum, baryon charge and (probably) of other conserving system charges. Usually the perfect relativistic hadronic fluid approximation is used. In this case as well as at zero baryon chemical potential, the equations have the form

$$\partial_\mu T^{\mu\nu} = 0$$

$$T^{\mu\nu} = (\varepsilon + p)u^\mu u^\nu - p g^{\mu\nu}, \quad g^{\mu\nu} = diag(+1,-1,-1,-1),$$

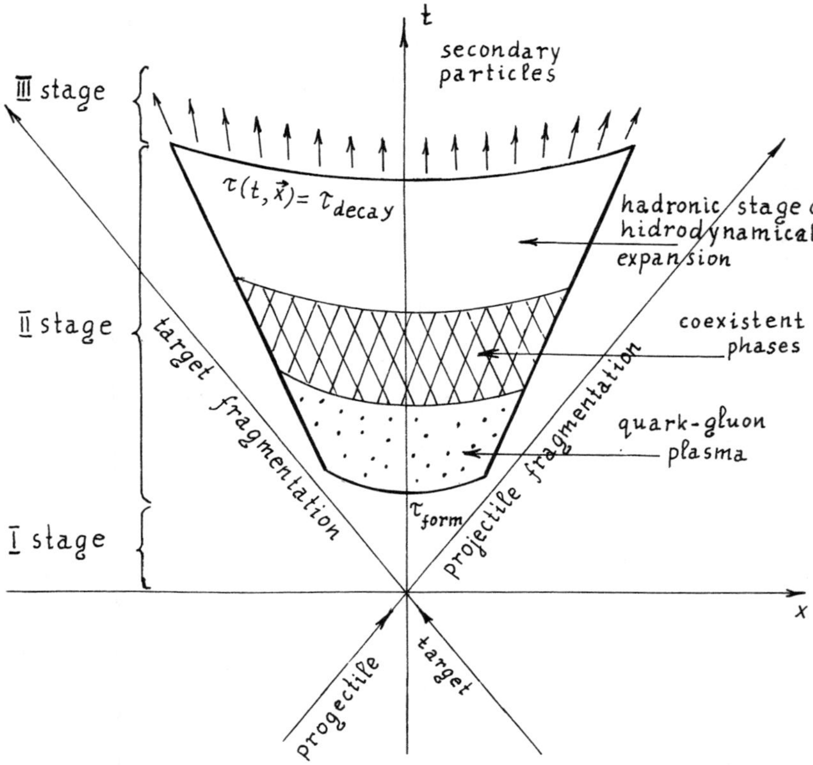

Fig.2.5. Three stages of the hydrodynamic theory of multiple production

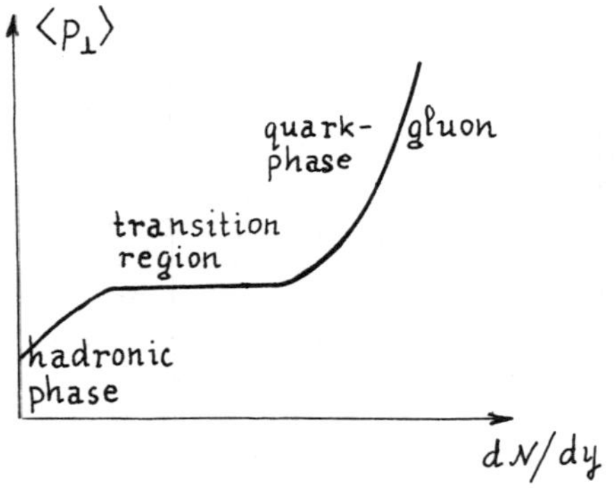

Fig.2.6. Average transverse momentum versus rapidity

where $T^{\mu\nu}$ - energy-momentum tensor, ε - energy density, p - pressure, $u^{\mu} = \{(1-v^2)^{-1/2}, \quad (1-v^2)^{-1/2}\}$ - 4-vector of fluid element velocity. In the expansion process, the quark and gluon mean free path should remain small as compared to the characteristic system size which is indispensable for hydrodynamics to apply. These equations describe the system in terms of pressure, energy density, 4-velocity of fluid element, and they are unclosed as their number is one less than that of the unknown quantities. To close the equations it is necessary to set the equation of state which would represent the relation of pressure versus energy density $p = p(\varepsilon)$. By means of the equation of state included into the hydrodynamics would be the effects of interaction between substructures forming the macroscopic system. It is the equation of state which the hydrodynamic multiparticle production theory demands from QCD thermodynamics, and it is in this that their intimate relationship finds its expression . It is possible in principle, just like in the non-relativistic case, to introduce effects of dissipation into the system, but even here the knowledge of kinetic coefficients (viscosity,heat conductivity) of QCD non-equilibrium statistics is necessary. For the first time the study of viscosity effects was performed more than twenty years ago by Feinberg and independently by Emelianov.

All quantities included in the equation of relativistic hydrodynamics, and also those (temperature, entropy density) which are related to them through thermodynamic equations are functions of the time and space coordinates $x^{\mu} = (t, \vec{x})$ .Thus solving hydrodynamic equations with the use of the equation of state describing the presupposed hadron matter phase transition into quark-gluon plasma we obtain unique information on space-time regions taken up by the plasma and hadron phases of strongly-interacting matter (Fig.2.5).Using this information allows to mark out the physical effects due to plasma phase. It is worthwhile to note that solutions of hydrodynamic equations essentially depend on phase transition character as well as even the details of the equation of state which is qualitatively manifested in the possibility of different shock waves being produced. Thus in the realisation of the first-kind phase transition different scenarios are possible: it is either deflagration or detonation shock wave formation, or else shock-free transition.Different versions have absolutely different sequels verifiable in experiment. As noted before, under natural conditions quark-gluon plasma could exist just for the first $10^{-5} \div 10^{-6}$ seconds after Big Bang, and the first-kind phase transition with great latent energy could spell substantial slowing down of the rate at which the Universe cools down, and its concrete realisation could throw light on small-scale inhomogeneity of the Universe (Olive,1981;Suhonen,1982; Kolb,Turner,1982; Crawford, Schramm,1982;

De Grand, Kajantie, 1984; Witten, 1984). Related to this, upcoming laboratory experiments with high-energy heavy ions could provide us with unique information on the primaeval state of matter many billion years before our time.

With the onset of hydrodynamic expansion the mean free path increases, and when it becomes equal to the linear dimensions of the system, the third evolution stage sets in - the decay of relativistic hydrodynamic fluid elements into secondary particles. The crucial moment of hydrodynamic theory is introducing a certain criterion of the third stage coming about, on which the predicted spectra of the secondary particles observed are dependent. In the Landau model, such a criterion is the temperature reaching the magnitude of order of π-meson mass $T(x) = T_{decay} \simeq m_{\pi} \simeq 140$ MeV (Fig.2.5) as the system cools down due to expansion. Its physical motivation is simple enough and consists in that at temperatures inferior to T_{decay} energy density concentrated in the fluid element is less than the hadronic (nucleonic) which the short-range interaction of nuclear forces taken into account (interaction radius $\sim m_{\pi}^{-1}$) means that the fluid element is the practically ideal hadron gas. This statement by Pomeranchuk and Landau was supplemented by Gorenstein, Pavlenko, Shelest and Zinoviev (1974-1975) who introduced for consideration an additional mechanism of secondary particle formation, that is, surface evaporation of hadrons. Besides, the volume effects of photon and lepton pair emission, first considered by Feinberg, the evaporation mechanism is a very important instrument in hydrodynamic system diagnostics early in its evolution. In particular, thanks to this mechanism the important conclusion was made stating there is no ultimate temperature in strongly-interacting matter. Detailed qualitative analysis of secondary particle spectra on the basis of the hydrodynamical model with evaporation, in contrast to the standard approach, calls for taking into account as well the influence of the evaporation process on system evolution. Such evolution should be described by hydrodynamic equations with energy-momentum 'drain' which corresponds to energy and momentum loss by the system in secondary particles evaporation.

Besides, the Landau model did not consider influence of the leading (rapid) particles which could be very important indeed in the hydrodynamic model. Although by their very significance leading particles are set apart in the hydrodynamic system at the expansion stage energy and momentum exchange takes place between the fluid and the leading particles. In this case boundary conditions for the expanding relativistic fluid undergo radical change as compared to the usual problem of expansion in vacuum. As first noted by Gorenstein, Sinyukov and Zhdanov (1978), the condition for system energy momentum conservation points to the necessity of introducing particle-like states at the periphery of the relativistic fluid. These states are thus identified with the leading particles.

Constructionwise, spectra of secondary particles appearing at the last (third) stage in system evolution, are obtained through the natural assumption that secondary particle number $d\mathcal{N}(t,\vec{x})$ of the fluid hydrodynamic element be proportional to its entropy $d\mathcal{S}(t,\vec{x})$ on spacelike surface $T(t,\vec{x}) = T_{decay}$ where their formation takes place according to the criterion given above. Temporal moments of the decay of different fluid hydrodynamic elements do not coincide, for the system as a whole is highly inhomogenous. Thus obtaining the solution of hydrodynamic equations, the dependence of entropy density and fluid element velocity on spacetime coordinates and finding the decay surface $T(t,\vec{x}) = T_{decay}$, we could get the velocity distribution of secondary particles:

$$d\mathcal{N}(\vec{v}) \sim d\mathcal{S}(x^{\mu})\Big|_{T(x^{\mu}) = T_{decay}} = d\mathcal{S}(\vec{x})\Big|_{\vec{x} = \vec{x}(\vec{v})}$$

$$= d\mathcal{S}(\vec{v}) = f(\vec{v})\, d^3 v \;.$$

(Here it is presupposed that the velocity of particles produced coincides with that of a fluid element at the moment of generation). It is this distribution supplemented by the contributions formed due to surface evaporation at earlier stages of hydrodynamic system evolution that is susceptible to experimental verification. It is clear that different types of solutions exist which satisfy different boundary conditions. Here it must be noted that for a concrete solution the formula given above fixes the functional dependence of secondary particle distribution according to the momenta (the passage to which is easy from velocities) with accuracy up to finite parameter arbitrariness which is done away with by fixation of energy density, entropy density and so on at the initial moment of time $\tau_{form} \sim 1$ F in hydrodynamic system formation. But how true values of these would be known as they are obtained in experimental situations? The seemingly reasonable suggestion that all energy of colliding particles is transformed into thermodynamic system energy stands in stark disagreement with the available experimental data. Thus in real collisions about half the initial energy stays with the leading particles while the rest is very unevenly distributed among the secondary particles. Moreover, in different events of one and the same reaction different energy and entropy densities might be reached. In this connection the problem of mechanisms of hydrodynamic system generation appears to be crucial, which is essentially related to the difficult and unresolved problem of dynamical description of the first stage in system evolution. It seems a vicious circle is created and little use is to be expected from the distribution obtained. But the pessimistic conclusion is premature. The thing is we could use the available experimental information to register the initial data on which momentum distribution depends. Thus perhaps a unique opportunity may be obtained of selecting and studying events where high energy density was achieved indispensable for eventual formation of quark-gluon plasma (Anishetty, Koehler, McLerran, 1980; Bjørken, 1983; Gyulassy, matsui, 1984). Moreover, it may provide the basis for the analysis of necessary conditions (besides high initial energy of colliding particles) in order to generate sufficient dissipation of energy.

There are two ways for achieving quark-gluon plasma state. The first is selecting rare events with high multiplicity of secondary particles in hadron-hadron collisions. At the highest acceleration energies achievable in our time, and for the typical hadron-hadron collisions we could only envisage the formation of two-phase states (if the transition between the hadron and the quark-gluon states is the first-kind phase transition). The second way of increasing energy density at the initial state of the hydrodynamic stage is going over to the processes of collisions between heavy nuclei at ultrarelativistic energies. It is these processes that the hopes of many researchers are pinned on. At central collision of two ultrarelativistic nuclei consisting of A nucleons the initial energy densities at the moment $\tau_{form} \sim 1$ F come into being out of $\sim A^{1/3}$ pairs of nucleon-nucleonic collisions from the 'nuclear tube', that is, the initial value of energy is $A^{1/3}$ times higher as compared to the respective value in the case of hadron-hadron collisions, and at a high enough A could correspond to the state of the quark-gluon plasma. This estimation is overvalued as some nucleon-nucleon collisions out of the 'nuclear tube' are somewhat shifted in time.

2.4. Experiments to detect signals of quark-gluon plasma formation

2.4.1. Disregarding the difficulties inherent to the hydrodynamic theory of elementary particle multiple production, let us consider on this basis one of the most intriguing questions fascinating now theoreticians as well as experimenters: what is the way of looking for quark-gluon plasma formation signals? As noted before, the difficulty lies in the fact the information on the time span during which the quark-gluon plasma was in existence, has been obliterated in the course of system evolution: secondary hadrons are formed after expansion and cooling of the strongly -

interacting matter down to the temperature T_{decay}.

It is well known that a macroscopic system mainly irradiates from surface that is less heated than its interior. So looking at the Sun we see only about $T \simeq 6 \times 10^3$ C although it is much hotter on the average. But observing solar neutrinos freely emitted from its interior we could pass judgment on the temperature of its core. Following this reasoning, Feinberg (1976) advanced the idea that the photons and leptons that do not take part in strong interaction and are emitted from the depths of strongly-interacting matter may provide direct information on the hot plasma phase. Later Shuryak (1978) within the framework of QCD kinetics made an evaluation of the generation rate of these particles in the process of quark-gluon plasma hydrodynamic expansion. It turned out that because photons and leptons quasi freely quit quark-gluon plasma, their spectra are the ideal 'thermometre' - by comparing theoretical and experimental spectra the temperature could be reconstituted as the parameter on which they are dependent. However, two reasons hinder using this. First the background has to be separated: hadronic processes (dominating at small transversal photon momenta and small invariant masses of lepton pairs) and hard quark-gluon collisions. Second, the possibility of picking out just this signal is essentially dependent upon sizes of the space-time region occupied by quark-gluon plasma in its expansion process. Unfortunately, at the colliding particle energies available at present this region turns out to be too small in typical events.

Another 'thermometre' could be fashioned by studying the composition of secondary particles. Thus c and s quarks are produced in the hot plasma phase (at the process of chemical equilibrium being established) at a relatively fast rate as compared with the rate of u and d quark production, and in quick cooling these may be 'tempered'. Thus increased yield of strange and charmed particles as compared to that in typical hadronic systems nay provide information on the stage of increased c and s quark production, the quark-gluon plasma. An especial role in the enhanced effect is played by the non-zero baryon chemical potential μ_B: at $\mu_B \neq 0$ in the system of light u and d quarks there is statistical suppression of $u\bar{u}$ and $d\bar{d}$ pair production due to the Pauli principle so that the share of $s\bar{s}$ and $c\bar{c}$ pairs is effectively increased at chemical equilibrium of quark flavours (inspite of the relatively high value of s and c quark masses at $\mu_B/T \gg 1$ we have $n_{\bar{s}}, n_{\bar{c}} > n_{\bar{u}}, n_{\bar{d}}$). As the quark-gluon plasma is cooled abruptly, the anomalous composition of secondary particles become 'frozen stiff'. Although this mechanism of strange and charmed particle formation directly from quark-gluon plasma is attractive, recently there have been indications casting doubt upon its validity as analogous increase of strange and charmed particle multiplicity could take place in purely hadronic systems at high enough temperatures.

2.4.2. An important direction in the search for the plasma matter formation signals is analysis of changes in behaviour of average transverse momentum $\langle p_\perp \rangle$ of the secondary hadrons (Hagedorn, 1965; Gorenstein, Shelest, Zinovjev, 1976; Shuryak, Zhirov, 1978; Van Hove, 1982). One-dimensional hydrodynamic description of multiparticle production leads to universal dependence of the inclusive spectra on transverse momentum in the form $\exp[-(p_\perp^2 + m^2)^{1/2}/T]$. Thence it may be seen that to higher $\langle p_\perp \rangle$ correspond higher T . But to higher initial temperatures correspond higher energy and entropy densities, and consequently higher secondary particle multiplicities per unit of the rapidity interval which are proportional, as discussed above, to entropy density. Thence thermodynamic arguments originating with Van Hove lead, in case first-kind phase transition is realised , to the characteristic correlation between the average transverse momentum $\langle p_\perp \rangle$ and the multiplicity per unit rapidity dN/dy in the central rapidity region (rapidity obtained using the formula $y = (1/2)\ln[(E + p_\parallel)/(E - p_\parallel)]$) (Fig.2.6). When the initial state is in the hadronic phase, entropy density s_0 increase and consequently that of dN/dy goes hand in hand with the increase of temperature and hence of $\langle p_\perp \rangle$. First-kind phase transition means possibility of growth for s_0 and dN/dy at constant temperature $T = T_c$ and its respective average transverse momentum $\langle p_\perp \rangle = \langle p_\perp \rangle_c$ (Fig.2.3). This causes flattening of the dependence of $\langle p_\perp \rangle$ on dN/dy. Further growth of dN/dy at $T > T_c$ causes new $\langle p_\perp \rangle$ increase. Unfortunately, the possibility cannot be excluded that the qualitative dependence $\langle p_\perp \rangle$ versus dN/dy and consequently on entropy density s_0 and energy density ε_0 could be due to action by simpler mechanisms. It is clear , for instance, that ε_0 growth increases hydrodynamic expansion time until system decay into secondary particles. The more the expansion time, the greater the role that could be performed by transversal hydrodynamic motion effects which will cause $\langle p_\perp \rangle$ growth. Another mechanism which may be responsible for $\langle p_\perp \rangle$ increase is the possibility for shock wave formation in the process of expansion and hadronisation of the quark-gluon plasma. However, qualitatively different mechanisms may bring about different curve shapes. To understand the growth effect of $\langle p_\perp \rangle$ correctly it is necessary first of all to make clear if they are caused by high temperatures in the system or are a manifestation of collective hydrodynamic movement in the direction transverse to the collision axis. With this aim in view, detailed theoretical calculations are necessary based on different mechanisms, and, of course, experimental data which theoreticians expect very eagerly.

At present a lot of different signals of quark-gluon

plasma formation are discussed but research in this direction is far from being completed.

LECTURE 3. Confinement, thermodynamics and deconfinement as seen by quantum field theory at finite temperature

3.1. Does absolute confinement contradict thermodynamics of quark-gluon ideal gas?

3.1.1. Inspite of great efforts by theoreticians directed at the studies of the phenomenology of strongly-interacting matter at high energy densities, stimulated in particular by the upcoming experiment, we still do not know very much about its properties which strictly follow from QCD. That is why it is hard to speak of reliable signals of its formation in ultrarelativistic heavy ion collisions, as these signals are to a very great extent model-dependent. The most fundamental information needed in this connection is the information on its structure, and in particular on excitations propagating in it and their distribution. What has just been said may look perplexing if confronted with that said about it earlier in Lecture 2, as well as concerning the name 'quark-gluon plasma' itself. In fact somebody well versed in science associates the word 'plasma' with a neutral state of a substance conducting a certain charge, while the phrase 'quark-gluon plasma' - with a colour charge as that charge. Thus the impression is created that in this state quasi-particles are the quasi-free quarks and gluons distributed with good accuracy according to the Fermi-Dirac and the Bose-Einstein laws respectively (and that is initially assumed in most semi-phenomenological calculations for detecting quark-gluon plasma formation signals). Moreover, as already discussed in lecture 2, this picture stands in agreement with the asymptotic freedom of QCD, with the presence of the Debye screening of the static chromoelectric fields, as well as the results of computations done within lattice QCD when it was found at high temperatures energy density is very close to what is obtained using the Stephan-Boltzmann formula with the number of degrees of freedom as in ideal gas of massless quarks and gluons.

But still this all is not enough to make definite conclusions on excitation spectrum in the system.

3.1.2. Let us remind that historically the name quark-gluon plasma (deconfinement of quarks and gluons) for the state of strongly-interacting matter at high energy densities was brought into being and struck root only due to QCD asymptotic freedom and the Debye screening of static chromoelectric fields (in analogy to the situation obtaining in quantum electrodynamics) and reflects nothing else but the presence of these properties. However, they, strictly speaking, do not mean that quarks and gluons are quasi-particles capable of propagating in the system . As to the thermodynamic arguments it is appropriate to note that generally speaking it is hard to draw unambiguous conclusions on the excitation spectrum from thermodynamics. Let us give a fairly obvious example to illustrate the latter. Thus having the expression for thermodynamic potential of the ideal quantum gas of an unknown kind of particles $\Omega = -7(\pi^2/90)T^4$ we may state only the following: there are contributions to the thermodynamic potential either by 7 massless bosonic degrees of freedom, or 8 massless fermion degrees of freedom (the gas consisting of either 2 kinds of fermions; or of 7 kinds of neutral scalar particles;or else of 1 kind of neutral vector, 1 kind of charged scalar and 3 kinds of neutral scalar particles, and so on). One more example could be the exactly solvable massless Schwinger model, quantum electrodynamics in 1 + 1 dimensions (QED_2). It is well known that there is absolute fermion confinement in QED_2 , and in the excitation spectrum there is a neutral scalar particle which is a quantum of the free massive Bose field. The exact answer is also known for the thermodynamic potential and its high-temperature asymptotics. The latter might be interpreted as corresponding to the free fermion gas (in the 1+1 dimensions the contribution of the two fermionic degrees of freedom imitates that of one bosonic degree of freedom).With this as the starting point we could proclaim at our will (if of course the correct answer is not known) that there is fermion deconfinement at high temperatures. But this is untrue as known from the exact excitation spectrum.

In other words, today convincing direct proof is lacking to show at high energy densities excitations arise in the system which have a colour charge, and therefore are capable of ensuring colour conductivity, an indispensable attribute of the plasma state.

In Lecture 2 the term 'quark-gluon plasma' was extensively used, and after what we have just said in this one, we are indeed obliged to explain the reason why - better late than never. We use the term 'quark-gluon plasma' to designate a state of strongly-interacting matter at high energy densities as it is widely used by physicists working on the problem, although we do not think it a fortunate designation, for we do not share the accepted point of view that the existing set of data indicates an eventual formation of a plasma state (in the conventional meaning). Moreover, it is not to be excluded that this state is not formed (that is to say, there is an absolute dynamic confinement) which in its turn does not contradict the information available. Further on, following the work by Zeitlin and the lecture's authors (1986) we will illustrate this eventuality within the framework of a simpler theory with confinement - quantum electrodynamics in 2+1 dimensions (QED_3) where more advance could be made in analytic calculations as compared to QCD. This admits of a more detailed analysis of the influence by confinement on thermal characteristics of the system.

3.1.3. Indirect arguments for this possibility based on dimensional reduction (Gross,Pisarski,Yaffe,1981;Appelquist, Pisarski,1981;Nadkarni,1983) of QCD to QCD in 2+1 dimensions or QCD_3 possessing confinement (Feynman,1981;D'Hoker, 1981) and computer simulation data were given in the interesting and rich work by De Tar (1985).In it, confinement was taken to mean that excitation generated in the physical vacuum of QCD by a colour source (for example,a bare quark) is indistinguishable from that generated by a colour singlet (white) source (for example,a bare quark-antiquark pair).

Accepting the assumption of dynamic confinement being absolute, the following qualitative description of the structure of strongly-interacting matter could be given for high energy densities. At $T \gg \Lambda_{QCD}$ due to the asymptotic freedom of QCD there is the following hierarchy of scales $1/T \ll 1/gT \ll 1/g^2T$. At small distances as compared to $1/T$ the image of the medium as a quark-gluon gas continues to hold.At distances of order of the Debye radius ($r_D \sim 1/gT$) perturbation theory works taking into account the Debye screening and the medium is a gas of quasi-quarks,plasmons and dressed transverse gluons. At distances $\sim 1/g^2T$ (g^2T-dimensional QCD_3 coupling constant) non-perturbative effects caused by confinement grow in importance. At distances greatly superior to the characteristic confinement scale ($\sim 1/g^2T$) there are bleached non-perturbative excitations. As in thermodynamic equilibrium medium the typical excitation wavelength $\sim 1/T$ (momentum $\sim T$) and the confinement scale is much greater than this wavelength ($1/g^2T \gg 1/T$) a good approximation in calculating thermodynamic potential is the representation of the high-temperature plasma as the gas of quasi-free quarks and gluons (for the dominant contribution to the thermodynamic potential is caused by small distances as compared to the confinement scale).Thus at very high temperatures non-perturbative contributions to the thermodynamic potential caused by confinement could be negligibly small, although at intermediate temperatures attainable, for instance, in collisions of ultrarelativistic heavy ions,they may be essential.

With in the framework of this qualitative picture many evaluations could be justified <u>a posteriori</u> based on considering the high-temperature phase of the strongly-interacting matter as a gas of quasi-free quarks and gluons (as a plasma) if , of course, the dominant contribution to the calculated quantity is really determined by small distance. It is evident that such a macroscopic characteristic as conductivity could not be such a quantity. Absolute confinement of course does not mean that the complex dynamic system under consideration could not experience phase transition that could, for instance, have chiral character or could not be subject to abrupt changes in their thermodynamic characteristics (even if phase transition in the rigorous sense is absent) when a certain energy density value is attained.In fact it is at the study of these peculiarities in distribution of secondary particles formed in heavy ion collisions that these investigations are directed which have been schematically treated in Lecture 2. Whatever the theoretical solution of the crucial question on the possibility or otherwise of deconfinement (that is, the absolute confinement question) these investigations contribute to a large extent to a more profound understanding of strong interaction physics and would provide more confirmation that QCD could work.

Which version (with absolute confinement or deconfinement) of strongly-interacting matter evolution is realised as the advance along the energy density scale takes place in reality, that is, in QCD (and at present everyone believes in QCD),is unknown now. From the fact deconfinement of the static (infinitely heavy) quarks was discovered in gluodynamic using computer calculations, liberation of dynamic quarks in QCD does not follow. This makes the problem all the more intriguing.On the whole the existing situation could not but cause some degree of internal discomfort, although it is natural in coming to terms with the new non-traditional physics, all the more so as only the first steps have been taken in this direction.

3.2.Critical appraisal of confinement criteria

3.2.1.At present confinement is the most important and the least studied, from the theoretical point of view, the phenomenon of quantum field theory.This is explicable both by technical difficulties and those of principle in a theory where asymptotic states do not correspond to the original fields.This latter just accounts for the back of the definition of confinement, couched in a rigorous formula,and consequently the lack of understanding as to the nature of the hypothetical deconfinement.

Intuitively, confinement means that quanta of the fields included in the Lagrangian, are not excitations of the physical (true) vacuum of theory. The physical vacuum essentially is not different from the macrophysical material media, and in particular from solids being a special kind of them. Just like such media vacuum is an extensive physical system having multiple degrees of freedom,is characterised by various internal processes and responds in a certain way to different external influences.On the basis of such a conception it could also be said that field quanta in confinement could not propagate in the physical vacuum as free particles.

In quantum field theory the quantity responsible for a quantum of a certain field $\varphi(x_0, \vec{x})$ propagates in time to some other point in space is the one-particle Green function (or the propagator) of the field $\varphi(x_0, \vec{x})$

$$G(x_0 - x_0', \vec{x} - \vec{x}') = i\langle 0|T(\varphi^*(x_0, \vec{x})\,\varphi(x_0', \vec{x}'))|0\rangle$$

(T stands for chronological ordering, $|0\rangle$ - vacuum state wave function).In momentum representation the one-particle Green function

$$G(k_0, \vec{k}) = \int G(x_0, \vec{x})\, exp(-i\vec{k}\cdot\vec{x} + ik_0 x_0)\, d^3\vec{x}\, dx_0$$

considered as a function of the energy variable k_0 would have a peculiarity at a certain value $k_0 = \varepsilon(\vec{k})$ (dispersion relation or the mass shell condition) which is indeed the energy of a real quantum of the field $\varphi(x_0, \vec{x})$ with the momentum $\vec{k}$ placed in the physical vacuum.Pole peculiarities of the retarded and advanced Green functions are similarly interpreted.

In many quantum field models studied, and in particular in quantum electrodynamics in 3+1 dimensions (QED4) it is this situation that really takes place. This is one of the main conditions needed for the standard (naïve) perturbation theory to apply in the infrared region (at great distances), where as the bare approximation the results of quantum free-field theory are used(with the coupling constant assumed to be zero).With perturbation theory corrections taken into account (radiation corrections),e.g. those brought about by virtual processes taking place in the physical vacuum,original field quanta properties are not radically changed in this case.It is usually said they are dressed or take on a cloud of similar but virtual (away from the mass shell) quanta.

A different situation holds in quantum field theories with confinement one of the characteristic manifestations of which seem to be strong infrared divergencies of the terms in the naïve perturbation theory series. Roughly speaking, in this case the contributions due to non-trivial vacuum structure are great and may change radically the excitation spectrum of a theory. Sometimes a meaningful answer could be obtained if at least a partial summation of leading terms in perturbation theory series is carried out. At present a number of procedures are known and are being intensively studied which are used to achieve this objective (non-perturbative methods), but as a rule it proves impossible to make at least an estimate of the contribution by the discarded terms.

But what answer could be expected from the theories with confinement? Or in other words, what are the confinement criteria?

3.2.2 The most well-known and popular criterion is the increase at great distances R of the static potential of quark-antiquark interaction V(R) determined by the gluon propagator. This behaviour by the interaction potential corresponds to a certain Wilson loop behaviour

$$W(C)=\langle 0|\mathrm{tr}\,P\exp\left(\frac{i}{2}g\oint_C A^a_\mu(t,\vec{x})\,dx_\mu\right)|0\rangle$$

(P denotes ordering along the contour C of the non-commuting Gell-Mann matrices λ) related to V(R) for timelike rectangular loops with the area R x t in the limit of large t, by the relation

$$W(R,t)\underset{t\to\infty}{\sim}\exp\left(-V(R)t\right)$$

Essentially, the Wilson loop is the quantitative measures of influence by the gluon fields on the external quark source transferred along the contour C perimeter. In the case of a timelike loop it corresponds to the creation of a quark-antiquark pair at the initial moment of time and then to its transfer along the world lines corresponding to the lateral parts of the loop, and finally, the annihilation of the pair at the final moment of time. For the potentials $V(R)=\sigma R$, linearly growing at large distances, there must be an exponential decrease of the Wilson loop $W(R,t)\sim\exp(-\sigma Rt)$ with the growth of the loop area (the area law), with the coefficient coinciding with the potential linear growth coefficient (string tension). It is this kind of behaviour by the Wilson loop and was found using computer simulation for pure gluodynamics.

In the case of non-zero temperature the modified Wilson loop -Polyakov thermal line is the value under analysis

$$L(\vec{x})=\langle\mathrm{tr}\,P\exp\left(\frac{i}{2}g\int_0^{1/T}A^a_0(\tau,\vec{x})\,\lambda^a\,d\tau\right)\rangle\ ,$$

where $\langle\ \rangle$ means averaging over the statistic ensemble of the gluon field configurations (because of the periodic boundary conditions according to the imaginary time τ ($\tau = it$), the line $0 \leq \tau \leq 1/T$ becomes a closed circuit and for this reason L ($\vec{x}$) is a gauge invariant quantity). The physical sense of L ($\vec{x}$) follows from the relationship $L(\vec{x}) \sim \exp(-\Delta F/T)$ with the change of free energy ΔF of the gluon system at the temperature T when an external static quark source is introduced into it. If L(x) = 0, it follows $\Delta F = \infty$, there is static confinement , if L(x) $\neq$ 0, ΔF is a finite quantity, there is static deconfinement (such a change in the behaviour L($\vec{x}$) was obtained for pure gluodynamics at the temperature $T_c \sim 200$ MeV using computer simulations).

The Wilson and the Polyakov criteria which work when there are static (infinitely heavy) quarks, are weak when applied to the systems of dynamical quarks as in this case the generation of virtual quark-antiquark pairs produced by vacuum fluctuation occurs, and consequently screening appears.Thus for example it is well known that static potential does not grow with distance in the massless Schwinger model, although fermions are in confinement (bare linearly increasing potential/ Coulomb potential on a one-dimensional line/ is screened by the virtual quark-antiquark pairs of the physical vacuum). In fact, in this case the Wilson loop determines the potential of the interaction between two mesons, and not two bare quarks.

Other criteria have been also discussed in literature but at present it is still unclear whether they are sufficient and/or necessary conditions for dynamic confinement to take place. One of them is the absence of peculiarities (including non-pole ones) of the quark propagator in the finite part of the energy plane (quark propagator being an entire function). If this condition is held, the impossibility is ensured of an isolated quark with finite energy existing in the physical vacuum of the theory, which is the indispensable sign of dynamic confinement. As the quark propagator is gauge dependent, criterion verification is needed at least in all relativistic gauges (with subsequent appeal to the physical level of the proof's strictness).

QCD confinement criteria analysis causes considerable difficulty and is successful after strong model assumptions are introduced (such as manmade infrared asymptotics of the gluon propagator). Studying temperature dependence of the results obtained is an even more hard problem. The difficulties there were in QCD study at high energy densities which are direct sequel to the vacuum infrared problems,stimulate consideration of the simpler field theories with a number of specific properties of QCD, those proved (asymptotic freedom) or those just assumed (confinement). Studying the temperature version of such theories could throw light upon the nature of the hypothetic deconfinement and the relation between system thermal properties and confinement. In subsequent sections we will try to analyse these rather difficult questions within the framework of the 2+1 dimensional electrodynamics.

3.3. Quantum field model with confinement - quantum electrodynamics in 2+1 dimensions

3.3.1. QED_3 as a model with singular infrared behaviour has attracted attention a considerable time ago (Cornwall, Tiktopoulos, 1977; Cornwall, 1980; Jackiw, Templeton, 1981). It is an example of a theory satisfying static confinement criterion: the growth of the Coulomb potential is logarithmic. As distinct from QED_2 there is a real degree of freedom corresponding to the gauge field (photon), and in this sense the situation is closer to the 4-dimensional. Further, as the interaction constant squared has mass dimensionality, QED_3 has naïve asymptotic freedom, is superrenormalisable in the ultraviolet region, but has problems in the infrared region. However, effective systematic methods could be used here to sum up the leading infrared contributions of the Feynman graphs developed in QED_4 (Fradkin; Barbashov, 1965).

Recently the theory attracted great interest in connection with the so-called topological mass (Deser, Jackiw, Templton, 1982). The thing is that in 2+1 dimensions as different from 3+1 dimensions, there is an additional gauge Lorentz-invariant photon mass term $\int d^3x\, \varepsilon_{\lambda\mu\nu} A_\nu F_{\lambda\mu}$, odd in time inversion (T) and inversion of one of the spatial coordinates (P). The presence of the photon mass term at the tree level (in bare action of the theory, that is with radiative corrections ignored) breaks the naïve static confinement and consequently we do not introduce it into bare action. But care should be taken so that the structure corresponding to the photon mass term should not emerge in theory effective action which is bare action plus the corrections caused by virtual processes in the physical vacuum of the theory. The danger of the photon mass term springing up in the effective action could be avoided if the initial bare action be invariant as to the discrete P- and T- transformations. Then photon mass term generation is prohibited for symmetry considerations (it should be reminded that it is P- and T- odd). The minimal possibility of attaining this is doubling the fermion representation, or in other words, introducing not just one, but even two kinds (flavours) of fermions (Jackiw, Templton, 1981). As a result the action has the form

$$S=\int d^3x\left[-\frac{1}{4}F^2_{\mu\nu}+\bar{\Psi}\left(i\gamma_\mu\partial_\mu-g\gamma_\mu A_\mu-m\right)\Psi\right],$$

where

$$\gamma_\mu=\begin{pmatrix}\sigma_\mu & 0\\ 0 & -\sigma_\mu\end{pmatrix},\quad \sigma_\mu=\left(\tau_3,\, i\tau_1,\, i\tau_2\right),\quad g_{\mu\nu}=diag\left(+1,-1,-1\right)$$

and τ_1, τ_2, τ_3 - the Pauli matrices.

3.3.2. Now we shall be considering the fermion and the photon one-particle Green functions to find an answer to the question of whether there are the corresponding excitations in QCD3 vacuum. In momentum representation in the arbitrary relativistic α -gauge they have the following form

$$G(p)=\left(\gamma_\mu p_\mu-m-\Sigma(p)\right)^{-1},\quad \left(p^2\equiv p_\mu p_\mu+i\varepsilon\right) \tag{3.1}$$

$$D_{\mu\nu}(k)=\left(g_{\mu\nu}-\frac{k_\mu k_\nu}{k^2}\right)\left(k^2+\Pi(k)\right)^{-1}+\alpha\frac{k_\mu k_\nu}{k^4},$$

$$\Pi_{\mu\nu}(k)=\left(g_{\mu\nu}-\frac{k_\mu k_\nu}{k^2}\right)\Pi(k),\quad k_\mu\Pi_{\mu\nu}(k)=0, \tag{3.2}$$

$$\left(k^2\equiv k_\mu k_\mu+i\varepsilon\right),$$

where $\Pi_{\mu\nu}(k)$ and $\Sigma(p)$-polarisation tensor and fermion self-energy, respectively, taking into account the medium response (in this case, of the QED3 physical vacuum) at an externally introduced perturbation brought about by interaction of the bare virtual fermion (photon) with medium constituents. If the interaction is switched off (g=0) it is evident that $\Pi_{\mu\nu}(k)$ and $\Sigma(p)$ are equal to zero, and the propagator $G(p)$(1) ($D_{\mu\nu}(k)$ (2)) has the pole at $p^2=m^2$ ($k^2=0$) corresponding to the bare fermion (photon). At the lowest perturbation theory order in coupling constant g (one-loop approximation) $\Pi_{\mu\nu}(k)$ and $\Sigma(p)$ are pictured by the following Feynman diagrams (solid line stands for a fermion, wavy line, for a photon)

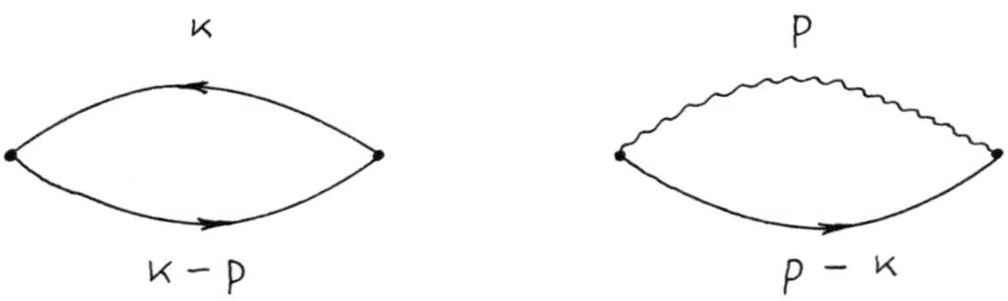

Calculating according to the standard Feynman rules, we have

$$\Sigma(p)=\frac{g^2}{8\pi p}\left[(2+\alpha)m\ln\frac{m+p}{m-p}+\alpha\gamma_\mu p_\mu\left(\frac{m}{p}-\frac{p^2+m^2}{2p^2}\ln\frac{m+p}{m-p}\right)\right], \qquad (3.3)$$

$$\Pi(k)=\frac{g^2 k}{4\pi}\left[\left(\frac{1}{4}+\frac{m^2}{k^2}\right)\ln\frac{2m+k}{2m-k}-\frac{m}{k}\right]. \qquad (3.4)$$

Let us analyse the expressions obtained.

The first thing that catches the eye is the divergency of fermion self-energy on mass shell that would not be eliminated at any α. To be convinced, put p=m ($\gamma_\mu p_\mu$ =m) everywhere except for the vanishing logarithm argument. Then we obtain

$$\Sigma_{p\simeq m}(p)=\frac{g^2}{8\pi}\left(2\ln\frac{2m}{m-p}+\alpha\right),$$

that is, the coefficient at the logarithm has no α and could not be vanished by any choice of gauge. Thence the crucial difference from QED$_4$ is seen in which self-energy is finite on mass-shell, and the divergency in its derivative is eliminated at $\alpha=3$. Consequently, to determine the fermion propagator a $p\sim m$ it is not enough to stick to the lowest order of perturbation theory and all leading contributions should be taken into account. At a momentum far from the naïve mass shell we could trust the one-loop approximation at $g^2/m \ll 1$ as in this case the correction to the tree fermion propagator due to fermion self-energy is not large. It could be convincingly seen that the fermion propagator (1) with $\Sigma(p)$ determined by formula (3), has no poles at p far from the naïve mass shell though the dispersion equation root could be found at momentum near the naïve mass shell. The root has a real part Rep - m ~

$g^2 \ln(m/g^2)$ and the imaginary one $\sim g^2$. From the preceding discussion it is clear we cannot trust this root. As it will be seen later, it is a one-loop approximation artefact.

As to the polarisation tensor, it could be seen from formula (4) that its real part diverges at the point $k = 2m$ which corresponds to the naïve threshold of fermion-antifermion pair creation and the imaginary one makes a jump (compare with QCD_4 where the peculiarity is radical, the real part is finite and the imaginary vanishes at the threshold). Such singular behaviour is connected to the peculiarity at the mass shell of the fermion self-energy. Inspite of this we cannot believe expression (4) in the vicinity of the point $k = 2m$, though far from it it is well defined, and we can trust the one-loop approximation (the condition is essential $g^2/m \ll 1$). At small k from (4) we have

$$\Pi(k)_{k\to 0} = \frac{g^2}{2\pi m} k^2 + O(k^4),$$

whence it is clear that the photon propagator (2) with $\Pi(k)$ determined by formula (4) has the pole at $k^2 = 0$ which corresponds to the photon and that in QED_3 as well as in QED_4 radiative corrections do not change infrared asymptotics of the tree photon propagator which causes singular behaviour of one-loop fermion self-energy in the vicinity of the naïve mass shell.

3.3.3. Let us now go back to the consideration of a fermion propagator, as the question remained unanswered as to the fate of the pole discovered in one-loop approximation in the vicinity of the naïve mass shell. In order to determine the propagator in this region all diagrams should be summed that provide leading contributions at $p \sim m$:

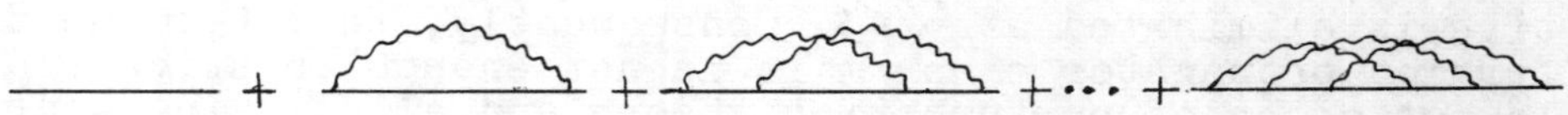

The way of consecutively summing them (Fradkin, Barbashov, 1965) is based on calculating the fermion Green function in the external field A_μ using the self-time representation and perturbation theory in the exponent, and averaging the result obtained over all configurations of the field A_μ with the Gaussian measure defined by the photon propagator infrared asymptotic (as we showed earlier it is given by the tree approximation). As a result we obtain the following expression for the fermion propagator

$$G(p)=i(\gamma_\mu p_\mu+m)\int_0^\infty ds\, \exp\left[is(p^2-m^2+i\varepsilon)-f(s)\right],$$

where the leading term of the asymptotics of the function f(s) at large s determining the character of the peculiarities G(p) at $p \sim m$ is independent from gauge, having the form

$$f(s)\sim i\,\frac{g^2 p s}{2\pi}\ln(sp^2).$$

At such asymptotics of the function f(s) the fermion propagator obtained has no peculiarities near the naïve mass shell.This is what diagram summation leads to: both the bare and the doubtful pole discovered earlier in the one-loop approximation , just vanish. Considering that G(p) has no peculiarities and at p far from the naïve mass shell, it could be stated that the fermion propagator is an entire function of the momentum p and there are no asymptotic states that correspond to the original fermions. Hence there is dynamic confinement in the model.

3.4. Thermal properties of QED_3. Absolute confinement

3.4.1. Now let us go over to consider QED_3 properties at non-zero high temperature $T \gg m$. Let us pick out the charge symmetry case when chemical potential is equal to zero. In the imaginary time formalism the theory is determined by the Euclidean version of QED_3 action where integration over $x_3(x_3 = ix_0)$ takes place from 0 to $\beta = 1/T$ and the (Fermi-) Bose-fields are (anti) periodical according to x_3 with the period β .From such a definition the Feynman rules immediately coinciding with the rules of the Euclidean field theory with the difference that the third momentum component accepts the values $2\pi nT$ for bosonic lines and $(2n+1)\pi T$ for the fermionic ones ($n=0,\pm1,\pm2...$) and the integration over these components is replaced by summation.

System specifics with confinement at non-zero temperature is determined by infrared divergencies arising in the multiparticle situation due to the same long-range forces that provide for confinement. In contrast to QCD in QED_3 such divergencies appear already at lowest order of perturbation theory. However, as there is one non-selfinteracting gauge degree of freedom (massless vector field spin being zero in 2 + 1 dimensions), these divergencies may be eliminated through consistent accounting for the corresponding static mode screening (this situation would correspond in QCD to the screening of not just the static chromoelectric fields, but the static chromomagnetic ones as well).That is why in QED3 long-range interaction influence upon thermal characteristics of the system could consistently be observed without involving additional model assumptions.

3.4.2.It is well known that photon propagator behaviour at small momenta is responsible for long-range forces.Let us analyse this behaviour.

At non-zero temperature the photon propagator and the polarisation tensor in momentum representation in arbitrary relativistic α -gauge have the following structure

$$D_{33}(k)=\frac{\vec{k}^4}{k^4}\,\frac{1}{\vec{k}^2+\Pi_{33}(k)}+\alpha\frac{k_3^2}{k^4},\ \left(k^2\equiv\vec{k}^2+k_3^2,\ \vec{k}=(k_1,k_2)\right),$$

$$D_{ij}(k)=\left(\delta_{ij}-\frac{k_i k_j}{\vec{k}^2}\right)\frac{1}{k^2+A(k)}+\frac{k_i k_j}{\vec{k}^4}k_3^2 D_{33}(k)+\alpha\frac{k_i k_j}{\vec{k}^4}\,\frac{\vec{k}^2-k_3^2}{k^2},$$

$$D_{i3}(k)=D_{3i}(k)=-\frac{k_i k_3}{\vec{k}^2}D_{33}(k)+\alpha\frac{k_i k_3}{\vec{k}^2 k^2},\qquad (3.5)$$

$$\Pi_{ij}(k)=\left(\delta_{ji}-\frac{k_i k_j}{\vec{k}^2}\right)A(k)+\frac{k_i k_j}{\vec{k}^2}\,\frac{k_3^2}{\vec{k}^2}\Pi_{33}(k),$$

$$\Pi_{i3}(k)=\Pi_{3i}(k)=-\frac{k_i k_3}{\vec{k}^2}\Pi_{33}(k),\ k_\mu\Pi_{\mu\nu}(k)=0\ \ (i,j=1,2).$$

At the same time let us pay attention to the Lorentz non-invariance of these expressions (compare with the corresponding expressions in the case T=0) which is caused by the presence of the 3-vector u_μ, the medium (thermal bath) velocity. That is why the polarisation tensor $\Pi_{\mu\nu}(k)$ is defined by two independent functions $\Pi_{33}(k)$ and A(k) (cf.(2)). By using u_μ formulas (5) could be rewritten in the Lorentz

-covariant form, but we work in a frame where $u_\mu=(0,0,1)$.

In calculating a separate diagram of the naïve perturbation theory expressions (5) with $\Pi_{\mu\nu}(k)=0$ are used as the photon propagator $D_{o\mu\nu}(k)$. As a result (in the QED3 case) already at lowest order of perturbation theory infrared divergencies arise, as for example in the one-loop fermion self-energy (in QED4 it is infrared-finite). Moreover, they arise even if an attempt is made to find the photon propagator $D_{o\mu\nu}(x)$ in the coordinate space starting with its form in the momentum space ((5) with $\Pi_{\mu\nu}(k)=0$)

$$D_{o\mu\nu}(\vec{x}, x_3)=\sum_{k_3=2\pi nT}\int\frac{d^3k}{(2\pi)^2}\, e^{-i\vec{k}\cdot\vec{x}-ik_3x_3}\, D_{o\mu\nu}(\vec{k},k_3) \quad (3.6)$$

because of the infrared divergency of the integral in the term with $k_3=0$ $(n=0)$. No wonder about that if it is noted the solution of the original equation for the propagator in the coordinate space with the x_3 periodically boundary conditions

$$(\partial_3^2+\vec{\partial}^{\,2})D_o(\vec{x}, x_3)=\delta(x_3)\,\delta(\vec{x})$$

(for simplicity's sake we disregard the tensor structure) could be shown as

$$D_o(\vec{x}, x_3)=\sum_{k_3\neq 0}\int\frac{d^2k}{(2\pi)^2}\, e^{-i\vec{k}\cdot\vec{x}-ik_3x_3}\,\frac{1}{k^2}+\frac{1}{2\pi}\ln(|\vec{x}|/const),$$

where the last term responsible for long-range interaction in the 2+1 dimension (logarithmic static bare potential of two charged bare fermions interaction) causes difficulty in a correct definition of the propagator in the momentum space. This is so for perturbation theory series terms divergencies as well. Thus to render sense to perturbation theory calculation it is needed to define additionally the momentum representation (that is, prescribe to it a certain procedure of passage to the limit). It is clear that such additional definitions should be reduced to introducing long-range interaction screening. We may choose the gauge-invariant infrared regularisation of the static mode which imitates the screening, that is, propagator (5) could be used with the polarisation tensor , with the only non-zero component $\Pi_{33}(\vec{k},0)=a$; $a\neq$ a certain constant. In physical terms this corresponds to screening the dangerous logarithmic potential at distances superior to $1/\sqrt{a}$, which could be verified by finding the Fourier representation of $D_{33}(\vec{k},$ The introduction of regularisation eliminates infrared divergency in the D_{33} component of formula (6), while the spatial components remain divergent. However, in the class of α-gauges there is a marked gauge $\alpha=-1$ where due to angle integrations there are no such divergencies. Therefore for infrared finiteness of gauge-invariant quantities it is sufficient to use the introduced static mode regularisation, but the gauge non-invariant quantities would be determined only in the gauge $\alpha=-1$.

At the end of calculation infrared regularisation should be removed, that is to say, the constant should be directed at zero. But in this case we obtain the same infrared divergencies which we struggled against. We seem to be in an impasse. But there is still a way round: the system's true polarisation tensor could act as the infrared regulator. The trick is in fact known since the times of Debye and Hückel (1923) and has been used many times over (Gell-Mann, Brueckner,1957;Fradkin,1965;Akhiezer,Peletminsky,1960,and others). Such perturbation theory was nicknamed the skeleton theory (cf.,e.g.Freedman,McLerran,1977). The thing is that the polarisation tensor satisfies demands imposed on the infrared regulator $\Pi_{33}(\vec{k},0)$ does not vanish with momentum tending to zero ($\vec{k}\to 0$).With the one-loop accuracy we have at $T \gg m$

$$\Pi_{33}(\vec{k},0)_{\vec{k}\to 0}=\frac{2g^2T}{\pi}\ln 2+O\left(\frac{g^2}{m}\vec{k}^2\right)\equiv r_D^{-2}+O\left(\frac{g^2}{m}\vec{k}^2\right),$$

where r_D - the Debye radius of screening of the bare Coulomb potential which increases logarithmically with distance. This dependence r_D on temperature and the coupling constant should be expected for dimensional reasons (compare with $r_D^{-2} \sim g^2T^2$ in 3+1 dimensions).

From the behaviour by the other components of the polarisation tensor in the one-loop approximation

$$\Pi_{33}(\vec{k},k_3\neq 0)_{\vec{k}\to 0}=O\left(\frac{g^2}{m}\vec{k}^2\right),\quad A(\vec{k},0)_{\vec{k}\to 0}=O\left(\frac{g^2}{m}\vec{k}^2\right) \qquad (3.7)$$

it follows we may trust this approximation in the infrared region (corrections being small).In this connection we should note its difference from QCD where $A(\vec{k},0)$ at $\vec{k}\to 0$ is proportional to $|k|$ (Kalashnikov,Klimov,1982),which is connected in its turn to gluon contribution and changes infrared asymptotics of the gluon propagator component D_{ij} which leads on to the as yet unresolved problems of perturbation theory. It needs also to be reminded that in QCD as distinguished from QED the polarisation tensor depends on the gauge that introduces additional difficulties into the proof of the Debye screening of static chromoelectric fields being gauge-independent.

Thus infrared asymptotics of the static mode of the photon propagator component D_{33} changes due to screening, and asymptotics of the spatial component D_{ij} coincides with the free one.

3.4.3. Thus we saw for certain that at non-zero temperature the Coulomb potential providing for static confinement at zero temperature is screened and static deconfinement arises - infinite energy is no longer needed to carry into infinity two external static charges placed in QED_3 medium. But it remains unclear whether dynamic deconfinement arises, that is, whether at high enough temperatures charged excitations arise in the spectrum ensuring, for instance, conductivity. The latter may only be collective and be manifested as peculiarities in the corresponding Green functions at finite temperature. Intuitively, it is clear that at high enough temperature (energy density) correlations between bound fermion and antifermion weaken because of screening (speaking in term of the classic Coulomb gas on plane, in the volume occupied by any pair, many identical pairs are found which screen the original one). But this weakening could be insufficient for a dressed fermion to propagate in the system.

To analyse this question, let us investigate behaviour of a fermion propagator in the high temperature regime $T \gg m$. In this connection we need information on fermion self-energy. At lowest order of naïve perturbation theory that is given by the formula

$$\Sigma(p)=g^2 T \sum_{k_3=2\pi n T} \int \frac{d^2 k}{(2\pi)^2} \gamma_\mu G_0(p-k)\gamma_\nu D_{0\mu\nu}(k) \qquad (3.8)$$

where G_0 - the Euclidean analogue of the fermion propagator (1) with $\Sigma(p)=0$ and the external momentum component p_3 accepts the values $(2n+1)\pi T$ (n=0, ±1, ±2...). As already discussed above, infrared divergencies of this expression are eliminated with the screening accounted for, which is reduced to the replacement $\vec{k}^2 \mapsto \vec{k}^2 + r_D^{-2}$ in the denominator of the photon propagator component $D_{33}^D(\vec{k},0)$ and the change-over to the gauge $\alpha=-1$ (potentially dangerous terms containing $\vec{p}\cdot\vec{\gamma}$ are cancelled after angle integration, and the remaining infrared divergency is proportional to $\alpha+1$). The unproper character of the fermion propagator (self-energy) in other α-gauges is not surprising as it may be obtained through gauge transformation which is an evident generalisation of the Euclidean vacuum field theory formula

$$G^{\alpha}(x) = \exp\left[-(\alpha+1) g^2 T \sum_{k_3 = 2\pi n T} \int \frac{d^2 k}{(2\pi)^2} \frac{e^{i k_\mu x_\mu} - 1}{k^4}\right] G^{\alpha=-1}(x),$$

where the term with $k_3 = 0$ $(n=0)$ is infrared-divergent.

An attempt could be made to use this modified expression (8), analytically continued on the variable p_3 in order to extract information on the excitation spectrum (on the poles of propagator $G(p)$ (1) with the self-energy $\Sigma(p)$ which is analytical continuation upon the whole complex plane of expression (8) which has been defined in the discrete set of Euclidean momenta $p_3 = (2n+1)\pi T$. This is what we are going to do.

Expression (8) could be represented as

$$\Sigma(p) = \Sigma^{vac}(p) + \Sigma^{T}(p),$$

where $\Sigma^{T}(p)$ -temperature contribution to self-energy and $\Sigma^{vac}(p)$ -the Euclidean vacuum self-energy which after analytical continuation diverges at $p \sim m$ (cf.3.3.2. and formula (3)).As at p m the contribution of $\Sigma^{T}(p)$ could be neglected as compared to $\Sigma^{vac}(p)$ $\left(\Sigma^{vac}(p) \gg_{p\sim m} \Sigma^{T}(p)\right)$, and taking into account the results in 3.3.3, it may be concluded that at $p \sim m$ the fermion propagator has no peculiarities. However,it cannot be excluded that they might appear in the region $p \gg m$ at high temperature $T \gg m$ where $\Sigma^{vac}(p) \ll_{p\gg m} \Sigma^{T}(p)$. In this case , even if the fermion excitation does appear, its mass m_T which is defined by the position of the pole of the fermion one-particle Green function at $\vec{p} = 0$, is mainly brought about by the thermal cloud, that is, by the polarisation cloud of real medium particles. This is the situation found in QED_4 where at high temperatures in the one-loop approximation $m_T^2 \sim g^2T^2$ (Klimov;Weldon,1982) and is gauge-independent. For dimensional reasons it could be expected in QED_3 that $m_T^2 \sim g^2 T$ and contributions from the higher loops are suppressed by the degrees of g^2/T (it should be reminded that in QED_3 g^2 has the dimensions of energy, and for this reason g^2/T is dimensionless). However,there are grounds for mistrusting this picture. Thus in calculating by contour integration the sum which is there in a formula for self-energy and determining the analytical continuation of the Green function at $\vec{p}=0$

$$G^{-1}(0, p_3) = -ip_3\gamma_3 - \Sigma(0, p_3)$$

from discrete points $p_3 = (2n+1)\pi T$ upon the whole complex plane on the basis of the condition

$$G(0, p_0) \xrightarrow[|p_0| \to \infty]{} |p_0|^{-1}$$

and finding in the region $m \ll |p_0| \ll T$ (p_0- energy variable) we are interested in the leading contribution s, we get the following dispersion equation

$$1 + \frac{g^2 T}{8\pi p_0^2} \ln\left(-\frac{p_0^2}{bT^2}\right) - \frac{g^2 T}{4\pi(p_0^2 + r_D^{-2})} \ln(-p_0^2 r_D^2) = 0, \quad (3.9)$$

where b - numerical constant of order of one. Let us analyse this.

The appearance of a large absolute value logarithm $\ln(-p_0^2/bT^2)$ at $p_0^2 \sim g^2T$ means in this region the equation may not be trusted until we are convinced such logarithmic contributions are absent from higher orders of perturbation theory. However, if equation (9) is taken literally, there are no solutions at $p_0^2 \sim g^2T$ though there is a solution with real part $\sim g^2T \ln(T/g^2)$ and the imaginary part $\sim g^2T$. Note the appearance of a large imaginary part which coincides with the real one with logarithmic accuracy. This shows the dispersion equation root could not be interpreted as responding to excitation (it should be recalled that for this interpretation to be justified it is necessary that the imaginary part to be considerably less than the real one, in this case the inverse imaginary part characterises excitation lifetime). The situation here reminds very much that with the vacuum where we found a doubtful complex root of a dispersion equation (cf.3.3.2.).The latter is an artefact of the one-loop approximation we verified in 3.3.3 by summing up the leading logarithms.

Equation (9) differs from the analogue in QED4 which in the main approximation has the form $1-g^2T^2/8p_0^2 = 0$ and includes neither logarithms,nor an imaginary part, and is gauge-independent.

Thus the one-loop calculation within its framework of the skeleton perturbation theory and(as noted earlier) the unproper charater of the temperature fermion propagator in relativistic gauges different from $\alpha = -1$ casts doubt on the existence in QED3 at high temperature of a quasi-particle corresponding to the fermion, and of the dynamical deconfinement.

3.4.4.As there are grounds for supposing that confinement in QED3 is absolute, it is interesting to analyse its influence on QED3 medium thermodynamics.It is not so easy bringing this analysis about as we, of course, do not have an exact answer for the thermodynamic potential and use (albeit improved) only perturbation theory.In this connection the guiding thread we choose will be observing characteristic difference of the skeleton perturbation theory in QED3 as compared to its analogue in the cognate theory of QED4 which, of course, has no confinement.

Well, let us get on and calculate the thermodynamic potential (the respective calculations in QED4 done by Akhiezer and Peletminsky,1960).The starting point will be the expression of the thermodynamic potential through the exact polarisation tensor and the photon propagator

$$\Omega=\Omega_0+T\int_0^g \frac{dg'}{g'}\sum_{k_3}\int\frac{d^2k}{(2\pi)^2}\Pi_{\mu\nu}(k)D_{\mu\nu}(k). \qquad (3.10)$$

Here Ω_0 - thermodynamic potential of photon-fermion ideal gas in 2+1 dimensions.

$$\Omega_0=1\cdot T\int\frac{d^2k}{(2\pi)^2}\ln\left(1-e^{-\beta k}\right)-4\cdot T\int\frac{d^2p}{(2\pi)^2}\ln\left(1+e^{-\beta\sqrt{m^2+p^2}}\right)$$

$$\underset{T\gg m}{=}-(1+3)\frac{\xi(3)}{2\pi}T^3+o(T^3), \qquad \xi(3)=1,202\dots$$

(we remind that the photon here has one degree of freedom so that its spin is zero, and the massive fermion has two degrees of freedom (particle-antiparticle) with the spin 1/2 ;in the model we are concerned with having double fermion representation (cf.3.3.1) the number of fermion degrees of freedom is four).High-temperature asymptotics ($T\gg m$) of Ω_0 demonstrates 2+1 dimension specifics which is that contribution by the four fermion degrees of freedom into the thermodynamic potential imitates that of threebosonic degrees of freedom, while in the 3+1 dimensions numerical coefficients are such that the contribution by the four fermion degrees of freedom , corresponds to the fractional numeral 7/2 of the neutral bosonic degrees of freedom. Later we shall go back to this remarkable fact.

Substituting $\Pi_{\mu\nu}(k)$ and $D_{\mu\nu}(k)$(5) into (1o) and integrating over g' we obtain

$$\Delta\Omega \equiv \Omega - \Omega_0 = \frac{T}{2}\sum_{k_3}\int\frac{d^2k}{(2\pi)^2}\left[\ln\left(1+\frac{A(k)}{k_3^2+\vec{k}^2}\right)+\ln\left(1+\frac{\Pi_{33}(k)}{\vec{k}^2}\right)\right]. \quad (3.11)$$

Later we shall limit ourselves to the one-loop approximation for the temperature polarisation tensor (recall that in this approximation $\Pi_{33}(k)$ and A(k) include g^2 as a multiplier). Although expansion of (11) over the parameter g^2 allows to obtain in a self-consistent way contributions up to order g^4 below we stick to accuracy g^2. This is justified as it will be shown that leading over g^2 non-analytic contributions have the structure $g^2\ln g^2$. There are also the contributions $g^4\ln g^2$ but to fix logarithm scale it is necessary to know the polarisation tensor with accuracy up to and including g^4 and therefore these transcend calculation accuracy.

The expansions over coupling constant of the integrand for the modes with $k_3 \neq 0$ and of the first bracket term for the mode $k_3 = 0$ with accuracy up to g^4 are easy due to the properties of $\Pi_{\mu\nu}(k)$ expressed by formulas (7). As a result of this expansion functions are obtained integrable in the infrared region. The expansion of the remaining term in formula (11) leads to a non-integrable peculiarity of the type $\int_0^\Lambda dk/k$. This peculiarity springs up due to screening, that is, to the fact that $\lim_{\vec{k}\to 0}\Pi_{33}(\vec{k},0)\neq 0$. Thence it is evident that this term's contribution is g^2-nonanalytical, and our task is to extract the leading part from it. Taking into account what has just been said, integrating by parts and using the fact that $\Pi_{33}(\vec{k},0) \xrightarrow[|\vec{k}|\to\infty]{} 0$ and $\Pi_{33}^{vac}(0,0)=0$ ($\Pi_{\mu\nu}^{vac}(k)$ - the Euclidean polarisation tensor at zero temperature) as a result we get

$$\Omega \equiv \Omega_1 + \Omega_2 = \frac{T}{2}\left[\sum_{k_3\neq 0}\int\frac{d^2k}{(2\pi)^2}\left(\frac{A(k)}{k_3^2+\vec{k}^2} - \frac{\Pi_{33}(k)}{\vec{k}^2}\right) + \right.$$

$$\left. + \int\frac{d^2(k)}{(2\pi)^2}\;\frac{A(\vec{k},0)+\Pi_{33}^{vac}(\vec{k},0)}{\vec{k}^2}\right] +$$

$$+\frac{T}{2}\left[\int\frac{d^2k}{(2\pi)^2}\,\frac{\Pi^T_{33}(\vec{k},0)}{\vec{k}^2+r_D^{-2}}+\frac{1}{4\pi}\,r_D^{-2}\right], \qquad (3.12)$$

where $\Pi^T_{\mu\nu}(k)\equiv\Pi_{\mu\nu}(k)-\Pi^{vac}_{\mu\nu}(k)$ .Even if we disregard the structure of $\Pi_{\mu\nu}(k)$ the following difference from QED4 is evident:the term with $k_3=0$ in the exchange part Ω_1 is modified so as to ensure infrared finiteness, and the corresponding addition is referred to the correlation part Ω_2 . On the other hand, in this latter due to the arising ultraviolet divergency, it is impossible to replace $\Pi_{33}(\vec{k},0)$ at the integrand by $\Pi_{33}(0,0)$ .For this reason in our case the division of $\Delta\Omega$ into the exchange and the correlation parts is conventional and shows only that the first does not yield a g^2-nonanalytic contribution.

Substituting A(k) and $\Pi_{33}(k)$ (we do not supply full expressions for these, clumsy as they are) into (12) and calculating the sum by contour integration, subtracting temperature-independent vacuum contributions and finding the leading term of the high-temperature asymptotics of the expression obtained with accuracy up to and including g^2 we have

$$\Delta\Omega\equiv\Omega_1+\Omega_2\underset{T\gg m}{=}\frac{2(\ln 2-1)\ln 2}{(2\pi)^2}g^2T+\frac{(3+\ln T^2 r_D^2/2)}{(2\pi)^2}g^2T^2+$$

$$+o(T^2)=\left(\frac{gT}{2\pi}\right)^2\ln\left(\frac{e\pi T}{g^2\ln 2}\right)\ln 2+o(T^2)$$

where e is the natural logarithm basis. In graphic terms the formula obtained corresponds to the leading over g contribution from an infinite sum of infrared-divergent diagrams of the type

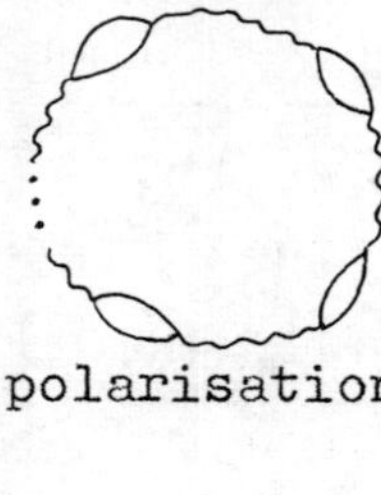

with any number of polarisation inserts.

Let us pay attention to the fact the expression obtained is non-analytic over coupling constant. For dimensional reasons it shpuld follow that corrections to Ω_0 at high temperature have a power character over the parameter g^2/T. The logarithm appears due to infrared divergencies of the naïve (non-skeleton) perturbation theory. It should also be noted that in contrast to QED_4 the correlation contribution $\sim g^2T^2\ln(T/g^2)$ prevails over the exchange one $\sim g^2T^2$.

The value of $\Delta\Omega$ is small at $T \to \infty$ as compared to the term Ω_0 which provides the main contribution to the thermodynamic potential. The latter means that inspite of absolute fermion confinement in QED_3 for which we cited arguments, at lowest orders of the skeleton perturbation theory the leading term of the high-temperature asymptotics of thermodynamic potential is the same as in photon-fermion ideal gas thermodynamic potential. However, although it is very hard to pass judgment on excitation spectrum starting from thermodynamics, in the model studied the above-noted coincidence of contributions by four fermionic and three neutral bosonic degrees of freedom into the thermodynamic potential introduces additional ambiguity into interpreting the high-temperature asymptotics obtained as that corresponding to the gas of free photons and fermions (deconfinement of fermions). For example, on the same basis it may be interpreted as corresponding to the gas of the bosons of the four types (the photon and, for example, of the three excitations which are fermion and antifermion bound states). This demonstrates once again that the attempt to get an answer to the question of whether confinement does or does not exist based on thermodynamic potential asymptotics, if not quite devoid of sense, could not, in any case, lead to an unambiguous conclusion.

So with QED_3 consistent accounting for screening allows to eliminate infrared divergencies and in this respect advance further than with QCD where the question on the way to eliminate them remains open. As a theory with confinement, QED_3 could not be the proving ground for testing its various criteria. The relative simplicity of the model permits to hope for an advance in such an important issue on the question of the excitation spectrum. Further study of QED_3 could deepen our understanding of theories with confinement.

4. CONCLUDING REMARKS

We hope that our retrospective exposition made it clear that physics of strongly-interacting matter is a rapidly developing branch of knowledge. Research done in its framework could be conventionally divided into three interconnected groups:

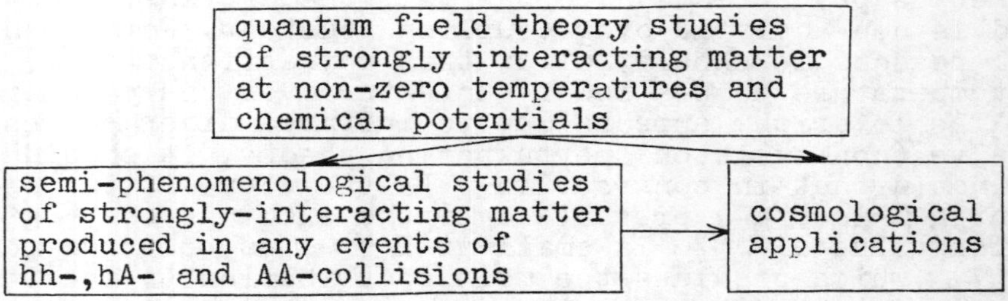

It is natural that at different times different approaches in research prevailed. Now it is the semi-phenomenological approach in research that dominates if we take into consideration the number of researches and of the works published. This seems to be brought about by experiments on heavy ions high energy collisions that began in late 1986. However, as the diagram shows the first approach has the fundamental importance, and the more so as it is only there that the crucial question of the structure of strongly-interacting matter (excitation spectrum) at high energy densities could be elucidated. At present there is no exact solution to it which of course is due to the complexities of QCD. Nevertheless the point of view has gained wide acceptance according to which its state is a quark-gluon plasma.One of the objectives which did much in stimulating us to write these Lectures is attempting to attract attention to the status of such a viewpoint(Lecture 3) as it is not to be excluded that it might have little to do with reality, although permitting in certain cases correct calculations to be made.

5.ACKNOWLEDGMENTS

We express our gratitude to V.Ya.Chernyak,M.I.Gorenstein, and G.M.Zinovjev for discussion of some of the questions concerned. Especial thanks go to V.Yu.Zeitlin for fruitful cooperation and constructive discussion in which our view was shaped on some problems of physics of high energy densities. Finally we thank S.I.Kazilo for assistance in translating and preparing the typescript.

DETECTORS OF LABORATORY GRAVITATION EXPERIMENTS AND A NEW METHOD OF MEASURING G

C.C. Speake and T.J. Quinn
Bureau International des Poids et Mesures (BIPM)
92312 Sèvres Cedex
France

ABSTRACT. In the three most recent experimental determinations of G a torsion balance was used as a detector and furthermore, it appears that the torsion balance is chosen systematically for experiments such as the Eötvös-Dicke-Braginski tests of the equivalence principle and for searches for the fifth force and other forces outside general relativity. In this article the sensitivity, thermal noise and susceptibility to seismic disturbance of torsion balances and alternative types of detectors are compared and a new method of determining G which uses a vertical dumb-bell balance is proposed.

1. INTRODUCTION

The recent progress in theoretical physics towards a unification of the gravitational interaction with the other three known forces in physics and the possibility of the existence of other forces in nature has prompted a revival of interest in macroscopic mechanical experiments. Despite the enormous progress in theoretical physics, the techniques available for such experiments have remained essentially unchanged since the time of Cavendish. Further work is clearly required to develop Cavendish-type experiments and alternative measurement schemes. It is ironic that one of the most exciting ideas of modern elementary particle physics, the fifth force[1], is now being tested with experimental devices[2] developed two hundred years ago !

V. De Sabbata and V. N. Melnikov (eds.),
Gravitational Measurements, Fundamental Metrology and Constants, 443–457.

The torsion balance was conceived by the Rev. John Mitchell, who was responsible for the major part of the design of Cavendish's apparatus, and, independently, by Ampere towards the end of the 18th century. It has, historically, been the most frequently used device for testing our knowledge of gravitation physics. In fact, the three most recent determinations of the gravitational constant, G, (the references to these G determinations and the other historical G measurements can be found in the bibliography compiled by Gillies[3]) and the Eötvös-Dicke-Braginski tests of the equivalence principle were all performed with the torsion balance. The purpose of this article is to list some performance criteria for choosing weak force detectors and to apply these to compare the torsion balance with some other, new or little employed, detectors. In the final part of the paper a brief outline is given of a proposal for a new measurement scheme for a precision determination of G using a vertical dumb-bell balance.

Von Jolly was the first individual to use a beam or common balance to measure G and his work was improved by Poynting and by Richarz and Krigar-Menzel. These determinations were superseded by the work of Boys (the results and uncertainties are listed in Table I).

TABLE I

author	year	$G \times 10^{-11}$ $(m^3 kg^{-1} s^{-2})$
von Jolly	1878	6.47 ± 0.11
Wilsing	1889	6.594 ± 0.015
Poynting	1892	6.70 ± 0.04
Boys	1895	6.658 ± 0.007
Richarz and Krigar-Menzel	1898	6.685 ± 0.011

The measurements of G performed at the end of the last century using beam balances. Boys' torsion balance measurement is also included and is the most precise.

The disadvantage of the beam balance is that the gravitational accelerations applied to the test masses by the attracting masses are parallel to the direction of the earth's own gravitational field; thus a G determination amounts to measuring the extremely small change in weight of the test masses on the application of the laboratory gravitational field. Typically the fractional change in weight of the test masses which can be produced by reasonable attracting masses is about 10^{-9}, so the precision of a beam balance relies heavily on the effective rejection of changes in the earth's field and on the stability of the torques due to the weight of the test masses about the pivot axis. In particular the most serious limitation arises from the differential thermal expansion of the balance arms. The most precise balances designed for G experiments achieve a sensitivity which is equivalent to a fractional weight change of the order of 10^{-11} [4].

In general, it is advisable to choose a detector which is sensitive to accelerations perpendicular to the direction of the earth's field. This condition was satisfied by Wilsing's vertical dumb-bell balance (see Table I). His device has received very little attention over the past one hundred years but, as will be shown below with the application of modern technology shows promise of being a viable alternative to the torsion balance. Other types of detector are being developed at present: De Boer[5] has developed a mercury bearing for a G determination. Faller has built a detector which comprises a cylindrical vessel floating in water[6]. Koldewyn and Faller[7] developed a magnetic suspension to replace the torsion fibre of a torsion pendulum and currently various groups are working on superconducting magnetic suspensions[8]. We will not discuss in detail these systems but will mention them briefly where appropriate in the list of performance criteria which follows.

2. PERFORMANCE CRITERIA

2.1. Sensitivity

A major difficulty with early beam balances was their lack of sensitivity. If we define the sensitivity, S, of a detector as the angular deflection produced by a gravitational field of 1 ms^{-2} acting on a single test mass (m) (we will suppose that the field at the location of the second mass is negligible) then,

$$S \simeq \frac{m\,R}{\tau_0} = \left(\frac{P}{2\,\pi}\right)^2 \cdot \frac{1}{2\,R} \quad \text{rad/ms}^{-2} \tag{1}$$

where τ_0 is the restoring torque, R is the radius of gyration, and P is the natural period of oscillation. We can deduce that a long period is required of a sensitive detector and that, if two detectors were to have equal periods, then the instrument with the smallest moment of inertia has the greatest sensitivity. This was first pointed out by Boys. Using equation 1 it is easy to show that Boys quartz fibre torsion balance was about 7×10^3 more sensitive than Richarz and Krigar-Menzel's beam balance and 2×10^5 more sensitive than the balance used by Poynting. Despite this colossal difference in sensitivities Boys barely gained a factor of 2 improvement in his uncertainty in G compared with Richarz and Krigar-Menzel. A result which clearly shows that sensitivity, although important, is by no means the only factor that limits the accuracy of such experiments. Nevertheless the work of Boys marked the end of beam balance G measurements.

TABLE II

	simple "hangdown" suspension or flexure strip	knife edge	torsion balance
restoring couple (τ_0)	$(MgEH)^{\frac{1}{2}}/\sinh(\alpha\ell)$ (per suspension element) M = load per element E = Young's modulus H = second moment of area of section $\alpha = (Mg/EH)^{\frac{1}{2}}$ ℓ = length of strip	$Mg\delta$ δ = radius of curvature of knife M = total suspended mass	$\pi\eta r^4/\ell$ η = shear modulus r = radius of fibre ℓ = length of fibre
typical practical sensitivity	10^2 (10^5)	10^2	10^8

Comparison of the restoring couples due to flexure strip, knife edge and torsion fibre suspensions.

Table II lists the sensitivity of various instruments. The sensitivity given for the torsion balance is that of the instrument used by Braginski and Panov[9] in their experimental test of the equivalence principle. Also listed in Table II are the sensitivities which can be expected from balances or compound pendulums suspended from flexure pivots and knife edges. It is extremely important to note that these expressions are those which apply when the centre of mass of the suspended object lies precisely at the axis of rotation of the pivot. It will be shown in the section below on seismic coupling why this particular case has been considered. The parameters chosen for the flexure pivot are those of the beryllium-copper flexures used in the BIPM prototype flexure strip balance[10] and the value in parenthesis is that which should be obtained in the near future using single crystal silicon. The sensitivity of the knife edge is limited by the radius of the fulcrum under load. The sharpest loaded ($Mg \sim 3$ kg) knife edge has a radius of about 10 μm.

It is difficult to assess quantitatively the sensitivity of the fluid and magnetic suspensions as they are nominally astatic devices i.e. of infinite sensitivity having no fixed equilibrium position. However deviations from axial symmetry usually impose a restoring torque and a prefered orientation at some level.

In order to calculate the resolution of the detector set by its sensitivity it is necessary to know the noise at the input of the transducer which is being used to monitor angular displacement. If the transducer has a white noise of amplitude $\Delta\Theta_t$ rad/√Hz then, provided the measurement takes place over timescales, $t_m \gg P$, the gravitation acceleration sensitivity, $\Delta\ddot{x}$, is simply

$$\Delta\ddot{x} \simeq \frac{\Delta\,\Theta_t}{S}\ \mathrm{ms^{-2}/\sqrt{Hz}}. \tag{2}$$

This expression is also valid for a servoed device under similar conditions (even though, of course, the servoed period can be much shorter than the natural value).

2.2. Gravitational and non-gravitational noise

Gravitational perturbations are spurious torques due to changes in the surrounding mass distribution (gravity gradient) and the apparatus cannot, in principle, be shielded from this noise. The ratio of the gravitational signal to gravitational noise for a G determination, for example, is independent of the size of the test masses. Noise due to ground vibrations could be considered to be a gravitational-type perturbation as it is also proportional to test mass size (for a given detector). However the influence of non-gravitational noise (electromagnetic and thermal effects, for example) is not in general proportional to test mass size and therefore large test masses are desirable to enhance the ratio of the gravitational signal to non-gravitational noise. This is a conclusion which is too easy to forget in an over enthusiastic search for high sensitivity.

A serious problem with the classical torsion balance is that a sensitive fibre cannot support large test masses. This follows obviously because the sensitivity of the fibre is proportional to r^{-4} (see Table II) whereas the supportable load increases as r^2. In other words the most sensitive torsion balance has an infinitely fine fibre and infinitely light test masses ! A great advantage of the flexure pivot is that it need not be circular in section and the restoring torque is largely defined by the thickness whereas the breadth can be increased to enable reasonable loads to be supported. Typical test mass values for flexure suspensions that we have examined[10] are about 1 kg, compared with a few grams for a torsion balance. Magnetic and fluid suspensions can also be designed to have high sensitivity and reasonably large test masses.

All the above mentioned noise sources (perhaps with the exception of seismic noise) can be reduced to low levels by choosing a sufficiently isolated experimental site and by careful electromagnetic and thermal shielding. However, the most serious limitation to measurement resolution is due to thermal or Brownian fluctuating torques and these are clearly an example of non-gravitational noise.

2.3. Thermal noise

Viscous damping due to collisions of gas molecules with the detector gives rise to Brownian noise, whose characteristics are well understood[11,12]. The spectral density of fluctuations is given as

$$f_B(\nu) = (4\ k_B\ T\ b)^{\frac{1}{2}} \quad \nu \leqslant \nu_B \tag{3}$$

where k_B is the Boltzmann constant, T is the thermodynamic temperature and b is the magnitude of the damping torque due to viscosity. The spectral density of the noise torque given in equation 3 is constant up to the mean frequency of molecular collisions, ν_B, (typically 10^{16} Hz in ultra-high vacuum). Above this characteristic frequency $f_B(\nu)$ drops to zero.

Braginski[13] derives one type of limit to the possible integration time or bandwidth reduction which can be employed to reduce the rms Brownian noise ($\Delta\tau_B$) where

$$\Delta\tau_B \geqslant \left(\frac{4\ k_B\ T\ b}{t_m}\right)^{\frac{1}{2}} \tag{4}$$

and t_m is the measurement time. The largest value of t_m must be much less than the damping time of the detector, t^*, where

$$t^* = 2\ I/b \tag{5}$$

with I as the moment of inertia.

Thus in order to reduce Brownian noise as much as possible we require high vacuum and the use of a detector of high Q. However, in general, if the parameters of the measurement system remain stationary, then the integration time can be increased without limit. It should be stressed that these arguments are completely independent of any electronic feedback employed.

Recent studies[15] have shown that the dominant energy loss mechanism in flexure pivots comes from anelasticity in the pivot material. The bulk elastic compliance of the flexure can be considered to be a generalised susceptibility. The imaginary component of the compliance then gives rise to damping which is, in general, frequency dependent. We have

$$\varepsilon = (K' + iK'')\ \sigma \tag{6}$$

where ε is the strain, K' and K" are the real and imaginary parts of the compliance and σ is an applied stress. The associated non-thermodynamic noise is given as[14]

$$\langle \varepsilon^2 \rangle = 4\ k_B T \int_0^\infty K'' \frac{d\nu}{\nu}\ . \tag{7}$$

Experiments which are described elsewhere[15] have determined that in many polycrystalline materials (Be-Cu and aluminium alloy) and in agate knife edges K" is independent of frequency at low frequencies. Equation 7 then leads to 1/f noise. Work is continuing on this interesting problem and measurements will be made on the frequency dependence of the damping of single crystal silicon flexures.

Equation 7 predicts rms torque uncertainties of about 10^{-11} Nm for Be-Cu flexures which represents a severe limitation to the performance of such a pivot in a gravitational physics experiment. Anelasticity must also exist in torsion fibres but has been little studied.

Brownian noise is the ultimate limitation to the resolution of room temperature fluid suspensions. However, the signal to noise ratio can sometimes be increased by increasing as necessary the test mass values[6]. The advantages of cryogenic detectors are obvious from the point of view of the thermal noise.

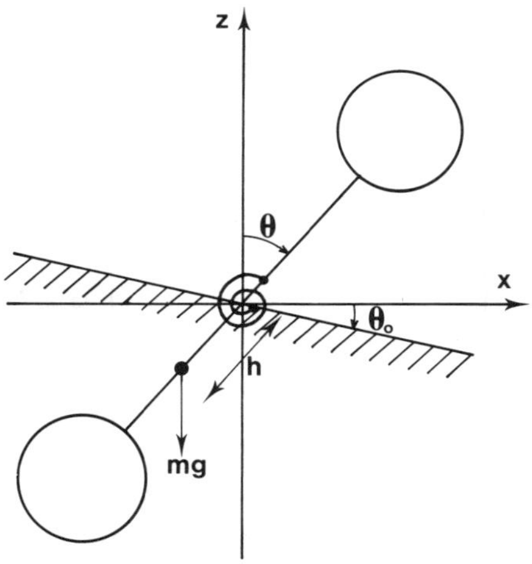

Figure 1. A schematic of an idealised dumb-bell balance subject to horizontal and vertical ground accelerations $\ddot{x}$ and $\ddot{z}$, and ground tilt Θ_0.

2.4. Seismic noise

Figure 1 shows a schematic diagram of a vertical dumb-bell balance suspended on a perfect astatic pivot. The rigidity of a real pivot (τ_0) is given by the schematic "watch" spring i.e. the rigidity of the balance when the centre of mass is at the point of suspension. Note, with the centre of mass suitably positioned above the axis of suspension the restoring torque τ_0 can be cancelled and the period made infinite. We can write down the equation of motion of the unperturbed balance as follows,

$$I\ddot{\Theta} + mg_0h\Theta + \tau_0\Theta_r = 0 \qquad (8)$$

g_0 is the mean value of the earth's gravity field. The angle (Θ_r), which is measured by a displacement transducer, is the difference between the absolute motion of the pendulum and the tilt of the ground reference,

$$\Theta_r = \Theta - \Theta_0 \;. \qquad (9)$$

Now, if we include vertical and horizontal ground acceleration, $\ddot{z}$ and $\ddot{x}$ respectively and ground tilt Θ_0, we find the following equation

$$I\ddot{\Theta}_r + (mg_0h + \tau_0)\,\Theta_r = -\,I\ddot{\Theta}_0 - m(g_0 + \ddot{z})\,h\,\Theta_0 - m\ddot{x}h \;. \qquad (10)$$

By choosing that the centre of mass of the balance lies on the pivot axis (h = 0) the two terms containing the ground movements can be suppressed. In the limit of the low frequencies $\nu \quad (\tau_0/I)^{\frac{1}{2}}$ we find

$$I\ddot{\Theta}_r + \tau_0\,\Theta_r = 0 \qquad (11)$$

and the pendulum is vibrationally isolated ! Thus, even though the sensitivity of dumb-bell balances (and beam balances[16]) can be tuned to be as large as desired, equation 10 implies that the balance with its centre of mass at the pivot axis is the best choice from the point of view of ground displacement noise coupling. An "over-sensitive" balance is a highly sensitive seismometer and tiltmeter.

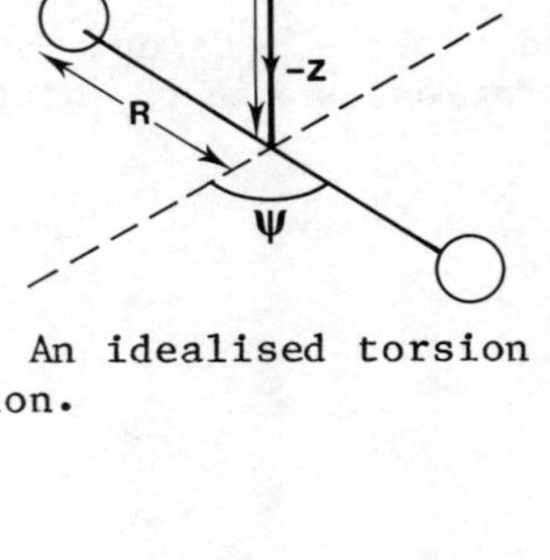

Figure 2. An idealised torsion balance subject to a horizontal acceleration.

The dynamical properties of torsion balances are much more complicated and the general equations of motion have been developed elsewhere[17]. We will simply quote here the result of a simplified analysis (see reference 18) which gives an approximate relationship between horizontal ground acceleration of spectral density $x(\nu)$ and the induced torque. A horizontal acceleration will excite a simple pendulum oscillation of the torsion pendulum bob (see Figure 2). The bob will then try to reduce its angular kinetic energy by aligning itself parallel to the axis of rotation of the simple pendulum motion. Depending on the direction of the horizontal acceleration (ψ), the torsion balance will experience a torque,

$$\Delta\tau_s \simeq \ddot{x}(\nu)^2 \cdot \frac{m R^2 d^{3/2}}{4 g_0^{\frac{1}{2}}} \cdot \frac{Q \sin 2\psi}{(d^2 + R^2 \sin^2 \psi)^{3/2}} \qquad (12)$$

where Q is the quality factor of the simple pendulum oscillation and the other parameters are defined in Figure 2. It is important to note that the above expression describes the behaviour of a torsion balance in the form of a rigid inverted 'T' supported by a universal joint. Equation 12 indicates that it is necessary to damp the simple pendulum oscillation as in Luther and Towler's experiment but that, even then, ground vibration severely limits the precision of a G determination.

The model has a rigid coupling between the bob and "fibre" and this is fairly good representation of actual torsion balances. Perhaps much could be gained by using a symmetric flexure joint (the fibre itself) and carefully balancing the torsion bob. This technique would decouple the modes as the bob would behave like a point mass for simple pendulum oscillations and its kinetic energy would therefore be independent of the direction of motion.

Another way of reducing this coupling is by increasing the symmetry of the torsion bob itself as was done by Dicke and Braginski, however this technique reduces the sensitivity of the apparatus to changes in gravity gradients which, of course, provide the signal in a G determination.

Fluid suspensions are extremely good at rejecting ground vibration noise as, at low frequencies, the fluid behaves like a stiff spring. The result is that, at frequencies low compared with C_s/D, where C_s is the speed of sound in the liquid and D is a typical dimension of the apparatus, the float is rigidly coupled to the laboratory frame and associated measuring instruments. The effects of ground vibrations are therefore not seen. Superconducting suspensions also have associated stiffnesses which could couple ground vibration to the suspended mass system, so care must be taken in their design to avoid mode couplings and resonances.

3. A NEW METHOD FOR MEASURING G

3.1. An outline of the new method

We shall describe here a new method for determining G which uses a vertical dumb-bell balance mounted on a flexure strip.

The balance will be held in a fixed position by a servo system employing a superconducting coil and ball transducer. This type of transducer has already been developed for quite different types of experiments[19]. Gravitational torques are applied to the balance by moving appropriate attracting masses around the upper and lower balls of the balance. The measured quantity is the change in magnetic energy of the superconducting coil/ball system. Provided that the electromechanical system can be made sufficiently conservative, the measured change in magnetic energy can be equated with the change in rotational gravitational potential energy (see below) resulting from a measured change in position of the attracting masses. The reason for proposing the superconducting coil-ball system as a detector is that it represents the closest practical approach to an ideal conservative system.

Suppose that the energy E of the complete system can be written

$$E = E_g + E_m + E_e \tag{13}$$

where E_g is the gravitational potential energy which can be coupled to an angular rotation of the balance, E_m is the magnetic energy stored in the coil-ball system and E_e is the elastic potential energy stored in the flexure pivot. For a change in gravitational potential energy dE_g and provided that the system is conservative and that the servo-control system is sufficiently precise and sensitive, we can assume that both E and E_e remain constant and we can write

$$dE = dE_g + dE_m = 0 \ . \tag{14}$$

In the absence of a servo system we would write equation 14 in the following way :

$$dE = dE_g + dE_e = 0. \tag{15}$$

But a precise measurement of dE_e would require an ideal, elastic, flexure pivot. These problems are avoided by using the superconducting-transducer servo-control so that the measurement of a change in mechanical potential energy is replaced by the measurement of a change in magnetic potential energy.

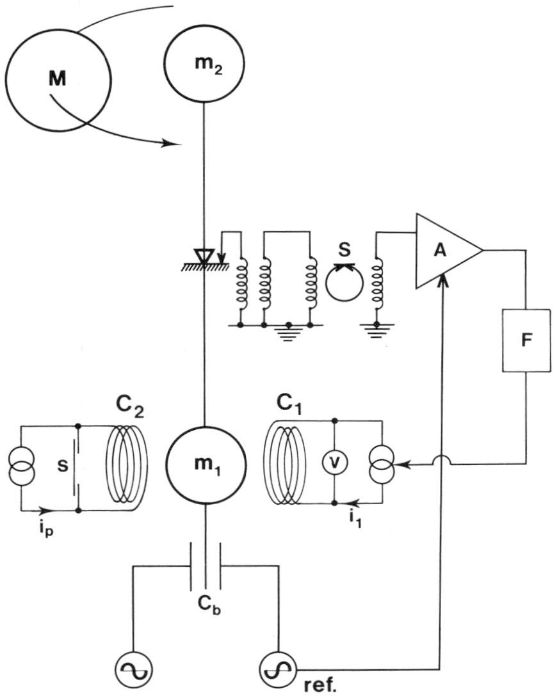

Figure 3. A schematic diagram of the new method for measuring G.

3.2. The experimental arrangement

Figure 3 shows a schematic outline of the proposed experimental arrangement. Masses m_1 and m_2 are mounted on an optimally tuned balance which uses a flexure suspension as a pivot and is maintained at liquid helium temperatures under ultra high vacuum. The lower mass is coated with, or is made of, lead and is positioned between two superconducting coils C_1 and C_2. The coil C_2 applies a continuous torque to the balance and is maintained in persistent mode with the switch s closed. As the attracting mass M rotates about the upper mass m_2 (which need not necessarily be superconducting), the balance will be deflected from a given reference rest position.

This movement is detected by the capacitance-bridge (C_b) lock-in system which, if necessary, could use a Squid (S) as an output detection element. The demodulated output signal from the Squid is proportional to the balance deflection angle and is integrated, differentiated and amplified in a suitable way by F. The square root of the output from F is used to control the current (i_1) flowing in coil C_1 to provide servo control of the balance.

The coil current $i(t)$ and voltage $v(t)$ can be measured using classical room temperature techniques but, of course, a Squid and a Josephson junction could be employed if a much higher sensitivity were required. The change in stored magnetic energy can be written

$$dE_m = \int i(t)\, v(t)\, dt \tag{16}$$

which can be evaluated by computer from series of measured values of $i(t)$ and $v(t)$.

3.3. Calculation of the energy stored in the rotational mode of the balance by displacements of an attracting mass

For a given position $\underset{\sim}{r}$ of the attracting mass (M) we can define the stored potential energy in the rotational mode of the balance such that it is zero when it lies parallel to $\underset{\sim}{r}$. A general position Θ of the balance is then characterised by an energy $E_g(r, \Theta)$ which can be calculated from the following expression,

$$E_g(r, \Theta) = \int_0^{\Theta} \underset{\sim}{\tau}(r, \Theta) \cdot \underset{\sim}{d\Theta}\ . \tag{17}$$

We can calculate the torque $\underset{\sim}{\tau}(\Theta)$ acting on the balance using Lagranges equations,

$$\underset{\sim}{\tau}(\Theta) = -\frac{\partial V}{\partial \Theta}(r, \Theta) \tag{18}$$

where $V(r, \Theta)$ is the potential energy of the two suspended masses in the field of the attracting mass. Combining equations 17 and 18 we find

$$E_g(r, \Theta) = V(r, 0) - V(r, \Theta). \tag{19}$$

The change in potential energy stored in the balance for a displacement of the attracting mass from (r_1, Θ_1) to (r_1, Θ_1) can be written,

$$\Delta E_g = E_g(r_2, \Theta_2) - E_g(r_1, \Theta_1)\ . \tag{20}$$

If we denote the gravitational potential distribution of the complete balance, including shaft etc. as $\Phi(r, \Theta)$, we can calculate the change in stored potential energy as

$$\Delta E_g = M\{\Phi(r_2, 0) - \Phi(r_2, \Theta_2) - \Phi(r_1, 0) + \Phi(r_1, \Theta_1)\} \tag{21}$$

where the direction $\Theta = 0$ corresponds to the angle, at a given r, at which the balance has a minimum potential energy. If the attracting mass were not spherical the energy stored could be calculated by integrating the potential over the attracting mass volume at each of the four required points. This procedure is considerably easier than the calculation of a gravitational torque.

From equations 4 and 9 we can therefore write

$$M\{\Phi(r_2, 0) - \Phi(r_2, \Theta_2) - \Phi(r_1, 0) + \Phi(r_1, \Theta_1)\}$$
$$= -\int i(t)v(t)dt. \qquad (22)$$

Therefore G can be determined from
a) the measurements of i(t) and v(t);
b) the calculated gravitational potential distribution of the balance;
c) the measured displacements of the attracting mass (or masses).

Taking the balance length to be 1 m, the suspended masses to be 1 kg and the attracting mass M to be 10 kg, a total energy change of 3×10^{-8} J occurs on rotating the attracting mass about the upper test mass at a convenient distance from m_2. If, as would probably be the case, an attracting mass moved around the lower superconducting ball as well, the total energy change would double to 6×10^{-8} J. For a relative precision of 10^{-5} in G we require an uncertainty of the order of 10^{-6} in electrical measurements. For example, choosing a coil of 10^3 turns having a self inductance of 1 mH gives a current change of 8 mA and a voltage of 3 μV. These changes are of such a magnitude as to pose no serious problem in measuring them to a good accuracy.

3.4. Discussion

The advantages of our proposed method over the torsion fibre method which have been given in detail in Section 2 can be summarised as follows

a) The stiffness of the flexure pivot is proportional to a low power of its thickness while the weight that it can support is proportional to its cross-sectional area. The latter can be increased to an arbitrary extent by making the flexure wide while maintaining a given thickness and hence a given stiffness. In the flexure balance described by Quinn et al.[15], a flexure of thickness 50 μm and width about 20 mm supported a mass of nearly 5 kg while maintaining the stress in the flexure relatively low.

b) The possibility of having large suspended masses allows much larger gravitational torques to be obtained (nearly 10^5 times those present in Luther and Towler's experiment). This should lead to a much larger ratio of the gravitational signal to the non-gravitational noise.

c) The measurement of the change in gravitational potential energy in terms of an equal but opposite change in magnetic energy using a superconducting system avoids the need for any detailed knowledge of the geometry of the magnetic system or magnitude of the stiffness of the flexure pivot and their evolution with time.

d) By proper choice of the natural period of the balance it can be made practically insensitive to tilt and accelerations due to microseismic ground movements which otherwise would be one of the principal sources of uncertainty.

In addition to the measurement of the electrical quantities it would, of course, also be necessary to measure various mechanical quantities. These include the masses of the suspended and attracting masses, the geometry and mass distribution in the balance and the distances of the attracting masses from the balance. Although by no means a trivial matter these measurements can be made using classical metrological methods and will not be discussed here.

In order to reduce noise torques due to nearby moving objects a quiet deserted site would be required.

The principal unknown quantity in the present proposal is the behaviour of the balance itself, its quality factor, general mechanical stability and mechanical noise. In experiments recently made at the Cavendish Laboratory in Cambridge[15], anelasticity of a Be-Cu flexure pivot was measured using a 1 m vertical dumb-bell balance of the type suggested here. Preliminary results were very encouraging and in fact the servo system of the balance was calibrated using the gravitational attraction of a 14 kg mass rotated about one of the extremities of the balance as in Fig. 3. Further experiments on anelasticity are now underway using both Be-Cu and single crystal silicon flexures, using the same balance which has now been transfered to BIPM.

REFERENCES

1. FISCHBACH, E., SUDARSKY, D., SZAFER, A., TALMADGE, C., ARONSON, S.H. Phys. Rev. Letters, 56 (1), 1986, pp. 3-6.

2. STUBBS, C.W., ALDELBURGER, E.G., RAAB, F.J., GUNDLACH, J.M., HECKEL, B.R., McMURRY, K.D., SWANSON, H.E. and WATANABE, R. Phys. Rev. Letters, 58 (11), 1987, pp. 1070-1073.

3. GILLIES, G.T. Metrologia, 24 (Suppl.), 1987, pp. 1-56.

4. SPEAKE, C.C. and GILLIES, G.T. Proc. R. Soc. A (in press).

5. AUGUSTIN, R., DE BOER, H., HAARS, H. and MICHAELIS, W. Feinwerktech. und Messtech., 89, 1981, pp. 280-283.

6. KEISER, G.M. and FALLER, J.E. Proc. 2nd Marcel Grossmann Meeting on Gen. Relativity, R. Ruffini ed., North Holland Pub. Co, 1982, pp. 969-976.

7. KOLDEWYN, W.A. and FALLER, J.E. Bull. Am. Phys. Soc., 17, 1972, pp. 472.

8. RITTER, R.C. Private communication.

9. BRAGINSKI, V.B. and PANOV, V.I. Soviet Phys. JETP, 34, 1972, pp. 463-466.

10. QUINN, T.J., SPEAKE, C.C. and DAVIS, R.S. Metrologia, 23 (2), 1986/87, pp. 87-100.

11. Mc COMBIE, C.W. Rep. Prog. Phys., 16, 1953, pp. 266-320.

12. RITTER, R.C. and GILLIES, G.T. Phys. Rev. A, 31, 1985, pp. 995-1000.

13. BRAGINSKI, V.B. and MANUKIN, A.B. 'Measurement of weak forces in physics experiments', D.M. Douglas ed., Chicago University Press, 1977.

14. LANDAU, L.D. and LIFSCHITZ, E.M. In Statistical Physics, London, Pergamon Press, 1959, pp. 391-402.

15. QUINN, T.J. and SPEAKE, C.C. BIPM Rapport BIPM-87/3.

16. SPEAKE, C.C. Proc. R. Soc. A (submitted for publication).

17. KARAGIOZ, O.V., VORONKOV, V.V., IZMAYLOV, V.P. and AGAFANOV, N.I. Izv., Physics of the Solid Earth, 11, 1975, pp. 69-74.

18. SPEAKE, C.C. and GILLIES, G.T. Zeitschrift für Naturforschung, 1987 (in press).

19. HORA, K., SHIOTA, F. and KUBOTA, T. In Precision Measurement and Fundamental Constants II, B.N. Taylor and W.D. Phillips eds., Natl. Bur. Stand. (U.S.) Spec. Publ. 617, 1984, pp. 479-482.

INTRODUCTION TO THE THEORY OF FIELDS IN FINSLER SPACES

Yoshiro Takano
Department of Physics
Yokohama National University
156 Tokiwa-dai, Hodogaya-ku
Yokohama 240, Japan

ABSTRACT. A brief survey of the theory of fields in Finsler spaces is presented. In the microscopic world, introducing the internal freedom into the elementary particles, the theory is expected to resolve two fundamental problems, that is, the divergence difficulty and the unified description of elementary particles. Abelian or non-Abelian gauge fields are necessarily introduced into the equation of spinor field. In the macroscopic world, effects originating from the directional dependence of metric will be expected to appear. Through the medium of torsion, the external and the internal energy-momentum seem to be connected with each other. Generalizations of Finsler spaces are also discussed.

1. INTRODUCTION

Roughly speaking, Finsler spaces are spaces whose metric tensor $g_{\mu\nu}$ depends on the directional variables y^λ as well as the coordinates x^λ. The directional variables y^λ transform like components of a contravariant vector under a coordinate transformation: $x^{\lambda'} = x^{\lambda'}(x^\lambda)$ Finsler geometry, however, is formulated so as to depend only on the direction of vector y^λ, but not its magnitude.

Accordingly, with respect to the infinitesimal changes of the directional variables and the coordinates, respectively, two kinds of affine connections $C_\lambda{}^\mu{}_\nu$ and $\Gamma^*_\lambda{}^\mu{}_\nu$ are defined, and further, three kinds of curvature tensors $S_\kappa{}^\lambda{}_{\mu\nu}$, $P_\kappa{}^\lambda{}_{\mu\nu}$ and $R_\kappa{}^\lambda{}_{\mu\nu}$ are defined, in the analogous way as Riemannian geometry.

The absolute differential and the covariant derivatives of vector field $V(x,y)$, which is assumed to be homogeneous of degree zero in y^λ, are defined as follows:

$$DV^\mu = V^\mu{}_{|\lambda}\, dx^\lambda + V^\mu{}_{\|\lambda}\, Dl^\lambda, \tag{1.1}$$

$$\begin{cases} V^\mu{}_{|\lambda} = \dfrac{\partial V^\mu}{\partial x^\lambda} - \dfrac{\partial V^\mu}{\partial y^\kappa}\Gamma^{*\kappa}_{\iota}{}_\lambda y^\iota + V^\kappa \Gamma^*_\kappa{}^\mu{}_\lambda, & (1.2) \\ V^\mu{}_{\|\lambda} = F\dfrac{\partial V^\mu}{\partial y^\lambda} + V^\kappa F C_\kappa{}^\mu{}_\lambda. & (1.3) \end{cases}$$

V. De Sabbata and V. N. Melnikov (eds.),
Gravitational Measurements, Fundamental Metrology and Constants, 459–466.

The metric fundamental function

$$F^2 = e g_{\mu\nu} y^\mu y^\nu , \tag{1.4}$$

where e is $+1$ or -1 according as y^λ is time-like or space-like, is multiplied to make any geometrical quantities to be homogeneous of degree zero in y^λ . For example, $l^\lambda = y^\lambda / F$.

The metric tensor is, inversely, given as follows:

$$g_{\mu\nu} = \frac{1}{2} \frac{\partial^2 (e F^2)}{\partial y^\mu \partial y^\nu} . \tag{1.5}$$

The absolute differential and the covariant derivatives of the metric tensor vanish identically:

$$D g_{\mu\nu} = 0 , \tag{1.6}$$

$$\begin{cases} g_{\mu\nu|\lambda} = 0 , & (1.7) \\ g_{\mu\nu\|\lambda} = 0 , & (1.8) \end{cases}$$

that is, such spaces as above are metrical.

Since new freedom is introduced, the theory of fields in Finsler spaces is expected to play several important parts both in the microscopic and the macroscopic worlds. In the microscopic world, introducing the internal freedom into the elementary particles, the theory is expected to resolve two fundamental problems, that is, the divergence difficulty and the unified description of elementary particles. It is also found that Abelian or non-Abelian gauge fields are necessarily introduced into the equation of spinor field. In the macroscopic world, effects originating from the directional dependence of metric will be expected to appear. Through the medium of torsion, the external and the internal energy-momentum seem to be connected with each other.

For the general references, see (1).

2. FIELDS OF ELEMENTARY PARTICLES

2.1. Scalar Field and Internal Freedom

In the theory of elementary particles in Lorentz spaces, to overcome the above-mentioned difficulties originating from the point particle, an attempt of nonlocal field theory has been proposed. In the nonlocal field theory, besides the usual external coordinates, the internal, or relative coordinates are used to describe the internal freedom of elementary particles.

Since any geometrical quantities in Finsler spaces depend on the directional variables as well as coordinates, the internal freedom is introduced in the fields of elementary particles and the effects equivalent to those of the nonlocal field theory can be expected.

For example, the equation of scalar field $\varphi(x,y)$ in Finsler spaces can be assumed as follows:

$$[\Box_{xx} + f(F\frac{\partial}{\partial y^\mu}, l^\mu, g_{\mu\nu}, \lambda)] \varphi(x,y) = 0 , \tag{2.1}$$

where $\Box_{xx}$ is the generalized d'Alembertian and the operator f is a certain invariant function of $F\,\partial/\partial y^\mu$, $l^\mu = y^\mu/F$, $g_{\mu\nu}$ and a constant λ with the dimension of length. This equation is nothing but the generalized equation of nonlocal scalar field. When Finsler spaces are reduced to Minkowski spaces and further to Lorentz spaces, the equation of scalar field may be separated to the usual Klein-Gordon equation and the eigenvalue equation concerning the internal freedom:

$$\begin{cases} (\Box + \kappa^2)\,\varphi^{(e)}(x) = 0, & (2.2)\\ [f(F\frac{\partial}{\partial y^\mu}, l^\mu, \lambda) - \kappa^2]\,\varphi^{(i)}(y) = 0, & (2.3)\end{cases}$$

$$\varphi(x,y) = \varphi^{(e)}(x)\,\varphi^{(i)}(y). \tag{2.4}$$

As to the divergence difficulty, the singularity of propagators might be removed by virtue of the average with respect to the directional variables.

In passing, it must be mentioned that the strings and the superstrings are closely connected with the nonlocal field theory.[2]

2.2. Spinor Field and Associated Gauge Fields

Spinor field $\psi(x,y)$ is naturally assumed to be homogeneous of degree zero in y^λ, and its absolute differential and covariant derivatives can be defined as follows:

$$D\psi = \psi_{|\lambda}\,dx^\lambda + \psi_{\|\lambda}\,Dl^\lambda, \tag{2.5}$$

$$\begin{cases} \psi_{|\lambda} = \frac{\partial\psi}{\partial x^\lambda} - \frac{\partial\psi}{\partial y^\kappa}\Gamma^{*\kappa}_{\iota\lambda}y^\iota - \Gamma^*_\lambda\psi, & (2.6)\\ \psi_{\|\lambda} = F\frac{\partial\psi}{\partial y^\lambda} - FC_\lambda\psi, & (2.7)\end{cases}$$

where Γ^*_λ and C_λ are spin affine connections and the latter is assumed to satisfy the condition:

$$C_\lambda y^\lambda = 0. \tag{2.8}$$

These spin affine connections are determined by the requirement that the covariant derivatives of γ^μ matrices vanish identically:

$$\begin{cases} \frac{\partial\gamma^\mu}{\partial x^\lambda} - \frac{\partial\gamma^\mu}{\partial y^\kappa}\Gamma^{*\kappa}_{\iota\lambda}y^\iota + \gamma^\nu\Gamma^{*\mu}_{\nu\lambda} - \Gamma^*_\lambda\gamma^\mu + \gamma^\mu\Gamma^*_\lambda = 0, & (2.9)\\ F\frac{\partial\gamma^\mu}{\partial y^\lambda} + \gamma^\nu F C_\nu{}^\mu{}_\lambda - FC_\lambda\gamma^\mu + \gamma^\mu F C_\lambda = 0, & (2.10)\end{cases}$$

where $\Gamma^{*\mu}_{\nu\lambda}$ and $C_\nu{}^\mu{}_\lambda$ are affine connections. The γ^μ matrices are related to the metric tensor as follows:

$$\{\gamma^\mu, \gamma^\nu\} = \gamma^\mu\gamma^\nu + \gamma^\nu\gamma^\mu = 2g^{\mu\nu}. \tag{2.11}$$

Spin affine connections Γ^*_λ and C_λ, however, are not determined unambiguously and these arbitrarinesses can be referred to the gauge transformations, in the same way as the case of Riemannian spaces.

It must also be emphasized that such gauge fields as above can be regarded as either Abelian or non-Abelian. One of these gauge fields may be identified to electromagnetic field and the other to Yang-Mills field.

The condition (2.8) is a kind of radial gauge and might be called indicatory gauge, since the above relation means that C_λ is indicatory in Finsler geometry.

The tangent space M_x at each point x of an n-dimensional Finsler space M with a metric fundamental function $F(x,y)$ is regarded as an n-dimensional Riemannian space with the metric tensor $g_{\mu\nu}(x,y) = \frac{1}{2}\cdot\partial(F^2)/\partial y^\mu \partial y^\nu$, where x is fixed. The indicatrix I_x at a point x is a hypersurface of the Riemannian space M_x, which is defined by the equation $F(x,y) = 1$, where x is fixed. Consequently, I_x is thought of as an (n-1)-dimensional Riemannian space with the Riemannian metric induced from the one of M_x. Namely, the indicatrix is a generalization of the unit sphere and a representation of anisotropy of space.

The behavior of any indicatory tensors is proved to be governed completely by their projection on indicatrix.[3] Indicatory tensors, therefore, can be regarded as those on indicatrix.

In the theory of fields in Finsler spaces, the internal physical quantities depend on the internal coordinates, that is, on the directional variables and, therefore, are expected to be represented by the geometrical quantities on indicatrix. The gauge field satisfying the indicatory gauge condition is regarded as the field on indicatrix, and, consequently, it must be essentially associated with the internal freedom of elementary particles.

This situation might be imaged in such a way that the indicatrix is an elastic membrane bounding the elementary particle and a gauge transformation of spinor field generates a strain on this membrane, being accompanied by a gauge field as a stress.

3. GRAVITATIONAL FIELD

3.1. Equations of Gravitational Field --- Directional Dependence

To determine the structure of space-time, the equations of gravitational field in Finsler spaces must be proposed.

Regarding the y-dependence of the metric tensor $g_{\mu\nu}$, in analogy with its x-dependence, we propose the following equation:

$$\hat{S}_{\mu\nu} - \frac{1}{2} g_{\mu\nu} \hat{S} = -\kappa^{(i)} T^{(i)}_{\mu\nu}, \tag{3.1}$$

where $\kappa^{(i)}$ and $T^{(i)}_{\mu\nu}$ might be termed the internal gravitational constant and the internal energy-momentum tensor.

The first curvature tensor $\hat{S}_\nu{}^\mu{}_{\lambda\kappa}$ is given as follows:

$$\hat{S}_\nu{}^\mu{}_{\lambda\kappa} = F^{-2} S_\nu{}^\mu{}_{\lambda\kappa} = C_\alpha{}^\mu{}_\lambda C_\nu{}^\alpha{}_\kappa - C_\alpha{}^\mu{}_\kappa C_\nu{}^\alpha{}_\lambda, \tag{3.2}$$

$$C_{\nu\mu\lambda} = \frac{1}{2}\frac{\partial g_{\nu\mu}}{\partial y^\lambda} = \frac{1}{4}\frac{\partial^3(F^2)}{\partial y^\nu \partial y^\mu \partial y^\lambda}. \tag{3.3}$$

The internal energy-momentum tensor $T^{(i)}_{\mu\nu}$ is symmetric and is assumed that its divergence with respect to y vanishes:

$$T^{(i)\mu\nu}{}_{\|\mu} = 0. \tag{3.4}$$

In other words, the conservation of the internal energy-momentum in the internal space is assumed. Since the left-hand side of eq.(3.1) is shown to be divergence-free by means of Bianchi identity, eq.(3.1) is consistent with the condition (3.4).

On the indicatrix eq.(3.1) is reduced to Einstein's equation in three-dimensional Riemannian spaces, and the internal cosmological term is automatically introduced in this equation.

For $T^{(i)}_{\mu\nu} = 0$, eq.(3.1) is reduced to

$$\hat{S}_{\mu\nu} = 0. \tag{3.5}$$

In four-dimensional Finsler spaces, eq.(3.5) is proved to lead necessarily

$$\hat{S}_{\nu}{}^{\mu}{}_{\lambda\kappa} = 0, \tag{3.6}$$

whether the metric is definite or indefinite.[3]

Although Brickell's theorem holds in Finsler spaces with positively definite metric, it has not been concluded whether indefinite Finsler spaces satisfying the condition (3.6) are Riemannian. However, if Brickell's theorem is supposed to hold also in the case of the indefinite metric, our physical space-time will be pictured in such a way that it is Finslerian in the region with internal source and is Riemannian in the empty region.

It seems quite natural to suppose that the internal energy-momentum is given by the fields of elementary particles. Then, the elementary particles inspire space-time with the directional dependence and, conversely, in Finsler spaces the elementary particles are inspired with the internal freedom.

3.2. Equations of Gravitational Field

With regard to the x -dependence of the metric tensor, it is natural to assume, as Riemannian spaces, the following equation:

$$R_{\mu\nu} - \frac{1}{2} g_{\mu\nu} R = \kappa T_{\mu\nu}, \tag{3.7}$$

i.e.

$$\left\{ \begin{aligned} & R_{\underline{\mu\nu}} - \frac{1}{2} g_{\mu\nu} R = \kappa T_{\underline{\mu\nu}}, && (3.8) \\ & R_{\underset{\vee}{\mu\nu}} = \kappa T_{\underset{\vee}{\mu\nu}}. && (3.9) \end{aligned} \right.$$

The antisymmetric part of the contracted third curvature tensor is given as below:

$$R_{\underset{\vee}{\mu\nu}} = \frac{1}{2} (C_{\mu}{}^{\lambda}{}_{\alpha} R_{\nu}{}^{\alpha}{}_{\lambda} + C_{\nu}{}^{\lambda}{}_{\alpha} R_{\lambda}{}^{\alpha}{}_{\mu} + C_{\lambda}{}^{\lambda}{}_{\alpha} R_{\mu}{}^{\alpha}{}_{\nu}), \tag{3.10}$$

where

$$R_{\lambda}{}^{\alpha}{}_{\kappa} = R_{\beta}{}^{\alpha}{}_{\lambda\kappa} y^{\beta} \tag{3.11}$$

is a torsion tensor.

The third curvature tensor $R_{\nu}{}^{\mu}{}_{\lambda\kappa}$ in Finsler spaces consists, in general, of three parts, that is, the first part originating from the torsion, the second originating in the y -dependence of the metric tensor $g_{\mu\nu}$ and the connection coefficient $\Gamma^{*}{}_{\nu}{}^{\mu}{}_{\lambda}$, and the third which has the same form as one of Riemannian spaces, but depends still on the directional variables, different from Riemannian spaces.

$$R_\nu{}^\mu{}_{\lambda\kappa} = K_\nu{}^\mu{}_{\lambda\kappa} + C_\nu{}^\mu{}_\alpha K_\beta{}^\alpha{}_{\lambda\kappa} y^\beta, \tag{3.12}$$

$$K_\beta{}^\alpha{}_{\lambda\kappa} y^\beta = R_\beta{}^\alpha{}_{\lambda\kappa} y^\beta = R_\lambda{}^\alpha{}_\kappa, \tag{3.13}$$

$$K_\nu{}^\mu{}_{\lambda\kappa} = \left(\frac{\partial}{\partial x^\kappa} - \Gamma^{*\alpha}_{\beta\kappa} y^\beta \frac{\partial}{\partial y^\alpha}\right)\Gamma^{*\mu}_{\nu\lambda} - \left(\frac{\partial}{\partial x^\lambda} - \Gamma^{*\alpha}_{\beta\lambda} y^\beta \frac{\partial}{\partial y^\alpha}\right)\Gamma^{*\mu}_{\nu\kappa} + \Gamma^{*\mu}_{\iota\kappa}\Gamma^{*\iota}_{\nu\lambda} - \Gamma^{*\mu}_{\iota\lambda}\Gamma^{*\iota}_{\nu\kappa}, \tag{3.14}$$

$$\Gamma^*_{\nu\mu\lambda} = \gamma_{\nu\mu\lambda} - C_{\lambda\mu\kappa}\Gamma^{*\kappa}_{\beta\nu} y^\beta - C_{\nu\mu\kappa}\Gamma^{*\kappa}_{\beta\lambda} y^\beta + C_{\nu\lambda\kappa}\Gamma^{*\kappa}_{\beta\mu} y^\beta, \tag{3.15}$$

$$\gamma_{\nu\mu\lambda} = \frac{1}{2}\left(\frac{\partial g_{\nu\mu}}{\partial x^\lambda} + \frac{\partial g_{\mu\lambda}}{\partial x^\nu} - \frac{\partial g_{\lambda\nu}}{\partial x^\mu}\right). \tag{3.16}$$

Contracting Bianchi identity, we obtain

$$\left(R^{\mu\nu} - \frac{1}{2} g^{\mu\nu} R\right)_{|\mu} = U^\nu, \tag{3.17}$$

where

$$U_\nu = \frac{1}{2} g^{\iota\lambda}\left(\hat{P}_\iota{}^\kappa{}_{\lambda\alpha} R_\kappa{}^\alpha{}_\nu + \hat{P}_\iota{}^\kappa{}_{\kappa\alpha} R_\nu{}^\alpha{}_\lambda + \hat{P}_\iota{}^\kappa{}_{\nu\alpha} R_\lambda{}^\alpha{}_\kappa\right). \tag{3.18}$$

Hence, we have to assume

$$T^{\mu\nu}{}_{|\mu} = \frac{1}{\kappa} U^\nu. \tag{3.19}$$

The vector U_ν acts as a source or a sink of the energy-momentum tensor $T_{\mu\nu}$, or acts as a force density. Namely, the energy-momentum inflows to or outflows from the internal space and the torsion (the dislocation, so to speak) of the external space.

Such non-conservation of the energy-momentum as above is not anything but the reflection of the inhomogeneity of space-time due to the existence of the torsion.

As to the second curvature tensor $P_\nu{}^\mu{}_{\lambda\kappa}$, its geometrical meaning has not yet been clarified and, therefore, the question how the equation is constructed from the second curvature tensor still remains unsettled.

3.3. Finsler Spaces of Constant Curvature

In order to clarify the significance of the non-conservation of the external energy-momentum in Finsler spaces, let us take up spaces of constant curvature.

In Finsler spaces of constant curvature the torsion tensor is given as follows:

$$R_\lambda{}^\alpha{}_\kappa = K\left(y_\lambda g^\alpha{}_\kappa - y_\kappa g^\alpha{}_\lambda\right), \tag{3.20}$$

where K is the scalar curvature and a constant. In these spaces the antisymmetric part of the contracted curvature tensor $R_{\mu\nu}$ is easily shown to be zero:

$$R_{\underset{\vee}{\mu\nu}} = 0, \tag{3.21}$$

and the vector U^ν can be reduced to the form of the divergence of the symmetric tensor as below:

$$U^\nu = \left(\frac{1}{2} K \hat{S} y^\mu y^\nu\right)_{|\mu}, \tag{3.22}$$

in virtue of the relation:

$$\hat{P}_\nu{}^\mu{}_{\underset{\vee}{\lambda\kappa}} - F^{-1} P_\nu{}^\mu{}_{\underset{\vee}{\lambda\kappa}} = -\frac{1}{2}\hat{S}_\nu{}^\mu{}_{\lambda\kappa|\beta} y^\beta. \tag{3.23}$$

Hence, eq.(3.17) is written as follows:

$$(R^{\mu\nu} - \frac{1}{2} g^{\mu\nu} R - \frac{1}{2} K \hat{S} y^\mu y^\nu)_{|\mu} = 0. \tag{3.24}$$

According to eq.(3.1), $\hat{S}$ can be determined from the internal energy-momentum tensor $T^{(i)}_{\mu\nu}$ as

$$\hat{S} = \kappa^{(i)} T^{(i)\ \lambda}_{\ \ \lambda}. \tag{3.25}$$

Therefore, we have to assume

$$(T^{\mu\nu} - \frac{\kappa^{(i)}}{2\kappa} K T^{(i)\ \lambda}_{\ \ \lambda} y^\mu y^\nu)_{|\mu} = 0. \tag{3.26}$$

It means that the total energy-momentum is conserved, if we regard a combination of the internal and the external energy-momentum tensors: $T_{\mu\nu} - \frac{\kappa^{(i)}}{2\kappa} K T^{(i)\lambda}_{\lambda} y_\mu y_\nu$ as the total energy-momentum tensor: $T^{total}_{\mu\nu}$. Consequently, the external and the internal energy-momentum are connected with each other, through the medium of torsion.

3.4. Geodesic Line and Vertical Path

By the variation principle we can derive the equation of the geodesic line in Finsler spaces as follows:

$$\frac{d^2 x^\lambda}{ds^2} + T^{*\ \lambda}_{\mu\ \nu}(x, \frac{dx}{ds}) \frac{dx^\mu}{ds} \frac{dx^\nu}{ds} = 0, \tag{3.27}$$

choosing the arc-length s as the parameter. Though it can be reduced to the same form as in Riemannian spaces, it remains the y -dependence of the connection coefficient, showing the directional dependence of the force acting on the particle.

The analogue of the geodesic line regarding y will be called the vertical path and is also derived by the variation principle.

Choosing the angle φ as the parameter, we obtain

$$\frac{d^2 y^\lambda}{d\varphi^2} + C_{\mu\ \nu}^{\ \lambda}(y) \frac{dy^\mu}{d\varphi} \frac{dy^\nu}{d\varphi} = 0, \tag{3.28}$$

where the angle $d\varphi$ between the vectors y^λ and $y^\lambda + dy^\lambda$ can be defined as below:

$$d\varphi^2 = e\, \varphi_{\mu\nu} dy^\mu dy^\nu, \tag{3.29}$$

$$\varphi_{\mu\nu} = \frac{1}{F} \frac{\partial^2 F}{\partial y^\mu \partial y^\nu}. \tag{3.30}$$

As a whole, a free particle in Finsler spaces moves along the geodesic line (3.27), being accompnied by the rotation of y with the angular acceralation (3.28).

As the next problem, studying the geodesic deviation and the conjugate point, we have to examine the singularity of Finsler spaces, in anticipation of the different situation from Riemannian spaces.

It must be mentioned here that Tavakol and van den Bergh, recently, proved one of special Finsler spaces, Berwald space to have the conformal and the projective structures.[14]

4. GENERALIZATIONS OF FINSLER SPACES

To increase the degrees of internal freedom, it will be necessary to generalize Finsler spaces in such a way that the geometrical quantities depend not only upon the direction of the vector y^λ but also upon its absolute value. The theory of fields in these generalized Finsler spaces can be formulated.

It must also be noted that the above-mentioned generalization is suited for the formulation of the geometry of Finsler spaces with indefinite metric.

In order to make the internal freedom to admit the symmetry SU(3), moreover, we have to define further generalized Finsler spaces, in which the geometrical quantities depend on plural, not single, directional vectors.

It is noticed that our theories of fields in Cartan's Finsler spaces, in generalized Finsler spaces, and in further generalized Finsler spaces correspond to the three stages of Yukawa's nonlocal field theory, respectively.

From a different point of view, Finsler spaces may be regarded as the superposition of Riemannian spaces corresponding to various values of the directional variables.

The differential geometry of spaces whose metric tensor $g_{\mu\nu}$ depends on the spinor variables ξ and $\bar{\xi}$ as well as the coordinates x^λ can be formulated in analogy with the differential geometry of Finsler spaces.

Such spaces as above may be regarded as spinor bundles in the same way that Finsler spaces may be regarded as tangent bundles.

The connection coefficients $\Gamma^{(*)}{}_\nu{}^\mu{}_\lambda$, $C_\nu{}^\mu$ and $\bar{C}_\nu{}^\mu$ and the spin connection coefficients $\Gamma^{(*)}_\lambda$, C and $\bar{C}$ are determined in such a way that the absolute differentials and the covariant derivatives of the metric tensor $g_{\mu\nu}$ and the generalized Dirac matrices γ^λ vanish identically.

The spin connection coefficients $\Gamma^{(*)}_\lambda$, C and $\bar{C}$, however, are not determined unambiguously and these arbitrarinesses can be referred to the gauge transformations, in the same manner as in Riemannian and Finsler spaces. Although the arbitrary term of $\Gamma^{(*)}_\lambda$ is interpreted as a vector gauge field, the arbitrary terms of C and $\bar{C}$ are interpreted as spinor gauge fields, which may intermediate between fermions and bosons.

REFERENCES

(1) Y. Takano: *Prog. Theor. Phys.*, 32, 365 (1964); 40, 1159 (1968); *Lett. Nuovo Cimento*, 10, 747 (1974); 11, 486 (1974); *Proceedings of the International Symposium on Relativity and Unified Field Theory, Calcutta*, p.17 (1975-76); *Lett. Nuovo Cimento*, 35, 213 (1982); *Tensor*, 40, 249 (1983).

(2) J. Rayski: *Prog. Theor. Phys.*, 76, 1393 (1986).

(3) M. Matsumoto: *Foundations of Finsler Geometry and Special Finsler Spaces* (Kaiseisha Press, Otsu, Japan, 1986).

(4) R. K. Tavakol and N. van den Bergh: *Phys. Lett.*, 112A, 23 (1985); *General Relativity and Gravitation*, 18, 849 (1986).

NEUTRINOS, GRAVITONS, METROLOGY, AND GRAVITATIONAL RADIATION

J. Weber
University of Maryland, College Park, MD 20742
and
University of California, Irvine CA 92717

ABSTRACT. New methods have been developed for observation of neutrinos and antineutrinos. These represent improvement in detection sensitivity by more than 20 orders, and provide a method for precise measurements of the Fermi weak interaction constant.

Theory and experiments are given for elastic scattering by quarks in a crystal.

A solar Eötvös experiment with equal masses having very different Debye temperatures, gives a positive result due to the solar neutrinos. We are observing the solar neutrinos.

The theory of scattering by strongly coupled scatterers is reviewed and extended for coherent scattering of MeV antineutrinos.

Antineutrinos from a reactor are scattered by a crystal, on one arm of a torsion balance. A 5 kilogram single crystal of sapphire placed in the line of sight between reactor and torsion balance, is employed as a shield for the antineutrinos. A force $3.9 \pm 0.4 \times 10^{-5}$ dynes is observed as a result of coherent scattering.

The theory and experiments have stimulated a new approach to the theory of gravitational radiation antennas. Some solutions of the antenna response as a many body problem, give much larger cross sections for pulses. Elastic solid antenna sensitivity for pulses is much greater than can be achieved by interferometer antennas in the forseeable future. A method is outlined for greatly increasing elastic solid antenna bandwidths.

Results of antenna observations during the rapid evolutionary phase of supernova 1987A are presented.

V. De Sabbata and V. N. Melnikov (eds.),
Gravitational Measurements, Fundamental Metrology and Constants, 467–499.

TABLE OF CONTENTS

A. NEW METHOD FOR OBSERVATION OF NEUTRINOS AND ANTINEUTRINOS

A new method for observation of neutrinos and antineutrinos employs a nearly perfect single crystal, with high Debye temperature.[1,2,3] Total cross sections are proportional to the square of the number of scatterers. We consider first the traditional approach to scattering problems, which does not give large cross sections.

We imagine a square array of N scatterers, which are delta function potentials, all in the x - y plane. Scatterer spacing is b, and the length L of each side of the square is given by $L = N^{1/2}b$. A scatterer located at $x = n_x b$, $y = n_y b$ has potential

$$U(\bar{r}) = B\,\delta(x - n_x b)\,\delta(y - n_y b)\,\delta(z) \tag{A-1}$$

The interaction Hamiltonian is

$$H' = \frac{B}{V}\sum_{n_x=1}^{n_x=N^{1/2}}\;\sum_{n_y=1}^{n_y=N^{1/2}} e^{i(\bar{p}_{IO}-\bar{p}_{IF})_x n_x b/\hbar + i(\bar{p}_{IO}-\bar{p}_{IF})_y n_y b/\hbar} \tag{A-2}$$

In (A-2) V is the volume, incident particles move in the Z direction with momentum $\bar{p}_{IO}$, and elastically scattered particles have momentum $\bar{p}_{IF}$.

First order perturbation theory and Fermi's golden rule give for the forward scattering part of the cross section

$$\sigma_F = \frac{\pi B^2 N}{(b\hbar c)^2} \tag{A-3}$$

(A-3) is proportional to N, and this occurs in the following way.

$$\sigma_F = \left(\frac{d\sigma}{d\Omega}\right)_{\text{FORWARD SCATTERING}} \Delta\Omega \qquad \text{(A-4)}$$

In (A-4) $\Delta\Omega$ is the solid angle of the forward scattering peak, $\frac{d\sigma}{d\Omega}$ is the differential cross section, given by

$$\frac{d\sigma}{d\Omega} = \frac{|p_{IO}|^2 B^2 N^2}{4\pi^2 \hbar^4 c^2} \qquad \text{(A-5)}$$

(A-5) is proportional to N^2, because all constributions to scattering in the forward direction arrive in phase. We compute the solid angle $\Delta\Omega$, by carrying out the sums in (A-2). The magnitude is

$$|H'| = \frac{B \sin[\frac{1}{2} N^{1/2} b(\bar{p}_{IO} - \bar{p}_{IF})_x/\hbar] \sin[\frac{1}{2} N^{1/2} b(\bar{p}_{IO} - \bar{p}_{IF})_Y/\hbar]}{V \sin[\frac{1}{2} b(\bar{p}_{IO} - \bar{p}_{IF})_x/\hbar] \sin[\frac{1}{2} b(\bar{p}_{IO} - \bar{p}_{IF})_Y/\hbar]} \qquad \text{(A-6)}$$

The forward peak minima are given by the first values of $(\bar{p}_{IO} - \bar{p}_{IF})_x$, $(\bar{p}_{IO} - \bar{p}_{IF})_Y$ which give zero in the numerator of (A-6).

$$\Delta\Omega = \frac{2\pi^3 \hbar^2}{(b|p_{IO}|)^2 N} \qquad \text{(A-7)}$$

(A-7) is inversely proportional to N, and the product of (A-5) and (A-7) is directly proportional to N.

This analysis, representing scatterers as potentials, is incorrect in principle, to describe scattering. Nuclei are composed of elementary particles, requiring wavefunctions for analyses of their interactions. Elementary particle scatterers, unlike model potentials, may exchange momentum and energy.

Suppose momentum $\Delta\bar{p}$ can be exchanged by a single scatterer which is very tightly coupled to all other scatterers. $\Delta\bar{p}$ is immediately shared, with the crystal recoiling as a single entity. If the particle scattering site cannot be identified after scattering, (A-6) becomes

$$|H'| = \frac{B \sin[\frac{1}{2} N^{1/2} b(\bar{p}_{IO} - \bar{p}_{IF} - \Delta\bar{p})_x/\hbar] \sin[\frac{1}{2} N^{1/2} b(\bar{p}_{IO} - \bar{p}_{IF} - \Delta\bar{p})_Y/\hbar]}{V \sin[\frac{1}{2} b(\bar{p}_{IO} - \bar{p}_{IF} - \Delta\bar{p})_x/\hbar] \sin[\frac{1}{2} b(\bar{p}_{IO} - \bar{p}_{IF} - \Delta\bar{p})_Y/\hbar]} \qquad \text{(A-8)}$$

Consider the differences between (A-6) and (A-8). In (A-6) the first diffraction minima occur for the incident and scattered momenta nearly parallel to each other, with resulting small solid angle (A-7), proportional to 1/N. This follows because the de Broglie wavelength λ is small compared with length L. Phase factors in the sum of exponentials in (A-6) are terms such as

$$\frac{N^{1/2} b\,(\bar{p}_{IO} - \bar{p}_{IF})_x}{\hbar} \tag{A-9}$$

Let $\hat{\iota}_I$ and $\hat{\iota}_F$ be unit vectors in the direction of incident and final momenta. (A-9) is written in the form

$$\frac{2\pi L}{\lambda}\left(\hat{\iota}_I - \hat{\iota}_F\right)_x \tag{A-10}$$

For the large $(\frac{L}{\lambda})$, $\hat{\iota}_I$ must be very nearly parallel to $\hat{\iota}_F$ in order for (A-10) to have the value π for the forward peak minimum, leading to the very small solid angle (7) . However if a single unidentifiable scatterer exchanges momentum $\Delta\bar{p}$ and $\hat{\iota}_{\Delta p}$ is a unit vector in the direction of $\Delta\bar{p}$, (A-10) is replaced by

$$\frac{2\pi L}{\lambda}\left(\hat{\iota}_I - \hat{\iota}_F - \hat{\iota}_{\Delta p}\frac{|\Delta p|}{|p|}\right)_x \tag{A-11}$$

In (A-11) if $|\Delta\bar{p}|$ is comparable with $|\bar{p}|$, $\hat{\iota}_I$ need not be nearly parallel to $\hat{\iota}_F$. There are three vectors involved in the sum, which can be small even if $\hat{\iota}_I$ and $\hat{\iota}_F$ make a large angle with each other, leading to solid angles approaching 4π. Analysis based on (A-8) then gives a total forward scattering cross section

$$\sigma_F \approx \frac{|p_{IO}|^2 B^2 N^2}{\pi \hbar^4 c^2} \tag{A-12}$$

proportional to N^2.

These arguments are valid for weakly interacting particles such as neutrinos, but not for strongly interacting particles. If a strongly interacting particle exchanges a large fraction of its momentum at a single scatterer, that close encounter will modify the wavefunction of the scattering site so much that subsequent measurements would reveal the scattering site. This is equivalent to setting up apparatus to measure through which slit of a multiple slit diffraction grating a photon passes. This destroys the usual interference pattern.

B. ELASTIC SCATTERING OF NEUTRINOS AND ANTINEUTRINOS BY QUARKS IN A NEARLY PERFECT CRYSTAL

The S matrix for interaction of two four current densities is given by

$$S = \frac{1}{\hbar c}\int \langle F|\bar{\psi}_S \Gamma \psi_S \bar{\psi}_I \Xi \psi_I |0\rangle \, d^4x \qquad \text{(B-1)}$$

In (B-1), $|0\rangle$ is the original state, $|F\rangle$ is the final state, $\bar{\psi}_S$ is a creation operator for a scatterer. $\bar{\psi}_I$ is a creation operator for an incident particle. ψ_S and ψ_I are the corresponding annihilation operators. Γ and Ξ are position independent operators.

For localized scatterers in a cubic crystal, in harmonic oscillator ground states, scatterer spin zero, the total cross section for N scatterers is given by[2]

$$\sigma = \frac{|U_{IF} \Xi U_{I0}|^2 E_{IF}^2 N^2}{\pi \hbar^4 c^4} \qquad \text{(B-2)}$$

U_{IF} and U_{IO} are incident particle spinors, for final and incident states, respectively. E_{IF} is the incident particle final state energy.

For N identical tightly bound quarks coupled to neutrinos via neutral current interactions, the cross section is given by[1,2]

$$\sigma_{\text{NEUTRAL CURRENTS}} = \frac{G_W^2 N^2 E_\nu^2}{32\pi^2 \hbar^4 c^4}\int |\langle F|\bar{U}_{\nu F}\gamma^\alpha(1+\gamma_5)U_{\nu 0}\bar{U}_{SF}\gamma_\alpha(C_{1q}+C_{2q}\gamma_5)U_{S0}|0\rangle|^2 d\Omega_I \qquad \text{(B-3)}$$

C_{1q}, C_{2q} are constants which differ for different quarks, Ω is the solid angle into which neutrinos are scattered.

For unpolarized scatterers, these relations give, for either neutrinos or antineutrinos, the total cross section

$$\sigma = \frac{G_W^2 E_\nu^2 (.387 N_U - .693 N_D)^2}{4\pi \hbar^4 c^4} \qquad \text{(B-4)}$$

In B-4, N_U is the total number of up quarks, N_D is the total number of down quarks. E_ν is the neutrino energy. B-4 does not include contributions to the cross section for processes in which 2 or more unidentified scatterers exchange momentum with incident particles. Contributions of these are included in section C. For the experiments to be discussed B-4 already exceeds the crystal area. Multiple scattering is therefore expected.

(B-4) assumes that the quarks are in nuclei of a nearly perfect crystal which is "infinitely stiff." The stiffness of available crystals is described by the Debye temperature. Clearly (B-4) would be expected to be accurate in the low energy limit, where the recoil energy of a single atom is small compared with the energy of the highest frequency phonons.

B.a. Observations

Experiments were carried out with the lowest available energy antineutrinos and a sapphire crystal. A titanium tritide source was employed, with initial activity approximately 3000 Curies. The torsion balance, Figure B-1, measured the momentum transferred by the antineutrinos to a single crystal sapphire target.

Torsion balance forces were measured by observing an electrical force required to balance forces associated with the scattering process. This was done in the following way. The force due to scattering rotates the balance disc. This produces an unbalance of a radiofrequency bridge, and provides an error signal. The error signal is amplified, and the amplified output voltage nearly restores the equilibrium position, by electrostatic forces.

The antineutrino source container mass produced a gravitational interaction, and the Beta electrons produced heat. A "dummy" without tritium, with a heat source, balanced gravitational and thermal effects. The balance was calibrated by replacement of source and dummy by a small lead mass.

Antineutrino source and dummy were secured to rods. A motor driven oscillator raised and lowered the rods cyclicly to vary the distance between source and target. The expected force law implied Green's solutions for the torsion balance response, as follows. For the antineutrino source and "dummy", the Green's solution for a harmonic oscillator displacement $X_{\bar{\nu}}(t)$ is given by

$$X_{\bar{\nu}}(t) = K'_\nu \int_{t_e}^{t} \frac{e^{-\alpha(t-t')} \sin\omega_0(t-t')\,dt'}{\left[a^2 + v^2(t'-t_0)^2\right]^2} \qquad \text{(B-5)}$$

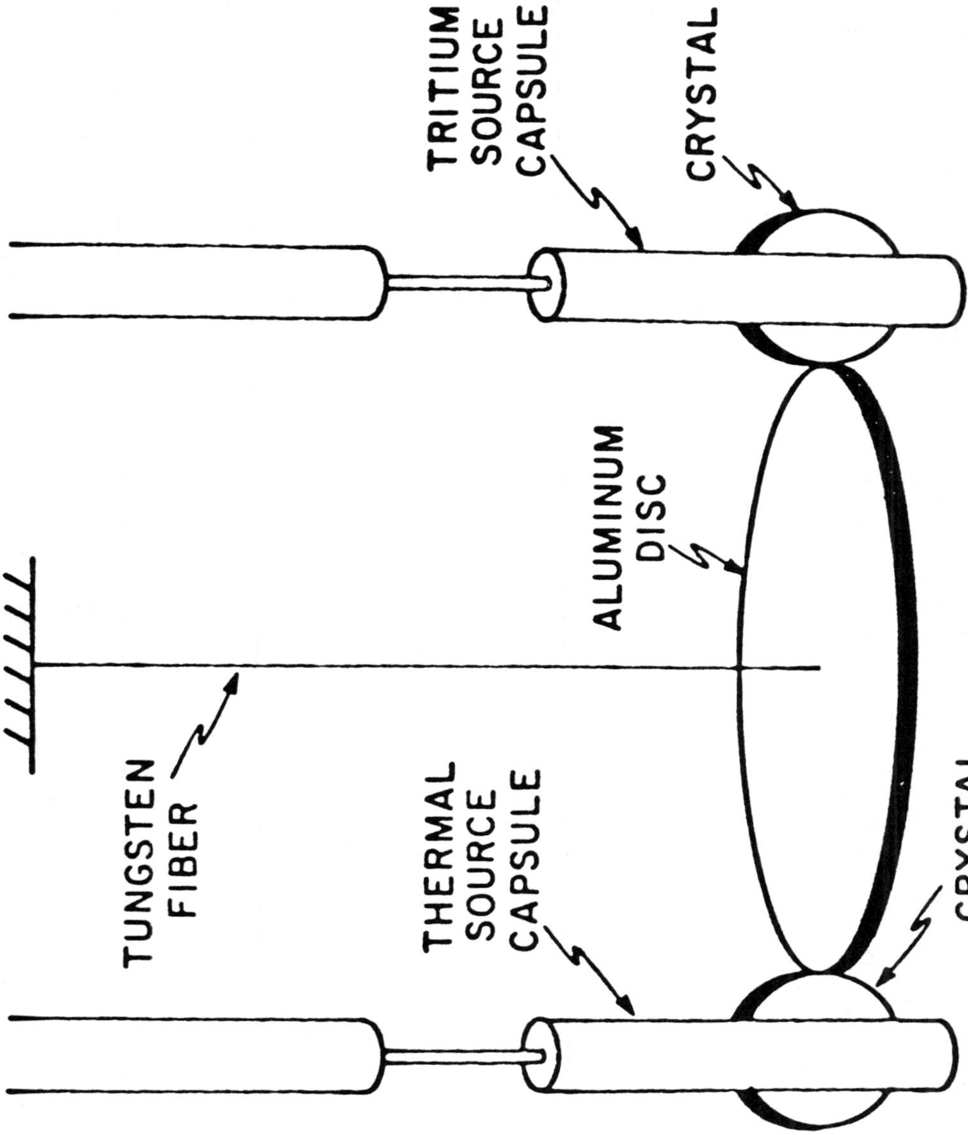

Figure B-1. Torsion Balance for Observation of Coherent Elastic Scattering of Antineutrinos.

(B-5) gives displacement at time t for oscillator initial displacement and initial velocity zero, at some earlier time t_e. K'_ν is a constant, α is the oscillator damping constant, ω_0 is the normal mode frequency, a is the distance of closest approach of the source to the target mass, v is the velocity of the source, and t_0 is the time at which the source reaches its closest approach point.

For the calibration lead mass, there is a different force law and

$$X_g(t) = -K'_m \int_{t_e}^{t} \frac{e^{-\alpha(t-t')} \sin\omega_0(t-t')\,dt'}{\left[a^2 + v^2(t'-t_0)^2\right]^{3/2}} \qquad \text{(B-6)}$$

Data were recorded on magnetic tape, which was read "blind" by our computer programmer, Mr. Gregory Wilmot.

Figure B-2 is the computer drawn predicted response for cycling a 27 gram lead mass, with predicted gravitational signal response based on equation (B-6). The system period was 13.4 seconds. The observed output voltage for signal averaging over 3013 cycles of 64 seconds is shown on Figure B-3.

Figure B-4 is a computer drawn predicted response for cycling the tritium source and "dummy" based on equation (B-5), again with torsion balance servosystem period 13.4 seconds. Figure B-5 is the observed response for tritium source and "dummy", signal averaging over 9826 cycles of 64 seconds.

Figure B-6 is for the parameters of Figures B-4 and B-5, but signal averaging over 359 cycles of 64 seconds. Figure B-7 is for the same parameters as Figures B-4, B-5, B-6, but signal averaging over 552 cycles.

Figure B-8 is the observed response to a pair of "dummies," for signal averaging over 995 cycles of 64 seconds, for torsion balance servosystem periods of 13.4 seconds.

Figure B-9 is the response, again drawn by computer for an antineutrino source, for torsion balance servosystem period of 7.0 seconds.

Figure B-10 is the observed response for tritium source and "dummy" for torsion balance servosystem period of 7.0 seconds, signal averaging over 533 cycles of 64 seconds.

Figure B-11 is the observed response of two heated "dummies," for torsion balance servosystem period of 7.0 seconds, signal averaging over 114 cycles.

There were more than 200 separate runs, each at least several hundred cycles. About half employed the 64 second cycling described earlier, and about half employed 32 second cycling. Two discs with different moments of inertia were employed. Linearity, thermal response, effects of differing source strengths, and effects of different calibration masses in the range 8 to 27 grams were studied.

Data of Figures B-5, B-6, B-7, B-10 are evidence that the antineutrino source is exerting a force opposite in sign to the gravitational force of the lead mass.

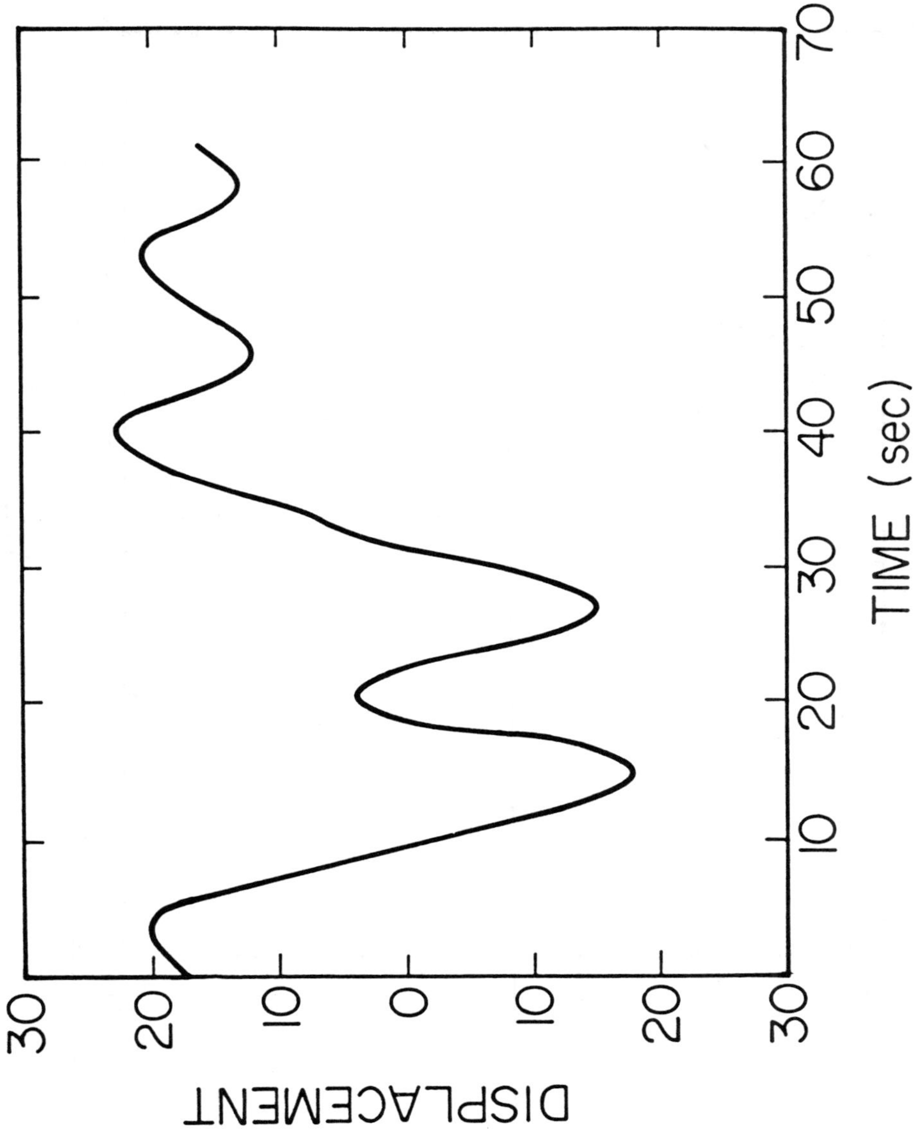

Figure B-2. Predicted Response of Torsion Balance-Servosystem to Gravitational Force due to a 27 gm mass moved vertically with a 64 second period. The Torsion Balance-Servosystem period is 13.4 seconds.

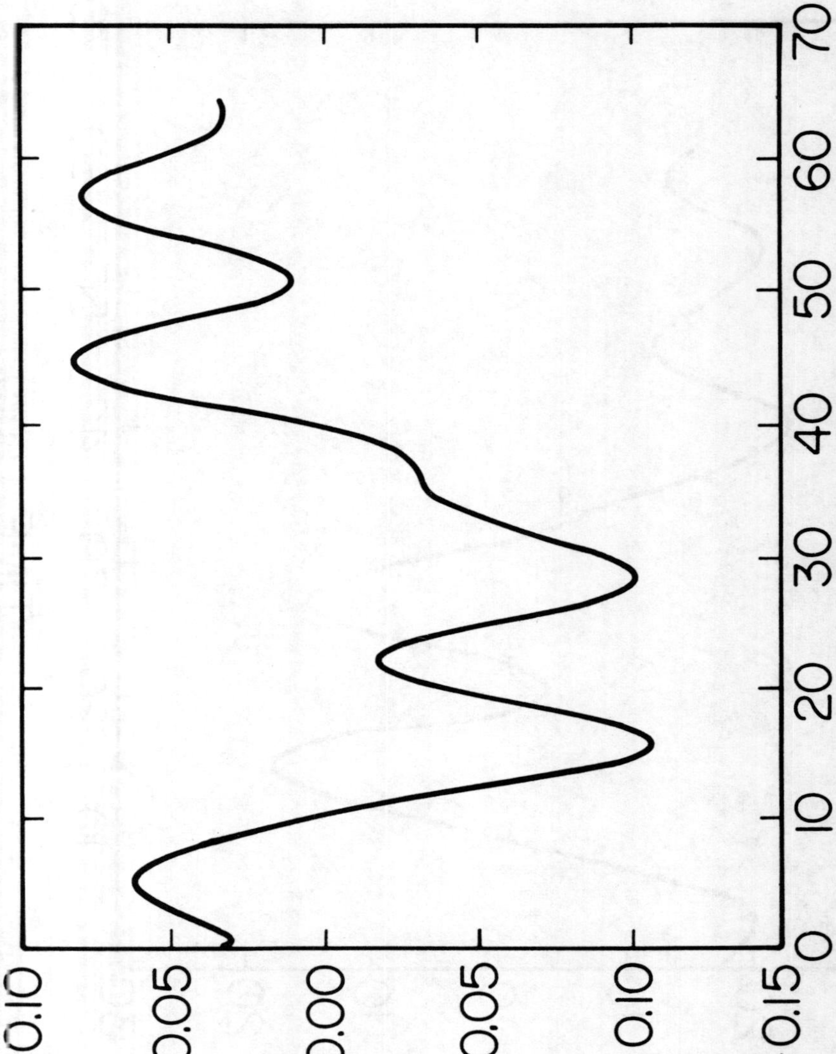

Figure B-3. Observed Response to the Modulated Gravitational Force, with the parameters of Figure 2, averaging over 3013 cycles.

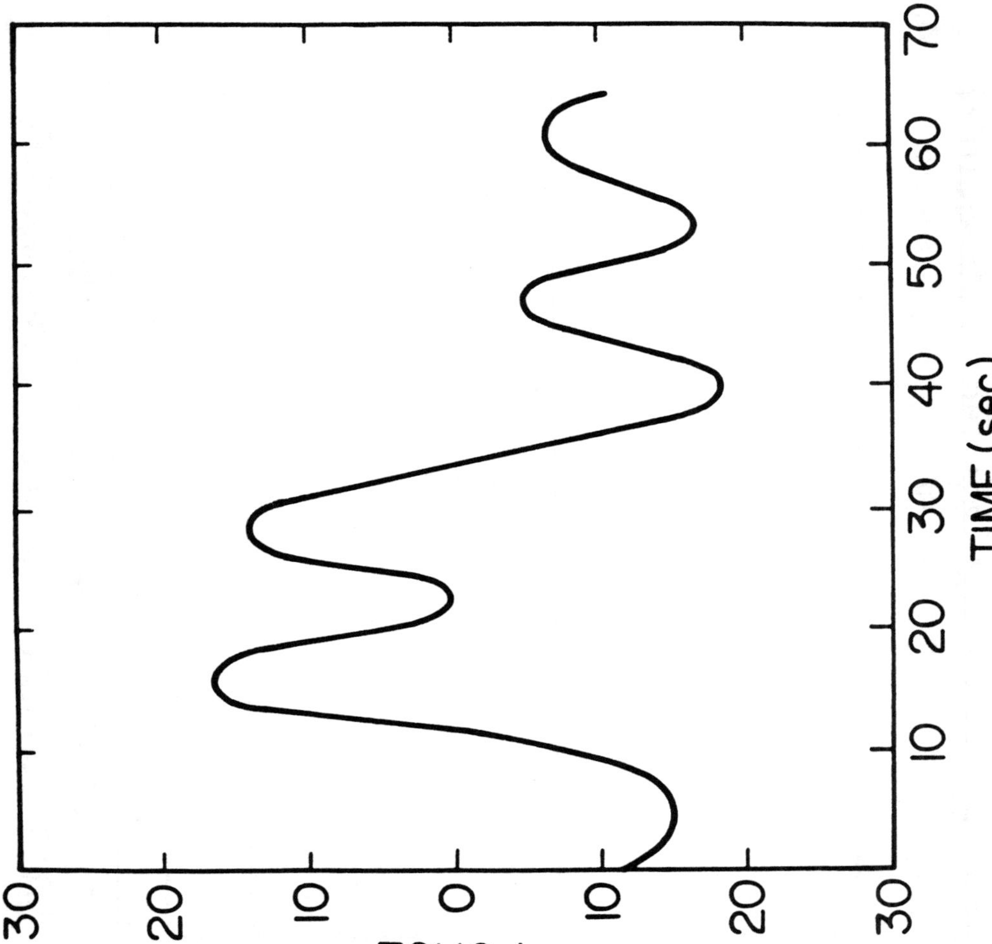

Figure B-4. Predicted Response of Torsion Balance-Servo-system to Antineutrino Source moved vertically with a 64 second period. The Torsion Balance-Servosystem period is 13.4 seconds.

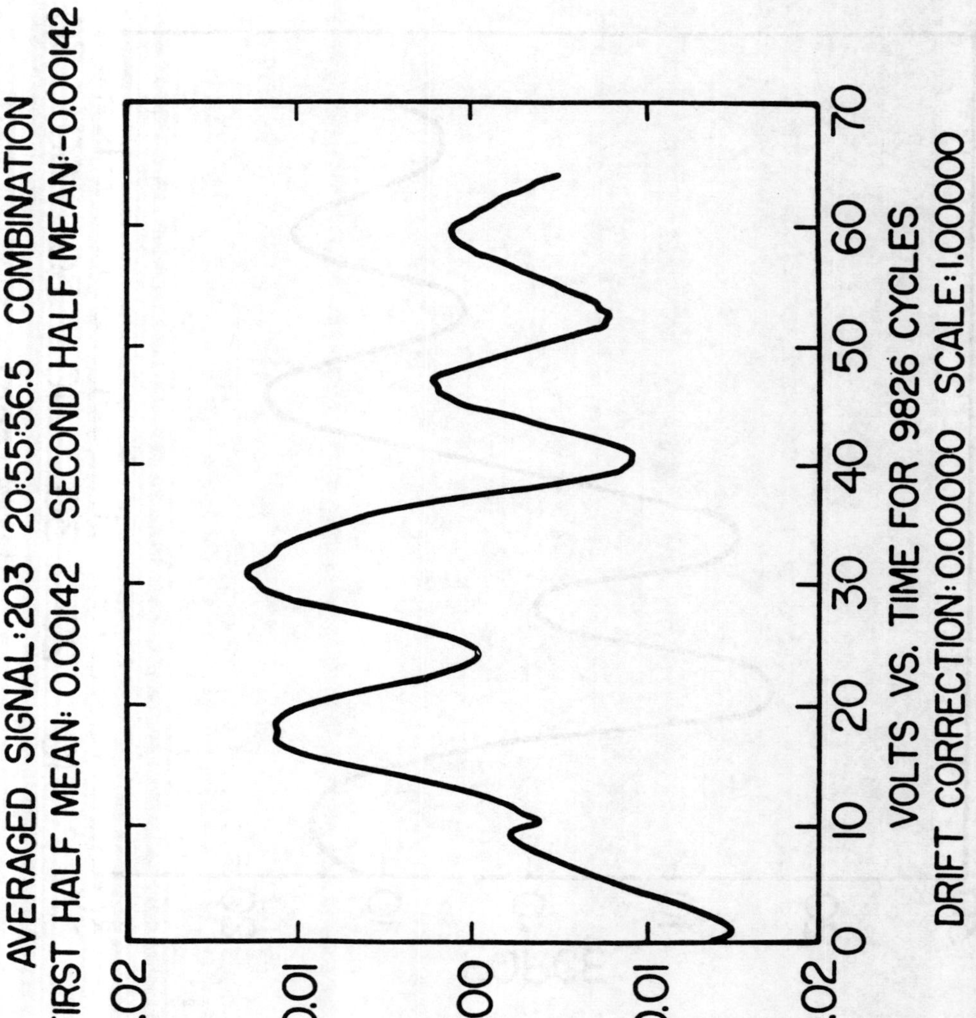

Figure B-5. Observed Response to the Antrineutrino Source and "Dummy" for the parameters of Figure 4, signal averaging over 9826 cycles.

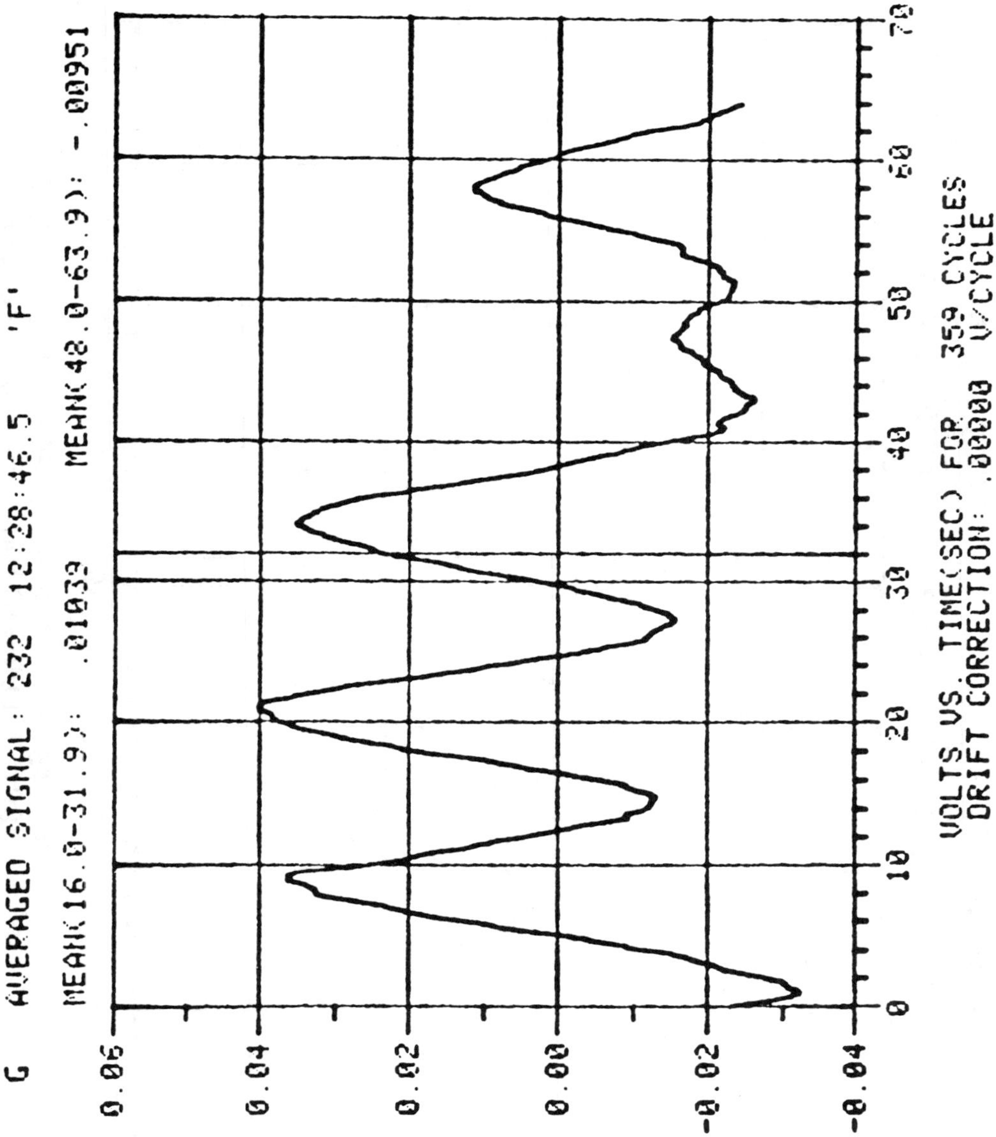

Figure B-6. Observed Response to the Antineutrino Source and "Dummy" for the parameters of Figure 4, signal averaging over 359 cycles.

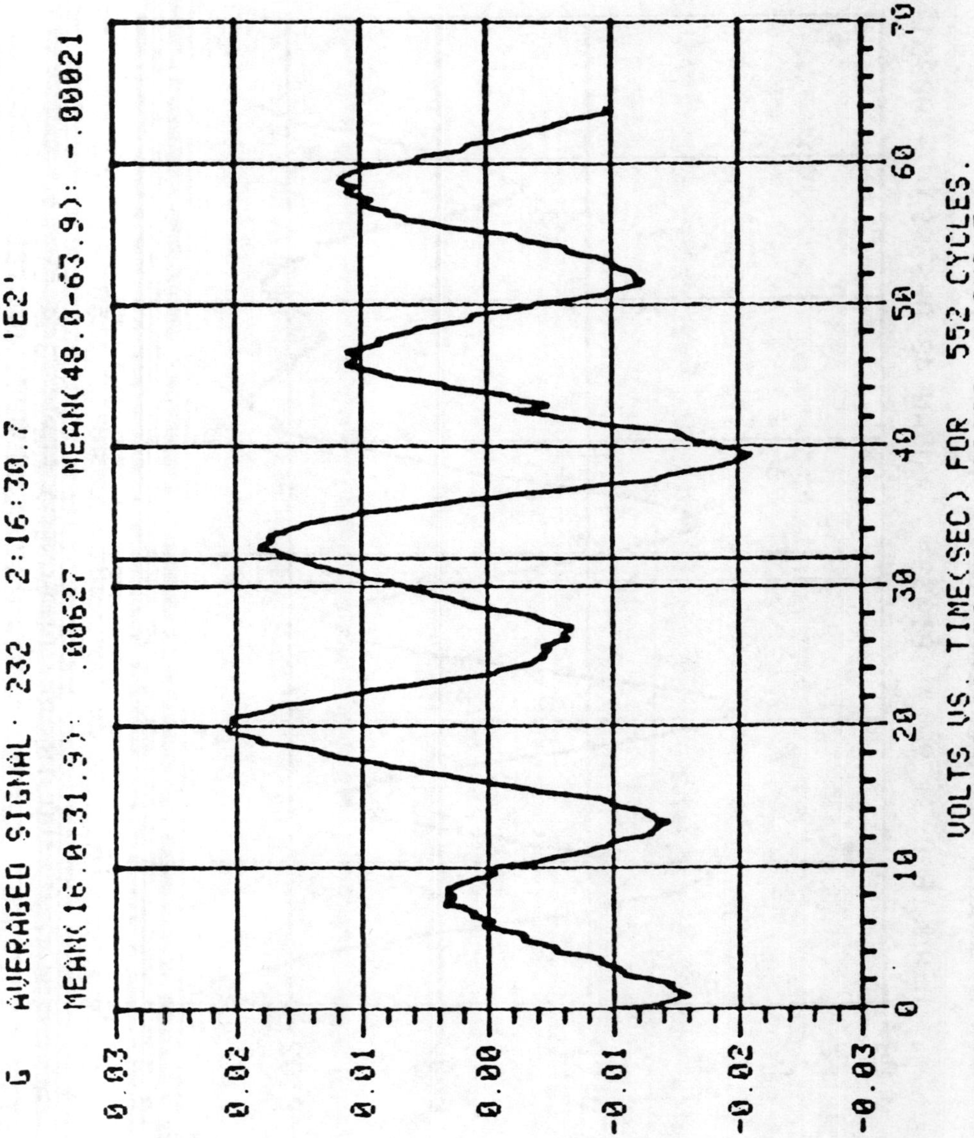

Figure B-7. Observed Response to the Antineutrino Source and "Dummy" for the parameters of Figure 4, signal averaging over 552 cycles.

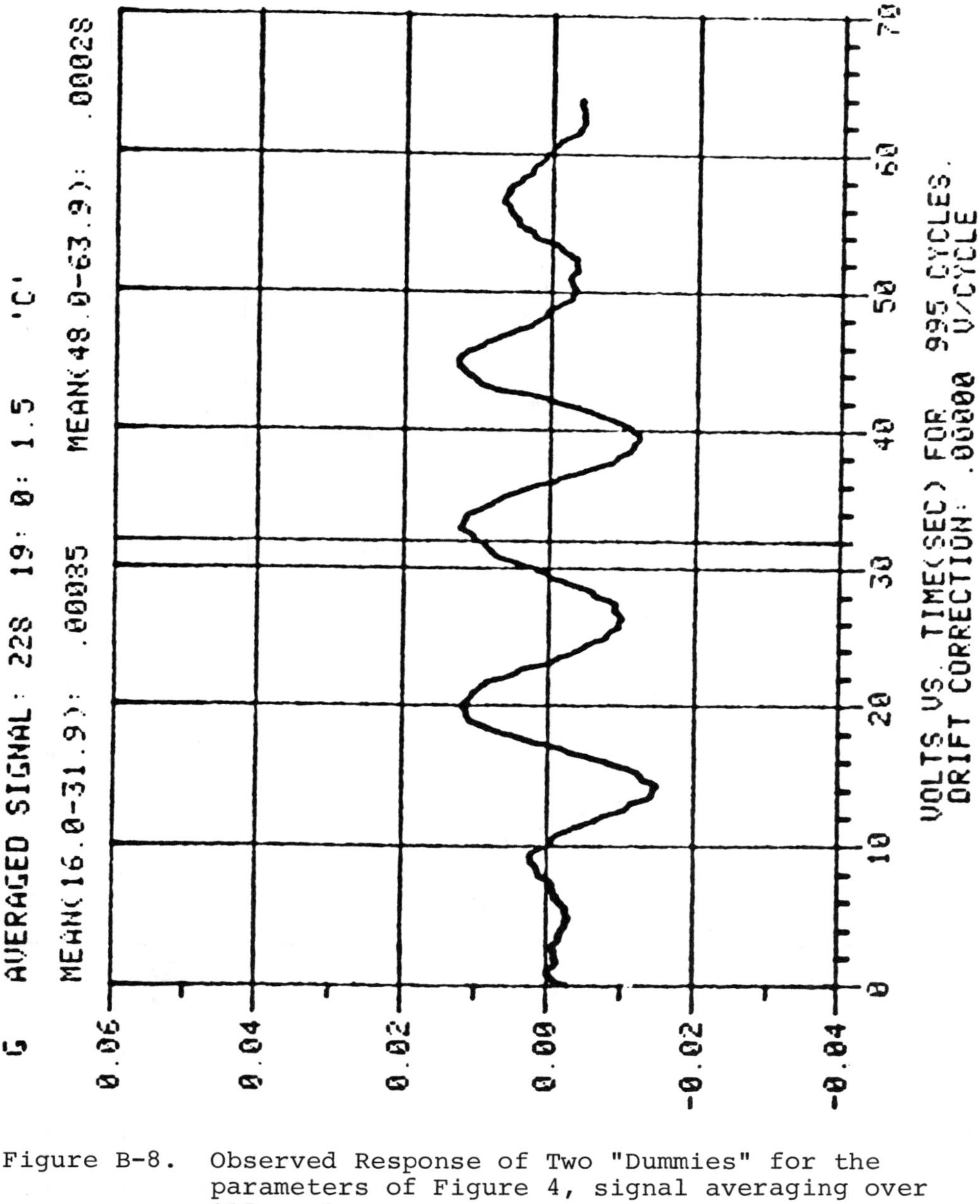

Figure B-8. Observed Response of Two "Dummies" for the parameters of Figure 4, signal averaging over 995 cycles.

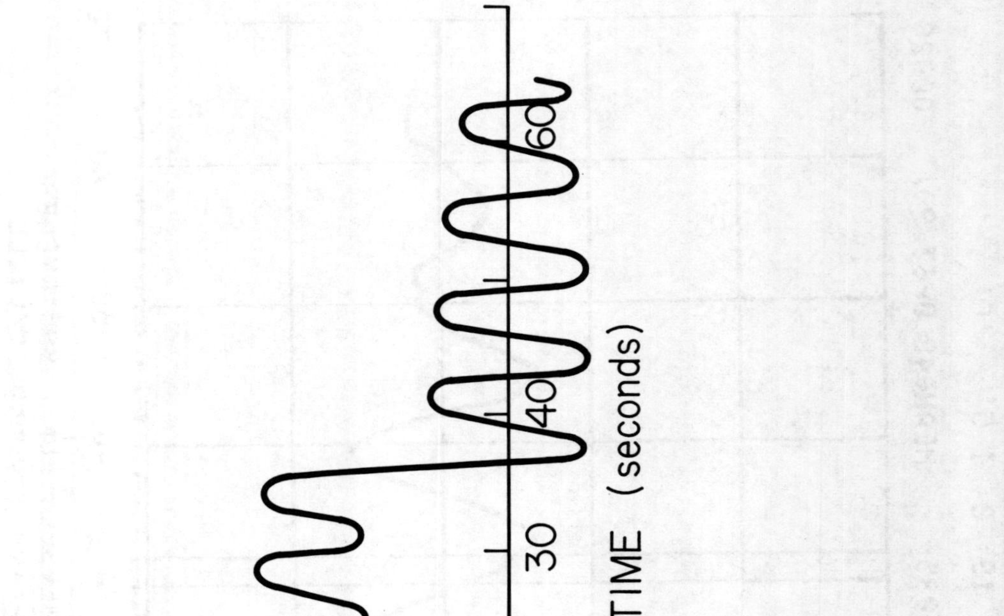

Figure B-9. Predicted Response of Torsion Balance-Servosystem to Antineutrino Source moved vertically with a 64 second period. The Torsion Balance-Servosystem period is 7.0 seconds.

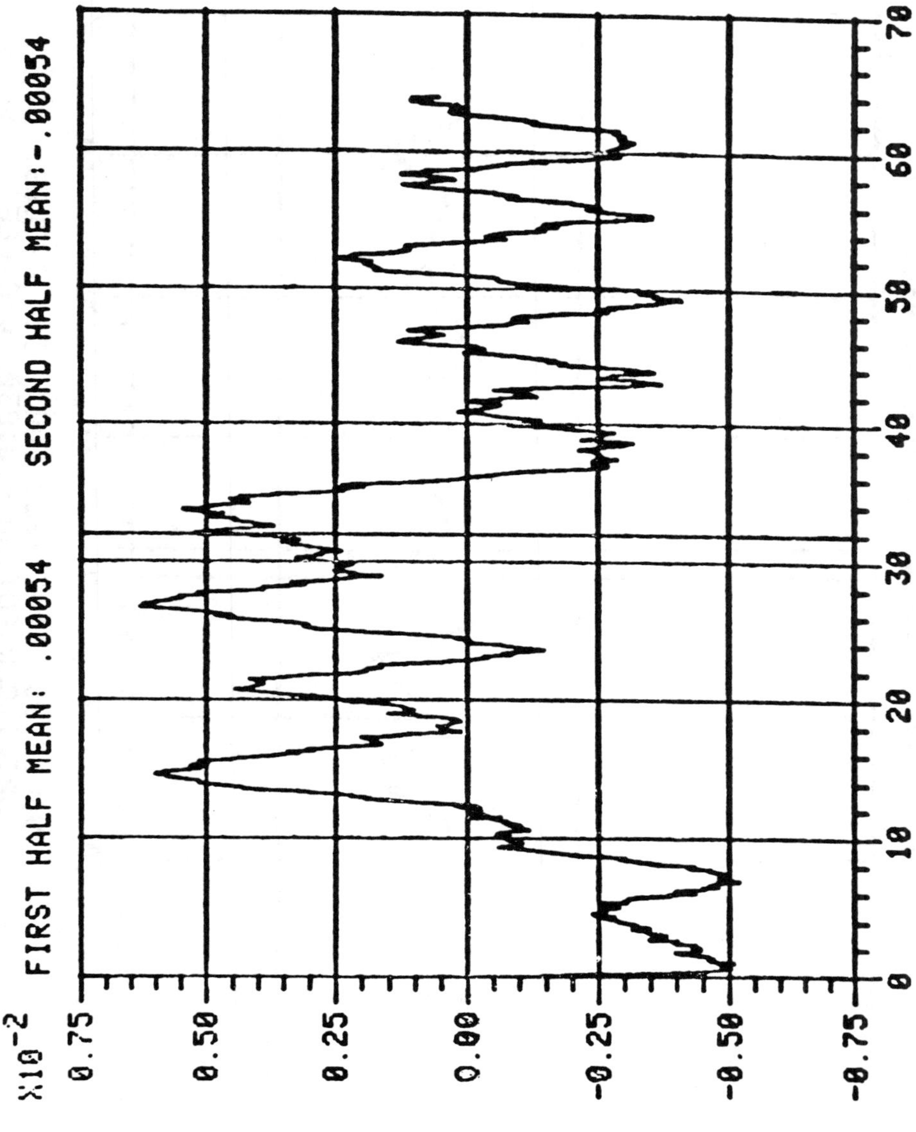

Figure B-10. Observed Response to the Antineutrino Source and "Dummy" for the Parameters of Figure 9, signal averaging over 533 cycles.

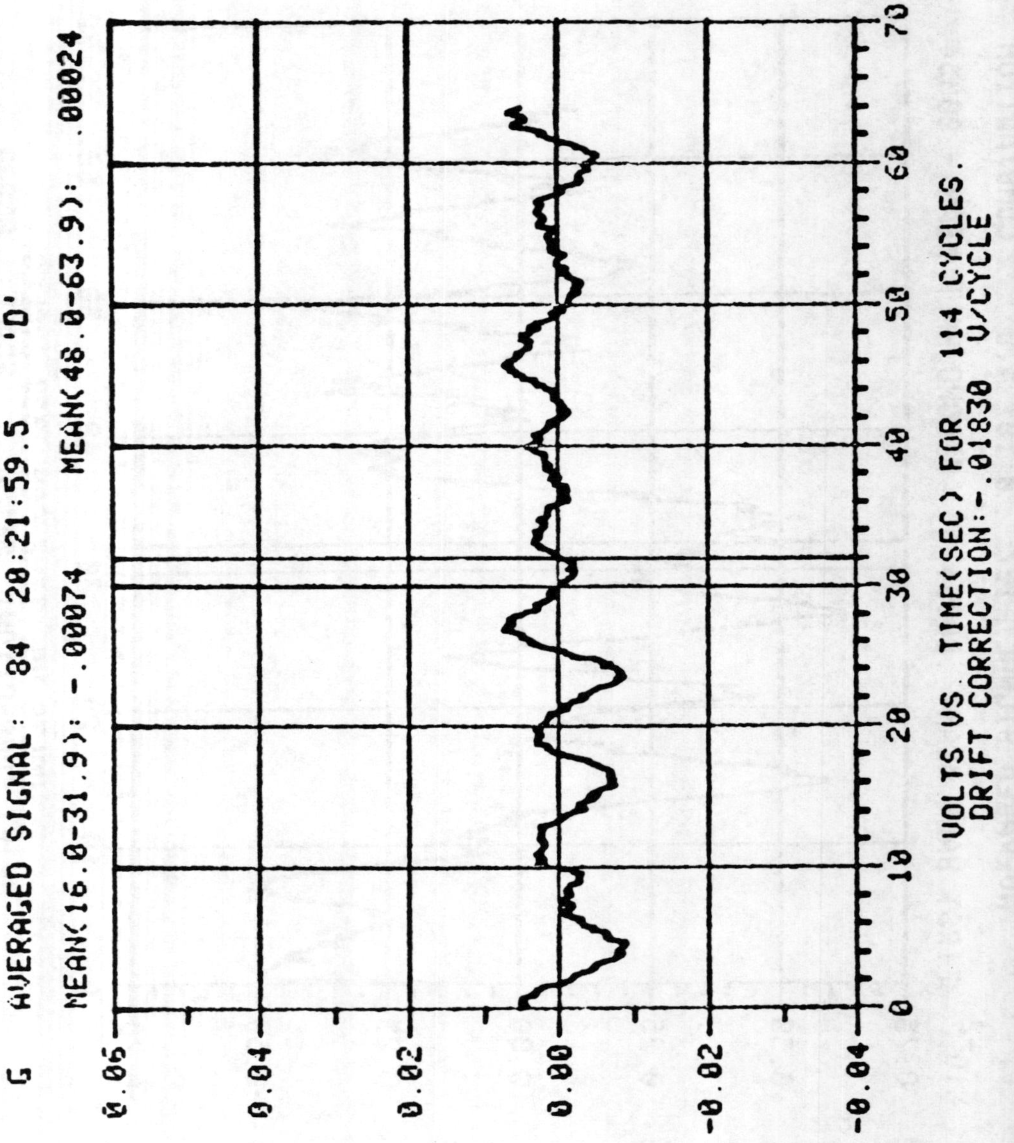

Figure B-11. Observed Response of Two Heated "Dummies" for the Parameters of Figure 9, averaging over 114 cycles.

In order to reduce magnetic effects to a minimum, no magnetic damping was employed, and no ferromagnetic materials were in the torsion balance. Magnetic contamination could produce a repulsive force between a ferromagnet and diamagnetic impurities. Magnetization measurements indicated that magnetic forces were very much smaller than those observed in these experiments.

No changes were observed in the data as a result of "degaussing" operations.

The tritium source was enclosed in a stainless steel capsule with wall thickness .08 cms. Measurements showed no significant β or ν emission outside the capsule. The capsule moved in an aluminum tube 1.5 mms. in thickness and 20 mms. in diameter. The maximum kinetic energy of the β electrons is 18.1 kilovolts and the mean value is 6.2 kilovolts. The range of 20 kilovolt electrons in aluminum is one milligram cm^{-2}. The aluminum tube represents 405 milligrams cm^{-2}. Therefore no significant number of Beta electrons penetrate both the stainless steel capsule cover and aluminum tube, to interact with the torsion balance.

The force which is observed corresponds to approximately 10^{14} tritium antineutrinos cm^{-2} sec^{-1} exchanging a large fraction of their incident momentum. The accuracy is not significantly affected by extremely small numbers of Beta electrons or gamma rays.

The ratio of force to antineutrino flux was measured as $\frac{F}{\Phi}$ = $1.05 \pm 0.12 \times 10^{-18}$.

The corresponding cross section is

$$\sigma = 2.05 \pm 0.23 \ CM^2$$

Different crystals of the same material may have differing cross sections. Orientation of the crystallographic axes was not considered and effects of dislocation defects have not been studied. For these reasons, agreement of measured cross sections by other observers, to better than an order of magnitude is not expected.

B.b. Independent Measurements of Energy and Flux

It is desirable to measure both the total neutrino flux and the neutrino energy.

Force measurements give a product of flux and cross section if the momentum distribution function is known.

The cross section, for a given crystal will be a function of the Debye temperature, crystal structure, and imperfections. Two crystals with very different Debye temperatures may be calibrated, employing sources with known flux, and momentum distribution. Measurements of the forces, for an unknown source, will provide two sets of data from which the flux and average source momentum can be determined.

B.c. Experiments in Progress and the Eötvös Experiment

Search is being carried out for solar neutrinos, in several ways. One employs the torsion balance with a single crystal of sapphire on one side and an equal mass of lead sheets on the other side.

This is the Eötvös experiment if equal masses of different materials are employed. If the different materials have different Debye temperatures, the solar neutrinos will give a positive result.

At present we do report a positive result, observing approximately 6×10^{10} solar neutrinos $cm^{-2}\ sec^{-1}$ with energy 0-400 keV, Fig. B.c.1.

B.d. Applications to Metrology

The experiment shown in Figure B-1 measures the Fermi weak interaction constant G_W. The cross section (B-4) is valid in the low energy limit after including the correction given by C-2 for a crystal which scatters a fraction of the incident neutrinos. Corrections for finite stiffness have been computed and these enable direct measurement of G_W, from the observed force. If a known gravitational force is employed for calibration, the experiment measures the ratio of G_W to the Newtonian gravitation constant. A knowledge of electrode dimensions and potentials permits writing G_W in terms of the fine structure constant.

C. SHIELDING AND SCATTERING OF MeV ANTINEUTRINOS BY NEARLY PERFECT CRYSTALS WITH HIGH DEBYE TEMPERATURES

As noted earlier, for crystals with infinite stiffness, the cross section for neutrinos, for the single unidentifiable scatterer momentum exchange, with neutral current scattering by up and down quarks, is given by

$$\sigma = \frac{G_W^2 E_\nu^2 (.387 N_U - .693 N_D)^2}{4\pi(\hbar c)^4} \qquad \text{(C-1)}$$

N_U is the number of up quarks, N_D is the number of down quarks, and G_W is the weak interaction coupling constant.

The total cross section includes (C-1), and other processes in which any number of scatterers may exchange momentum-energy with an incident particle. These may be included by multiplication of (C-1) by the factor p_Σ, with[2]

$$P_\Sigma = \left(\frac{\pi^2}{3} - 1\right) \qquad \text{(C-2)}$$

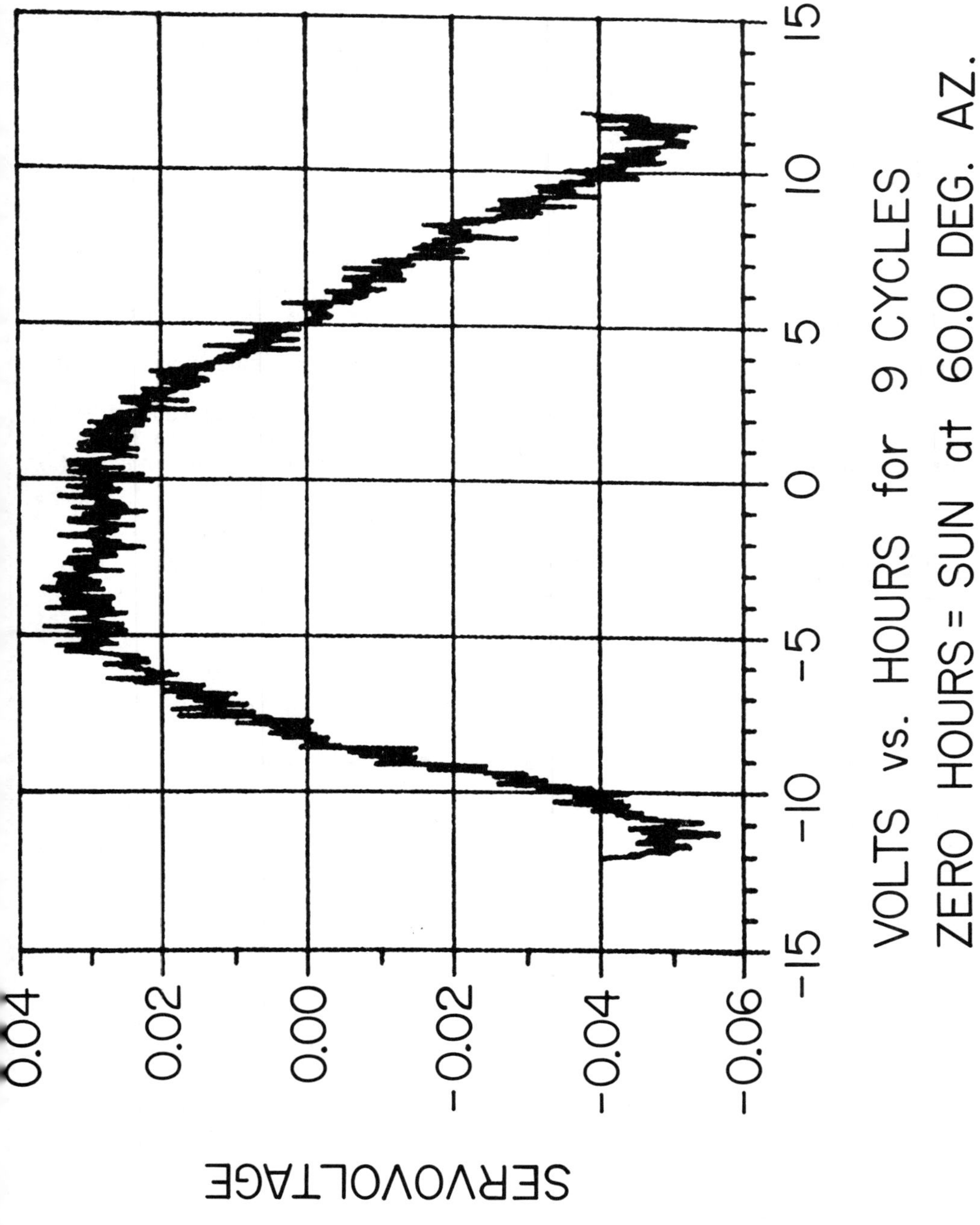

Figure B.c.1. Solar Neutrino Signal for 26 Gram Mass on One Arm, Lead Sheets on Other Arm of Torsion Balance.

Available crystals have finite stiffness described by the Debye temperature. Mossbauer effect theory states that a fraction f of the scatterings will involve recoil of the entire crystal, without excitation of phonons, for

$$f = e^{(-E_R/kT_D)\left(\frac{3}{2} + \pi^2 T^2/T_D^2\right)} \tag{C-3}$$

In (C-3) E_R is the recoil energy given in terms of the momentum transfer $\Delta\bar{p}$, and appropriate scatterer mass m, as $\frac{(\Delta p)^2}{2m}$. T is the temperature of the experiment and T_D is the Debye temperature of the crystal.

(C-1), (C-2), and (C-3) are employed to compute the expected force and cross section for the MeV region. The force associated with elastic scattering for incident flux Φ of neutrinos with energy E_ν and momentum p_ν is

$$F = \frac{\left(\frac{\pi^2}{3}-1\right)\Phi\, p_\nu E_\nu^2 (.387N_U - .693N_D)^2 \int_0^\pi \cos^2\frac{\theta}{2}(1-\cos\theta)\sin\theta\, e^{-4p_\nu^2 \sin^2\frac{\theta}{2}\left[\frac{3}{2}+\pi^2\frac{T^2}{T_D^2}\right]/2mkT_D}\, d\theta}{4\pi(\hbar c)^4 \int_0^\pi \cos^2\frac{\theta}{2}\sin\theta\, d\theta} \tag{C-4}$$

Incident particles are again in the Z direction and θ is the angle into which incident particles are scattered. The factor $\cos^2\frac{\theta}{2}$ in the integrand has its origin in the spinor transformation law. The exchanged momentum $|\Delta\bar{p}| = 2|\bar{p}|\sin\frac{\theta}{2}$.

Carrying out the integration gives

$$F = \frac{\left(\frac{\pi^2}{3}-1\right)\Phi\, G_W^2 (.387N_U - .693N_D)^2}{\pi \hbar^4 c^5 E_\nu}\left(\frac{mc^2 kT_D}{3 + 2\pi^2 T^2/T_D^2}\right)^2 \tag{C-5}$$

(C-5) is an approximation for (C-4) valid at MeV energies, but not at very low energy. In a similar way an appropriate scattering cross section may be defined, by writing

$$\sigma = \frac{\left(\frac{\pi^2}{3}-1\right) G_W^2 E_\nu^2 (.387 N_U - .693 N_D)^2 \int_0^\pi \cos^2\frac{\theta}{2} \sin\theta \, e^{-4p_\nu^2 \sin^2\frac{\theta}{2}\left[\frac{3}{2}+\pi^2 T^2/T_D^2\right]/2mkT_D} \, d\theta}{4\pi(\hbar c)^4 \int_0 \cos^2\frac{\theta}{2} \sin\theta \, d\theta} \quad (C6)$$

For experiments described here, (C-6) gives a cross section considerably greater than the crystal area. This implies that most antineutrinos are scattered by the first portions of the crystal. However the exponential factor implies that the scattering is in a narrow cone in the forward direction. Thus the antineutrinos, while multiply scattered, continue to move mainly in the forward direction.

C.a. Experiments

The torsion balance arrangement shown in Figure (C-1) was employed. A 100 gram sapphire crystal was on one side, and a 100 gram lead mass was on the other side of the balance.

An AC bridge was employed for sensing and measurements. The bridge is balanced for the equilibium position of the torsion balance. A force applied to one of the masses will rotate the torsion balance, producing an error voltage from the AC bridge. This voltage is amplified and employed to produce electrostatic forces to restore the torsion balance to equilibrium by a closed loop servosystem. In final equilibrium, the force resulting from elastic scattering of antineutrinos is therefore balanced by an electrical force. The system is calibrated by moving a lead brick into position, and measuring the servovoltage for the known gravitational interaction between lead brick and torsion balance.

The apparatus was located about fifteen meters from a 20 megawatt nuclear reactor, which was maintained at relatively constant output for other experiments.

If the large cross sections predicted by the theory are correct, a larger crystal should serve as an antineutrino shield. Switching is carried out by moving the shield into the line of sight between reactor and balance. Torsion balance output voltages were recorded on magnetic tape, with and without shield in place. Similar experiments were carried out with reactor off. Magnetic tapes were read "blind" by computer programmer G. Wilmot. The observed force was

$$F_{OBSERVED} = 3.9 \pm 0.4 \times 10^{-5} \text{ DYNES} \quad (C\text{-}7)$$

The predicted value, averaging (C-5) over the momentum distribution is

$$F_{PREDICTED} = 3 \times 10^{-5} \text{ DYNES} \quad (C\text{-}8)$$

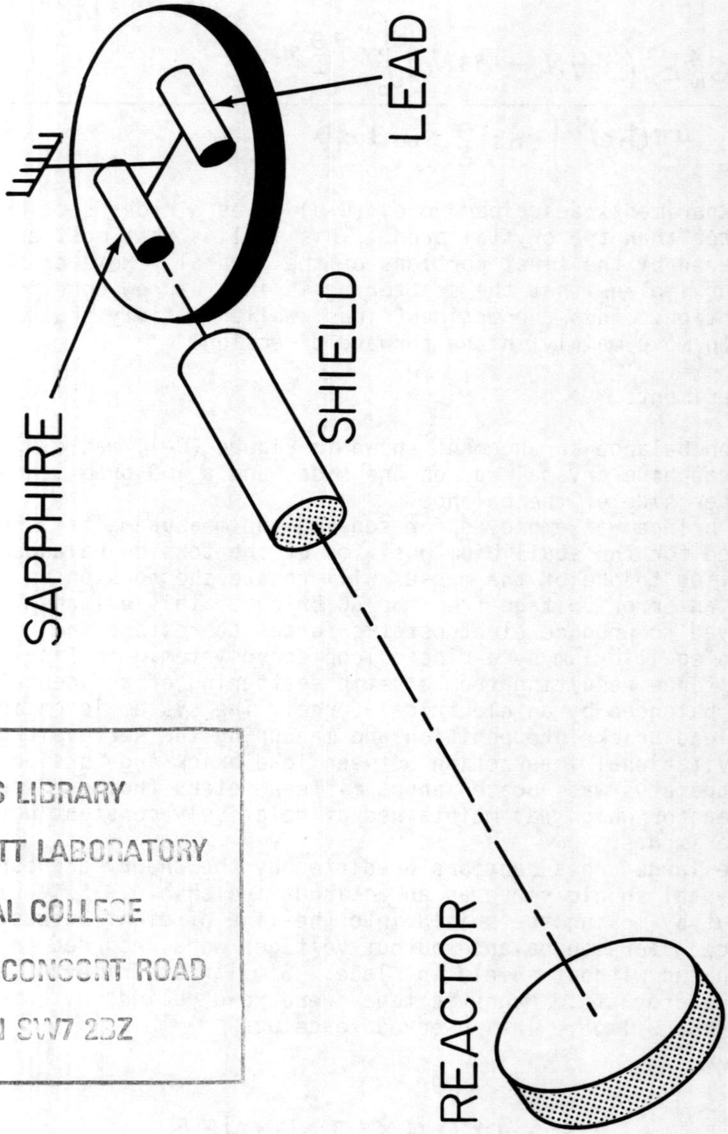

Figure C-1. Shielding, and Elastic Scattering of MeV Antineutrinos, Employing a Torsion Balance.

(C-7) and (C-8) are in fair agreement. We are not stating that the theory presented here is complete, and agreement of observation with the theory is not claimed. Considerable refinement of the theory is being studied, to properly consider anistropy, effects of dislocations...

C.b. Solar Neutrinos

As already noted, we are observing the solar neutrinos. (C-5) and the reactor calibration provide measurement of the solar neutrino flux.

D. GRAVITATIONAL RADIATION ANTENNAS

D.a. Cross Sections and Bandwidths

We will consider the quantum theory, beginning with the Hamiltonian

$$H = \frac{p^2}{2M} + \frac{K\xi^2}{2} + Mc^2 R_{o\ell oj} r^\ell \xi^j = H_o + H' \qquad \text{(D-1)}$$

The first two terms on the right of (D-1) are the kinetic and potential energies of a mass quadrupole harmonic oscillator, reduced mass M, momentum p, displacement ξ, force constant K. The third term is the interaction with the Riemann tensor of incident radiation. We will employ the semiclassical approximation. $R_{o\ell oj}$ is the Riemann tensor in a normal coordinate system, r^ℓ is the equilibrium position vector of the second mass relative to the first. The radiation interaction is treated by perturbation theory, with H' the perturbing potential.

For the wavefunction we employ

$$\Psi = \sum_q a_q(t) U_q e^{-iE_{Aq}t/\hbar} \qquad \text{(D-2)}$$

In (D-2) the U_q are eigenfunctions of the unperturbed oscillator with energies E_{Aq}. The expansion coefficients $a_q(t)$ are time dependent. (D-2) satisfies the Schroedinger equation

$$i\hbar \frac{\partial \Psi}{\partial t} = (H_o + H')\Psi \qquad \text{(D-3)}$$

The a_q are written as a series in successive orders of approximation

$$a_q = a_q^0 + a_q^1 + a_q^2 + \cdots \quad \text{(D-4)}$$

These equations give

$$i\hbar \sum \dot{a}_q^1 U_q e^{-iE_{Aq}t/\hbar} = Mc^2 R_{0\ell 0j} r^\ell \xi^j a_q^0 U_q e^{-iE_{Aq}t/\hbar} \quad \text{(D-5)}$$

At t = -T/2 assume the antenna is initially in an eigenstate, $a_q^{(0)} = \delta_{qn}$. Multiply (D-5) by $U_k^* e^{iE_{Ak}t/\hbar}$, and integrate over configuration space to obtain

$$i\hbar \dot{a}_q^1 = Mc^2 R_{0\ell 0j}(t) r^\ell \xi^j_{kn} e^{i\omega_{kn}t} \quad \text{(D-6)}$$

In (D-6) ξ^j_{kn} are the harmonic oscillator matrix elements. The time dependent Riemann tensor is written as a Fourier integral, and the integral of (D-6) is

$$a_k = \frac{Mc^2}{i\hbar\sqrt{2\pi}} \int_{-\infty}^{\infty}\int_{-T/2}^{T/2} r^\ell \xi^j_{kn} \tilde{R}_{0\ell 0j}(\omega) e^{i(\omega+\omega_{kn})t} \, d\omega \, dt \quad \text{(D-7)}$$

The integral of (D-7) over time gives

$$a_k = \sqrt{\frac{2}{\pi}} \, \frac{Mc^2}{\hbar} \int_{-\infty}^{\infty} \tilde{R}_{0\ell 0j}(\omega) r^\ell \xi^j_{kn} \left[\frac{\sin(\omega_{kn}+\omega)T/2}{(\omega_{kn}+\omega)} \right] d\omega \quad \text{(D-8)}$$

In (D-8), the mass M is a measure of the coupling to the gravitational field, and is properly the gravitational mass. We will therefore write it as M_G.

Integration and squaring of (D-8) gives

$$|a_k|^2 = 2\pi \left| M_G c^2 \tilde{R}_{0\ell 0j}(\omega_{kn}) r^\ell \xi_{kn} \right|^2 / \hbar^2 \quad \text{(D-9)}$$

The harmonic oscillator squared matrix elements are

$$\xi^2_{kn} = \frac{\hbar}{M_I \omega_{kn}} \left[\frac{n+1}{2}\right] \quad , \quad k = n+1 \tag{D-10}$$

$$\xi^2_{kn} = \frac{\hbar}{M_I \omega_{kn}} \left[\frac{n}{2}\right] \quad , \quad k = n-1 \tag{D-11}$$

The mass here is properly the inertial mass, and it is labeled M_I.
The energy change ΔU is given by

$$\Delta U = \sum_j \Delta |a_j|^2 E_{Aj} = \frac{\pi M_G^2 c^2}{M_I} \left| R_{o\ell oj}(\omega_{n,n\pm 1}) r^\ell V^j \right|^2 \tag{D-12}$$

V^J is a unit vector in the oscillator displacement direction.
For a pulse with spectral width $\sim \omega_{kn}$

$$\left| \tilde{R}_{o\ell oj}(\omega_{kn}) \right|^2 = 4\pi\omega G U_{INCIDENT} / A c^7 \tag{D-13}$$

In (D-13), G is the Newtonian constant of gravitation, $U_{INCIDENT}$ is the incident gravitational radiation energy, over area A.
The total cross section is given by (D-12), (D-13) then as

$$\sigma = \frac{8\pi^3 G M_G^2 r^2}{c^2 \lambda M_I} \longrightarrow \frac{8\pi^3 G M r^2}{c^2 \lambda} \tag{D-14}$$

On the far right of (D-14), the gravitational and inertial masses are taken to be the same. In this analysis, the heat bath interactions were not considered, and the result (D-14) is the limit for antennas with infinite quality factor Q.

D.b. Antenna Composed of Many Quadrupoles

An elastic solid antenna is composed of atoms coupled by chemical forces. The model as a single large quadrupole is not realistic. For a large number of quadrupoles coupled to one detector

$$H = \sum_{\alpha} \frac{p_\alpha^2}{2m_\alpha} + \frac{k_\alpha \xi_\alpha^2}{2} + m_\alpha c^2 R_{0\ell 0j} r_\alpha^\ell \xi^j \tag{D-15}$$

Let us write the system wavefunction as

$$\Psi_{SYSTEM} = \sum a_s U_s e^{-iE_{As} t/\hbar} \tag{D-16}$$

U_S is a product of individual quadrupole wavefunctions.
For N quadrupoles

$$a_k' = \sqrt{\frac{2}{\pi}} \sum_{\alpha} \frac{m_{G\alpha} c^2}{\hbar} \int_{-\infty}^{\infty} \tilde{R}_{0\ell 0j}(\omega) r_\alpha^\ell \xi_{kn}^j \left[\frac{\sin \frac{(\omega_{kn}+\omega)T}{2}}{\omega_{kn}+\omega} \right] d\omega \tag{D-17}$$

For N quadrupoles in a small region, there are no phase shifts in $R_{0\ell 0j}(\omega)$

$$|a_k'|^2 = 2\pi N^2 \left| m_G c^2 \tilde{R}_{0\ell 0j}(\omega_{kn}) r^\ell \xi_{kn}^j \right|^2 / \hbar^2 \tag{D-18}$$

The cross section is then

$$\sigma_N = \frac{8\pi^3 G m_G^2 r^2 N^2}{c^2 \lambda m_I} \tag{D-19}$$

The factor N^2 is also present in the corresponding electromagnetic many element problem, when the entire array linear dimensions are small compared with a wavelength.

If the antenna length L is given approximately by

$$L \approx N^{1/3} r \tag{D-20}$$

$$M \approx Nm \tag{D-21}$$

(D-19) becomes

$$\sigma_N = \frac{8\pi^3 G M L^2 N^{1/3}}{c^2 \lambda} \qquad \text{(D-22)}$$

D.c. Slab Model

Present gravitational antennas employ the lowest frequency longitudinal wave mode. This is a one dimensional wave, with propagation in the length direction.

A reasonable model is to imagine the antenna composed of slabs. Each slab consists of the cross sectional area of the antenna, one atom in thickness.

It is important that detailed analyses show that individual slabs can be externally driven. The reaction of the rest of the slabs on front and back slab faces has resonance at the normal mode frequency.

We may imagine therefore that an individual atom in the slab couples to the Riemann tensor with coupling strength proportional to the mass of the atom. That atom drives the entire slab, which has the inertia of the entire slab. The total cross section is therefore given, from (D-19) by

$$\sigma_N = \frac{8\pi^3 G m_{ATOM}^2 L^2 N_{ATOM}^2}{c^2 m_{SLAB}} \qquad \text{(D-23)}$$

$$M = m_{ATOM} N \qquad \text{(D-24)}$$

$$m_{SLAB} = m_{ATOM} N^{2/3} \qquad \text{(D-25)}$$

$$\sigma_N = \frac{8\pi^3 G M L^2 N^{1/3}}{c^2 \lambda} \qquad \text{(D-26)}$$

(D-26) has again assumed no losses due to heat bath and other interactions. Analysis shows that the quality factor of a driven slab may be small, given by

$$Q_{MODE} = Q_{SLAB} N^{1/3} \tag{D-27}$$

For the case where Q_{SLAB} is less than one, taking losses into account, the cross section is

$$\sigma_N = \frac{8\pi^3 G M L^2 Q_{SLAB} N^{1/3}}{c^2 \lambda} \tag{D-28}$$

D.d. Discussion

The result (D-28) for pulses is important. The total cross section is orders larger than for an elastic solid represented by a single mass quadrupole interaction.

The pulses observed during the period 1968 - present therefore represent a much smaller gravitational energy flux than assumed on the basis of the continuous elastic solid arguments. Furthermore the present bar antennas are orders more sensitive for pulses, than presently projected designs for interferometer antennas.

D.e. Bandwidth Issues

Let us imagine an antenna with piezoelectric crystals in the center. Many modes couple energy to the crystal output terminals. If a wideband amplifier is employed, all of this information may be recorded. A computer analysis can then perform, for each spectral portion, the analysis normally done by a fixed frequency analogue filter detector, which has a narrow frequency output recorded on tape. In this way, the available bandwidth is increased by orders.

E. OBSERVATIONS DURING RAPID EVOLUTIONARY PHASE OF SUPERNOVA 1987A

Gravitational antennas have been operating at the University of Maryland for about 20 years. Two antennas have operated continuously for the past 16 years. One has a mass of 3000 kilograms, and is suspended by cables on acoustic filters in a large, thick walled vacuum chamber, about four meters in diameter. The pulse noise temperature was recently measured and found to be 11 Kelvin. The second antenna is much smaller, with a mass about 1000 kilograms, and it is mounted in a small - one meter diameter thin walled chamber. The electromagnetic and acoustic isolation are much smaller for the small antenna, and the cross section for absorption of gravitational radiation is also much smaller. The small antenna is a useful control experiment because its gravitational radiation response is much smaller than that of the large antenna, while its response to local seismic and electromagnetic disturbances is much larger than that of the large antenna.

Magnetic tapes for the period February 23 were studied by G. Wilmot. Figure E-1 consists of the first computer plot, for the large antenna. Eleven second averages were employed because of available software. Times of reported neutrino events were not known until after Figure E-1 was prepared. The largest pulse, close to the center of the Figure, is preceded by two somewhat smaller pulses. The 23 pulses projecting above the highest horizontal line exceed the mean by 2.8 standard deviations. The largest pulse exceeds the mean by 4.5 standard deviations. The time of the peak of the largest pulse is 0252 hours and 5 seconds, approximately 30 seconds before the first neutrino pulse observed by the Mont Blanc Tunnel group. There was no significant response on the small, control antenna. This makes it probable that the largest pulses of Figure E1 are not of local seismic or electromagnetic origin. The two large pulses which just precede the largest one, are the third and fourth largest of the entire set.

It is appropriate to compute the energy required in gravitational radiation, for production of these pulses.

In 1960 the theory of the antenna was considered, with the aluminum bar assumed to be a continuous elastic solid. The total cross section is given by

$$\sigma_{1960} = \frac{8\pi^3 GML^2}{c^2\lambda} \tag{E-1}$$

G is Newton's constant of gravitation, M is the reduced mass, L is the "effective quadrupole length," C is the speed of light, λ is the wavelength. However, we have seen that the more realistic model of an antenna, as a large number of atoms coupled by chemical forces, gives a much larger cross section:

$$\sigma_{1984} = \frac{8\pi^3 GML^3 Q_s}{c^2\lambda L_a} \tag{E-2}$$

Here Q_s is the quality factor of a cross sectional slab of the antenna, with plane normal to the axis of acoustic propagation, of thickness equal to the length L_a occupied by a single atom. That value would require about 4×10^{-4} solar masses to produce the largest pulse observed on Figure E1.

It is important to stress that time differences of 30 seconds in arrival times, for neutrinos, for different frequency gravitational pulses, could be caused by several factors. Propagation is in the gravitational field of galaxies, with matter present, not in vacuum.

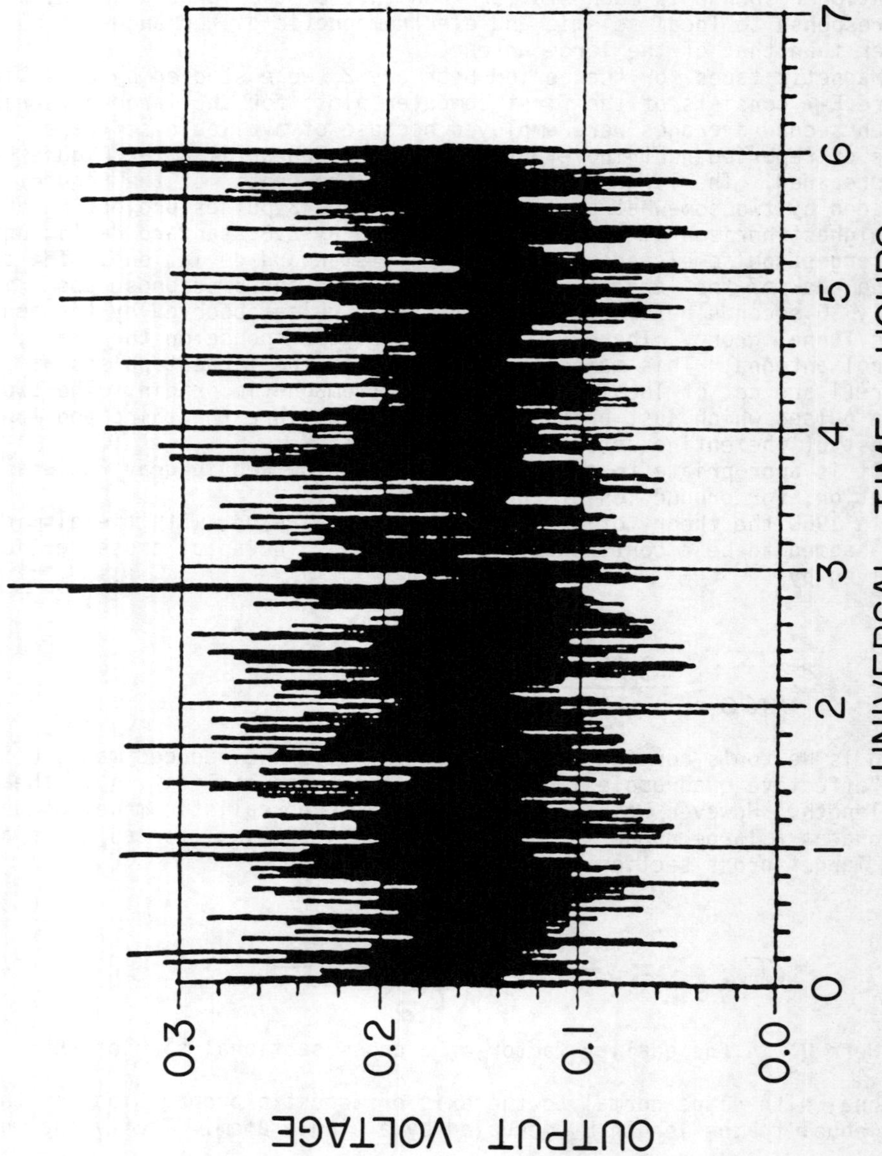

Figure E-1. Gravitational Antenna Pulses During Rapid Evolutionary Phase of Supernova 1987A.

Effects of dispersion, gravitational lensing, and possible neutrino masses, all require careful investigation.

These pulses and those observed by the Rome group are probably gravitational radiation from the supernova.

ACKNOWLEDGEMENTS

We are pleased to acknowledge support of the National Science Foundation, DARPA, and the Harry Diamond Laboratories SDI SDIO.

REFERENCES

1. J. Weber, Foundations of Physics 14, 1185 (1984).
2. J. Weber, Physical Review C 31, 1468 (1985).
3. A number of authors have commented on references 1 and 2. These include T.H. Ho, Phys. Lett. 168B, 295 (1986); R. Casella in Intersections Between Particle and Nuclear Physics 1986 AIP Conference Proceedings 150 pages 1038; G.F. Bertsch and S.M. Austin, Phys. Rev. C 34, 361 (1986); Y. Aharonov, F.I. Avignone, III, A. Casher, S. Nussinov, Phys. Rev. Lett. 58, 1173 (1987); H. Lipkin, Phys. Rev. Lett. 58, 1176 (1987).

APPENDIX

ON QUIRINO MAJORANA'S PAPERS REGARDING GRAVITATIONAL ABSORPTION

G.Dragoni
Dipartimento di Fisica
Via Irnerio 46
40126 Bologna
Italy

ABSTRACT. After a brief presentation of the figure of Quirino Majorana, his work on gravitational absorption is presented on the basis of two of his papers. A renewal of interest for these long forgotten contributions arises from the possibility of reading them as checks *ante litteram* of the hypothesis of a fifth force depending on the chemical nature of substances.

1. INTRODUCTION

Nowadays there is considerable scientific debate as to the existence of a possible new force: the fifth force, the intermediate-range composition -dependent force. This being so, we are glad to accept Professor Venzo De Sabbata's invitation to present some of the experimental research work carried out by Quirino Majorana (1) on the absorption of gravitation. Indeed, he carried out some of the first most important and complete experiments on the subject during the first decades of the 20th century.

Now Majorana's experiments have mainly a historical interest. However, we think that the presentation of some papers of his might be useful.

In order to present a brief introduction to the works by Majorana on gravitational absorption, let us choose among his papers (see note 2, References on Gravitation by Q.Majorana) a general review: 'Quelques Recherches sur l'Absorption de la Gravitation par la matière', *Journal de Physique*, **9**, 314-324 (1930), and a scientific paper, 'On gravitation ,Theoretical and Experimental Researches', *Philosophical Magazine*, **39**, 408-504 (1920).

First of all, we think it necessary to give some biographical information about Quirino Majorana.

2. BIOGRAPHICAL NOTE

Quirino Majorana (1871-1957), the uncle of the famous Ettore, was an

V. De Sabbata and V. N. Melnikov (eds.),
Gravitational Measurements, Fundamental Metrology and Constants, 501–539.

outstanding personality in the field if Italian Physics from the end of the 19th century up to the middle of the 20th century. He was born in Catania in 1871 and graduated in Engineering at the University of Rome in 1891 and in Physics in 1893.

From 1894 to 1904 he was assistant to Professor Pietro Blaserna (the Director of the Institute of Physics in Rome); from 1904 to 1914 he was appointed Director of the Istituto Centrale Telegrafico in Rome. Then he obtained the chair for experimental physics (university level) at the Politecnico of Turin (1914-1920). After the death of Augusto Righi (1850-1920) he moved to Bologna where he obtained Righi's chair, and the Direction of the Physics Institute up to 1941. He continued his research activity in Rome until he died (at Rieti in 1957).

He remained at the summit of Italian Physics for several decades, acting as President of the Italian Physical Society, while Enrico Fermi was Vice-President. For years he edited "Il Nuovo Cimento", the official journal of Italian Physics. He was also acclaimed in the international field and received numerous awards.

An outstanding experimentalist, he offered many original contributions to the knowledge of physics. In short, he carried out studies on photoelectric phenomena (interpretation of Volta effect), magnetic double refraction (Majorana effect), electric discharges in gases, X-rays, diamond reproduction,... His studies in the field of radiotelephony are most important.

It must be remembered that in 1901 Marconi had transmitted signals through headphones between Europe (Poldhu, Cornwall) and America (Signal Hill, Newfoundland), whereas, in 1903, Majorana transmitted the spoken word by radiotelephony through the Institute of Physics in Rome. In 1907 he broadcast on a distance of 500 km from Monte Mario (Rome) to Erice (Trapani).

In the 20's, he was engaged on secret broadcasts by ultraviolet and infrared rays, obtaining important results.

After the 2nd world war his fame declined owing to the negative opinion that many physicists had of his numerous attempts to deny Einstein's theories.

Like many experimental physicists of his time, he distrusted theoretical speculations not based upon sound experimental results, and, as a general attitude, he always tried to envisage experiments capable of testing them. As an example pf this attitude Majorana put the 2nd principle of Einstein's restricted relativity to test, but recognized that his 1914, 1917-1918 and 1934 experiments favoured Einstein's theories (3).

These steps of his remain in the scientific literature as the very first reliable experiments confirming Einstein's special relativity.

Later he carried out research on gravitational absorption.

3. AN OUTLOOK ON MAJORANA'S PAPERS

Majorana intended to check experimentally the hypothesis of gravitational absorption by matter interposed between two interacting masses. Majorana thought that it was reasonable to assume that the presence of this matter interposed would modify Newton's gravitational law. This hypothesis was probably made in analogy with the photons in electromagnetism, however his hypothesis was that: "la force de gravitation doit, en l'absence d'un milieu quelconque dépendre d'une sorte de bombardement... de particules 'sui generis'". And besides: "Il suffira de dire que, comme consequence du passage des particules à travers un milieu matériel, le flux des particules, c'est-a-dire le nombre de celles qui, dans l'unité des temps, traversent l'unité de section, diminue par l'absorption" (Quelques Recherches, p.316). Anyway, in his experiments Majorana admitted as a hypothesis the law of absorption for gravitational force given by a modification of Newton's law attributed to Laplace:

$$F=k\frac{M_1M_2}{R^2}\exp(-hdr) \qquad (1)$$

in which H=hd is the quenching factor for density d; h represent a universal constant upon which the measure of the gravitational absorption depends; r measures the thickness of the absorbing medium.

From theoretical calculations, Majorana decided that h must be between 0 and $8*10^{-12}$. His numerous experiments led to values such as $h=6.7*10^{-12}$ (in Hg) or $h=2.8*10^{-12}$ (in Pb) (see articles already mentioned).

He noticed: i) that if such an effect existed it should be five orders of magnitude smaller with respect to the limits available in the scientific literature of his time; ii) that his results implied a non-vanishing effect for the gravitational absorption: h was of the order of 10^{-12} $cm^2 g^{-1}$.

His results give rise to some hesitation even for the author himself, who, for more than 10 years, carried out very exacting experiments, almost beyond expectation, concerning all the sources of error. These errors were satisfactorly considered, calculated and corrected.

Several considerations can be made:

i) In the worst of hypotheses, Majorana's results provide a highly accurate confirmation of Newton's law.

ii) Majorana's results might depend on the different atomic numbers A, Z for Hg e Pb (anyway Majorana thought that h was a universal constant, see 'On Gravitation, p.495)

iii) Majorana was simply anxious to observe and measure the screening effect of matter interposed between interacting masses. The validity of Newton's law in a vacuum does not seem to come under discussion. In this sense, his measurements appear exclusively those regarding the presence or absence of interposed matter. At present, however, it might be interesting to consider them but a possible generalization of Newton's law according to a relation like this:

$$F=k\frac{M_1 M_2}{R^2}\exp(-gR) \qquad (2)$$

where g parametrizes the deviation of Newton's law even in a vacuum. An experimental investigation of this kind requires absolute measurements in a vacuum. The existence of a non vanishing g could be interpreted as a non zero mass for gravitons in a modern field theory. On the other hand if g is sufficiently small, as one can infer from astronomical effects, one can expand the equation given by an addictional force such as:

$$F=k\frac{M_1 M_2}{R^2} - gk\frac{M_1 M_2}{R} + \ldots \qquad (3)$$

In any case, whatever the theoretical reason accepted, the last word is left to the experimental evidence, a fact of which Majorana was well aware: "C'est sur les faits que doivent se fonder les théories nouvelles ou les vieilles prudemment modifiées. Le contraire serait inadmissible" ('Quelques Recherches, p.324).

Lastly, let us recall that Q.Majorana **constantly** insisted that his measurements be taken up and repeated by other experimentalists using more suitable methods than he had at his disposal.

This paper of ours furthers this invitation bearing in mind particularly the ideas expressed by a well-known phylosopher of science, Imre Lakatos, who maintained that one is never justified in saying a programme of scientific research has degenerated beyond hope.

Acknowledgement

Published with permission from the Philosophical Magazine.

Notes:

(1) Q.MAJORANA, 'Un sessantennio di ricerca scientifica nel campo della fisica', *Istituto Superiore delle Poste e Telecomunicazioni, Piccole note, recensioni e notizie*, **3**, 3-10 (1954)

E.PERUCCA, 'Commemorazione del Socio Quirino Majorana', *Accad. Lincei, Rendiconti*, **25**, 354-362 (1958)

D.GRAFFI, 'Commemorazione dell'Accademico benedectino Quirino Majorana', *Accad. Scienze Istituto di Bologna*, Serie XI, Tomo VI, **247**, 175-188 (1959). Con ampia bibliografia dei lavori di Majorana.

(2) References on Gravitation by Q.Majorana:

Q.MAJORANA, 'Sulla gravitazione', *Atti Accad.Lincei*, **28**, 165-174 (1919)

Q.MAJORANA, 'Sulla gravitazione', *Atti Accad.Lincei*, **28**, 221-223 (1919)

Q.MAJORANA, 'Sulla gravitazione', *Atti Accad.Lincei*, **28**, 313-317 (1919)

Q.MAJORANA, 'Sulla gravitazione', *Atti Accad.Lincei*, **28**, 416-421 (1919)

Q.MAJORANA, 'Sulla gravitazione', *Atti Accad.Lincei*, **28**, 480-489 (1919)

Q.MAJORANA, 'Sur la gravitation', *C.R.Acad.Sci.*, **169**, 646-649 (1919)

Q.MAJORANA, 'Expériences sur la gravitation', *Comptes Rendus Acad. Sci.*, **169**, 719- 721 (1919)

Q.MAJORANA, 'Sulla gravitazione', *Atti Accad.Lincei*,**29**, 23-32 (1920)

Q.MAJORANA, 'Sulla gravitazione', *Atti Accad.Lincei*, **29**, 90-99 (1920)

Q.MAJORANA, 'Sulla gravitazione', *Atti Accad.Lincei*, **29**, 163-169 (1920)

Q.MAJORANA, 'Sulla gravitazione', *Atti Accad.Lincei*, **29**, 234-240 (1920)

Q.MAJORANA, 'Sulla gravitazione', R.Accad.Scienze, **69**, (1919) Torino

Q.MAJORANA, 'On gravitation, Theoretical and experimental researches', Philos.Mag., **39**, 488-504 (1920)

Q.MAJORANA, 'Sull'assorbimento della gravitazione', Atti Accad. Lincei, **30**, 75-79 (1921)

Q.MAJORANA, 'Sull'assorbimento della gravitazione', Atti Accad.Lincei, **30**, 289-294 (1921)

Q.MAJORANA, 'Sull'assorbimento della gravitazione', Atti Accad.Lincei, **30**, 350-354 (1921)

Q.MAJORANA, 'Sull'assorbimento della gravitazione', Atti Accad.Lincei, **30**, 442-446 (1921)

Q.MAJORANA, 'Sur l'absorption de la gravitation', C.R. Acad.Sci., **173**, 478-479 (1921)

Q.MAJORANA,'Sull'assorbimento della gravitazione', Atti Accad.Lincei, **31**, 41-45 (1922)

Q.MAJORANA, 'Sull'assorbimento della gravitazione', Atti Accad.Lincei, **31**, 81-86 (1922)

Q.MAJORANA, 'Sull'assorbimento della gravitazione', Atti Accad.Lincei, **31**, 141-146 (1922)

Q.MAJORANA, 'Sull'assorbimento della gravitazione',Atti Accad.Lincei, **31**, 221-226 (1922)

Q.MAJORANA, 'Sull'assorbimento della gravitazione',Atti Accad.Lincei, **31**, 343-346 (1922)

Q.MAJORANA, 'Sur l'absorption de la gravitation', Revue Générale des Sciences, **15**, 70 (1922)

Q.MAJORANA, 'Sull'assorbimento della gravitazione', Atti Soc. Ital. Progr. Scienze, Riunione di Trieste (1921)

Q.MAJORANA, 'A proposito di alcune mie ricerche sulla gravitazione', Rend.R. Acc. delle Scienze di Bologna, **26**, 145 (1921-1922)

Q.MAJORANA, 'Quelques recherches sur l'absorption de la gravitation', J. Phys. Radium, **9**, 314-324 (1930)

Q.MAJORANA, 'Su di un'ipotesi cosmogonica', Atti Accad. Lincei, **17**, 150-157 (1954)

Q.MAJORANA, 'Su di una nuova teoria della gravitazione', Atti Accad.Lincei, **19**, 95-101 (1955)

Q.MAJORANA, 'Sull'ipotesi dell'assorbimento gravitazionale', Atti Accad.Lincei, **22**, 392-397 (1957)

Q.MAJORANA, 'Ipotetiche conseguenze dell'assorbimento gravitazionale' Atti Accad.Lincei, **22**, 397-402 (1957)

(3) See:

Q.MAJORANA, 'Sul secondo postulato della teoria della relatività', Accad.Lincei, **26**, 118-122 (1917)

Q.MAJORANA, 'Dimostrazione sperimentale della costanza di velocità della luce riflessa da uno specchio in moto', Accad. Lincei, **26**, 155-160 (1917)

Majorana pubblicherà questi suoi risultati anche su riviste internazionali: Philosophical Magazine, **35**, 163 (1918) e **37**, 145 (1919) e Archives des Sciences Physique et Naturelles de Genéve, **53**, 793 (1918)

Q.MAJORANA, 'Sulla propagazione della luce riflessa da uno specchio mobile nel vuoto' Accad.Lincei, **19**, 754-759 (1934) Furono inviati resoconti anche a Riviste Internazionali. Nel 1934 la nota di Majorana su questo esperimento apparve sul Physik Zeitschrift, **35**, 740 (1934)

LE

JOURNAL

DE

PHYSIQUE

Fondé par J. Ch. d'ALMEIDA

ET LE

RADIUM

Fondé par Jacques DANNE

PUBLICATION de la SOCIÉTÉ FRANÇAISE de PHYSIQUE

Subventionnée par la Confédération des Sociétés scientifiques françaises
à l'aide des fonds alloués par le Parlement

ANNÉE 1930

TOME I — SÉRIE VII

REVUE GÉNÉRALE DE L'ÉLECTRICITÉ, ÉDITEUR

Administration :
12, place de Laborde, PARIS VIIIe

Rédaction :
10, rue Vauquelin, PARIS V[e]

QUELQUES RECHERCHES SUR L'ABSORPTION DE LA GRAVITATION PAR LA MATIÈRE

par M. Q. MAJORANA

J'ai été très flatté par l'aimable invitation du Conseil de la Société Française de Physique, de faire une conférence à l'occasion de son assemblée annuelle, bien que le choix du sujet m'embarassât quelque peu. Mais puisque, comme d'habitude, il est permis au conférencier de parler de ses recherches personnelles, je me suis enfin décidé à choisir le sujet annoncé. Il s'agit là d'une question bien délicate à laquelle j'ai dédié une dizaine d'années de recherches qui ne sont pas encore achevées et dont les résultats, s'ils sont définitivement confirmés ne manqueront pas d'acquérir une importance remarquable (¹).

Cependant je vous prierai de considérer le choix de ce sujet, principalement comme un désir d'attirer votre attention sur la possibilité de sonder expérimentalement une propriété que l'on soupçonne dans la manifestation d'une des forces les plus imposantes de la nature : celle de la gravitation.

Je ne prétends, en effet, nullement affirmer que mes expériences, d'ailleurs très variées, dont je vais vous entretenir brièvement, sont à coup sûr concluantes. Mais il serait, à mon avis, opportun, que mes expériences pussent être reprises par d'autres collègues plus habiles que moi et disposant de moyens plus perfectionnés. Il pourrait certes se faire que ces recherches éventuelles se concluent par la constatation que *l'effet* que j'ai trouvé reste réduit dans une mesure plus ou moins grande, ou que les limites de sensibilité ou les erreurs d'observation ne permettent réellement pas de préciser cet effet d'une façon sûre. Mais, même dans ce cas, le physicien ferait une œuvre utile au progrès de la science.

Le titre de ma conférence pourrait tout simplement faire croire à l'existence d'une absorption de la force gravifique dont on voudrait effectuer une détermination quantitative. Or, il ne s'agit pas précisément de cela.

La force gravifique est, peut-on dire, dans ses manifestations, la plus parfaite des forces naturelles. Exprimée par la loi extrêmement simple de Newton, aucune action ou force physique concomitante ne peut la modifier. Cela est vrai dans les limites de précision des appareils que les physiciens peuvent réaliser, ou des observations astronomiques qu'ils sont à même de faire. Mais il y a tout lieu de penser que de même que la matière ne laisse pas absolument inaltérée une action ou force quelconque s'exerçant à travers elle, elle peut amener une modification quelconque dans la gravitation, dont les lignes de forces la traverseraient.

Ce raisonnement, fondé sur des considérations de caractère analogique, ne nous assure pas de l'existence d'un affaiblissement d'une force gravifique, lorsqu'elle traverse la matière. Nous pourrions tout au plus admettre qu'elle subit une variation. Cette variation pourrait concerner, d'une façon générale, soit l'intensité de la force (augmentation ou diminution), soit la distribution de ses lignes. Le raisonnement analogique ne tend d'ailleurs nullement à préciser si la force de gravitation doit être attribuée à coup sûr à une déformation statique d'un milieu hypothétique (l'éther cosmique), comme c'est le cas du champ électrique ou magnétique, ou de caractère dynamique, comme dans le cas de l'énergie rayonnante,

(¹) On pourra trouver un exposé plus détaillé de mes premières recherches sur les *Rendiconti della R. Accademia dei Lincei*, 1920, 1921, 1922.

comme cela a lieu pour les ondes électromagnétiques ou la lumière, ou pour les ondes élastiques (mécaniques ou acoustiques).

En parlant d'absorption de la gravitation, j'ai voulu faire allusion, d'une façon générale, à une action du milieu matériel existant entre deux masses agissantes ; sans me préoccuper *a priori* du sens de la variation de la force même.

Je rappellerai en outre, que l'idée d'une véritable absorption de la force de gravitation n'est ni nouvelle ni récente. Laplace avait effleuré la possibilité d'une modification de la loi de Newton suivant la relation :

$$F = k \frac{M_1 M_2}{r^2} e^{-hr},$$

où h est l'affaiblissement de la force par unité de longueur à travers une masse de densité unitaire.

Seeliger fit constater que ce coefficient d'amortissement h pourrait donner la clé de la rotation du périhélie de Mercure s'il était égal à 4.10^{-7}. Mais, outre le fait qu'une absorption de cette valeur serait facilement constatable par des expériences directes, comme celles que j'ai faites moi-même, on observe que, pour le périhélie de mars, cette valeur serait excessive.

Au surplus, il faut noter que la formule de Newton, modifiée de la sorte, présenterait l'avantage de détourner la difficulté de l'application de la loi de Newton aux masses infiniment grandes, telles qu'on devrait les admettre dans le cas où l'univers serait infini.

L'existence de l'influence sur la force de gravitation a été recherchée expérimentalement avant moi-même par d'autres physiciens. Je citerai Austin et Thwing, Kleiner, Laager, Crémieu, Erisman et, tout récemment Schlomka. Cependant, exception faite pour Schlomka, ces expériences ont été dépourvues d'un résultat positif. Quant aux recherches de Schlomka lui-même, une confirmation s'impose avant leur acceptation.

Je soulignerai encore le fait que, suivant Botlinger, la valeur de h pourrait être de 3.10^{-15} ce qui expliquerait certaines perturbations au cours des éclipses lunaires.

Il faudrait également rappeler, en nous plaçant à un point de vue purement théorique, que la théorie de la relativité d'Einstein porte aussi à admettre un certain coefficient d'absorption d'un ordre de grandeur bien plus réduit, et précisément 10^{-23} environ.

Si donc le problème de la recherche des modalités et de la mesure de l'influence du milieu sur la gravitation a été entrevu de plusieurs côtés, il n'a pas été abordé et résolu définitivement et sûrement.

Et cependant il mérite, comme je viens de le dire, toute l'attention des physiciens.

C'est ainsi que je tâche depuis dix ans de mettre au point des dispositifs expérimentaux pouvant m'aider dans la solution du problème. De là, j'ai tiré le sujet de ma conférence, et j'effleurerai ensuite les quelques critiques auxquelles ces recherches ont donné lieu ; je parlerai enfin de mon avis à l'égard de ces mêmes critiques.

Recherches expérimentales. — Une étude préalable quantitative et soignée nous porterait facilement à la constatation que la façon de vérifier, avec la plus grande sensibilité, les variations éventuelles de la force gravifique traversant un milieu matériel, est de se servir d'une balance oscillant dans un plan vertical. Celle-ci nous permettrait d'étudier l'influence que ce milieu exerce sur la force de gravité terrestre. Il est, à mon avis, impossible de penser à réaliser avec une sensibilité pareille une expérience tendant à la recherche d'une action que le milieu exercerait éventuellement sur une balance horizontale du type Cavendish. Toutes mes recherches ont donc été menées au moyen d'expériences dans lesquelles je me servais de balances à fléau rigide munies d'un couteau d'acier ou de pierre dure et oscillant dans un plan vertical.

Avant de penser à réaliser un appareil de ce genre, il fallait se rendre compte de l'ordre de grandeur probable, de l'effet à rechercher moyennant des hypothèses subsidiaires.

A ce propos, qu'il me soit permis de me placer dans une attitude agnostique en ce qui concerne quelques théories modernes et leurs conséquences.

En effet, je crois que si l'on s'occupe d'étudier la nature intime du phénomène gravifique, la formulation d'hypothèses nouvelles (pourvu qu'elles ne soient pas illogiques), peut être consentie.

Bien entendu, ces hypothèses n'auront d'autre valeur que celle *d'hypothèses de travail*, moyennant lesquelles on peut établir le plan d'une recherche expérimentale. On doit en outre être prêt à les abandonner ou à les modifier si l'expérience ou bien des raisonnements irréfutables montrent qu'elles ne sont plus soutenables.

Cela établi je me reporte à ce que je disais tout au début de mon discours, et je rappelle le double aspect sous lequel nous pourrions considérer ce phénomène : aspect statique ou dynamique. Le premier nous porte nécessairement à admettre l'existence d'un milieu (l'éther cosmique), tandis qu'on peut s'en passer pour le second. Conformément à ce second aspect, on peut penser plus vraisemblablement que la force de gravitation doit, en l'absence d'un milieu quelconque dépendre d'une sorte de bombardement, pour ainsi dire centrifuge ou centripète, de *particules « sui generis »*.

L'hypothèse centripète est, au fond, celle de Lesage et on connaît les difficultés auxquelles elle donne lieu lorsqu'on veut en approfondir les particularités. Par contre, l'hypothèse centrifuge ne représente pas un modèle ayant un caractère mécanique ; car il faudrait admettre que, au choc de particules émises par un corps, correspond une force de sens contraire à ce choc même. Je ne me suis pas arrêté cependant à perfectionner une de ces hypothèses ; mais en choisissant parmi les deux, la seconde, j'ai essayé d'en déduire quelques conséquences.

J'ai déjà montré par ailleurs, qu'en suivant cet ordre d'idées, on peut arriver à préciser l'ordre de grandeur probable du facteur d'amortissement h. Il me faut, pour la clarté de mon exposé, résumer ici ce que j'ai dit à cet égard.

En formulant ladite hypothèse, on peut supposer que quelques particules, dans leur mouvement à travers un milieu matériel, viennent à être arrêtées. Pour mieux comprendre encore le phénomène gravifique, on pourrait même dire que la force attractive dépend, d'une manière certaine, des particules qui s'arrêtent ; mais cette hypothèse n'est pas rigoureusement nécessaire, et elle ne nous conduit pas, au moins pour le moment, à entrevoir des contrôles expérimentaux plus précis. Il suffira de dire que, comme conséquence du passage des particules à travers un milieu matériel, le flux des particules, c'est-à-dire le nombre de celles qui, dans l'unité de temps, traversent l'unité de section, diminue par l'absorption. Et, puisque la force attractive ne doit dépendre que de ce flux, elle ne peut que diminuer par l'interposition d'un milieu matériel entre deux corps qui s'attirent. Je reviendrai après sur les contrôles expérimentaux entrepris par moi dans ce sens.

Mais en attendant, d'après les hypothèses faites, une masse considérable comme celle d'un astre (le Soleil, par exemple), ne pourrait exercer à l'extérieur une action gravifique correspondant à la somme de ses actions élémentaires ; elle se comportera comme si, le phénomène d'absorption n'existant pas, elle était plus petite. L'astronome, qui déduit de la troisième loi de Kepler ou de la théorie des perturbations, les masses des astres, ne déterminerait donc que des masses *apparentes* ; les masses *réelles* pourraient être notablement plus grandes.

J'ai pu déduire de ces considérations une limite supérieure du coefficient h (coefficient d'absorption gravifique). L'existence de cette limite est la conséquence évidente du fait, que si l'on doit admettre l'existence d'une masse *apparente* et d'une masse *réelle* pour tout astre déterminé, la masse apparente aura précisément la valeur de laquelle dépendent les forces attractives des astres mêmes. Si le coefficient d'absorption était extrêmement grand, la masse apparente serait évidemment négligeable, tout son flux attractif étant absorbé par les couches externes de l'astre. Par un calcul préalable, j'ai pu établir que la valeur de h doit donc avoir la limite supérieure d'environ 8×10^{-12}. Certainement, donc, la vraie valeur de cette constante est comprise entre 0 et 8×10^{-12}.

Ce point de départ une fois posé, il fallait que mes appareils eussent une sensibilité telle qu'elle me permît de vérifier si la force de gravitation peut s'affaiblir d'environ un billionième (10^{-12}) en traversant un centimètre de masse de densité unitaire, ou d'environ un milliardième (10^{-9}) pour une masse de densité 10 et une distance de quelques dizaines de centimètres. Une bonne balance possède réellement cette sensibilité, pourvu que l'on adopte des dispositifs spéciaux. Il faut noter en outre qu'il ne s'agit pas d'établir une mesure absolue, mais de mesurer une variation du poids d'un corps entouré d'un milieu matériel qu'on peut, à volonté, éloigner.

Il y avait lieu de croire, selon toute probabilité, que ce milieu matériel peut être d'autant plus efficient qu'il est plus dense. J'ai donc constamment adopté dans mes expériences, l'artifice de peser une sphère de plomb entourée ou non de mercure ou de plomb.

La distribution de ces masses doit être en tout cas telle, qu'elle donne lieu à des forces attractives dont la résultante soit parfaitement nulle.

Les dispositifs que j'ai adoptés et expérimentés sont au nombre de quatre.

Première expérience. — L'appareil est indiqué figure 1. Avec cet appareil, j'ai

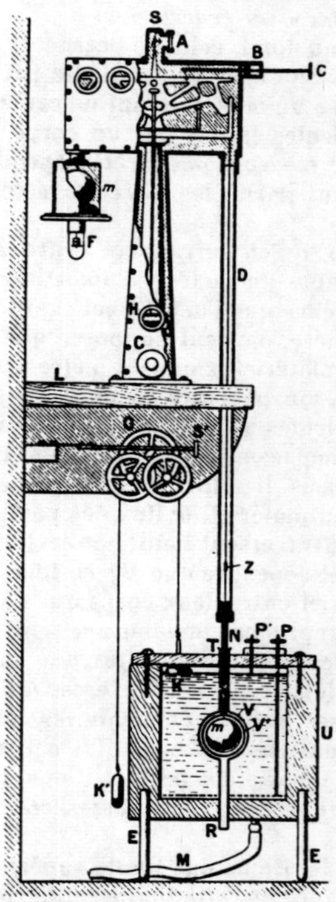

Fig. 1.

réussi à peser une sphère de plomb de 1 274 gr, libre, puis entourée symétriquement de 104 kg de mercure. La balance employée avait une telle sensibilité, qu'elle pouvait déceler une variation de poids correspondant à un amortissement de la force gravifique de 10^{-9}. En effet, la déviation provoquée dans le rayon lumineux réfléchi par le fléau, sous l'action d'une surcharge de 1 mg, était environ de 400 à 500 mm, c'est-à-dire d'environ 1/2 mm par 1/1000 de mg. Puisqu'on pouvait apprécier le 1/10 de mm, la détermination du 1/1000de mg.

était suffisamment sûre. Sans discuter ici de nombreuses causes d'erreur (thermiques, mécaniques et magnétiques), que j'ai eu à éliminer au cours de cette expérience, je dirai simplement que j'ai cru trouver une diminution du poids de la sphère provenant de la présence du mercure égale à 1/1000 de mg. Autrement dit, conformément aux idées annoncées, la force gravifique (la gravité dans ce cas), en traversant « le manteau de mercure » de 10 cm d'épaisseur environ, disposé autour de la sphère, s'est trouvée affaiblie. La valeur de h, c'est-à-dire l'affaiblissement subi par l'unité de poids pour l'unité de densité du milieu et pour l'unité de distance parcourue par la force gravifique, est, par conséquent, de :

$$h = \frac{1 \text{ millième de mgr}}{1.274 \text{ gr} \times 13{,}6 \times 8{,}5 \text{ cm}} = 6{,}66 . 10^{-12}$$

où 13,6 est la densité du mercure ; cette valeur confirmerait donc l'idée de l'absorption non seulement comme signe, mais encore comme ordre de grandeur.

Deuxième expérience. — Plus tard, j'ai répété l'expérience dans des conditions notablement différentes. La sphère de plomb et la balance sont les mêmes que dans la première expérience: la masse absorbante est constituée par 9 603 kg de plomb disposés en

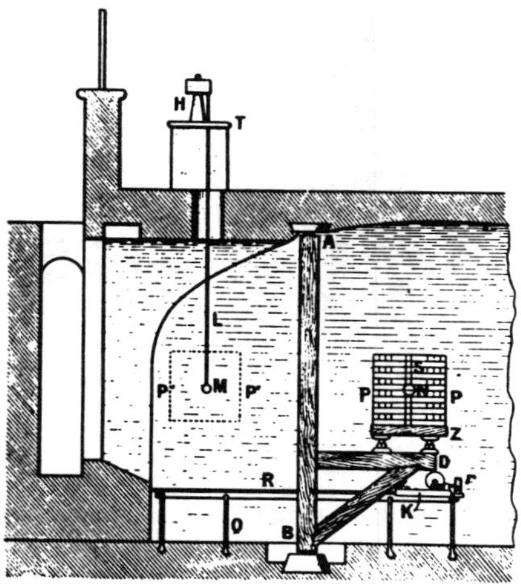

Fig. 2.

forme de cube de 95 cm de côté (fig. 2). Le cube peut être décomposé en deux prismes de 95 cm de hauteur et de 95 × 47,5 cm² de base.

Ces prismes sont mobiles et on peut les réunir de manière que, dans une première série d'expériences, la sphère de plomb soit emprisonnée au centre du cube ainsi formé, et suspendue sans le toucher par un fil à la balance extérieure. Cette balance est disposée à l'étage supérieur de l'édifice. Dans la figure 3 on voit les détails de la balance avec les artifices adoptés pour corriger les déformations produites par le mouvement des plombs dans le souterrain, etc. Les prismes de plomb peuvent être écartés jusqu'à 3 mètres de la sphère. La diminution de poids de la sphère qu'on peut prévoir avec ce nouveau dispositif et qui dépendrait du parcours rectiligne de la gravité dans la masse du cube (ce parcours, malgré la masse environ 100 fois plus grande, n'est que 4 ou 5 fois plus grand que dans le

cas du mercure), aurait dû être précisément 4 ou 5 fois plus forte, c'est-à-dire égale à 4 ou 5 millièmes de mg, étant donné la faible différence de densité entre le mercure et le plomb. En fait, après avoir lutté plus d'une année contre les difficultés résultant de la nécessité de faire mouvoir une masse de 10 tonnes dans le voisinage d'une balance qui devait déceler des variations de poids de l'ordre du milliardième de la masse attachée à ses bras (c'est-à-dire 1/1000 de mg pour 1 kg environ), j'ai trouvé que la variation de poids, quoique se manifestant dans le sens voulu, n'est que de 2 millièmes de mg, c'est-à dire d'environ la moitié de la différence prévue. Ceci conduit pour h à une valeur de $2{,}8 . 10^{-12}$, au lieu de $6{,}7 . 10^{-12}$. Je ne saurais pas dire si la différence entre ces deux résultats provient de quelque

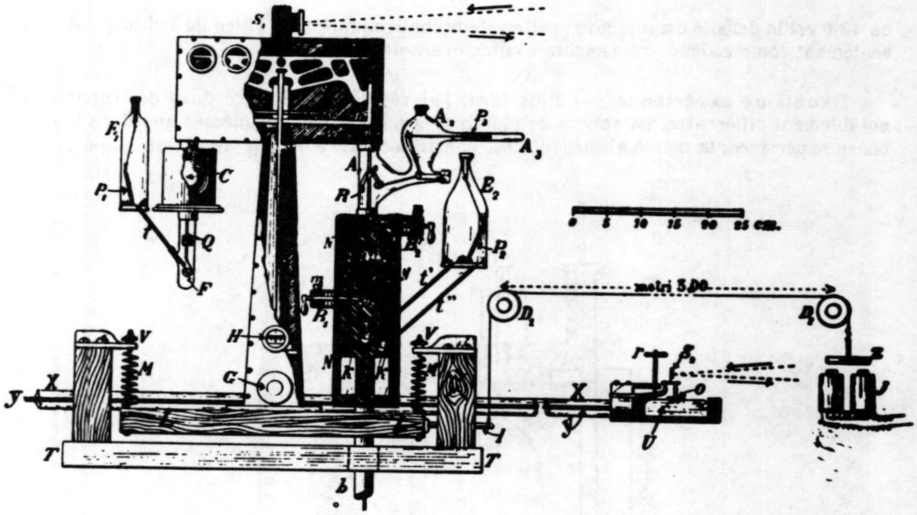

Fig. 3.

erreur dans la première détermination, ou (c'est moins probable) dans la seconde; ou bien d'un défaut de ma théorie qui serait valable de cette façon, seulement en première approximation. D'autres expériences pourront répondre à cette question; mais la concordance du signe et, dans une certaine mesure, de l'ordre de grandeur des deux déterminations est déjà un résultat satisfaisant.

Dans une seconde série d'expériences exécutées avec le même appareil, j'ai disposé la sphère de plomb immédiatement au-dessus du cube protecteur, en déterminant de cette façon l'attraction newtonienne correspondante, égale à peu près à 2/10 de mg. Les figures 4, 5, 6, se réfèrent à cette série d'expériences : dans la figure 4, les prismes sont écartés de la sphère; dans 5 ils sont en marche; dans 6 ils sont sous la sphère.

Et enfin, dans une troisième série, j'ai placé la sphère immédiatement au-dessous du cube; la nouvelle attraction, de direction contraire à la précédente, se trouve quelque peu plus grande en valeur absolue (c'est-à-dire supérieure à deux dixièmes de mg) d'environ 4/1000 de mg. La figure 7 indique en M, M_1, M_2, les positions de la sphère dans les trois séries d'expériences.

Cette différence est donc le double de la diminution de poids que l'on avait observée

lorsque la sphère était au centre du cube. L'explication la plus simple de ce résultat est que, dans la deuxième série d'expériences, l'absorption gravifique est double, justement parce que les rayons gravifiques terrestres traversent toute la largeur du cube, et non la moitié, comme dans le cas de la sphère placée au centre ; au contraire, dans la troisième série, l'absorption est nulle.

Ce résultat, non seulement montrerait l'accord entre l'absorption trouvée dans le cas de la sphère placée au centre du cube et dans le cas de la sphère placée au-dessus, mais nous permettrait, étant définitivement confirmé, de décider entre l'hypothèse des particules locales, avancée par moi, et celle des particules ultramondaines de Lesage En effet, d'après cette dernière théorie, lorsque la sphère se trouve au-dessous du cube, elle aurait dû subir un allégement à cause de la présence du cube même, parce que des particules, venant de

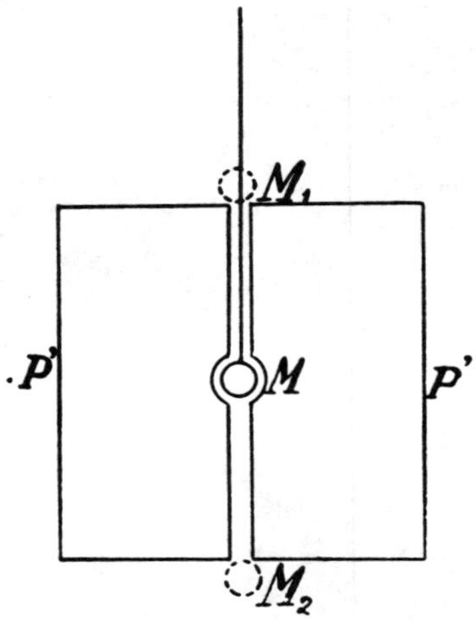

Fig. 7.

l'extérieur, s'y seraient arrêtées. Au contraire, mon observation montre que l'allégement a lieu lorsque la sphère est au-dessus du cube. Il me semble donc que la théorie des particules locales soit préférable.

Troisième expérience. — J'ai repris, à Bologne, les expériences sur la gravitation et en vue des difficultés que j'avais rencontrées avec l'usage de grandes masses, je me suis décidé à me servir encore d'appareils relativement plus petits. En outre (puisque je doutais si la différence entre les résultats de la première et de la seconde expérience pouvaient dépendre de la qualité du milieu absorbant employé), je me suis encore servi de masses de plomb comme dans l'appareil dernièrement décrit,

La figure 8 indique le schéma de la disposition adoptée. Une balance porte aux extrémités de ses bras deux sphères : S_1 S_2 en plomb ayant comme précédemment une masse d'environ 1 000 gr. Les deux sphères peuvent être entourée par un cylindre en plomb divisé en deux parties. Le poids de chacune de ces parties est de 180 kg environ. Le cylindre est soutenu d'une manière convenable (fig. 9) : il peut être porté dans l'une ou l'autre des posi-

PLANCHE I.

Fig. 4.

Fig. 5.

Fig. 6.

Fig. 9.

PLANCHE III.

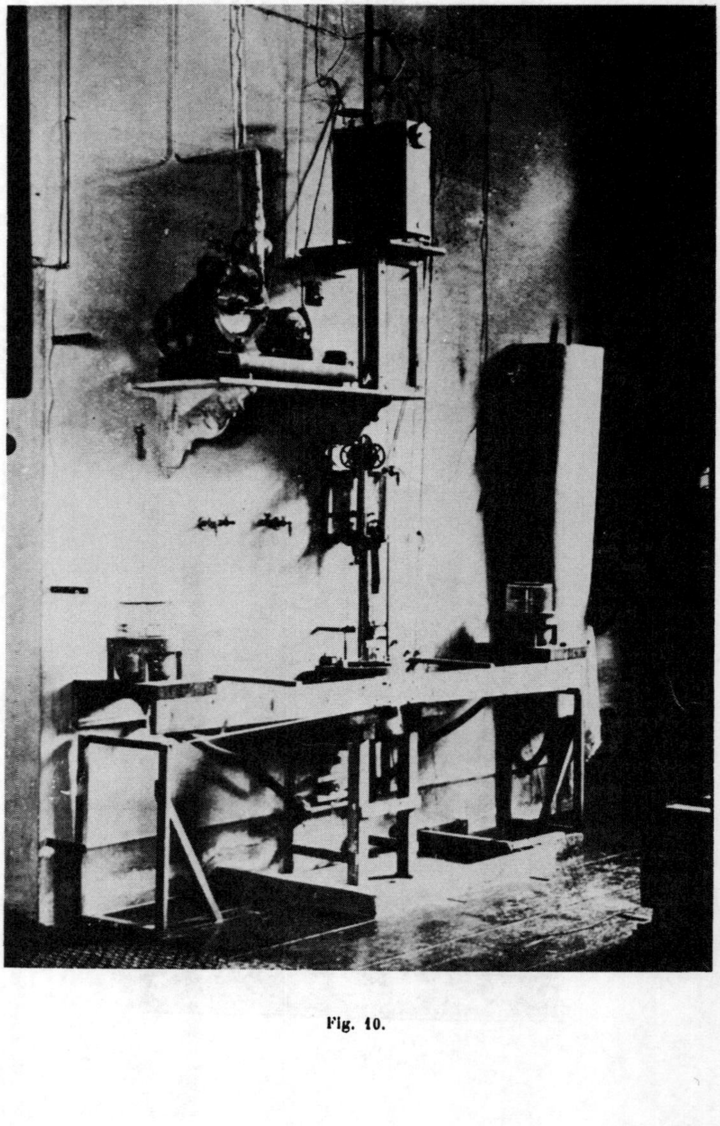

Fig. 10.

tions indiquées par la figure 8 ; de la sorte, l'effet observable sur une des deux sphères est doublé. Bien que j'aie travaillé longtemps à la réalisation de cette disposition expérimentale plutôt compliquée, j'ai dû reconnaître qu'elle engendre beaucoup de causes d'erreur dans la mesure de l'effet, au point qu'on ne peut pas déduire par son usage des valeurs certaines pour sa grandeur.

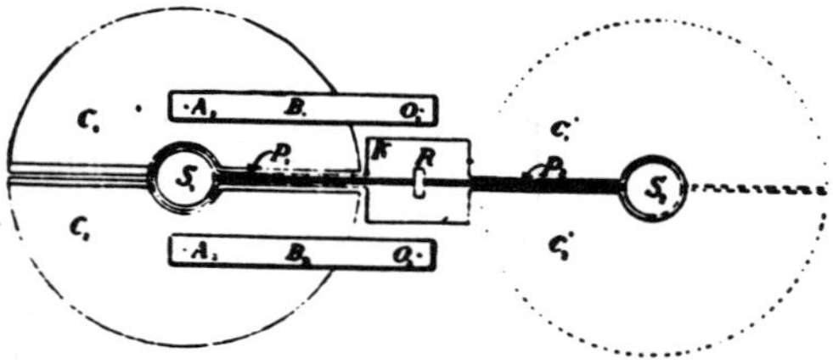

Fig. 8.

Quatrième expérience. — En vue de cela, je me suis décidé, encore une fois, à expérimenter avec le mercure (matière absorbante) moyennant une disposition semblable à celle déjà employée la première fois à Turin. Cet appareil est indiqué dans la figure 10. Je me borne à fixer votre attention sur le perfectionnement plus important que j'ai apporté à cet appareil, dans le but d'éviter un inconvénient qui se présentait presque toujours dans tous les appareils de ce genre réalisés par moi. C'est-à-dire, qu'il arrive souvent, lors d'une mesure, que le déplacement de masses très pesantes qu'on doit opérer pendant la marche d'une expérience, conduit nécessairement à des déformations, très petites cependant, soit du pavé, soit des murailles du bâtiment de la salle dans laquelle on travaille. Ces déformations ont pour effet de faire tourner d'un très petit angle, le plan d'appui du couteau central de la balance.

Une méthode simple pour éviter cette déformation est illustrée par la projection que je vous montre (fig. 10). Le mercure qui se déplace de sa position autour de la sphère que l'on pèse, se trouve dans deux récipients latéraux soutenus par deux bras en bois fixés rigidement avec le tonneau qui le contenait dans ladite position. On comprend que le poids de tout ce système agit toujours sur le même point du pavé.

Les observations avec ce dernier dispositif n'ont pas encore été achevées à cause d'une étude minutieuse que je conduis depuis les deux dernières années sur les meilleures conditions de fabrication et de fonctionnement des couteaux de la balance et qui n'est pas encore terminée. Les quelques mesures déjà faites semblent donner des résultats qui confirment le sens de l'effet autrefois établi, c'est-à-dire l'absorption de la force de gravitation. Bien que je ne puisse donner aujourd'hui des résultats quantitatifs sur l'effet recherché, j'ai confiance qu'avec le dernier appareil en cours d'expérimentation je serai à même, dans quelque temps, de dire, pour mon compte, un mot définitif *sur le sujet.*

Après avoir fait la description des dispositifs expérimentaux dont je me suis servi pour contrôler l'existence d'une absorption de la force gravifique, je vous parlerai brièvement des objections dont mes recherches ont été l'objet. Je vous dirai tout d'abord que, si j'en suis bien informé, trois savants seulement, Lippmann, Russell, Eddington, ont soulevé, il y a quelques années, des objections aux conclusions auxquelles j'avais abouti ; mais ces objections touchent le seul point de vue théorique. Au point de vue expérimental, le sujet méritait la plus grande attention de la part des physiciens, et son importance aurait dû induire quelques-uns d'entre eux à répéter mes recherches. Je vous rappellerai, d'ailleurs, que tout de suite après la publication de mon premier mémoire, le célèbre physicien Michelson m'écrivit en me demandant de l'autoriser à répéter mes expériences. Il est certain que personne n'aurait pu, avec plus d'autorité que Michelson, dire le dernier mot sur un sujet si délicat. N'ayant pu m'empêcher, cependant, de déclarer à M. Michelson que j'entendais poursuivre mes recherches avec d'autres dispositifs bien que j'eusse été heureux de sa collaboration sur le sujet, notre éminent collègue s'abstint, par délicatesse, d'insister sur sa demande. La première objection émana du regretté G. Lippmann. Il m'écrivait : « ... du moment que la matière altère l'attraction newtonienne, ne s'ensuit-il pas que cet effet doit être plus marqué pour les grosses planètes que pour les petites? Ne s'ensuit-il pas que la loi de Kepler $\frac{a^3}{T^2}$ = const. ne se vérifie pas exactement, le quotient $\frac{a^3}{T^2}$ devant être plus petit pour les grosses planètes que pour les petites? »

J'eus l'honneur de répondre à M. Lippmann au sujet de cette objection. Mais, celle ci concordant dans une certaine mesure avec les remarques de l'astronome Russell, je reviendrai plus loin là-dessus.

Les considérations de M. Russell sont plus détaillées et plus minutieuses. Je ne lui ai jamais répondu de façon à épuiser le sujet et je suis heureux que l'assemblée de la Société Française de Physique m'en donne l'occasion, quoique peut-être un peu tard. Je justifierai ce retard en observant que mes recherches avaient un caractère purement expérimental et il ne m'a jamais semblé opportun d'entraver par des « à priori » une expérience possible, à moins que celle-ci ne fût tout simplement absurde.

M. Russell divise ses objections en les appuyant sur trois hypothèses distinctes. « Supposons, dit-il, tout d'abord, que les masses inertes des planètes restent inaltérées, tandis que les masses pesantes varient suivant le coefficient d'absorption h de Majorana. On voit immédiatement qu'il devrait découler de cette supposition que les orbites des grosses planètes, calculées suivant les temps de leurs révolutions, devraient être notablement plus petites qu'elles ne le sont en réalité. Ainsi, la distance entre Jupiter et le Soleil devrait être d'environ 1 pour 100 plus petite qu'elle n'est en réalité. » M. Russell dit, en conclusion, que, pour admettre le principe de l'absorption de la gravitation, il faudrait ramener la valeur d'h à un maximum d'1/5000 du résultat. Au fond, cette première objection n'est autre que celle de M. Lippmann et elle est sans nul doute fondée, sans préjudice cependant de ce qui suit.

M. Russell fournit toutefois par la suite le moyen d'éliminer son objection. Il suppose, maintenant, que la masse inerte d'un corps suit exactement les mêmes variations que sa masse pesante; ou, en d'autres termes, que les accélérations subies, par exemple, par les planètes correspondent à leurs masses, réduites celles-ci par suite de l'absorption. M. Russsell reconnaît de la sorte, que l'absorption de la gravitation ne peut amener de perturbation sensible dans le mouvement des planètes.

Il remarque en troisième lieu cependant que le phénomène des marées solaires et lunaires devrait s'accomplir d'une façon considérablement différente de celle qui a été constatée. Selon M. Russell, l'hypothèse satisfaisante de la proportionnalité entre la masse inerte et la masse apparente elle-même ne suffirait plus à expliquer ce que l'on observe sur la terre même.

Voulant cependant, par une bienveillance dont je le remercie, motiver mes résultats expérimentaux, M. Russell est porté à croire que, s'ils étaient fondés, il faudrait admettre que la présence d'un corps dans le voisinage d'un autre ne donne pas lieu à l'existence d'une masse totale égale à leur somme. En d'autres termes, les propriétés newtoniennes de la matière ne seraient pas parfaitement additives.

Répondant à M. Russell, je dirai que je ne peux le suivre dans ses raisonnements que jusqu'à un certain point, précisément jusqu'à la conclusion de sa première objection.

Je suis, en d'autres termes, parfaitement d'accord avec lui lorsqu'il pense que la masse inerte doit varier dans une proportion exacte à la variation de la masse pesante. Mais il me semble que les raisonnements qu'il fait en calculant les perturbations éventuelles des marées, ne peuvent être acceptés comme définitivement concluants. En effets, ces raisonnements se basent sur la vérification de la loi de Newton des marées, faite par Michelson. Or, il faut rappeler que selon les déterminations du physicien américain, on peut conclure qu'en Amérique les marées d'Est à Ouest sont seulement 0,7 de celles prévues par la dite loi. Pour justifier cette différence, on peut attribuer à la croûte terrestre un manque de rigidité, quoiqu'elle reste toujours aussi rigide que l'acier. Supposons maintenant que soit vrai le fait de l'absorption gravifique. Le problème de la détermination de la rigidité de la terre par l'observation d'une des marées (par exemple, la lunaire) ne serait plus déterminé. Il pourrait arriver par exemple que la terre fût beaucoup moins rigide que l'acier, au point que la marée fût seulement 0,3 de la théorique (terre rigide), et l'absorption gravifique pourrait porter cette valeur à 0,7. Or puisqu'on peut faire aussi des déterminations sur les marées solaires, le problème n'est plus indéterminé, et il faudrait, selon les considérations de M. Russell, arriver à la réfutation de l'hypothèse de l'absorption.

Mais il y a lieu d'observer que M. Russell arrive à ces conclusions moyennant l'hypothèse suivant laquelle l'action du soleil ou de la lune sur une parcelle de l'océan s'exerce exactement et exclusivement suivant la résultante géométrique.

Or, dans l'ignorance où nous sommes de la véritable essence du phénomène gravifique, il me semblerait déplacé d'admettre avec toute rigueur cette hypothèse et de vouloir, surtout, en tirer des conséquences incontestables; je dis cela bien que moi-même, dans certains cas, j'aie fait usage de cette hypothèse.

L'objection de M. Russell sur les marées ne me semble donc pas aussi grave qu'elle le parait; et je me permets de dire qu'avant d'abandonner définitivement l'hypothèse de l'absorption, il faudrait discuter plus complètement la question des marées. Après, on pourrait voir si l'autre hypothèse avancée par lui, sur la propriété non additive des masses matérielles peut être, examinée plus attentivement.

Je parlerai, maintenant, des objections de M. Eddington. Elles se basent sur la théorie de la relativité. Celle-ci pourrait, en premier temps, motiver le désaccord qui existerait entre mes résultats et les observations astronomiques (désaccord qui a déjà fait l'objet de l'examen de M. Russell), mais serait, tout de suite après, en contradiction flagrante avec ces résultats mêmes.

M. Eddington part de l'application du principe fondamental de la relativité générale : un corps qui subit un mouvement accéléré perd son champ de gravitation, générateur du dit mouvement. Pour M. Eddington, un corps (une planète) se comporte, dans son mouvement, comme s'il ne se trouvait pas immergé dans un champ de gravitation : l'absorption doit donc être nulle.

Je me permettrai de faire observer que ce raisonnement néglige complètement le côté physique du phénomène. Supposons en effet qu'un corps A, attiré, tombe d'un mouvement accéléré dans la direction d'un autre corps, B. Supposons le corps A divisé en deux parties : A_1 et A_2, la première faisant face à B, l'autre, un peu plus éloignée. Admettons encore que la distance entre A et B soit considérable par rapport aux dimensions d'A. Il s'ensuivra que les deux moitiés A_1 et A_2 tomberont de conserve et unies dans la direction de B. Les accélérations subies par A_1 seront exactement identiques à celles que subira A_2. Or, en dépit de Mr. Eddington, qui affirme que le champ de gravitations engendré par B n'existe plus pour le corps A, A_1 et A_2 poursuivent leur chemin en vertu de la force gravifique qu'elles subissent continuellement. A proprement parler, la partie A_2 devra subir l'action d'une force qui aura dû traverser au moins en partie la moitié A_1 et il n'y a pas de raisonnement qui puisse expliquer pourquoi cette force ne doit pas être partiellement absorbée.

J'ai naturellement donné, de l'idée de Mr. Eddington, un schéma plutôt simple pour expliquer qu'il n'est pas admissible d'appliquer sans jugement la théorie de la relativité, fût-ce même au problème de l'absorption, du moins de la façon dont procède cet auteur.

Se plaçant cependant à un point de vue diamétralement opposé à celui de Mr. Russell, Mr. Eddington, après m'avoir offert le moyen de défendre mes résultats en les étayant sur les théories d'Einstein, m'objecte carrément le fait que ces théories font prévoir une absorption d'un ordre de grandeurs incomparablement plus petit que celui que j'ai déterminé.

Vu que les conclusions de la relativité se basent, dit-il, sur le principe de la conservation de l'énergie, l'admission de l'absorption de Majorana serait la négation de ce principe.

Il me semble que cette conclusion manque, elle aussi, de rigueur. L'idée que je me suis faite est celle-ci : si l'absorption de la gravitation existe dans la mesure que j'ai observée, elle dépend d'une propriété intime de la matière. Cette propriété, en admettant qu'elle soit vraie, n'est pas prévue par la théorie de la relativité.

Me voici à la fin de cet exposé, rapide en dépit de sa longueur, et je m'empresse de conclure.

Comme je l'ai dit, je vous ai exposé le résumé de mes recherches laborieuses principalement pour attirer votre attention sur la possibilité réelle de mesurer une variation de poids de l'ordre du milliardième d'une masse donnée. Cette possibilité est naturellement soumise à l'éventualité d'erreurs d'observation. Il me semble cependant difficile que quelques-unes de ces erreurs, insuffisamment étudiées ou profondément cachées, puissent rendre illusoires toutes les déterminations.

Ceci posé, j'ai montré comment on peut appliquer ce genre de mesure à l'étude de la gravitation et entrevoir, au moins, l'existence du phénomène de son absorption, dont je viens de parler.

Je poursuivrai ces recherches dans l'espoir que mes résultats futurs puissent enfin me dire d'une façon sûre si le phénomène existe réellement.

Mais s'il est vrai que je dédierai tous mes efforts à l'obtention de ce but, il n'en reste pas moins avéré qu'une matière si délicate exige la collaboration de plusieurs expérimentateurs. Il faudra cependant que ces messieurs, après avoir considéré avec soin le pour et le contre des objections soulevées contre le programme de recherches actuelles, admettent que le véritable progrès scientifique s'appuie d'abord et avant tout sur l'observation des faits.

C'est sur les faits que doivent se fonder les théories nouvelles ou les vieilles prudemment modifiées. Le contraire serait inadmissible.

Manuscrit reçu le 25 juin 1930.

THE

LONDON, EDINBURGH, AND DUBLIN

PHILOSOPHICAL MAGAZINE

AND

JOURNAL OF SCIENCE.

CONDUCTED BY

SIR OLIVER JOSEPH LODGE, D.Sc., LL.D., F.R.S.
SIR JOSEPH JOHN THOMSON, O.M., M.A., Sc.D., LL.D., F.R.S.
JOHN JOLY, M.A., D.Sc., F.R.S., F.G.S.

AND

WILLIAM FRANCIS, F.L.S.

"**Nec aranearum sane textus ideo melior quia ex se fila gignunt, nec noster vilior quia ex alienis libamus ut apes.**" JUST. LIPS. *Polit.* lib. i. cap. 1. Not.

VOL. XXXIX.—SIXTH SERIES.

JANUARY—JUNE 1920.

LONDON:

TAYLOR AND FRANCIS, RED LION COURT, FLEET STREET.

SOLD BY SMITH AND SON, GLASGOW;—HODGES, FIGGIS, AND CO. DUBLIN;—
AND VEUVE J. BOYVEAU, PARIS.

Characters of Newton's Law.—In its simplicity this law seems the most perfect of the physical laws. Up to now no influence of the nature of the medium has been detected in the propagation of the attractive force between two material masses. The researches of Austin and Thwing *, Kleiner †, Laager ‡, Cremieu §, Erisman ‖, and others having in view the discovery of an action of that kind, failed utterly. In consequence of Laager's experiment, in which he studied the weight of a silver ball, covered with a thick coat of lead, it may be thought that the lack of effect has been until now verified up to the approximation of about 5.10^{-5}. The result was obviously a confirmation of the exactitude of Newton's Law.

Doubts on the exactitude of Newton's Law.—It does not seem right to me to deduce from an experiment, like Laager's for instance, that what has been verified in the Laboratory, might repeat itself with the same appearance, even in astronomical cases. Hence it would not be right to deduce therefrom that the mass of the silver ball would appear the same if in the centre of the earth, or in the centre of the sun (333,000 times the mass of the earth).

Let us admit as an hypothesis, that the mass may appear smaller if surrounded by other masses, that is, that there is a diminution of the force of gravitation on account of its propagation across a material medium. This diminution might be owing to a property of the material medium, to be compared with electric or magnetic permeability, or to the progressive absorption of the force. In the first case, if the analogy with the electrical and magnetical phenomena could be proved, small thicknesses of the medium would be sufficient to allow of the verification of the hypothetical permeability of gravitation; this has not been done in the experiments known up to now. In the second case the absorption would occur for very great thicknesses of medium, and therefore escape the researches of the laboratory, and yet manifest itself in the celestial bodies. In consequence, this second hypothesis of absorption seems more probable, and its conception would be more easy, if the force of gravitation could be explained by a kind of energical flux, continually emanating from ponderable matter. This flux

* Phys. Rev. v. (1897).

† *Arch. sc. phys. et nat.* xx. p. 420 (1905).

‡ *Dissert. Zürich* (1904).

§ *C. R.* cxl. p. 80 (1905); cxli. pp. 653, 713 (1905); cxliii. p. 887 (1906).

‖ *Vierteljahrschr.* liii. p. 157 (1908).

would be gradually absorbed, as in the case of light in its propagation through a not perfectly transparent medium: Newton's Law would only be exact in the first approximation.

Consequences of the Hypothesis of Absorption.—The first consequence would be the knowledge of the *true mass*, and of the *apparent mass*. The true mass would be that property of the mass to which the attractive force is proportional, if it is reduced in very small particles. The apparent mass is, on the other hand, the apparent value that the true mass assumes in consequence of the gradual absorption.

In order to respect the principle of the conservation of energy, it would be required, furthermore, to admit that any kind of matter is gradually transforming itself. This would be analogous, so to speak, to what happens with radium, with the difference that for this substance the transformation lasts some thousands of years, whilst for every one of the other known substances, the transformation would last enormously longer.

Another consequence might be deduced: since the force of gravitation is produced by an energical flux that has been absorbed, and as the energy cannot be destroyed, it ought to transform itself, into heat, for instance. Therefore matter subjected to gravitation is heated; this fact would give a new explanation of all the solar heat or of a part of it at least. Furthermore, the hypothesis might be kept in mind when considering the fact, which is nearly certain, that there are no *non-luminous* stars of great dimensions in the sky; ponderable matter ought to heat itself if condensed into little space.

It is, however, with the utmost reserve that I state the hypothesis of the energetic character of the force of gravitation; and if considerations that I do not detect at present should arise in the future and make me judge it erroneous, I should withdraw it.

I rather consider as probable, and in fact, after some experiments that I shall describe as certain, the hypothesis of gravitational absorption.

Analytical Researches.—In order to be able to establish a plan of experiments with the purpose of verifying the hypothesis of absorption, it is necessary to consider this hypothesis analytically. A physical quantity, *sui generis*, can now be defined and called *flux of gravitational action*; it is not yet necessary to connect the notion of this quantity with the notion of energy. Let us suppose a material particle dm so small that its inner gravitational absorption might be valued as null. We can suppose, in consequence of the above stated hypothesis, that it would continually put forth a certain flux, proportional to dm, that is to say, $k\,dm$, uniformly irradiated in all directions. Let us suppose the particle to be in a vacuum; across a solid angle subtending the surface $d\omega$ at the distance of 1, the flux would only be

$$\phi = k\frac{dm\,d\omega}{4\pi}.$$

If the particle instead of being in a vacuum is in a medium of true density δ_r, the flux that will have arrived at the distance x from the particle, will be expressed by

$$\phi = k\frac{dm\,d\omega}{4\pi}e^{-\mathrm{H}x}. \quad . \quad . \quad . \quad . \quad . \quad (1)$$

This is equivalent to admitting a gradual absorption of the flux, proportional to its value at every point, to the thickness of the medium that has been crossed, and to the medium's density. It is supposed, in fact, that

$$\mathrm{H} = h\delta_r, \quad . \quad . \quad . \quad . \quad . \quad . \quad . \quad (2)$$

H being the *quenching factor* for the density δ_r, and h the quenching factor for density 1.

We will now consider a massive sphere, with the uniform density δ_r; and determine the flux that emerges from it. Let us call R its radius, O its centre (fig. 1). I consider an

Fig. 1.

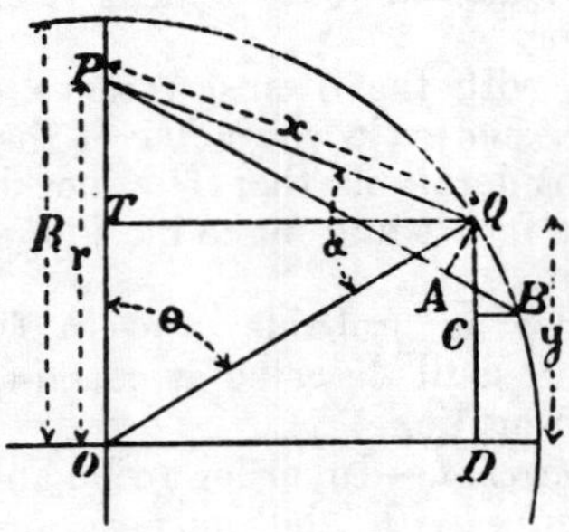

inner point P of it, in which the mass dm would be concentrated. I trace the PO radius of the sphere passing through P; and I trace QPB, an infinitesimal angle, with the vertex at P; I trace QA perpendicular to PQ. I say: $\mathrm{OP}=r$; $\mathrm{PQ}=x$; $\mathrm{QD}=y$. I make the triangle QPA turn around the PO axis; the AQ segment will trace the $2\pi . TQ . QA$ surface. It is possible in equation (1) to substitute for $d\omega$ this surface, reduced to unit distance from P, that is to say, dividing it by x^2. We have

$$\phi = k\frac{dm\,TQ\,.\,QA}{2x^2}e^{-Hx}.$$

QD is to be traced parallel to OP; B projected normally to QD on c; we have $Qc = dy$. I call $P\hat{Q}O = \alpha$; $P\hat{O}Q = \theta$; it appears from the figure that

$$dy = QB \sin\theta\ ;\ QA = QB \cos\alpha\ ;$$

therefore $$QA = \frac{dy}{\sin\theta}\cos\alpha.$$

Moreover,

$$TQ = R\sin\theta\ ;\ x = \sqrt{R^2 + r^2 - 2ry}.$$

Differentiating

$$dy = -\frac{x\,dx}{r}.$$

From the triangle OPQ we have

$$r^2 = x^2 + R^2 - 2xR\cos\alpha,$$

therefore $$\cos\alpha = \frac{x^2 + R^2 - r^2}{2xR}.$$

Hence we have

$$\phi = -k\frac{dm}{4r}\left(1 + \frac{R^2 - r^2}{x^2}\right)e^{-Hx}dx.$$

Let us call dF the total flux of action that is put forth by the dm particle, and that succeeds in emanating itself from the sphere; this flux will be given by the integral of ϕ, taken between $R + r$ and $R - r$ limits,

$$dF = -k\frac{dm}{4r}\int_{R+r}^{R-r}\left(1 + \frac{R^2 - r^2}{x^2}\right)e^{-Hx}dx\ ;$$

and carrying out the integration

$$dF = k\frac{dm}{4r}\left[\frac{e^{-Hx}}{H} + \frac{(R^2 - r^2)}{x}e^{-Hx} + H(R^2 - r^2)\int\frac{e^{-Hx}}{x}dx\right]_{R+r}^{R-r}.$$

Extending up to the limits, where it is possible,

$$dF = k\frac{dm}{4r}\left[e^{-H(R-r)}\left(\frac{1}{H}+R+r\right)-e^{-H(R+r)}\left(\frac{1}{H}+R-r\right)\right.$$
$$\left.-H(R^2-r^2)\int_{R-r}^{R+r}\frac{e^{-Hx}}{x}dx\right].$$

The integral remaining in this expression is transcendent, and one can only obtain its value by developing this expression in series; but this can be avoided by an opportune artifice. Let us call dm not only the mass that is contained in the point P, but all the mass of a spherical layer whose radius is r and thickness dr,

$$dm = 4\pi r^2\delta_v dr,$$

therefore

$$dF = k\pi\delta_v r\,dr\left[e^{-H(R-r)}\left(\frac{1}{H}+R+r\right)\right.$$
$$\left.-e^{-H(R+r)}\left(\frac{1}{H}+R-r\right)-H(R^2-r^2)\int_{R-r}^{R+r}\frac{e^{-Hx}}{x}dx\right].$$

To obtain the value of the total flux that emerges from all the points of the sphere, it is necessary to integrate this expression from O to R, and we have

$$F = k\pi\delta_v\int_0^R r\,dr\left[e^{-H(R-r)}\left(\frac{1}{H}+R+r\right)\right.$$
$$\left.-e^{-H(R+r)}\left(\frac{1}{H}+R-r\right)-H(R^2-r^2)\int_{R-r}^{R+r}\frac{e^{-Hx}}{x}dx\right]$$
$$= k\pi\delta_v\left[\frac{2R^2}{H}-\frac{2R}{H}+\frac{1}{H^3}-\frac{1}{H^3e^{2HR}}\right.$$
$$\left.-H\int_0^R r(R^2-r^2)dr\int_{R-r}^{R+r}\frac{e^{-Hx}}{x}dx\right].$$

We can carry out the double integration of the last term, changing the order of integration; but it is necessary to change also suitably the limits of integration. Proceeding in this way, and making $p = RH$, we have finally

$$F = k\pi\delta_v R^3\left[\frac{1}{p}-\frac{1}{2p^3}+e^{-2p}\left(\frac{1}{p^2}+\frac{1}{2p^3}\right)\right]. \quad (3)$$

The external action of gravitation can therefore be considered as the effect of this flux. Inasmuch as k is the coefficient of proportionality that gives the Newtonian force as a function of the apparent mass, calling this mass M_a, we have

$$F=kM_a;\; M_a=\pi\delta_v R^3\left[\frac{1}{p}-\frac{1}{2p^3}+e^{-2p}\left(\frac{1}{p^2}+\frac{1}{2p^3}\right)\right]. \quad (4)$$

If we put

$$\psi=\frac{3}{4}\left[\frac{1}{p}-\frac{1}{2p^3}+e^{-2p}\left(\frac{1}{p^2}+\frac{1}{2p^3}\right)\right], \quad . \quad . \quad (5)$$

we have

$$M_a=\frac{4}{3}\pi\delta_v R^3\psi=M_v\psi, \quad . \quad . \quad . \quad . \quad (6)$$

in which M_v represents the true mass of the sphere. Moreover we have

$$M_a=\frac{4}{3}\pi\delta_a R^3;\; \delta_v=\frac{\delta_a}{\psi};\; \delta_a=\delta_v\psi;\; \psi=\frac{\delta_a}{\delta_v}. \quad . \quad (7)$$

It is easy to see that

$$\lim_{p=0}\psi=1, \quad . \quad . \quad . \quad . \quad . \quad . \quad . \quad (8)$$

and hence

$$\lim_{p=0} M_a=M_v.$$

Fig. 2.

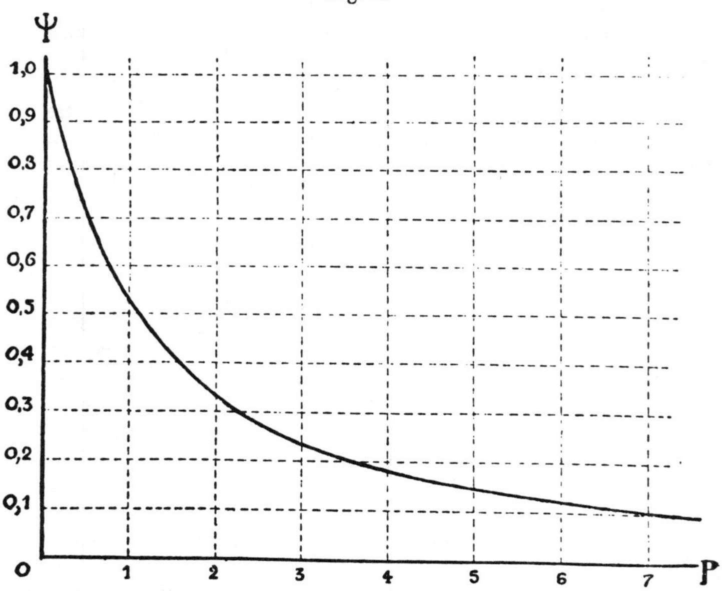

That is to say, that the apparent and true mass coincide if $p=0$, that is to say, if it is a question of a sphere with a very small radius, or if $H=0$. In fig. 2 the values of p are set down on the abscissa, and the ψ values on the ordinates; in this way the curve corresponding to equation (15) is obtained It touches the axis of the ordinates with a value 1 (see (8)) ; and it is asymptotic to the axis of the p's.

Application of the ψ function to the Sun.—The sun's density is assuredly not uniform ; but for a roughly approximate research I shall suppose this density to be constant, and equal to δ_v. Its apparent density is the astronomical one, and we have $\delta_a = 1{\cdot}41$. Several hypotheses can be made as to the real value of δ_v for the sun; for each of them one can calculate the value of ψ by means of (7); then, from the curve of fig. 2, the corresponding value of p can be deduced; and finally, since $p = \mathrm{RH} = \mathrm{R}_s \delta_v h$, we can deduce the value of h, seeing that the sun's radius is $\mathrm{R}_s = 6{\cdot}95 \,.\, 10^{10}$ cm. We can construct the following table :—

$\delta_v = 1{\cdot}41$	2	5	10	15	20
$\psi = \delta_a/\delta_v = 1$	0·705	0·281	0·141	0·094	0·070
$p = 0$	0·53	2·46	5·20	7·95	10·40
$p/\mathrm{R}_s\delta_v = 0$	$3{\cdot}81 \,.\, 10^{-12}$	$7{\cdot}08 \,.\, 10^{-12}$	$7{\cdot}49 \,.\, 10^{-12}$	$7{\cdot}63 \,.\, 10^{-12}$	$7{\cdot}64 \,.\, 10^{-12}$

Hence we see that if the true density is increasing, the h value increases rapidly, up to a density of about 2 ; and then more slowly, with a tendency towards a limit-value that we can see remains fixed at $7{\cdot}65 \,.\, 10^{-12}$.

Furthermore, we can see that, even admitting only a true density of the sun slightly greater than the apparent (*i. e.* 2), the order of magnitude for the h *quenching factor* remains fixed between 10^{-12} and 10^{-11}.

The h factor.—According to the already made hypothesis, the h factor represents a *universal constant*, upon which the measure of the gravitational absorption depends; and its probable value would be, as aforesaid, fixed between 10^{-12} and 10^{-11}; but its exact determination cannot be arrived at by considering the solar phenomenon. In fact, we lack the necessary elements to enable us to state the true density of the sun; perhaps we can believe it to be certainly greater than 1·41 (apparent or astronomical density), when we consider the great density of some heavier bodies. The sun's very high temperature, that would have the effect of dilating to an enormous extent such bodies, might be compensated by the enormous pressure in the solar mass interior. Anyway the value of the aforesaid true density of the sun cannot be established *a priori* with sufficient exactitude.

We can then imagine an experimental method for the investigation of the h constant. It would consist in finding the eventual variation of weight for a mass m, relatively small, surrounded by another mass M much bigger. In fact, as it is probable, according to the already made hypothesis, that the flux of the mass m must be partly absorbed by M, so also the gravitational flux that is put forth by the earth must weaken itself, before it reaches m, across M. I suppose this mass M shaped as a sphere, with radius r; and the mass m small, and situated at M's centre. If δ is the density of the substance that forms M, we shall have, according to (1) and (2),

$$f_m = kme^{-h\delta r};$$

this represents the flux of the mass m that succeeds in emerging from M. Taking in correspondence m_a and m_v as the true and apparent mass of m, we have

$$\frac{m_a}{m} = e^{-h\delta r}, \text{ or } m_a = m_v e^{-h\delta r};$$

r being very small (at the most a few decimetres), we have

$$m_a = m_v(1 - h\delta r).$$

This means that the mass m would undergo a variation (of diminution) of

$$\epsilon = m_v h\delta r. \quad . \quad . \quad . \quad . \quad . \quad . \quad . \quad (9)$$

From this the value of h could be deduced

$$h = \frac{\epsilon}{m_v \delta r}. \quad . \quad . \quad . \quad . \quad . \quad . \quad (10)$$

We can establish what would be the order of magnitude of ϵ, in a possible experiment of this kind. Let us put

$$m = m_v = 1 \text{ kg.}; \ \delta = 13{\cdot}60; \ r = 10 \text{ cm.}$$

This corresponds to the conditions that I realized in an experiment that I shall describe: I had taken for m a leaden ball; mass M was mercury, symmetrically distributed around m. Since h is probably equal to about 10^{-12} (on the hypothesis that the true density of the sun is a little greater than 2),

$$\epsilon = 1000 \text{ gr.} \, . \, 10^{-12} \, . \, 13{\cdot}60 \, . \, 10 \text{ cm.} = 1{\cdot}4 \, . \, 10^{-7} \text{ gr.}$$

This means that it would be necessary to be able to detect a variation of 1/10,000 of a mg. in 1 kg. The apparatus employed to carry out the planned experiment ought to fulfil such condition.

Description of the experimental disposition.—A Ruprecht balance of the bearing of about 1 kilogram was removed from its original protecting box and enclosed in a metallic box shaped like a T (fig. 3), able to resist the atmospherical

Fig. 3.

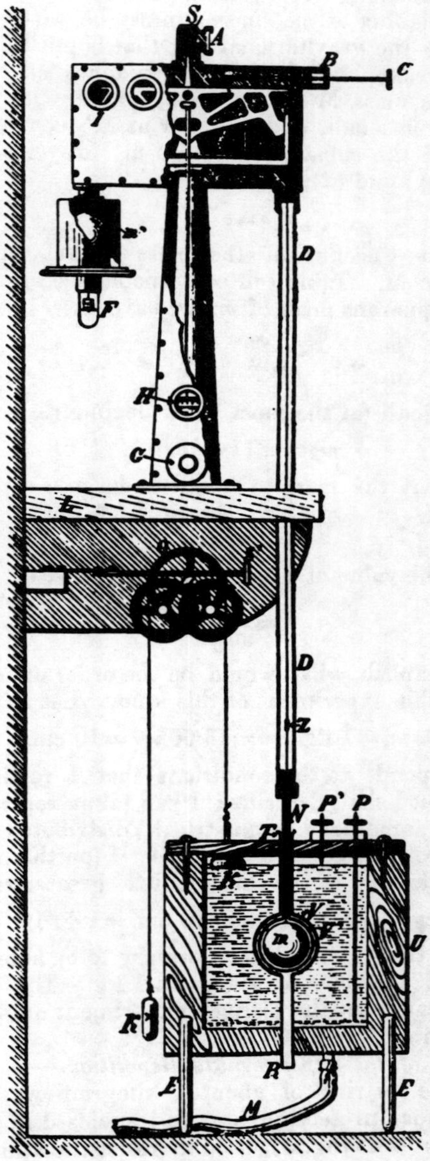

pressure when the vacuum was established in it. Special artifices have been devised in order to command from outside the beam and the little rider of a centigram upon it. The scales' original plates also have been taken off. Under the right arm of the balance a glass tube D has been fixed to the protecting box; this tube connects the box itself with the ambient in which is the mass *m*, as will be described further on. Under the left arm is a metallic protection that contains a leaden ball *m'*, which acts as tare of the mass *m*: this is also a leaden ball. On the beam at its middle point is a concave mirror S for the observation of the oscillations with a beam of light and vertical scale. The balance and the box are on the bracket table L fixed to the wall.

Under L and on the floor of the room is the recipient U, destined to receive the mercury that will surround the mass *m*. It is made out of very strong pieces of wood joined together; it is a cylinder whose diameter and inner height are about 22 cm. In the axis of the cylinder U are fixed two brass tubes R, T, of which one is the prolongation of the other, and these are joined by means of a hollow brass sphere V, whose diameter is 79 mm. This sphere can be separated in two pieces by means of a joint with screws in its diametral horizontal plane. In the interior of Y and concentric with it there is a second hollow brass sphere V', whose diameter is 70 mm. It is connected by means of a brass tube N to the glass tube D, that comes down from the balance. The sphere V' and the tube N do not come in contact with any point of the sphere V and tube T. The sphere V' can be separated, like V, into two hemispheres; so that it can contain the leaden ball *m*. This is suspended, by means of a brass wire, to the right arm of the balance, through the tubes D and N. The enlargement Z of this wire allows us to control with the cathetometer the position of the *m* ball, relatively to the recipient U. In this the mercury can flow upwards; this liquid can be removed at will by aspiration. The levels that the mercury reaches when U has been filled or emptied nearly completely are controlled rigorously by electric contacts P, P', that can be easily regulated. Moreover, a delicate system consisting of a float K and its counterpoise K' shows, by means of a mirror S', the position that the mercury takes every moment in the recipient U. All the fittings are made with a precision superior to 2/10 of a millimetre; within this limit it can be reckoned that the mercury has its centre of gravity coincident with the centre of gravity of *m*, the leaden ball. This has a mass of 1274 gr.; the mercury of 104 kg.

The balance with its accessories maintains the vacuum in

a way practically perfect. Even after twenty-four hours the inner pressure does not rise above 7/10 mm. of mercury, which represents the vapour tension of the mastics used to seal the balance. While the tests are going on, the mercury's rotatory pump must be kept going to reduce the pressure to less than 1/10 mm. of mercury. Under such conditions the perturbations of temperature that result from the mercury surrounding the V′ and V capsules are completely avoided.

The observations are made with a beam of light reflected by S on a scale at the distance of 12 m.; it is possible to detect 1/10 mm. on this scale. The sensibility of the balance can be brought in this way to a deviation of 170 mm. for the beam of light *per mg.* Hence it becomes possible to estimate about 1/1700 mg. in a direct reading, and reach a superior preciseness with many observations.

However, a doubt arises, is not such a preciseness illusive, and will not the slightest causes of perturbation hide completely a deviation of a few millimetres? I have nevertheless been able to remove all the more considerable causes of error. The most important of these consisted of the external mechanical shakes produced by the life of the town. I avoided them either by making my observations in the night or by availing myself of the days of the general strike, in this respect very useful.

Observation on the variation of weight.—The effect of the presence of mercury around the ball *m* was ascertained in the following way. In the recipient U determinations of

Fig. 4.

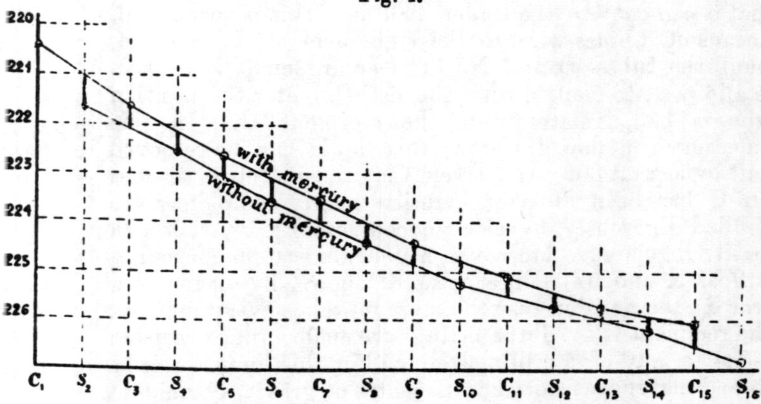

the balance position of rest were made rapidly and alternately with and without mercury. Fig. 4 shows the diagram of one of this series of observations, carried out on July 20–21, 1919.

As abscissæ I take the successive intervals of time $C_1 S_2$, $S_2 C_3$, $C_3 S_4$, $S_4 C_5$, ..., all equal, elapsing between the single observations *with mercury* and *without mercury* alternately. As ordinates I take the scale's positions of rest, determined each of them with three readings of oscillations. I join the points obtained in this way with two lines. These have a descending direction caused by the gradual displacement of the rest position of the balance, which displacement is a consequence of slight variations in the temperature. But the *with-mercury line* is always, with its points, above the *without-mercury one.* This means that the presence of the mercury *always* makes the leaden ball *m seem lighter.* In the same fig. 4 the 14 vertical segments represent the various successive means obtained from the represented series of observations. For briefness sake I shall not report all the other diagrams corresponding to five other series of observations that were made, together with the one described in fig. 4, on July 20 and 21. I shall say only that, taking the general mean of 57 partial means, I find as the value of the displacement of the balance position of rest, due to the presence of the mercury,

$$\text{mm. } 0{\cdot}358 \pm 0{\cdot}012\,;$$

the probable error 0·012 has been calculated by the method of least squares. The direction of the displacement indicates a diminution of weight, that is to say, *absorption of the terrestrial force of gravitation* on the leaden ball through the mercury.

The balance sensibility in the course of the aforesaid experiments was constantly equal to 171 mm. per mg. Therefore that displacement corresponds to a variation of

$$\text{mg. } \frac{0{\cdot}358 \pm 0{\cdot}012}{171} = \text{mg. } 0{\cdot}00209 \pm 0{\cdot}00007.$$

Correction of the observed effect.—However, it must be stated that in the experiment carried out in this way, several causes intervene, and superposing themselves with their own effects upon the phenomenon sought, modify the result notably. I cannot in this brief exposition discuss such causes in detail ; but among them I select those which have sensible effect, and I construct with them the following table (each bears its own sign) :—

Observed effect	+mg. 0·00209±0·00007	
Newtonian effect of the Hg on the tare	−mg. 0·00085	
" " " receptacle of Hg	+mg. 0·00007	
" " " floats K and K′	−mg. 0·00034	
" " " Hg on the beam	mg. 0·00000	
Correction for the displacement of the zero...	+mg. 0·00001	
Greatest error admissible for asymmetry......	mg.	±0·00009
Net effect	ϵ=+mg. 0·00098±0·00016	

The four corrections given here for the Newtonian effects of the mercury and float K with the K′ counterpoise have been rigorously calculated. Their probable error is very much inferior to the probable error of my observations. The greatest error admissible for asymmetry in the mercury's position relatively to the leaden ball, estimated at ±0·00009 mg., is certainly superior to the true one: it has been my purpose to exaggerate in its admission, to show that it cannot cover the phenomenon that has been discovered.

Hence *we have a net diminution of weight undergone by the leaden ball weighing* 1274 *gr., and this diminution is equal to* 0·00098 *mg.* (*that is to say,* $7\cdot7 \,.\, 10^{-10}$ *of its value*) *due to the fact of the ball being surrounded by the mercury.*

Possibility of other causes of error.—In the detailed relation of these experiments that will be published by the *Accademia dei Lincei* in Rome, I discuss minutely the possibility of other causes of error. Here I shall simply enumerate them :—

I. Perturbations of a mechanical character, such as the effects of the weight of the mercury on the balance of the projection lamp, the position of the scale, or the deformation of the recipient U, augmentation by compression of the density of the mercury, &c.;
II. Perturbations of a calorific character;
III. Radiometric actions;
IV. Electrostatic actions;
V. Magnetic actions;
VI. Electromagnetic actions.

I shall add only that such causes of error, if intervening, cannot modify sensibly the result that has been obtained.

Determination of the h constant.—The already verified variation of weight allows the valuation of the universal quenching constant h at least between certain limits of approximation. I make use of the relations (9) and (10). I must, however, introduce an hypothetical simplification in the experiment as performed if I want to avoid a very great difficulty of calculation; moreover, for a first research of the kind, it may be permitted. I shall suppose the leaden mass m, weighing 1274 gr. concentrated in one point; I shall suppose, moreover, the mass of mercury, weighing 104 kg. transformed from cylindrical to spherical shape, though containing still concentrically the V sphere (fig. 3). The radius of the mercury sphere so resulting will be equal to 12·35 cm. Finally, the thickness of mercury crossed by the single gravitational actions that the lead puts forth (or receives) can be rated, always in rough approximation, as equal to the difference between the radius of the mercury sphere and the sphere V. This corresponds to 12·35−3·95 =8·40 cm. Consequently we have in the formula (10):

$$\epsilon=9{\cdot}8.10^{-7}\text{ gr.}; \; m_v=1274\text{ gr.}; \; \delta=13{\cdot}60; \; r=8{\cdot}40;$$

and hence

$$h=\frac{9.10^{-7}}{1274.13{\cdot}60.8{\cdot}4}=6{\cdot}73.10^{-12}.$$

The order of magnitude in this determination coincides with the one predicted.

Application of the experimental result to the Sun's case.— The results already obtained are based chiefly upon the hypothesis that the sun's astronomical density, called here *apparent*, may be inferior to another density: the true density. Making always the simplification, deriving from the hypothesis of the constancy of the sun's true density, it can be considered as determined by the experiment described above. Let us call R_s the sun's radius, δ_{rs}, δ_{as}, its densities (apparent and true). Putting $p=RH=Rh\delta$, we have for the sun

$$p_s=h\delta_{vs}R_s.$$

To p_s's value corresponds a determinate value ψ_s of the ψ function, which might be deduced from fig. 2, if δ_{vs} were known to us. Now, from (7), we have

$$\delta_{vs}=\frac{\delta_{as}}{\psi_s};$$

therefore $$p_s\psi_s=hR_s\delta_{as}.$$

Since $R_s=6{\cdot}95.10^{10}$ cm. $\delta_{as}=1{\cdot}41$; $h=6{\cdot}73.10^{-12}$; we have further

$$p_s\psi_s=6{\cdot}18.10^{-12}.6{\cdot}95.10^{10}.1{\cdot}41=0{\cdot}660.$$

This condition must be satisfied. Considering the curve in fig. 2 we note that for the point $p=2{\cdot}0$, $\psi=0{\cdot}433$, and this really does happen. Hence I will put $p_s = 2{\cdot}0$, $\psi_s = 0{\cdot}433$, and from this we can deduce

$$\delta_{vs} = \frac{\delta_{as}}{\psi_s} = \frac{1{\cdot}41}{0{\cdot}433} = 4{\cdot}27.$$

That is to say, *it ensues that the sun's true density is the triple of what it is believed to be by the astronomers* (1·41).

But, although considering it possible to take as true the general result about a true density greater than the apparent, I do not attach exceeding importance to the determination I have been working out. The problem on the research of the true density so laid down, shows itself to be rather uncertain. In fact, it is sufficient to admit even a relatively slight error in the determination of ϵ, to have the value of δ_{vs} notably altered. That can be deduced from the following table:—

ϵ.	h.	δ_v.
0·0007	$4{\cdot}80 \,.\, 10^{-12}$	2·42
0·0009	$6{\cdot}18 \,.\, 10^{-12}$	3·27
0·00098	$6{\cdot}73 \,.\, 10^{-12}$	6·73 (experimental determination).
0·0011	$7{\cdot}55 \,.\, 10^{-12}$	10·04
0·0012	$8{\cdot}23 \,.\, 10^{-12}$	——

i. e., it is enough to admit $\epsilon = 0{\cdot}0011$, to have the true density rise to 10·04.

But, considering the ψ function, we come to note an interesting consequence: the h constant cannot be greater than $7{\cdot}55 \,.\, 10^{-12}$; because if it were, in the sun's case the following expression ought to be exact:

$$\frac{p_s \psi_s}{R_s \delta_{as}} > 7{\cdot}65 \,.\, 10^{-12}, \text{ or } p_s \psi_s > 0{\cdot}75,$$

and such a condition never can be verified from (5), which for great values of p, gives at most $p\psi = \frac{3}{4}$.

In other words we can also say *inasmuch as the sun has an apparent density equal to* 1·41, *the coefficient of absorption* h *cannot be greater than* $7{\cdot}65 \,.\, 10^{-12}$. The experiment gives $6{\cdot}73 \,.\, 10^{-12}$; therefore the facts agree up to now with the proposed theory.

I bring these considerations to a close, calling attention to the fact that if we admit the hypothesis of gravitational absorption, the calculation worked out for the sun with the simplification of constant density cannot lead us to very erroneous results. In fact, if we substitute for this hypothesis of constant density, another law of variable density, this will be, as a matter of course, greater at the centre than at the surface. Therefore, on the one hand, the fact that the matter would accumulate itself towards the centre would have for consequence that the absorption of the gravitational force of the greatest part of the matter would be accomplished across greater thickness, as the gravitational action would have to pass first chiefly from the deeper layers to the surface, then to the exterior ; but, on the other hand, the exterior mass has a reduced density, hence the absorption itself is diminishing. Therefore these are two contrary causes that in general will not balance each other, but the effect of one is subtracted from the effect of the other, leaving the mean true density not very different from the one established by my experiment and calculation.

Summary and Conclusion.—Examining Newton's Law, I have come to think that the force of gravitation can weaken itself by absorption due to ponderable matter. Following other arguments, I have come to suspect that the matter which shows the force of gravitation might heat itself. Although such conception might resolve in a new way the old controversy on the origin of the sun's heat, I state it with all reserve.

I have undertaken afterwards to treat theoretically the case of a spherical mass with constant density, subject to the absorption of its own force of gravitation, and from this work I have deduced the elements necessary to carry out an experimental control of my hypothesis. I have carried out this experiment by weighing *in vacuo* a leaden ball whose weight was 1274 gr. symmetrically surrounded by 104 kg. of mercury. Having previously avoided all the possible causes of error, I have been able to conclude that *the leaden ball loses* $7{\cdot}7 \,.\, 10^{-10}$ *of its weight by the presence of the mercury.* Such result causes the determination of the *quenching constant* (factor of absorption) per unit of density and length, as $6{\cdot}73 \,.\, 10^{-12}$.

Applying finally this result to the sun's case, I calculate its true density as 4·27.

The importance of this research is obvious, and I do not think that reasons for criticism can easily be found. Anyway, as I am the first to wish to test in all possible ways the results I am publishing, I may mention that it is my intention to repeat my experiments with far bigger apparatus. For the purpose, in the Laboratorio di Fisica del Politecnico di Torino (Italy), an apparatus is being built in such proportions as will render possible an experiment with 10,000 kg. of lead. On the results that I shall obtain with it I shall report in due time.

SUBJECT INDEX

A

B

E

F

I

J

K

L

M

N

Q

R

S

T

Y

Z